Valve Handbook

Other McGraw-Hill Handbooks of Interest

Avallone and Baumeister
MARKS' STANDARD HANDBOOK FOR MECHANICAL ENGINEERS

Bleier
FAN HANDBOOK

Brady et al.
MATERIALS HANDBOOK

Brink
HANDBOOK OF FLUID SEALING

Chironis & Sclater
MECHANISMS AND MECHANICAL DEVICES SOURCEBOOK

Czernik
GASKET HANDBOOK

Harris and Crede
SHOCK AND VIBRATION HANDBOOK

Hicks
HANDBOOK OF MECHANICAL ENGINEERING CALCULATIONS

Hicks
STANDARD HANDBOOK OF ENGINEERING CALCULATIONS

Lingaiah
MACHINE DESIGN DATA HANDBOOK

Parmley
STANDARD HANDBOOK OF FASTENING AND JOINING

Rothbart
MECHANICAL DESIGN HANDBOOK

Shigley and Mischke
STANDARD HANDBOOK OF MACHINE DESIGN

Suchy
DIE DESIGN HANDBOOK

Walsh
McGRAW-HILL MACHINING AND METALWORKING HANDBOOK

Walsh
ELECTROMECHANICAL DESIGN HANDBOOK

Valve Handbook

Philip L. Skousen
Valtek International

McGraw-Hill

New York   San Francisco   Washington, D.C.   Auckland   Bogotá
Caracas   Lisbon   London   Madrid   Mexico City   Milan
Montreal   New Delhi   San Juan   Singapore
Sydney   Tokyo   Toronto

Library of Congress Cataloging-in-Publication Data

Skousen, Philip L.
    Valve handbook / Philip L. Skousen.
        p.   cm.
    Includes index.
    ISBN 0-07-057921-0 (alk. paper)
    1. Valves—Handbooks, manuals, etc.   I. Title.
TS227.S55   1997
621.8′4—dc21   97-14832
                                                CIP

McGraw-Hill

A Division of The **McGraw·Hill** Companies

1 2 3 4 5 6 7 8 9 0   DOC/DOC   9 0 2 1 0 9 8 7

ISBN 0-07-057921-0

*The sponsoring editor for this book was Harold B. Crawford, the editing super-
visor was Frank Kotowski, Jr., and the production supervisor was Clare
Stanley. It was set in Palatino. It was composed by Priscilla Beer of McGraw-
Hill's Professional Book Group composition unit.*

Printed and bound by R. R. Donnelley & Sons Company.

McGraw-Hill books are available at special quantity discounts to use as
premiums and sales promotions, or for use in corporate training programs. For more information, please write to the Director of Special
Sales, McGraw-Hill, 11 West 19th Street, New York, NY 10011. Or contact your local bookstore.

 This book is printed on recycled, acid-free paper containing a minimum of 50% recycled, de-inked fiber.

Contents

Preface

The editors at McGraw-Hill first approached me about writing this handbook nearly three years ago, after reviewing an article that I authored for *The Valve Magazine*. They indicated that they were interested in having me author a valve handbook, using my common denominator writing style. I liked the challenge that they proposed, and I accepted. Now, after literally hundreds of hours, dozens of phone calls and facsimiles, and four drafts, the handbook is ready for the valve-using public.

When I began my career with Valtek International in 1975, I was like many who have started out in this industry: My only experience with a valve was taking a pair of pliers to a leaky faucet in the bathroom. But spending three years in the engineering department at Valtek and then some 18 years as a technical communicator cured me of the notion that a process valve is just a larger version of a simple faucet. True, the two are related in a number of ways and they both work by the same scientific principles. However, the engineering design and complexity of process valves can be immense. The process services they are installed in can sometimes be brutal—even capable of destroying a valve in hours if misapplied. To me this is an exciting and dynamic industry, especially with the advent of smart technology, which has lifted the science and application of valves to a whole new level.

Twenty years ago, when I first picked up a drawing pencil (yes, computer-aided design was still a couple years away), if one wanted to find basic information about valves, not much was readily available. Unfortunately, if a person starting out in the valve business or process industry wanted to learn the basics about valves, instead of turning to a good reference handbook, he or she had to ask questions of more senior engineers or technicians. More often than not, the individual might not have had the educational background or experience to even ask the right questions. Hence, learning the basics of valves often took months and maybe even years to fully understand certain designs and principles.

It's not that valve books didn't exist; they did then and still do today. The problem is that such books do not typically address the questions or level of understanding that most nonvalve experts ask. A few valve books existed, but none began with the simple concepts of fluid dynamics and valve design to build a foundation of common understanding between the reader and the author. Once this foundation is established, the more complex issues can then be addressed. Those valve books in existence were primarily authored by one or several of the industry "gurus" or experts. Many were product specific, such as those experts in severe service trims, packing boxes, or actuators. Others had studied the adverse effects of process flow, such as those who extensively studied cavitation or noise. Some handbooks were compiled from a series of "white papers" from a wide assortment of experts and brought together by an independent editor—a great method to put out a high level of knowledge quickly, but lacking in continuity and basics. While most books concentrated on design and severe services, little information was provided about selection options, or about installing, starting up, troubleshooting, and servicing valves.

Over the years, I've had the opportunity to meet and work with many of these valve experts. I have a great respect for their knowledge and pioneering efforts in the field of valves. The knowledge that such experts impart is important to the entire industry. Once understood by the user, it can help solve application and process control questions. A problem inherent with many authors and some industry experts is their basic assumption that the reader knows as much as the author—or that at least the author and reader have the same high level of understanding or engineering background. Although that may be fine for those experienced in the industry over some 10 or 20 years or with advanced engineering degrees, it leaves a great many people out of the loop of understanding. I decided to write this book for those who need to know the essence of valves and know it quickly. In this day of corporate turnarounds or internal reengineering, the ability to understand a particular segment of process control—such as valves—cannot wait for a decade of experience. Engineers and technicians responsible for valves need knowledge now—knowledge that is simple to understand and easy to apply. After the basics are understood, the finer parts of this business can then be explored, some of which are explained in this handbook, as well as other valve books.

As a certified business communicator, one with years of technical writing and editing about valves, I have learned that the best approach to communication is to begin simply and write to a common denominator. This means that if a principle or concept is written to a high school level, both the high school graduate and the engineer with a masters degree will understand it. But if that same concept is written to a higher level, such as a 16th grade level, only the university graduate will understand it. That's not to say that higher knowledge in this book is missing or "dumbed down." Rather, this higher degree of information is presented in a structured, simplified manner with no assumptions of knowledge made. Because of this style, if the user reads this handbook from cover to cover, he or she may find some duplication of information. This is because most handbooks, like encyclopedias and dictionaries, are used for reference purposes: they sit on a shelf until they are needed to answer a particular question or to explain a particular concept. With common denominator style, the reader will be able to turn to any section of the handbook, read it, and understand the concepts...without keeping a finger glued to the glossary or the index, or left wondering about a term.

With my experience in referencing valve books, I have learned one important fact: This industry is so large that no one book could ever hope to contain every fact, design concept, sizing equation, or principle about the thousands of different valve models available today. If possible, the book would become a set of 20 volumes, and then be so massive that the user would find it extremely cumbersome.

As I accepted this challenge of writing a valve handbook for McGraw-Hill, I took the approach that a dozen basic designs and a handful of scientific principles represent the foundation of the valve industry. Taking into account the dozens of valve manufacturers, each design can have literally hundreds of particular features. Rather than research and include them all, I have opted to take the most common features and have described them in detail. A number of statements are made in the book describing the general design of a particular valve design or feature. Because no one rule can be steadfast in this dynamic business, these general statements are by no means certain or definite. Exceptions can always be found to these general statements.

This also applies to any information in the handbook about installation, quick-checking, troubleshooting, and servicing of a valve. These sections are provided as general guidelines to the user, compiled from various users and manufacturers. In no way can they possibly apply to every type of valve and are certainly not intended to replace the manufacturer's technical and maintenance literature. By including this information, I hope that these tips and ideas will provide the user with a broader base of information than may be provided by the manufacturer's literature alone.

The terminology used in the book is based upon my experience and the advice of others. With the wide diversity represented by the valve industry, I found the same valve part or concept can be called by three or four different names. In the introduction of a new term, I have also included other common names for reference purposes. However, I use the first term consistently throughout the entire handbook. This is not to say that the other terms are incorrect. Rather, I believe that a consistency of terminology makes the concepts and designs easier to understand.

Some of the information contained in this book has come to me through technical materials, training manuals, or white papers that I have collected over the years. In addition, dozens of valve manufacturers graciously responded to my initial request for information and sent me boxes of material. In some cases, I have relied upon my own knowledge and experience with valves, as well as my interviews with dozens of users over the years. Overall, I was impressed with much of the recent material produced by valve manufacturers. Many have gone to great lengths to portray their products with simple, easy-to-understand concepts.

Because my primary focus in valves has been control valves, I am indeed grateful to those experts in the manual, check, and pressure relief valve industries who patiently explained the finer points of their products to me. I am also grateful for their review of my material, as well as their suggestions and criticisms. One thing I have learned from authoring this handbook is that a great number of opinions exist among the valve experts of today. Although I respect all opinions and arguments offered to me as part of this project, in some cases I had to act as referee when two opinions conflicted. In such situations, the decision to promote one idea over another was based upon my judgment and the opinions of several leaders whose judgment I have come to trust.

Philip L. Skousen

Acknowledgments

Over the past two years, a great number of individuals have assisted the author with the preparation of this handbook, sharing their knowledge of particular portions of the valve industry, including the design, operation, troubleshooting, and service of a wide range of process valves. These individuals have not only provided valuable input, but have also reviewed portions of the manuscript and recommended clarifications, which have been extremely valuable. Many of these individuals also provided the photography, artwork, illustrations, graphs, and table data—greatly adding to the content of the handbook.

Special thanks to: Mark Peters of Accord Controls (Cincinnati, Ohio), a subsidiary of the Duriron Company; Tim Martin of Adams (Houston, Texas); Peter Amos and John Stofira of Advanced Products Company (North Haven, Connecticut); Roland Larkin and C. H. Lovoy of the American Flow Control, a division of American Cast Iron Pipe (Birmingham, Alabama); Bill Knecht of Anchor/Darling (Williamsport, Pennysylvania); Chris Buxton and Michelle Strauss of Anderson, Greenwood & Co. (Houston, Texas), a subsidiary of Keystone International, Inc.; Richard H. Stern of the Automatic Switch Company (Florham Park, New Jersey); Richard Weeks of Automax (Cincinnati, Ohio), a subsidiary of The Duriron Company; Dan Wisenbaker of Betis Actuators and Controls (Waller, Texas); Fermo Gianesello, Robert Katz, Herb Miller, Andrew Noakes, and Nicole Woods of Control Components Inc. (Rancho Santa Margarita, California); Nancy Winalski of Conval Inc. (Somers, Connecticut);

Walter W. Mott of Copes-Vulcan (Lake City, Pennsylvania); Lew Babbidge and Cindy Sartain of the Daniel Valve Company, a division of Daniel Industries, Inc. (Houseton, Texas); Jean Surma of DeZURIK (Sartell, Minnesota); Rom Bordelon of Dresser Industries (Alexandria, Lousiana); Ken Senior of the DuPont Company–Polymers (Newark, Delaware); Dennis Garber of Durco Valve (Cookeville, Tennessee), a subsidiary of the Duriron Company; Philip R. Vaughn of DynaTorque Valve Actuators and Accessories (Muskegon, Michigan); Bob Sogge and John Wells of Fisher Controls (Marshalltown, Iowa); Susan Anderson of Flowseal (Long Beach, California), a division of Crane Valves; Lee Ann McMurtrie of the Groth Corporation (Houston, Texas); James D. Phillips of the Gulf Valve Company (Houston, Texas); Will Gavin of the Hydroseal Valve Company, Inc. (Kilgore, Texas); Lou Gaudio and Valerie D. Litz of ITT Engineered Valves; Ian W. B. Johnson of Kammer Ventile (Essen, Germany), a subsidiary of the Duriron Company; Domenic DiPaolo of Kammer USA (Pittsburgh, Pennsylvania), a subsidiary of The Duriron Company; Carter Hydrick of Keystone International, Inc. (Houston, Texas); Robert Hoffman of Mueller Steam Specialty (St. Pauls, North Carolina); Jime Holmes of Parker Electrohydraulics (Elyria, Ohio); Michael Fitzpatrick of Orbit Valve Company (Little Rock, Arkansas); Susan Anderson of Pacific Valves (Long Beach, California), a division of Crane Valves; Christ Letzelter of the Red Valve Company (Pittsburgh, Pennsylvania); Kevin Speed of Jordan Valve (Cincinnati, Ohio), a division of the Richards Industries Valve Group; Chris Warnett of Rotork Actuation (Rochester, New York); Pierre Brooking of Sereg Vannes (Paris, France), a subsidiary of The Duriron Company; Stephen R. Gow of Spirax Sarco (Allentown, Pennsylvania); Frank Breinholt, Fred Cain, Candee Ellis, Alan Glenn, and Craig Heraldson of Valtek International (Springville, Utah), a subsidiary of the Duriron Company; Bill Sandler of the Valve Manufacturers Association of America (Washington, D.C.); Deborah Lovegrove and Tom Velan of Velan Valve Corporation (Williston, Vermont); Gilbert K. Greene of the Victaulic Company of America (Easton, Pennsylvania); John J. Murphy of Yarway (Blue Bell, Pennsylvania), a subsidiary of Keystone International Inc.

I would also like to thank my employer, Valtek International, for its valuable assistance and support during this two-year project. Twelve years ago, as a technical communicator for Valtek, I was given the assignment to author a sizing and selection guide for control valves, working with a number of excellent engineers who helped guide me through that 200-page document. It was during that time that I first envisioned a handbook that would explain valves in a simple, straightforward manner. Valtek's parent company, The Duriron Company, Inc., took a special interest in this project—in partiular Duriron's chairman and CEO Bill Jordan. Bill supported this project from day one and encouraged me to complete it, for which I am grateful.

And, finally, I'd like to thank my wife, Patty, for her general support and assistance with proofreading the manuscript. Her insightful comments and objectivity helped make this handbook what it is. I am also thankful for my three daughters—Lindsay, Ashlee, and Kristin—who saw a little less of their dad during this project and were very understanding (well, mostly understanding) when I needed to use the home computer. It's all yours now, girls!

Valve Handbook

1
Introduction to Valves

1.1 The Valve

1.1.1 Definition of a Valve

By definition, *valves* are mechanical devices specifically designed to direct, start, stop, mix, or regulate the flow, pressure, or temperature of a process fluid. Valves can be designed to handle either liquid or gas applications.

By nature of their design, function, and application, valves come in a wide variety of styles, sizes, and pressure classes. The smallest industrial valves can weigh as little as 1 lb (0.45 kg) and fit comfortably in the human hand, while the largest can weigh up to 10 tons (9070 kg) and extend in height to over 24 ft (6.1 m). Industrial process valves can be used in pipeline sizes from 0.5 in [nominal diameter (DN) 15] to beyond 48 in (DN 1200), although over 90 percent of the valves used in process systems are installed in piping that is 4 in (DN 100) and smaller in size. Valves can be used in pressures from vacuum to over 13,000 psi (897 bar). An example of how process valves can vary in size is shown in Fig. 1.1.

Today's spectrum of available valves extends from simple water faucets to control valves equipped with microprocessors, which provide single-loop control of the process. The most common types in use today are gate, plug, ball, butterfly, check, pressure-relief, and globe valves.

Valves can be manufactured from a number of materials, with most valves made from steel, iron, plastic, brass, bronze, or a number of special alloys.

Figure 1.1 Size comparison between 30-in and 1-in globe valves. (*Courtesy of Valtek International*)

1.2 The History of Valves

1.2.1 Earliest Use of the Valve

Prior to the development of even simple irrigation systems, crops cultivated by early civilizations were at the mercy of whims of weather, water levels of rivers or lakes, or the strength of humans and animals to transport water in primitive vessels. Because of the unpredictability or hardship associated with these methods, early farmers sought a number of ways to control the flow of nearby water sources.

The primary ideal of a valve most likely arose when these simple farmers noticed that fallen trees or debris diverted, or even stopped, the flow of streams; thus the concept arose of using artificial barriers to divert water into nearby fields. Eventually, this idea expanded into simple irrigation using a planned series of ditches and canals, which by using gravity could transport, store, and widen the reach of the water source.

An important element of these early irrigation systems was a removable wooden or stone barrier, which could be placed at the entrance of each irrigation channel. This barrier was the early progenitor of what we now commonly call the gate valve and could be wedged between the walls of a canal to stop the flow or divert the flow to other channels, or when placed in a position between shut and fully open could regulate the amount of water entering the channel downstream.

As early as 5000 BC, crude gate valves were found in a series of dikes designed as part of ancient irrigation systems developed by the Egyptians along the banks of the Nile River. Archaeologists have found that other ancient cultures in Babylon, China, Phoenicia, Mexico, and Peru also used similar irrigation systems.

As early engineers examined these primitive process systems, they began to apply the technology to new uses. For example, as early as 1500 BC, the tombs of Egypt were equipped with extensive drainage systems, which included siphons, bellows, and simple plug valves carved from wood. Designed to bring water to the surface from underground wells, sophisticated saqqiehs in Egypt were equipped with simple wooden valves in the buckets used to transport the water.

The Romans, having conquered the Middle East, quickly saw the value of the Middle Eastern hydraulic engineering and expanded the concept into a series of aqueducts in Europe, which were used to sustain new cities that were located in areas away from major water sources. These aqueducts included early pumps, piping, and waterwheels, as well as gate and plug valves made of wood, stone, or lead.

1.2.2   Historical Development of the Valve

Generally, valves during the Middle Ages were crude, carved from wood, and used mainly as bungs in wine and beer casks. Valve design changed very little until the Renaissance when modern hydraulic engineering principles began to evolve. In an attempt to improve the performance of canal locks, Leonado da Vinci analyzed the stresses that would occur at different lock gates with varying heights of water on either side of the gate. These early studies of the concept of *pressure drop* helped determine the basis for modern fluid dynamics, which is essential to understanding and calculating the performance of valves.

In 1712, Englishman Thomas Newcomen invented his *atmospheric engine* (sometimes called a *heat engine*), which used low-pressure steam to drive a piston forward. When attached to a pivot beam, this simple engine could be used to lift water. As Newcomen improved his machine, he introduced a simple iron plug valve, which could be used to regulate the flow of steam to the piston—the first known application of a throttling valve.

In the late 1700s, the pioneering Scottish engineer James Watt looked for ways to improve Newcomen's atmospheric engine. Watt examined a number of ways to improve the Newcomen machine, which was slow and not very powerful because of the low-pressure steam. Also, because of the single-direction action, each stroke had to be returned to position by counterweights, which was extremely inefficient. Watt's final redesign of the inefficient Newcomen engine evolved into the first double-acting engine. Watt's engine introduced steam to both sides of the piston, driving both the upstroke and downstroke simultaneously. A piston rod was attached to both sides of a crank to produce rotary motion for driving wheels, which finally led to the development of steam locomotives and steamboats. Critical to Watt's steam engine were self-acting valves, which were used to introduce and vent steam from both sides of the piston. Although these iron valves were crude by today's standards, their function was critical to the success of the steam engine, which ushered in the Industrial Age.

During the 1800s the use of steam power in transportation and textile industries, as well as waterworks, accelerated the development of more sophisticated valves. With the obvious temperature considerations of steam service, valves could no longer be made of wood or soft metals. Instead, steam engineering led to iron valves, machined to close tolerances. Not only were these iron valves far more durable, they were able to withstand the high temperatures and excessive stroking associated with steam engines without excessive leaking.

The advent of steam power produced a greater need for coal, which led to the development of sophisticated underground pumping systems, including new types and styles of valves, such as gate valves. The Industrial Age also spurred the use of natural gas in cities as the fuel for lighting and heating, which required simple ball valves for this early gaseous application.

The Corliss steam engine (Fig. 1.2), unveiled in 1876, was designed with sophisticated self-acting control valves—including the first introduction of linear globe valves, some of which are similar to designs available today.

The discovery of crude oil as a plentiful and inexpensive form of power in the early nineteenth century spawned the creation of the petroleum refinery. From refineries, other process industries soon followed, which led to the development of chemical, petrochemical, pulp and paper, and food and beverage processing plants—creating the need for hundreds of process valves in each plant.

Electricity as a source of power led to the creation of coal-fired, hydroelectric, and, eventually, nuclear power plants, which involved

Figure 1.2 The Corliss steam engine was one of the first applications of the control valve. (*Courtesy of Valtek International*)

the use of valves in not only simple water and steam applications but also severe service applications that involved high pressure drops and subsequent cavitation, flashing, and choking.

1.2.3 The Valve and the Modern Era

Before the 1930s, nearly all valves in process plants were manually operated, which required workers to open and close the valves by hand according to the needs of the process. Obviously, this resulted in a slow response, since a worker had to run or pedal a bicycle from the control room to the valve, as well as poor accuracy in throttling situations where a worker had to estimate a certain valve position. For those reasons, during that decade, automated control valves made their first appearance. Control valves allowed the control room to send pneumatic signals directly to the valve, which then moved to the necessary position automatically, without the need of human involvement.

Today, the global valve industry involves hundreds of global manufacturers who produce thousands of designs of manual, check, pressure-relief, and control valves. Modern valve designs range from simple gate valves, similar in function to those used by early Egyptian

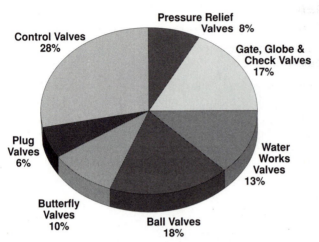

Figure 1.3 1993 Usage of valves, according to classification. (*Data courtesy of Valve Manufacturers Association of America*)

farmers, to control valves equipped with microprocessors for single-loop control. According to the Valve Manufacturers Association (VMA), valve manufacturing in 1993 was a US$2.7 billion industry. As shown in Figs. 1.3 and 1.4, control valves are the fastest growing segment of the valve industry, indicating the quickening pace of automation in process industry.

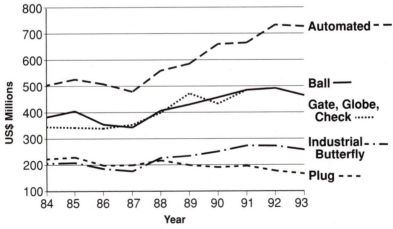

Figure 1.4 Ten-year sales of valves, according to classification. (*Data courtesy of Valve Manufacturers Association of America*)

1.3 Valve Classification According to Function

1.3.1 Introduction to Function Classifications

By the nature of their design and function in handling process fluids, valves can be categorized into three areas: *on–off valves,* which handle the function of blocking the flow or allowing it to pass; *nonreturn valves,* which only allow flow to travel in one direction; and *throttling valves,* which allow for regulation of the flow at any point between fully open to fully closed.

One confusing aspect of defining valves by function is that specific valve-body designs—such as globe, gate, plug, ball, butterfly, and pinch styles—may fit into one, two, or all three classifications. For example, a plug valve may be used for on–off service, or with the addition of actuation, may be used as a throttling control valve. Another example is the globe-style body, which, depending on its internal design, may be an on–off, nonreturn, or throttling valve. Therefore, the user should be careful when equating a particular valve-body style with a particular classification.

1.3.2 On–Off Valves

Sometimes referred to as *block valves,* on–off valves are used to start or stop the flow of the medium through the process. Common on–off valves include gate, plug, ball, pressure-relief, and tank-bottom valves (Fig. 1.5). A majority of on–off valves are hand-operated, although they can be automated with the addition of an actuator (Fig. 1.6).

On–off valves are commonly used in applications where the flow must be diverted around an area in which maintenance is being performed or where workers must be protected from potential safety hazards. They are also helpful in mixing applications where a number of fluids are combined for a predetermined amount of time and when exact measurements are not required. Safety management systems also require automated on–off valves to immediately shut off the system when an emergency situation occurs.

Pressure-relief valves are self-actuated on–off valves that open only when a preset pressure is surpassed (Fig. 1.7). Such valves are divided into two families: relief valves and safety valves. Relief valves are used to guard against overpressurization of a liquid service. On the other hand, safety valves are applied in gas applications where overpressurization of the system presents a safety or process hazard and must be vented.

Figure 1.5 Tank bottom valve used in a steel processing application. (*Courtesy of Kammer USA*)

Figure 1.6 Quarter-turn plug valve with rack and pinion actuation system in chemical service. (*Courtesy of Automax, Inc. and The Duriron Company, Valve Division*)

Figure 1.7 Pressure-relief valve being tested for correct cracking pressure. (*Courtesy of Valtek Houston Service Center*)

1.3.3 Nonreturn Valves

Nonreturn valves allow the fluid to flow only in the desired direction. The design is such that any flow or pressure in the opposite direction is mechanically restricted from occurring. All check valves are nonreturn valves (Fig. 1.8).

Nonreturn valves are used to prevent backflow of fluid, which could damage equipment or upset the process. Such valves are especially useful in protecting a pump in liquid applications or a compressor in gas applications from backflow when the pump or compressor is shut down. Nonreturn valves are also applied in process systems that have varying pressures, which must be kept separate.

1.3.4 Throttling Valves

Throttling valves are used to regulate the flow, temperature, or pressure of the service. These valves can move to any position within the stroke of the valve and hold that position, including the full-open or full-closed positions. Therefore, they can act as on–off valves also. Although many throttling valve designs are provided with a hand-operated manual handwheel or lever, some are equipped with actua-

Figure 1.8 Piston check valve in natural gas service. (*Courtesy of Valtek International*)

tors or actuation systems, which provide greater thrust and positioning capability, as well as automatic control (Fig. 1.9).

Pressure regulators are throttling valves that vary the valve's position to maintain constant pressure downstream (Fig. 1.10). If the pressure builds downstream, the regulator closes slightly to decrease the pressure. If the pressure decreases downstream, the regulator opens to build pressure.

As part of the family of throttling valves, *automatic control valves*, sometimes referred to simply as *control valves*, is a term commonly used to describe valves that are capable of varying flow conditions to match the process requirements. To achieve automatic control, these valves are always equipped with actuators. Actuators are designed to receive a command signal and convert it into an specific valve position using an outside power source (air, electric, or hydraulic), which matches the performance needed for that specific moment.

1.3.5   Final Control Elements within a Control Loop

Control valves are the most commonly used final control element. The term *final control element* refers to the high-performance equipment

Figure 1.9 Globe control valve with extended bonnet (left) with quarter-turn blocking ball valves (right and bottom) in refining service. (*Courtesy of Valtek International*)

Figure 1.10 Pressure regulator. (*Courtesy of Valtek International*)

needed to provide the power and accuracy to control the flowing medium to the desired service conditions. Other control elements include metering pumps, louvers, dampers, variable-pitch fan blades, and electric current-control devices.

As a final control element, the control valve is part of the *control loop*, which usually consists of two other elements besides the control valves: the *sensing element* and the *controller*. The sensing element (or sensor) measures a specific process condition, such as the fluid pressure, level, or temperature. The sensing element uses a transmitter to send a signal with information about the process condition to the controller or a much larger distributive control system. The controller receives the input from the sensor and compares it to the set point, or the desired value needed for that portion of the process. By comparing the actual input against the set point, the controller can make any needed corrections to the process by sending a signal to the final control element, which is most likely a control valve. The valve makes the change according to the signal from the controller, which is measured and verified by the sensing element, completing the loop. Figure 1.11 shows a diagram of a common control loop, which links a controller

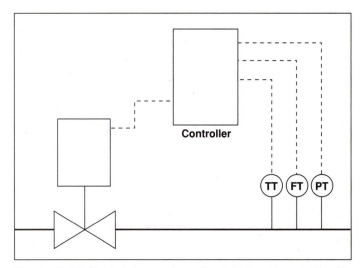

Figure 1.11 Control loop schematic showing the relationship among flow (FT), pressure (PT), and temperature (TT) transmitters, and the controller and control valve. (*Courtesy of Valtek International*)

with the flow (FT), pressure (PT), and temperature transmitters (TT) and a control valve.

1.4 Classification According to Application

1.4.1 Introduction to Application Classifications

Although valves are often classified according to function, they are also grouped according to the application, which often dictates the features of the design. Three classifications are used: *general service valves,* which describes a versatile valve design that can be used in numerous applications without modification; *special service valves,* which are specially designed for a specific application; and *severe service valves,* which are highly engineered to avoid the side effects of difficult applications.

1.4.2 General Service Valves

General service valves are those valves that are designed for the majority of commonplace applications that have lower-pressure ratings between American National Standards Institute Class 150 and 600 (between PN 16 and PN 100), moderate-temperature ratings between −50 and 650°F (between −46 and 343°C), noncorrosive fluids, and common pressure drops that do not result in cavitation or flashing. General service valves have some degree of interchangeability and flexibility built into the design to allow them to be used in a wider range of applications. Their body materials are specified as carbon or stainless steels. Figure 1.12 shows an example of two general service valves, one manually operated and the other automated.

1.4.3 Special Service Valves

Special service valves is a term used for custom-engineered valves that are designed for a single application that is outside normal process applications. Because of its unique design and engineering, it will only function inside the parameters and service conditions relating to that particular application. Such valves usually handle a demanding temperature, high pressure, or a corrosive medium. Figure 1.13 shows a control valve designed with a sweep-style body and ceramic trim to

Figure 1.12 Wedge gate valves used in a blocking service to bypass general service control valves in a gasification process. (*Courtesy of Valtek International*)

Figure 1.13 Sweep-style globe valve used in an erosive mining application involving high-pressure air and sand particulates. (*Courtesy of Valtek International*)

handle an erosive mining application involving sand particulates and high-pressure air.

1.4.4 Severe Service Valves

Related to special service valves are *severe service valves,* which are valves equipped with special features to handle volatile applications, such as high pressure drops that result in severe cavitation, flashing, choking, or high noise levels (which is covered in greater detail in Chap. 11). Such valves may have highly engineered trims in globe-style valves, or special disks or balls in rotary valves to either minimize or prevent the effects of the application.

In addition, the service conditions or process application may require special actuation to overcome the forces of the process. Figure 1.14 shows a severe service valve engineered to handle 1100°F (593°C) liquid-sodium application with multistage trim to handle a high pressure drop and a bonnet with special cooling fins. The electrohydraulic actuator was capable of producing 200,000 lb (889,600 N) of thrust.

Figure 1.14 Severe service valve designed to handle high-pressure-drop, high-temperature liquid-sodium application. (*Courtesy of Valtek International*)

1.5 Classification According to Motion

1.5.1 Introduction to Motion Classifications

Some users classify valves according to the mechanical motion of the valve. *Linear-motion valves* (also commonly called *linear valves*) have a sliding-stem design that pushes a closure element into an open or closed position. (The term *closure element* is used to describe any internal valve device that is used to open, close, or regulate the flow.) Gate, globe, pinch, diaphragm, split-body, three-way, and angle valves all fit into this classification. Linear valves are known for their simple design, easy maintenance, and versatility with more size, pressure class, and design options than other motion classifications—therefore, they are the most common type of valve in existence today.

On the other hand, *rotary-motion valves* (also called *rotary valves*) use a closure element that rotates—through a quarter-turn or 45° range—to open or block the flow. Rotary valves are usually smaller in size and weigh less than comparable linear valves, size for size. Applicationwise, they are limited to certain pressure drops and are prone to cavitation and flashing problems. However, as rotary-valve designs have matured, they have overcome these inherent limitations and are now being used at an increasing rate.

1.6 Classification According to Port Size

1.6.1 Full-Port Valves

In process systems, most valves are designed to restrict the flow to some extent by allowing the flow passageway or area of the closure element to be smaller than the inside diameter of the pipeline. On the other hand, some gate and ball valves can be designed so that internal flow passageways are large enough to pass flow without a significant restriction. Such valves are called *full-port valves* because the internal flow is equal to the full area of the inlet port.

Full-port valves are used primarily with on–off and blocking services, where the flow must be stopped or diverted. Full-port valves also allow for the use of a *pig* in the pipeline. The pig is a self-driven (or flow-driven) mechanism designed to scour the inside of the pipeline and to remove any process buildup or scale.

1.6.2 Reduced-Port Valves

On the other hand, *reduced-port valves* are those valves whose closure elements restrict the flow. The flow area of that port of the closure element is less than the area of the inside diameter of the pipeline. For example, the seat in linear globe valves or a sleeve passageway in plug valves would have the same flow area as the inside of the inlet and outlet ports of the valve body. This restriction allows the valve to take a pressure drop as flow moves through the closure element, allowing a partial pressure recovery after the flow moves past the restriction.

The primary purpose of reduced-port valves is to control the flow through reduced flow or through throttling, which is defined as regulating the closure element to provide varying levels of flow at a certain opening of the valve.

1.7 Common Piping Nomenclature

1.7.1 Introduction to Piping Nomenclature

Although a complete glossary is included in this handbook, the reader should be acquainted with the piping nomenclature commonly used in the global valve industry. Because the valve industry, along with a good portion of the process industry, has been driven by developments and companies originating in North America over the past 50 years, valve and piping nomenclature has been heavily influenced by the imperial system, which uses such terms as *pounds per square inch* (*psi*) to refer to pressure or *nominal pipe size* (*NPS*) to refer to valve and pipe size (in inches across the pipe's inside diameter). These terms are still in use today in the United States and are based upon the nomenclature established by the American National Standards Institute (ANSI).

Outside of the United States, valve and piping nomenclature is based on the International System of Units (metric system), which was established by the International Standards Organization (ISO). According to the metric system, the basic unit measurement is a *meter*, and distances are related in multiples of meters (kilometers, e.g.) or as equal units of a meter (centimeters, millimeters). Typically metric valve measurements are called out in millimeters and pressures are noted in *kilopascal* (kPa) (or *bar*). ISO standards refer to pipe diameter as *nominal diameter* (DN) and pressure ratings as *nominal pressure* (PN). Tables 1.1 and 1.2 provide quick reference for both ANSI and ISO standards.

Table 1.1 Nominal Pipe Size
vs. Nominal Diameter*

Nominal Pipe Size (NPS) (inches)	Nominal Diameter (DN) (millimeters)
0.25	6
0.5	15
0.75	20
1.0	25
1.25	32
1.5	40
2.0	50
2.5	65
3.0	80
4.0	100
6.0	150
8.0	200
10.0	250
12.0	300
14.0	350
16.0	400
18.0	450
20.0	500
24.0	600
36.0	900
42.0	1000
48.0	1200

Data courtesy of Kammer Valve.

Table 1.2 ANSI Pressure Class
vs. Nominal Pressure*

ANSI Pressure Class *pounds of force per square inch of surface area*	Nominal Pressure (PN) *allowable pressure in bar*
150	16
300	40
600	100
900	160
1500	250
2500	400
4500	700

Note: PN is an approximation to the corresponding ANSI pressure class, and should not be used as an exact correlation between the two standards. PN correlates to DIN (Deutsche Industrie Norme) pressure–temperature rating standards, which may vary significantly from ANSI pressure–temperature ratings.

*Data courtesy of Kammer Valve.

2

Valve Selection Criteria

2.1 Valve Coefficients

2.1.1 Introduction to Valve Coefficients

The measurement commonly applied to valves is the *valve coefficient* (C_v), which is also known as the *flow coefficient*. When selecting a valve for a particular application, the valve coefficient is used to determine the valve size that will best allow the valve to pass the required flow rate, while providing stable control of the process fluid. Valve manufacturers commonly publish C_v data for various valve styles, which are approximate in nature and can vary—usually up to 10 percent—according to the piping configuration or trim manufacture.

If the C_v is not calculated correctly for a valve, the valve usually experiences diminished performance in one of two ways: If the C_v is too small for the required process, the valve itself or the trim inside the valve will be undersized, and the process system can be starved for fluid. In addition, because the restriction in the valve can cause a buildup in upstream pressure, higher back pressures created before the valve can lead to damage in upstream pumps or other upstream equipment. Undersized C_v's can also create a higher pressure drop across the valve, which can lead to cavitation or flashing.

If the C_v is calculated too high for the system requirements, a larger, oversized valve is usually selected. Obviously, the cost, size, and weight of a larger valve size are a major disadvantage. Besides that consideration, if the valve is in a throttling service, significant control problems can occur. Usually the closure element, such as a plug or a disk, is located just off the seat, which leads to the possibility of creating a high pressure drop and faster velocities—causing cavitation, flashing, or erosion of the trim parts. In addition, if the closure element is closure to the seat and the operator is not strong enough to hold that position, it may be sucked into the seat. This problem is appropriately called *the bathtub stopper effect*.

2.1.2 Definition of C_v

One C_v is defined as one U.S. gallon (3.78 liters) of 60°F (16°C) water that flows through an opening, such as a valve, during 1 min with a 1-psi (0.1-bar) pressure drop. As specified by the Instrument Society of America (ANSI/ISA Standard S75.01), the simplified equation for C_v is

$$C_v = \text{flow} \times m \sqrt{\frac{\text{specific gravity at flowing temperature}}{\text{pressure drop}}}$$

A step-by-step process for calculating C_v is found in Chap. 9.

2.2 Flow Characteristics

2.2.1 Introduction to Flow Characteristics

Each throttling valve has a *flow characteristic*, which describes the relationship between the valve coefficient (C_v) and the valve stroke. In other words, as a valve opens, the flow characteristic—which is an inherence to the design of the selected valve—allows a certain amount of flow through the valve at a particular percentage of the stroke. This attribute allows the valve to control the flow in a predictable manner, which is important when using a throttling valve.

The flow rate through a throttling valve is not only affected by the flow characteristic of the valve, but also by the pressure drop across the valve. A valve's flow characteristic acting within a system that allows a varying pressure drop can be much different or can vary significantly from the same flow characteristic in an application with a constant pressure drop. When a valve is operating with a constant pressure drop without taking into account the effects of piping, the flow characteristic is known as *inherent flow characteristic*. However, if both the valve and piping effects are taken into account, the flow characteristic changes from the ideal curve and is known as the *installed flow characteristic*. Usually, the entire system must be taken into account to determine the installed flow characteristic, which is discussed further in Sec. 2.2.5. Some rotary valves—such as butterfly and

Figure 2.1 Characterizable quarter-turn plug. (*Courtesy of The Duriron Company, Valve Division*)

ball valves—have an inherent characteristic that cannot be changed because the closure element cannot be modified easily. For that reason, rotary control valves in a throttling application can modify this inherent characteristic using a characterizable cam with the actuator's positioner, or by changing the shape of the closing device, such as a V-notched ball valve. Quarter-turn plug and ball valves can modify the characteristic by varying the opening on the plug (Fig. 2.1). On the other hand, linear valves usually have a flow characteristic designed into the trim, by determining either the size and shape of the holes in a cage (Fig. 2.2) or the shape of the plug head (Fig. 2.3).

The three most common types of flow characteristics are *equal percentage, linear,* and *quick-open.* The ideal curves for these three flow characteristics are shown in Fig. 2.4. However, the inherent characteristic of these curves can be affected by the body style and design, and piping factors.

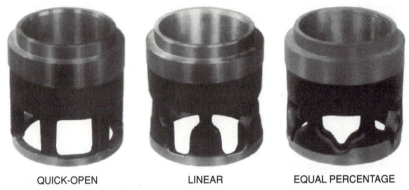

QUICK-OPEN   LINEAR   EQUAL PERCENTAGE

Figure 2.2 Characterizable cages. (*Courtesy of Fisher Controls International, Inc.*)

2.2.2 Equal-Percentage Flow Characteristic

Of the three common flow characteristics, the equal-percentage characteristic is the most frequently specified with throttling valves. With an equal-percentage characteristic, the change in flow per unit of valve stroke is directly proportional to the flow occurring just before the change is made. With an inherent equal-percentage characteristic, the flow rate is small at the beginning of the stroke and increases to a larger magnitude at the end of the stroke. This provides good, exact control of the closure element in the first half of the stroke, where control is harder to maintain because the closure element is more apt to be affected by process forces. On the other hand, an equal-percentage characteristic provides increased capacity in the second half of the stroke, allowing the valve to pass the required flow. An equal-percent-

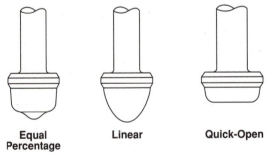

Equal   Linear   Quick-Open
Percentage

Figure 2.3 Characterizable linear plugs. (*Courtesy of Valtek International*)

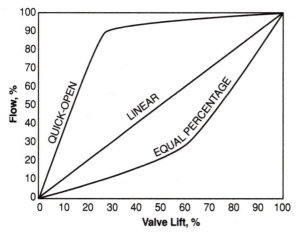

Figure 2.4 Typical inherent flow characteristics. (*Courtesy of Valtek International*)

age characteristic results in improved rangeability (Sec. 2.2.7) for a particular valve, as well as better repeatability and resolution in the first half of the stroke.

The mathematical formula for an equal-percentage characteristic is

$$Q = Q_0\, e^{nL}, \qquad \frac{dQ}{dL} = nQ$$

where Q = flow rate
L = valve travel
e = 2.718
Q_0 = minimum controllable flow
n = constant

Although the flow characteristic of the valve itself is equal percentage, the installed flow characteristic is closer to the linear flow characteristic. This is usually the case when the process system's pressure drop is larger than the pressure drop across the valve. Figure 2.5 shows two flow curves for an equal-percentage characteristic: the inherent flow characteristic and the installed characteristic that takes into account piping effects. The addition of the piping effects has a tendency to move the flow characteristic away from the ideal equal-percentage characteristic toward the inherent linear characteristic.

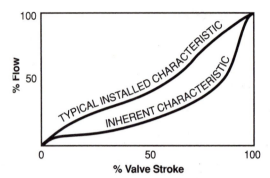

Figure 2.5 Typical inherent and installed equal-percentage flow characteristics. (*Courtesy of Valtek International*)

2.2.3 Linear Flow Characteristic

The inherent linear flow characteristic produces equal changes in flow per unit of valve stroke, regardless of the position of the valve. Linear flow characteristics are usually specified in those process systems where the majority of the pressure drop is taken through the valve. For the most part, linear flow characteristics provide better flow capacity throughout the entire stroke, as opposed to equal-percentage characteristics.

The mathematical formula for the linear characteristic is

$$Q = kL, \qquad \frac{dQ}{dL} = k$$

where Q = flow rate
L = valve travel
k = constant of proportionality

Figure 2.6 shows the inherent linear flow characteristic, as well as the installed characteristic (taking into account piping effects). As can be seen by this figure, the piping effects have a tendency to push the linear flow characteristic toward the quick-open characteristic.

2.2.4 Quick-Open Flow Characteristic

The quick-open characteristic is used almost exclusively for on–off applications, where maximum flow is produced immediately as the valve begins to open (Fig. 2.7). Because of the extreme nature of the

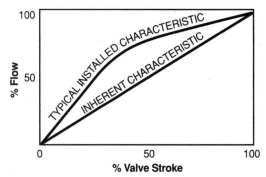

Figure 2.6 Typical inherent and installed linear flow characteristics. (*Courtesy of Valtek International*)

quick-open characteristic, the inherent and installed characteristics are similar.

2.2.5 Determining Installed Flow Characteristics

As discussed earlier, the inherent flow characteristic can change dramatically when the valve is installed in a process system. When the system's piping effects are taken into account, the equal-percentage characteristic moves toward linear, and the linear characteristic moves toward quick-open. Two examples of installed applications follow, one without piping effects and the other with piping effects.

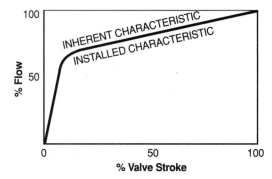

Figure 2.7 Typical inherent and installed quick-open flow characteristics. (*Courtesy of Valtek International*)

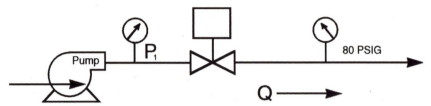

Figure 2.8 Typical flow schematic showing no piping losses. (*Courtesy of Valtek International*)

2.2.6 Flow Characteristic Example A (without Piping Effects)

Figure 2.8 shows a schematic of a process system that includes a centrifugal pump and a valve, which is used to maintain the pressure downstream to 80 psi or 5.5 bar. For illustration purposes, Fig. 2.9 provides the pump's relationship between the pump output (psi) and the flow (gal/min).

For this example, piping losses are assumed to be minimal. A total of 200 gal/min (757 liters/min) is required for the maximum flow rate. From Fig. 2.9, at 200 gal/min, the pump discharge pressure (P_1) is found to be 100 psi (6.9 bar) upstream of the valve, while 80 psi (5.5 bar) is required downstream (or, in other terms, a 20-psi or 1.4-bar

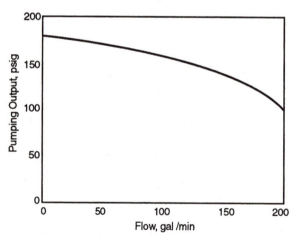

Figure 2.9 Flow chart of typical pump characteristics. (*Courtesy of Valtek International*)

pressure drop). Using the sizing formula for C_v (Sec. 2.1.2), we determine the C_v required for this application, which is

$$C_v = Q\sqrt{\frac{G_F}{\Delta P}} = 200\sqrt{\frac{1}{20}} = 45$$

Assuming that the C_v of 45 is the maximum C_v, several values of flow can now be estimated, along with the required valve C_v and the percent of maximum C_v the valve must have to control the process. These flow data are included in Table 2.1.

Using the definitions of both equal-percentage and linear characteristics, the installed characteristics can be plotted on a graph, using the data from Table 2.1, which is found in Fig. 2.10. This figure graphically illustrates the effect the installation has on the inherent flow characteristic. The linear characteristic moves away from the ideal linear line toward the quick-open characteristic. On the other hand, the equal-percentage characteristic moves toward the ideal linear line. In this example, either characteristic would provide good throttling control.

2.2.7 Flow Characteristic Example B (with Piping Effects)

For illustration purposes, Example A was simplified with a constant downstream pressure and a pressure drop only affected by the pump

Table 2.1 Flow Rate, C_v, and Pump Pressure (Without Piping Losses)†

Q Flow gpm	P₁ Pump Discharge Pressure psig	ΔP Across Valve psi	C_v Required	Percent of Valve Maximum C_v
50	170	90	5.2	11
100	150	70	12	27
150	125	45	22	49
200	100	20	45*	100

†Data courtesy of Valtek International.
*Maximum C_v.

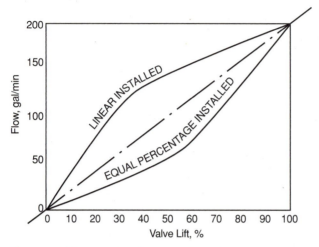

Figure 2.10 Installed linear and equal-percentage flow characteristics (without piping losses). (*Courtesy of Valtek International*)

characteristic. In Example B, the application is modified using a restriction downstream from the valve, as shown in Fig. 2.11. Note that the constant downstream pressure (80 psi or 5.5 bar) must be held constant after passing through the restriction.

Because of the restriction, the pressure drop must be distributed between the valve and the restriction (R). For this example, a 4-psi (0.3-bar) pressure drop across the valve is required at a flow rate of 200 gal/min (757 liters/min). Using the C_v equation, the maximum C_v for the valve is

$$C_v = Q\sqrt{\frac{G_f}{\Delta P}} = 200\sqrt{\frac{1}{4}} = 100$$

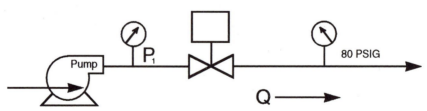

Figure 2.11 Typical flow schematic showing piping losses. (*Courtesy of Valtek International*)

Table 2.2 Flow Rate, C_v, and Pump Pressure (Without Piping Losses)†

Q Flow gpm	P_1 Pump Discharge Pressure psig	ΔP_R Across Restriction	ΔP Across Valve	C_v Required	Percent of Required Maximum Valve C_v
50	170	1	89	5	5
100	150	4	66	12	12
150	125	9	36	25	25
100	100	16	4	100*	100

†*Data courtesy of Valtek International.*
*Maximum C_v.

According to the square-root law ($Q = R\sqrt{\Delta P}$), the pressure drop across the valve's restriction will vary somewhat. Thus, using the pump characteristic, the available pressure drop across the valve can be estimated, which is shown in Table 2.2. Figure 2.12 shows the installed linear and installed equal-percentage characteristics from the data in

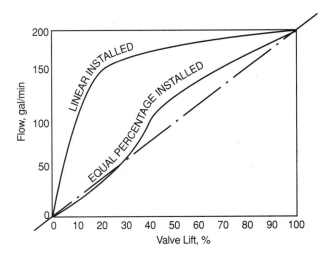

Figure 2.12 Installed linear and equal-percentage flow characteristics (with piping losses). (*Courtesy of Valtek International*)

Table 2.2. Note that the piping losses from the restriction have modified the installed equal-percentage characteristic to an inherent linear characteristic. In turn, the installed linear characteristic has become an inherent quick-open characteristic. Because of this effect of the piping losses, the use of a linear characteristic would create a highly sensitive system with a very small change in lift at the beginning of the stroke. On the other hand, using an equal-percentage characteristic would produce a more constant sensitivity throughout the entire stroke.

2.2.8 Choosing the Correct Flow Characteristic

When throttling valves are selected, a choice must be made between linear and equal-percentage characteristics. Two general rules apply that will simplify this choice. First, if most of the pressure drop is taken

Table 2.3 Recommended Flow Characteristics for Liquid Level Systems*

Constant Valve Pressure Drop	Recommended Inherent Flow Characteristic
Constant ΔP	Linear
Decreasing ΔP with increasing load: ΔP at maximum load > 20% of minimum load ΔP	Linear
Decreasing ΔP with increasing load: ΔP at maximum load < 20% of minimum load ΔP	Equal Percentage
Increasing ΔP with increasing load: ΔP at maximum load < 200% of minimum load ΔP	Linear
Increasing ΔP with increasing load: ΔP at maximum load > 200% of minimum load ΔP	Quick Open

Data courtesy of Valtek International.

through the valve and the upstream pressure is constant, a linear characteristic will provide the best control. However, such systems are rare, especially considering the complexities of today's process systems. A linear characteristic is also recommended when a variable-head flowmeter is installed in the system. Second, if the piping and downstream equipment provide significant resistance to the system, the equal-percentage characteristic should be chosen. This is usually the case with most process systems today, where a majority of all throttling valves have equal-percentage characteristics. The equal-percentage characteristic is also used for applications of high pressure drops with low flows and low pressure drops with high flows. When the valve is oversized as a precaution because limited data are available, the equal-percentage characteristic will provide the greatest range of control. Tables 2.3, 2.4, 2.5, and 2.6 provide more specific recommendations,

Table 2.4 Recommended Flow Characteristics for Pressure Control Systems*

Application	Recommended Inherent Flow Characteristic
Liquid process	Equal Percentage
Gas process, small volume, less than 10 feet (3 meters) of pipe between control valve and load valve	Equal Percentage
Gas process, large volume (process has a receiver, distribution system or transmission line exceeding 100 feet of nominal pipe volume), decreasing ΔP with increasing load, ΔP at maximum load > 20% of minimum load ΔP	Linear
Gas process, large volume, decreasing ΔP with increasing load, ΔP at maximum load < 20% of minimum load ΔP	Equal Percentage

Data courtesy of Valtek International.

Table 2.5 Recommended Flow Characteristics for Flow Control Processes†

Flow Measurement Signal to Controller	Location of Valve in Relation to Measuring Element	Wide Range of Flow Set Point	Small Range of Flow with Large ΔP Change at Valve with Increasing Load
Proportion to flow	Series	Linear	Equal Percentage
	By-pass*	Linear	Equal Percentage
Proportion to Flow Squared	Series	Linear	Equal Percentage
	By-pass*	Equal Percentage	Equal Percentage

†*Data courtesy of Valtek International.*
*When valve closes, flow rate increases in measuring element.

Table 2.6 Recommended Flow Characteristics for Miscellaneous Systems*

Application	Recommended Inherited Flow Characteristic
Three-way valves and two-way valves used as three-way valves (If characterized positioners are used, they must be calibrated by the valve manufacturer)	Linear
Gas compressor recycle control valve	Linear
Constant pressure drop service	Linear
Temperature control where control valve ΔP > 50% of System ΔP	Equal Percentage
pH control where control valve ΔP < 50% of system ΔP	Equal Percentage
pH control where control valve ΔP > 50% of system ΔP	Linear

Data courtesy of Valtek International.

depending on whether the system is for liquid level, pressure control, flow control, or another type of system, respectively.

For the most part, today's control instrumentation can make satisfactory signal adjustments to the throttling valve despite the flow characteristic. However, if manual control is ever required, having the correct flow characteristic allows such changes to be made easily.

2.2.9 Rangeability

Related to flow control and flow characteristics is the term *rangeability,* which is defined as the ratio of maximum to minimum flow that can be acted upon by a control valve after receiving a signal from a controller. Today's control valve applications often require a degree of *high rangeability,* which requires a valve to control flow from large to small flows. The rangeability of a control valve is affected by three factors.

The first factor is the valve's geometry (for example, the geometry of the plug and seat in globe valves), which has an inherent rangeability due to the design and configuration of the body and the regulating element. Sometimes the configuration can be modified, improving the rangeability as long as the valve's sensitivity is not affected. *Sensitivity* is defined as the specific change in flow area opening produced by a given change in the regulating element when compared to the previous position. In dealing with small flows when the regulating element is nearly closed, such as when a plug or a disk is close to the seat, oversensitivity can be a problem due to the small clearances involved.

The second factor, seat leakage, can also affect rangeability. Excessive seat leakage can cause instability as the valve lifts off the seat, especially with screwed-in seats that are not lapped, as opposed to floating clamped-in seats that are held in place by a retainer or cage.

Rangeability is also affected by the valve's actuation or actuator, which is the third factor. Some actuators are much more stiff at near-closure than others. For example, when a pneumatic spring diaphragm actuator is specified, a throttling valve is seldom accurate within the 5 percent of the valve closing. This is due primarily to the effects of the positioning spring, hysteresis, changing area of the diaphragm (as the actuator changes position), and the pressure drop itself. On the other hand, spring cylinder actuators use supply air pressure on both sides of a piston, which can provide control within less than 1 percent of valve lift, as well as a stiffness factor up to 10 times that of a comparable diaphragm actuator. Thus, a throttling valve equipped with a spring cylinder actuator would have a higher rangeability than the same valve with a diaphragm actuator.

Taking into account the effects of the valve geometry and the actuator, rangeability can be calculated in a simple manner. For example, if a valve is not accurate at less than 5 percent of stroke, then the rangeability is 20:1 (100 percent divided by 5 percent). As a common rule for common throttling valves, V-notched ball valves usually have the highest rangeability (up to 200:1), followed by eccentric plug valves (100:1), globe valves (50:1), and butterfly valves (20:1). Usually, the valves with the highest rangeability are those with the low sensitivity as the regulating element is nearly closed, but increases in sensitivity as the valve opens. Because the equal-percentage flow characteristic promotes increased sensitivity as the valve opens, it is usually chosen for most throttling applications. The term *clearance flow* is used to designate any flow that occurs between the lower end of the valve's rangeability and the actual closed position.

ISA Standard S75.11 ("Inherent Flow Characteristic and Rangeability of Control Valves") establishes guidelines for rangeability, sensitivity, and limits of deviation.

2.3   Shutoff Requirements

2.3.1   Shutoff Standards

Industry standards have been established for the control valve industry regarding the amount of permissible leakage of the process fluid through a valve's seat or seal. Usually this standard is applied to throttling valves, but may be applicable to other types of valves also. Specifically, ANSI Standard 70-2-1976 (reaffirmed in 1982) provides the outline for six classifications of shutoff.

2.3.2   Shutoff Classifications

Shutoff classifications are determined by a percentage of a test fluid (usually water or air) that passes through the valve, as part of the valve's rated capacity. This must take into account the predetermined pressure, temperature, and time limits. Shutoff classifications range from ANSI Class I, where the valve does not require tight shutoff, to ANSI Class VI, where shutoff must be complete or nearly bubble-tight. The following briefly describes each shutoff classification and maximum leakage rates for each.

The ANSI Class I shutoff is an open classification that does not require a test, while allowing for a specified agreement between the user and the valve manufacturer as to the required leakage. The ANSI Class II shutoff is 0.5 percent of the rated valve capacity and is associated with double-ported seats or pressure-balanced trims where metal piston rings and metal-to-metal seat surfaces are used. The ANSI Class III shutoff is 0.1 percent of rated valve capacity and is associated with the same types of valves listed in Class II, but is used for applications that require improved shutoff.

The ANSI Class IV shutoff is the industry standard for single-seated valves with metal-to-metal seating surfaces, which calls for a maximum permissible seat leakage of 0.01 percent of rated valve capacity. To achieve this higher classification with metal-to-metal seating surfaces, the load applied to the surfaces from the manual operator or actuator must reach certain levels. Table 2.7 provides a listing of typical required seat loads for Classes IV, V, and VI with metal and soft seating surfaces.

Both ANSI Classes V and VI were developed for throttling valves where shutoff is a primary focus. The ANSI Class V shutoff is defined as 0.0005 cm^2/min per inch of orifice diameter per pounds-per-square-inch (psi) differential. Class V is unique in that it is the only classification where the allowable seat leakage is allowed to vary according to the orifice diameter and the differential pressure (pressure drop). This classification is necessary for those applications where a throttling or control valve is used as a blocking valve that is required to stay closed for lengthy periods against a high pressure drop. It is applied to single-seat valves with either metal or soft seating surfaces or with pressure-balanced trim that requires extraordinary seat tightness.

The ANSI Class VI shutoff is commonly referred to as *bubble-tight shutoff* and is associated with metal-to-elastomer soft seating surfaces (such as using an elastomer insert in the seat ring or the plug head)—although with extremely high seating loads (as shown in Table 2.7), it is possible to achieve Class VI shutoff with a metal-to-metal seat. Class VI is independent of the pressure differential, but it does take into account milliliter per minute of leakage versus the seat orifice diameter. That means that valves with large seat diameters applied to a service with a low pressure drop can have a lesser leakage requirement than Class V. Figure 2.13 shows this relationship between Classes V and VI, taking into account the pressure differential for Class V and the lack of pressure differential for Class VI.

2.4 Body End Connections

2.4.1 Introduction to End Connections

A number of different end connections are available that allow the valve to be joined to the system's piping. In most cases, the valve's end

Table 2.7 Typical Seat Loads vs ANSI Classification for Shutoffs*

Seat Surface	ANSI Shutoff Classification	Valve Sizes	Required Seat Load (linear seating force)
Metal	Class IV	0.5 to 4-inch DN 15 to 100	50 pounds/inch 60 joules
Metal	Class IV	6-inch and above DN 150 and above	75 pounds/inch 91 joules
Metal	1% of Class IV	0.5 to 4-inch DN 15 to 100	100 pounds/inch 121 joules
Metal	1% of Class IV	6-inch and above DN 150 and above	150 pounds/inch 181 joules
Metal	Class V	0.5 to 4-inch DN 15 to 100	250 pounds/inch 303 joules
Metal	Class V	6 to 10-inch DN 150 to 250	400 pounds/inch 484 joules
Metal	Class VI	0.5 to 4-inch DN 15 to 100	250 pounds/inch 303 joules
Metal	Class VI	6 to 10-inch DN 150 to 250	400 pounds/inch 484 joules
Soft	Class V	0.5 to 4-inch DN 15 to 100	50 pounds/inch 60 joules
Soft	Class V	6-inch and above DN 150 and above	100 pounds/inch 121 joules
Soft	Class VI	0.5 to 4-inch DN 15 to 100	50 pounds/inch 60 joules
Soft	Class VI	6-inch and above DN 150 and above	100 pounds/inch 121 joules

Data courtesy of Valtek International.

connection is designed or specified to match the piping connection. In an ideal situation, end connections and materials between the valve and the piping would be identical; however, this is not always the case.

The general rule is that smaller-sized valves—smaller than 2-in (DN 50) valves—can use threaded connections (Fig. 2.14), while larger

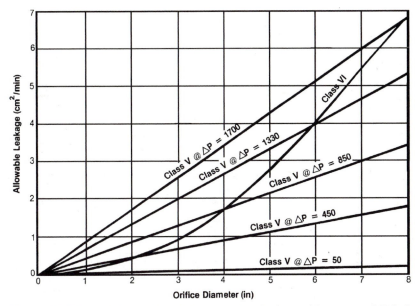

Figure 2.13 ANSI Class V and VI allowable leakage. (*Courtesy of Valtek International*)

sizes—2-in (DN 50) and larger—use flanged connections (Fig. 2.15). The refining industry uses such a standard, since it is very conscious of fugitive-emission mandates against leakage. Some process systems where fugitive emissions or process leakage is not a problem (such as water systems) will use threaded connections in sizes up to 4 in (DN 100).

Most process system applications require both ends of the valve to have identical connections. On some applications, such as vent and

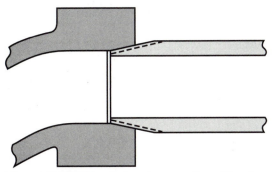

Figure 2.14 Threaded end connection. (*Courtesy of Valtek International*)

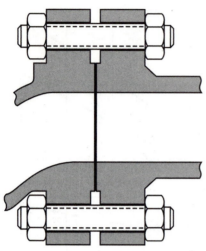

Figure 2.15 Integral flange end connection.
(*Courtesy of Valtek International*)

drain valves, one end may require one type of connection on the
upstream port and a different connection on the downstream port.

2.4.2 Threaded End Connections

As noted above, threaded connections are used in smaller sizes—1.5 in
(DN 40) and smaller. The standard end connection for valves smaller
than 1 in (DN 25) is a threaded connection. If leakage is not a concern,
threaded connections can be used in sizes up to 4 in (DN 100).

The valve's end connection is designed with a female National Pipe
Thread (NPT), which mates with the piping that uses a male NPT
thread. Because of the leakage and pressure limitations of threaded
ends, they are only rated up through ANSI Class 600. Also, threaded
ends should not be used with corrosive processes, since the threads
can either fail or become inseparable.

A National Pipe Thread is the most commonplace thread joint. One
exception is for fire management systems, which require the use of the
National Hose Thread (NHT), which matches connections used by fire
departments. Another thread occasionally seen in a process system is
the ordinary ¾-in Garden Hose Thread (GHT). Threads can be either
cut or molded in place, especially when precision moldings are used.
The molded threads do not have sharp edges (which are produced by
machining), but are more rounded at the peak of the thread.

When used in smaller sizes, threaded connections are easy to install since the valve is smaller and lightweight. This is important because the pipe and valve must be rotated to make the connection. In some cases, the pipe fitting will require piping tape or compound to ensure a tight seal.

Because the threaded design requires little machining and is commonplace among most valve manufacturers, it is the least expensive to specify.

2.4.3 Flanged End Connections

Flanges are commonly required on valves larger than 2 in (DN 50). Flanges are easier to install than threaded connections, because the valve's face is matched up with piping and bolted together without any rotation of the pipe or valve. Flanges can be applied in most temperatures, from absolute zero to 1500°F (815°C). As the temperature increases, some limitations are placed on high pressures.

Force generated by the flange bolting, coupled with the gasket between the flanges, is used to seal the connection. Flanges are built to the ANSI Standard B16.5 (or API 6A or similar standards), which addresses design criteria for the flat face, the height and diameter of the raised face, standard hole patterns, and the necessary dimensions for even rare joints, such as tongue and groove, and male and female designs. Flanges are rated according to the type of service, material requirement, maximum service temperature, and pressure. Although the main advantage of flanges is that the valve can be removed easily from the line, flanges are subject to thermal distortion and shock. If temperature cycles vary significantly, then a welded connection should be considered as an alternative.

Two types of flange designs exist. *Integral flanges,* as the name implies, are an integral part of the body. With integral flanges, the flange hole pattern is either machined or cast into the body casting. Integral flanges are commonplace since they are standard with many valve manufacturers and have been used from the earliest designs. On the other hand, *separable flanges* have been a relatively new addition to end-connection design. Separable flanges are individual flanges that slide over the hub ends of the body and are held in place by half-rings.

Integral flanges can be provided with a *flat face* (Fig. 2.16), which allows full contact between the two matching flanges and the flange gasket. Flat-face flanges are commonplace with low-pressure applications as well as brass and cast-iron valves. Because the flanges are in complete contact with each other, this design minimizes flange stresses

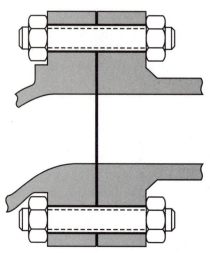

Figure 2.16 Flat-face end connection. (*Courtesy of Valtek International*)

as well as possible bending of the flange as the bolting is tightened. However, the flange faces must be completely flat to create an equal seal through the entire flange. When flat-faced flanges are specified, larger-diameter gaskets (same as the flange) are used to provide the seal.

Another common flange face is *raised face* (refer to Fig. 2.15), which is a circular area that physically separates the two flanges. The raised face is only a slight step. The inside diameter of the raised face is identical to the inside diameter of the pipe–valve port, while the outside diameter is smaller than the bolt circle. ANSI standards call for this raised face to be 0.06 in (1.5 mm) below ANSI Class 600 (PN 100) and 0.25 in (6 mm) in sizes above ANSI Class 600 (PN 100). The raised face separates the flanges themselves, preventing any incidental flange-to-flange contact that may result in decreased gasket sealing pressure, although some flange stress may be created when the bolting is tightened. This raised face may be serrated with concentric circular grooves when using simple sheet gaskets or may have a smoother surface if spiral-wound gaskets are used. The raised face is finished with a series of concentric circular grooves, which are designed to keep the gasket in place (preventing blow-out) and to provide a better seal. This type of flange is specified on ANSI Class 250 iron valves and all steel valves. It is recommended in pressures through 6000 psi (400 bar) and in temperatures to 1500°F (815°C).

The *ring-type joint* (also known as RTJ) is a modification of the raised-face design (Fig. 2.17). A U-shaped groove is cut into the face,

Figure 2.17 Ring-type joint end connection. (*Courtesy of Valtek International*)

which is concentric with the valve port. A soft metal gasket (commonly Monel or iron, but any soft metal can be specified) is then inserted in this groove, which is wedged in place as the flanges are tightened. RTJ flanges are specified for high-pressure applications—up to 15,000 psi (1000 bar)—although not with high-temperature applications.

As mentioned earlier, separable flanges (Fig. 2.18) are now accepted as an inexpensive, versatile alternative to integral flanges. Because the flange is not wetted by the process, it can be produced from simple carbon steel and be painted for atmospheric protection, which lowers the cost of a valve that requires a stainless-steel or alloy body. The separable flange is designed to slide over the body hub. To fasten the flange in place, two half-rings are inserted in a groove in the body, which act as mechanical stops. When the flange bolting is tightened, the flanges lock against the rings, holding the valve body securely in place. Although carbon steel is the most common (and inexpensive) material for separable flanges, stainless steel flanges are necessary for high-temperature–high-pressure applications.

One important advantage of separable flanges over integral flanges is their range of motion when dealing with misaligned pipe flanges. If the flange of an upstream pipe is fixed in place and is not exactly aligned with the flange of the downstream pipe, the misalignment will prevent the installation of a valve with integral flanges—unless the flange and

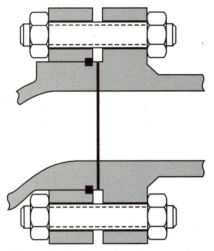

Figure 2.18 Separable flange end connection.
(*Courtesy of Valtek International*)

valve hole patterns are modified to align the holes. On the other hand, with separable flanges, the flange on either end of the valve can be rotated slightly to compensate for the misalignment. This ability to modify the alignment of the flanges also allows the valve to be rotated and fixed in a different position (especially if a space conflict exists).

Separable flanges can be designed to be interchangeable among low-pressure classes. They are rated to ANSI Classes 150–600 (PN 16–PN 100) in sizes of 4 in (DN 100) and smaller. With ANSI Classes 150–300 (PN 16–PN 40), flanges are available in 6- and 8-in sizes (DN 150 and DN 200). Separate flanges can also be used with ANSI Class 150 (PN 16) in sizes larger than 10 in (DN 250).

Although the separable flange design is less expensive and more versatile, one drawback is that if the flange bolting is not properly tightened, the valve could rotate accidentally because of gravitation forces or excessive line vibration—especially if the valve has a heavy actuator or other top-works. Following installation, this problem may be remedied by using tack welds to keep the flange or body from rotating.

2.4.4 Welded End Connections

When zero leakage is required—for environmental, safety, sanitary, or efficiency reasons—the piping can be welded to the valve, providing one-piece construction. Many users insist that high-pressure applications—ANSI Class 900 (PN 160) and higher—require a permanent end

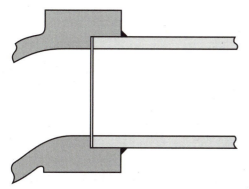

Figure 2.19 Socketweld end connection.
(*Courtesy of Valtek International*)

connection, especially if they involve high temperatures. Nearly all steam and water services in the power industry call for welded connections. The two most common welded connections are socketweld and buttweld connections.

The *socketweld connection* (Fig. 2.19) is specified in high-pressure–high-temperature fluids in sizes 2 in (DN 50) and smaller. The socketweld design for a valve involves boring into the valve's body end to a predetermined depth (according to ANSI Standard B16.11). The piping is then mated or inserted into the bore, and a weld is then applied between the pipe outside diameter and the face of the body. The welding standard for socketweld connections is the piping welding specifications according to the local or ANSI codes (B31.1 or B31.3).

For larger valve sizes 3 in (DN 80) and larger, a *buttweld connection* (Fig. 2.20) is specified for high-pressure–high-temperature applica-

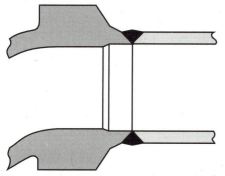

Figure 2.20 Buttweld end connection.
(*Courtesy of Valtek International*)

tions. Buttweld ends involve a lip that butts up against a similar lip on the pipe. Following the lip, both the pipe and valve use a single- or double-angle bevel to create a V-shaped butt joint that is filled with a full penetration weld. Some smaller industrial valves may incorporate a J-bevel or U-bevel in the design. These joints are harder to manufacture, but easier to inspect with radiology. Most buttweld ends are specified according to ANSI Standard B16.25, which calls for a 37.5° angle for wall thicknesses up to $\frac{7}{8}$ in (22 mm). If the wall thickness exceeds $\frac{7}{8}$ in, a compound buttweld of 37.5° and 10° is specified.

The user may also designate a special buttweld design according to individual specifications. For example, power applications sometimes require the use of a backing ring, which must be incorporated into the buttweld specifications. Backing rings are inserted to ensure proper alignment of the pipe and valve.

When considering socketweld and buttweld connections, material compatibility between the valve and piping must be a consideration to ensure proper welding and mating of the valve to the piping. Since carbon alloys or high-chrome steel have a tendency to air-harden, they should be avoided (or be heat-treated.)

2.4.5 Other End Connections

Nonmetallic valves, of which plastic is the most common, are equipped with other types of end connections. Small plastic valves can be manufactured with *union end connections,* which are used to join the plastic valve to plastic piping. Each end of the valve retains an external nut that can be threaded onto the pipe to make a solid connection. Small plastic or metal valves used in vacuum service can be equipped with an O-ring joint.

Valves made from polyvinylchloride (PVC) and chlorinated polyvinylchloride (CPVC) use a male–female socket arrangement, similar to the socketweld design, except that a solvent cement is used to fuse the two pieces together. Another method used to bond plastic piping and valves is heat fusion, in which an outside heating source melts the plastic and allows the two parts to fuse together.

Iron valves can be connected to piping using a clamp coupling that fits into special grooves cut into the ends of the valve and pipe. Stainless-steel sanitary valves may use special clamp joints, which allow the system to be disassembled regularly for cleaning (Fig. 2.21).

Some rotary valves have *flangeless* connections, where the valve body—which by its rotary design has a short face-to-face—is placed between two pipe flanges, which are then bolted together. This config-

Figure 2.21 Sanitary end connection. (*Courtesy of Jordan Valve*)

uration allows the valve to be bolted securely between the flanges and uses a simple flat gasket. The outside diameter of the body hub matches the outside diameter of the raised face on the pipe end. Some consideration should be given to thermal expansion, as the longer bolting can lengthen or shorten accordingly, causing leakage or crushing the gasket, respectively. Thermal effects can be modified by using a flexible gasket that can control the compression. However, this design is only recommended when there are no fire-safe considerations. During a fire, thermal expansion can cause the bolting to expand, causing process leakage that may feed the fire.

2.5 Pressure Classes

2.5.1 Introduction to Pressure Classes

A valve is designed to handle a certain range of internal pressure up to a certain point, which is called the valve's *pressure rating*. The higher

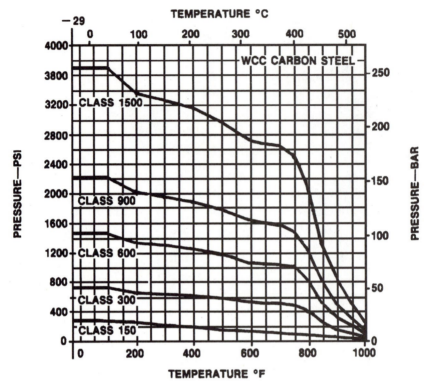

Figure 2.22 Pressure–temperature ratings for carbon steel. (*Courtesy of Fisher Controls International, Inc.*)

the pressure rating for a valve, the thicker the wall thickness must be so that the valve body subassembly will not rupture. The pressure rating is affected by the temperature of the service also: the higher the process temperature, the less pressure can be handled by the body subassembly, as shown in Fig. 2.22. ANSI Standard B16.34 is used to determine the pressure–temperature relationship, as well as applicable wall thickness and end connections.

An understanding of common pressure class ratings and pressure ratings is important, especially since a valve's pressure class can be designated as a standard class, a special class, or an intermediate class.

2.5.2 Standard Classification

The most common pressure class standard is ANSI B16.34, which specifies six *standard classes:* Class 150, 300, 600, 900, 1500, or 2500. (See Table

1.2 for nominal pressure designations.) These classes apply to valves with NPT threaded, flanged, socketweld, and buttweld end connections.

2.5.3 Special Classification

Special class ratings are available when nondestructive examination requirements are met for valves with buttweld end connections. ANSI Standard B16.34 allows buttweld valves to be upgraded to ANSI Special Classes 15, 300, 600, 900, 1500, 2500, and 4500.

2.5.4 Intermediate Classification

This ANSI standard also permits the use of *intermediate ratings* for valves with buttweld end connections, such as an ANSI Intermediate Class 3300. Using this class requires additional engineering time, but does allow a special service valve to be reduced in size, weight, and cost. For example, a carbon-steel valve is required for a 300°F (150°C) service at 6500 psi (450 bar). Normally, if using a conventional standard or special pressure class, the valve would require an ANSI Special Class 4500 pressure rating, which would increase the size, weight, and cost of the valve. However, if the ANSI Intermediate Class 3300 is chosen, a smaller valve could then be used. One point should be remembered, however. Unless the valve manufacturer has engineered this intermediate class, special design and casting patterns will be required, which may increase the cost of the valve. This added cost of new engineering should be weighed against the cost of the larger, existing valve design.

The ANSI intermediate classification can also be used to designate pressure classes larger than ANSI Special Class 4500, although one should not confuse a 6600 psi (450 bar) pressure rating for ANSI Intermediate Class 6600, which has a maximum pressure of 13,200 psi or 910 bar.

2.6 Face-to-Face Criteria

2.6.1 Introduction to Face-to-Face

The dimension between one pipe mating surface of the valve to the surface on the opposite end is called the *face-to-face* dimension. This physical dimension is always determined by the surface-to-surface measurement regardless of the type of end connection (threaded, flanged, or welded).

Most valves' face-to-face is determined by the industry standards, although some custom designs, such as Y-body valves, are determined by the manufacturer or restricted by the limitations of the design. In some cases, the user's process system layout may determine a special face-to-face. For example, some valves designed for the power industry come with buttweld end connections that are designed with custom face-to-faces.

A question often arises about the ring-joint end connection, where the sealing surface is the end of the ring and not the surface of the valve end. In this case, the face-to-face dimension is still considered to be the valve's face surfaces.

2.6.2   Common Face-to-Face Standards

Several standards for face-to-face valves are commonly used throughout the process industry, as outlined in Table 2.8. These standards have been set by the following organizations: American National Standards Organization (ANSI), Instrument Society of America (ISA), American Society of Mechanical Engineers (ASME), British Standards Institute (BSI), and Manufacturers Standardization Society of Valves and Fittings Industry (MSS).

2.7   Body Material Selection

2.7.1   Introduction to Body Materials

Normal practice calls for the control-valve user to specify the body material, especially with special service or severe service valves. Many general service valves are specified with commonly found materials, such as carbon or stainless steels. In most cases, the required body material is the same as the pipe material—which most likely is carbon steel, stainless steel, or chrome–molybdenum steel (commonly called *chrome-moly*).

Carbon steel is probably the most common material specified for valves. Overall, it is the ideal material for noncorrosive fluids. Carbon steel is also widely used for steam and condensate services. It does exceptionally well in high temperatures: up to 800°F (425°C) in continuous service, or even up to 1000°F (535°C) in noncontinuous service. Carbon steel is readily available in most common general service valves and generally inexpensive, especially when compared to other commonly used metals.

Stainless steel is very corrosion resistant, extremely strong, and is commonly specified for high-temperature applications—temperatures

Table 2.8 Common Face-to-Face Standards

Standard	Valve Type	Pressure Rating
ANSI/ISA S75.03	Globe valves	150 - 600 (valve is interchangeable between Class 150, 300 and 600)
ANSI/ISA S75.04	Flanged globe valves	125, 150, 250, 300, 600
ANSI/ISA S75.04	Flangeless globe valves	150, 300, 600
ANSI/ISA S75.08	Flanged clamp or pinch valves	All classes
ANSI/ISA S75.12	Socketweld and threaded end globe valves	150, 300, 600, 900, 1500, 2500
ANSI/ISA S75.14	Buttweld globe valves	4500
ANSI/ISA S75.15	Buttweld globe valves	150, 300, 600, 900, 1500, 2500
ANSI B16.10	Iron (ferrous), gate, plug, globe valves	All classes
BS 2080	Steel valves used in the petroleum, petrochemical and associated industries	All classes
MSS SP-67	Butterfly valves	All classes
MSS SP-88	Diaphragm valves	All classes
MSS SP-42	Stainless steel valves (gate, globe, angle and check)	All classes

of 1000°F (535°C) and higher. Its cost is somewhat higher than carbon steel, although less than other steel alloys.

Chrome–molybdenum steel is a good material that falls between the characteristics of carbon steel and stainless steel. It can handle higher pressures and temperatures than carbon steel, making it ideal for high-pressure steam or flashing condensate applications. Its strength surpasses carbon steel and is nearly equal to that of stainless steel. However, chrome–molybdenum steel is not as corrosion resistant as stainless steel.

Special alloys are specified for special service or severe service valves. For example, Hastelloy B and C or titanium may be selected to avoid fluid incompatibility, such as a highly acidic fluid. In another case, a Monel or bronze body may be selected for a pure oxygen service, where having a nonsparking material is critical for safety reasons.

Table 2.9 lists a number of common valve materials and their temperature limits. Valve bodies are manufactured from castings, forgings, or barstock, or can be fabricated from piping tees and flanges. Castings are the least expensive choice because of the process and the higher volumes run by the manufacturer. Forgings are required for special materials and/or higher-pressure ratings, such as ANSI Classes 1500 (PN 250), 2500 (PN 400), or 4500 (PN 700). Barstock bodies are required for critical deliveries where a cast or forged body is not readily available, or when structural integrity is essential. Fabricated bodies are required for large angle valves.

As a general rule, bonnets or bonnet caps (which are used to seal the upper portion of the body subassembly) are made from the same material as the body, although most are manufactured from barstock instead of castings. One exception to this rule is a low-pressure chrome–molybdenum valve, which often requires a stainless-steel bonnet as the standard for sizes 6 in (DN 150) and smaller.

2.7.2 Material Selection Standards

Since several parts of a valve are exposed to pressure, process fluid, corrosion, and other effects of the service, those parts are required by regulation to be manufactured from approved metals. These parts are usually specified as the body, bonnet, bonnet bolting, plug, ball, disk, wedge, and/or drainage plug. Although a plug stem or rotary shaft extends from the pressure vessel, they are not considered to be pressure-retaining parts by the leading quality- and safety-related organizations.

The American National Standards Institute publishes specific pressure and temperature limits for specified materials (Standard B16.34). This standard should be reviewed before any material is selected to ensure that it will fall within the correct pressure–temperature limits. Materials used in the construction of pressure-retaining parts are designated by codes formalized by the American Society for Testing and Materials (ASTM). ASTM provides specifications for materials as they are subjected to that organization's standard testing procedures, as well as acceptance criteria. ASTM codes are critical in that they ensure that a material is duplicated time and time again according to correct specification, regardless of the manufacturer. If the material is pro-

Table 2.9 Temperature Limits for Body Materials†

Material	Upper Limit (°F)	Upper Limit (°C)	Lower Limit (°F)	Lower Limit (°C)
Cast Iron	410	210	-20	-5
Ductile Iron	650	345	-20	-5
*Carbon Steel (Grade WCB)	1000	535	-20	-5
Carbon Steel (Grade LCB)	650	345	-50	-10
Carbon Moly	850	455	-20	-5
1-1/4 Cr - 1/2 Mo (Grade WC6)	1000	535	-20	-5
2-1/4 Cr - 1/2 Mo (Grade WC9)	1050	565	-20	-5
5 Cr - 1/2 Mo (Grade C5)	1100	595	-20	-5
9 Cr - 1 Mo (Grade C12)	1100	595	-20	-5
Type 304 (Grade CF 8)	1500	815	-425	-220
Type 347 (Grade CF8C)	1500	815	-425	-220
Type 316 (Grade CF8M)	1500	815	-425	-220
3-1/2 Ni (Grade LC3)	650	345	-150	-65
Aluminum	400	205	-325	-160
Bronze	550	285	-325	-160
Inconel 600	1200	650	-325	-160
Monel 400	900	480	-325	-160
Hastelloy B	700	370	-325	-160
Hastelloy C	1000	535	-325	-160
Titanium	600	315	N/A	N/A
Nickel	500	260	-325	-160
Alloy 20	300	150	-50	-10

†*Courtesy of Valtek International.*

*The carbon phase of carbon steel may be converted to graphite upon long exposure to temperatures above 775°F (415°C). Check applicable codes for maximum temperature ratings of various materials. Other specific data available in ANSI B16.34.

duced according to specification, its properties will be able to withstand or handle the application it was designed for, such as corrosive fluids, severe temperatures, or high pressures.

ASTM codes are not intended to cover all known materials but do cover all common materials used in known applications. Since a number of new materials are being introduced annually, ASTM has procedures that allow new materials to be submitted for acceptance, and sometimes even allowed to be used before being formally accepted by ASTM, as long as the procedures are followed exactly. Table 2.10 provides applicable body and bonnet material standards (ANSI Standards B16.34 and B16.24) for castings, forgings, and barstock.

Another organization associated with the manufacture and performance of pressure-retaining parts is the American Society of Mechanical Engineers (ASME), which oversees and publishes *The Boiler and Pressure Vessel Code.* Section II of that code covers material selection for equipment that is under pressure, which includes valves. A comparison of the materials outlined in Sec. II with ASTM-specified materials shows that nearly all are covered by both standards. The

Table 2.10 Common ASTM Materials for Bodies and Bonnets*

Body Type	Material	Body Standard	Bonnet Standard
Castings	Stainless Steel	A351-CF8M	A479-316
	Carbon Steel	A216-WCB	A675-70
	Chrome-moly	A217-WC6	A479-316
		A217-WC9	A479-316
		A217-C5	A479-316
Forgings	Stainless Steel	A743-CF8M	A479-316
	Carbon Steel	A105	A675-70
	Chrome-moly	A182-F11	A479-316
		A182-F22	A479-316
		A182-F5a	A479-316
Barstock	Stainless Steel	A182-F316	A479-316
		A479-316	A479-316
	Carbon Steel	A675-70	A675-70
	Chrome-moly	See Forgings	See Forgings

Data courtesy of Valtek International.

materials listed in Sec. II carry the same numerical designation as ASME, although ASTM uses a specification prefix "S" before the number. ASME also regulates procedures for welding, heat-treating, and preheating. Another organization, the American Welding Society (AWS), oversees procedures and regulations for welding rod and wire.

2.7.3 Material Testing Methods

Valve materials are often examined to verify that the material in question is indeed the material specified by the user or that a part is not flawed by small cracks, voids, laminations, or porosity. Porosity is characterized by small bubbles in the metal formed during the casting process. In many cases, valve users require such testing to bolster their confidence that the manufacturer's valve is corrosion-resistant and has the required hardness or pressure-retaining ability. In the case of certified or nuclear-certified valves, examination using certain testing methods is a requirement by the user and/or may be subject to government regulations. Larger valves or valves designed for higher-pressure classes, ANSI Class 900 (PN 160) and higher, are examined for flaws in the metal surfaces that may result in leaks or a burst vessel in high pressures. Such examinations and the procedures and methodology are established by ASME and AWS.

The term *examination* refers to any test performed by the machinist or technician producing the part, a quality assurance engineer or inspector, the end user, or a representative of the user. On the other hand, the term *inspection* is used when that same test is conducted and/or witnessed, and approved or rejected by an authorized representative of the user or the user's insurance carrier.

To verify any metal alloy that is not carbon-based, an x-ray spectrometer can be used (Fig. 2.23). In most cases, however, a *certified material test report* can be produced for the material, which traces the history of the metal back to its heat number and batch number from the foundry.

Five examinations are commonly used to verify a material and to test its integrity and ability to perform as required: visual, magnetic particle, liquid penetrant, radiology, and ultrasonic.

The most commonplace examination is *visual* (coded *VT* by the AWS), which involves looking closely at the part for any defects—such as cracks, surface porosity, etc.—outside those accepted by a written or reference photograph or illustration. Visual examination may also include looking at the surface finish or overall quality of workmanship.

Figure 2.23 X-ray spectrometer. (*Courtesy of Valtek International*)

The *magnetic particle examination* is a quick method to determine small cracks or porosity in the surface of the metal that cannot be seen with a visual examination. It works best with ferrous castings and forgings. Generally it is an easy procedure that can be accomplished quickly with little setup. However, special electronic equipment is required and it is only effective with ferrous material, such as carbon steel and certain ferrous alloys. It cannot be used with nonmagnetic metals, such as stainless steels. The magnetic particle examination involves sprinkling iron filings or a similar granular ferrous material over the metal surface to be tested and then passing an electrical current through the metal part. As the electrical current generates magnetic lines of force on the metal surface, the iron filings will cluster along those lines. If the metal is void of cracks or porosity, the lines will remain uninterrupted. However, if a crack or pore exists, the iron filings will cluster around that area and reveal the flaw to the examiner. In most cases, a magnetic particle examination is only used for surface flaws, although a large void just under the surface may also be detected.

The *liquid penetrant examination* is also used to find minute surface cracks or porosity that may not be as evident in a magnetic particle examination. It is used with cast parts with a great deal of curved surfaces, such as valve bodies, or with machined rounded surfaces. However, by virtue of its methodology, it is somewhat more involved and time-consuming than a magnetic particle examination. However, because it does not rely upon any special equipment or power supply, it can be done nearly anywhere. In a liquid penetrant examination, a red liquid is sprayed onto the metal surface. After the dye dries, any excess is wiped from the surface. A developer is then sprayed over the surface, which leaves a white film after drying. This white film has a tendency to draw out any red dye that may bleed from the cracks or porosity. Another option with a liquid penetrant examination is to use a fluorescent dye, which involves examining the part in a darkened room with a black light.

Radiography is similar to an x-ray procedure and is used to check for any cracks or voids under the surface of the part. In almost all cases, such testing requires a certified technician and prudent safety measures. A γ radiation source is placed on one side of the part and unexposed film on the opposite side. The radiation source is removed from its lead container for a certain amount of time, which exposes the film. The film is then processed, and using a penetrameter under-surface cracks and voids can be discovered. (A penetrameter is a tool placed behind the film to help determine the actual size and depth of the flaw.) Although radiography is usually thorough, some cracks that are parallel to the plane of the unexposed film may not show up when developed. In this case, an ultrasonic examination could be used to detect such cracks.

The last examination commonly used is *ultrasonic*. An ultrasonic examination involves generating a high-frequency wavelength that is directed toward the metal surface. By looking at the reflection of the returning waves in an oscilloscope, any cracks or laminations in the interior of the metal part can be seen. Although ultrasonic examination works well with finding cracks or laminations, it may have a problem with finding small voids or porosity. For this reason, ultrasonic and radiographic testings are conducted at the same time as a safeguard against missing both cracks and voids.

The term *hardness* refers to the ability of a metal to resist penetration, deformation, or denting. A number of hardness tests can be conducted to determine a metal's ability to resist penetration and can sometimes be used to determine tensile strength. Two tests are commonly used: Brinnel and Rockwell. The standard *Brinnel hardness test* uses a hardened ball—called the *indentor*—to apply a standard load to the metal. The indentation is then measured to determine the *Brinnel hardness number*, which can vary between 111 (soft) and 745 (hard). The *Rockwell hardness test* is similar in some aspects to the Brinnel hardness test in that it uses a steel ball or a diamond point to penetrate the metal. However, the depth of the penetration is measured to determine the Rockwell hardness number, which has two scales: B and C. The Rockwell B scale is for softer metals and ranges normally from 100 (soft) to 65 (hard). The Rockwell C scale is for harder metals and ranges normally from 65 (soft) to 20 (hard).

The *Charpy impact test* is used to determine how brittle a metal is at a specific temperature. Some metals are more brittle at lower temperatures. The test is conducted to see if the metal will stay intact even if a force is applied to it under abnormal temperatures. The Charpy impact test involves using a notched sample of metal, which is struck by a blow from a testing machine. The sample can be measured according to the amount of force needed to fracture the sample or by the amount of fracture if the force remains constant. Temperatures can also be varied during the test to determine the brittle behavior over that temperature range.

2.8   Gasket Selection

2.8.1   Introduction to Gaskets

A gasket is a malleable material, which can be either soft or hard, that is inserted between two parts to prevent leakage between that joint. It is designed to be placed in a predetermined space in a joint between the two parts. This space may be a counterbore, groove, or retainer plate (Fig. 2.24). Pressure is applied by bolting or using a clamp to compress the gasket firmly in place. As a general rule, to avoid damage to parts and to seal properly, gaskets must be softer in composition than the materials of the parts themselves.

Gaskets are made from all different types of materials, depending on the temperature, pressure, or fluid characteristics of the process. Some are designed to be resilient or self-energizing to allow for variations in temperature or pressure, which may require the gasket to expand or condense accordingly. Other gaskets, when used in more constant or severe service conditions, are made with harder materials (such as soft metals) that provide a strong seal, but are not self-energizing and once compressed may not be used again.

Gaskets are used in valves for three major purposes. First, as mentioned earlier, gaskets prevent leakage around the closure mechanism. For example, a gasket is used to seal the joint between the body and seat in a linear valve to prevent leakage from the upstream side of the valve to the downstream side. Without the gasket, the fluid would leak

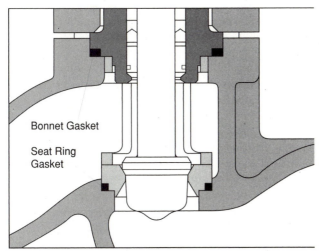

Figure 2.24 Gasket placement in typical globe valve design. (*Courtesy of Valtek International*)

past the seated plug. Second, gaskets are used to prevent leakage of fluid to atmosphere. For example, split-body and top-entry valves are designed with gaskets at the disassembly joints. Third, gaskets are used to allow the function of internal mechanisms that depend on separate fluid chambers, such as pressure-balanced trim.

Obviously, the ability for gaskets to function correctly is dependent on the correct seating load, which can vary widely according to the style of gasket, free height, wall thickness, material, and groove (or step) depth. Usually the valve manufacturer provides a torque specification for the associated bolting to ensure the proper seating load for the gasket. A common problem with such torque requirements is that if a torque wrench is not readily available, the risk may exist for a technician to overtighten the bolting, thus crushing the structure gasket, which can actually create a leak path. On the other hand, some valve designs prevent gasket crushing by using a metal-to-metal fit between the two mating parts, which ensures the proper gasket seating compression without a torque wrench. When the two parts are tightened so that they achieve a metal-to-metal connection, the height of the step and the gasket compression are assured. When the metal-to-metal connection is achieved, it can easily be felt through the wrench.

Gaskets come in a number of different styles, the most common being flat gaskets, spiral-wound gaskets, metal O-ring gaskets, metal C-ring gaskets, metal spring-energized gaskets, and metal U-ring gaskets. In some applications, the gaskets are coated with a rubber or plastic material to improve the self-energizing ability of the gasket or the corrosion resistance of the gasket. Some metal O-rings can be plated to improve the corrosion resistance.

To seal adequately, the gasket surfaces of the step or groove must be sufficiently smooth and flat. Ideally, surfaces should be finished to between 125 and 500 μin RMS (root mean squared) (between 3.2 and 12.5 μm).

Common specifications for these gasket styles are found in Table 2.11.

Table 2.11 Typical Gasket Specifications†

Type	Gasket Material	Maximum Temperature (°F/°C)	Minimum Temperature (°F/°C)	Maximum Pressure (psi/bars)
Flat	Virgin PTFE	350/175	-200/-130	6000 - 1000 psi 415 - 70 bar
Flat	Reinforced PTFE	450/230	-200/-130	6000 - 500 psi 415 - 35 bar
Flat	CTFE	200/95	-423/-250	6000 - 500 psi 415 - 35 bar
Flat	FEP	400/205	-423/-250	6000 - 500 psi 415 - 35 bar
Spiral-wound	AFG*	1500/815	-20/-30	6250 psi 430 bar
Spiral-wound	304 SS/Asbestos	750/400	-20/-30	6250 psi 430 bar
Spiral-wound	316 SS/Asbestos	1000/540	-20/-30	6250 psi 430 bar
Spiral-wound	316 SS/PTFE	350/176	-200/-130	6000 - 500 psi 415 - 35 bar
Spiral-wound	316 SS/Graphite	1500/815	-423/-250	6250 psi 430 bar
Hollow O-ring	Inconel X-750	1500/815	-20/-30	15,000 psi 1035 bar

†*Data courtesy of Valtek International*
*Asbestos-free gasket.

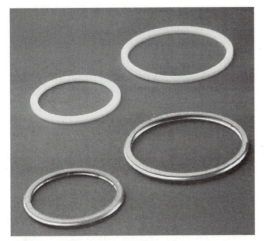

Figure 2.25 Flat (above) and spiral-wound (below) gaskets. (*Courtesy of Valtek International*)

2.8.2 Flat Gaskets

Of the different types of gaskets, the most simple and inexpensive are *flat gaskets,* which as the name describes are gaskets that are machined with a simple outside diameter, inside diameter, and a certain height (Fig. 2.25). For the most part, these gaskets adapt easily to any irregularities in metal surfaces of the joint due to its elasticity or plastic deformation.

Flat gaskets are best used for general service applications without severe temperature or pressure considerations. Flat gaskets can be made from industrial plastics, such as polytetrafluoroethylene (PTFE) or chlorotrifluoroethylene (CTFE), or soft metals, such as aluminum, copper, silver, soft iron, lead, or brass. Some metal flat gaskets are applied to high-temperature service, such as nickel [1400°F (760°C)], Monel [1500°F (815°C)], or Inconel [2000°F (1095°C)].

2.8.3 Spiral-Wound Gaskets

Spiral-wound gaskets are all-purpose, medium-priced gaskets that consist of alternate layers of metallic and nonmetallic materials wound together (Fig. 2.25). The metal strip winding is normally V-shaped and is set on edge with the filler material sandwiched between windings. Spiral-wound gaskets combine the elastic properties of flat gaskets with the inclusion of soft metal windings, which adds strength to pre-

62
Chapter Two

vent possible gasket blow-out high-pressure–high-temperature applications. The strength of spiral-wound gaskets can be varied by the materials specified. The strength is also determined by the number of windings: the higher the number of windings, the greater the pressure load handled by the gasket. When spiral-wound gaskets are compressed, the metal layers are crushed, providing an effective seal even with uneven gasket surfaces. However, because the metal strips are deformed during compression, spiral-wound gaskets can never be reused.

As a general rule, spiral-wound gaskets should never be used with soft-seat or soft-seal designs, where the closing device, such as a plug or disk, seats against a nonmetallic surface. The force needed to compress the spiral-wound gasket is partially transmitted through the soft-seat (or seal) insert, which is more compressible than the gasket. Therefore, the soft insert is likely to extrude before the spiral-wound gasket is fully compressed. Unfortunately, the outcome is usually a damaged soft insert or a valve that leaks.

In the past, a common filler material for high-temperature spiral-wound gaskets has been asbestos paper. However, due to the controversial health and legal aspects of this material, many valve manufacturers—especially those in North America—do not offer it as a standard option. In its place, newer (and safer) filler materials have been developed or used, such as a ceramic fiber paper. Gaskets with this new filler have been known by the generic term *asbestos-free gaskets* (AFG), which can be substituted for gaskets with asbestos filler in most high-temperature applications. Their ability to seal at high temperatures is very similar to a spiral-wound gasket that contains graphite. Safety controversies and legal issues aside, asbestos gaskets are occasionally specified by users, especially by those in the power generation industry. As noted earlier, because asbestos spiral-wound gaskets are used primarily for high-temperature applications, they are typically installed in stainless-steel, carbon steel, and chrome-moly valves. Besides asbestos, common filler materials include polytetrafluoroethylene, graphite, mica, or ceramic paper.

Graphite spiral-wound gaskets are used for high-pressure–high-temperature applications associated with valves in severe service. Either 316 stainless steel or Inconel can be used for the metal windings, depending on the process fluid.

Spiral-wound gaskets can be also custom-made depending on the process fluid and its interaction with the metal windings or filler. In addition to those noted earlier, windings can be made from the following materials: 304, 315, 347, or 321 stainless steels, Monel, nickel, titanium, Alloy 20, Inconel, carbon steel, Hastelloy B, Hastelloy C-276, phosphor bronze, copper, gold, or platinum.

2.8.4 Metal O-Ring Gaskets

For exceptional severe service, *metal O-rings* are very versatile and can be applied in a wide range of services. Instead of a flat gasket design, some metal gaskets are designed as a metal O-ring, which is a tube that is circular in nature with the ends welded together. Although most are circular in shape (Fig. 2.26), they can also be formed in custom nonround or irregular shapes. Like most specialized parts, metal O-rings are more expensive than flat or spiral-wound gaskets. The hollow nature of the metal O-ring gasket allows the gasket to be compressed as the bolting or clamp is tightened, providing a reliable seal especially with high-temperature–high-pressure applications. They are especially effective in applications that involve reversing pressures. The inside volume of the rings can be pressurized for certain high-temperature–low-pressure applications.

A chief advantage of using metal O-rings is their ability to conform to the mating gasket surfaces despite any minor variations in flatness or concentricity. Like spiral-wound gaskets, once a metal O-ring has been compressed it cannot be reused but must be replaced every time disassembly takes place.

2.8.5 Metal C-Ring Gaskets

Metal C-ring gaskets are characterized by their unique shape, which is C shaped with the slot facing the inside diameter (Fig. 2.27) and the

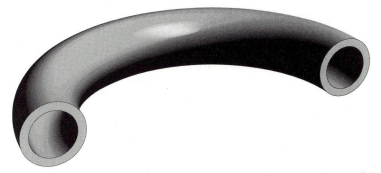

Figure 2.26 Metal O-ring. (*Courtesy of Advanced Products Company*)

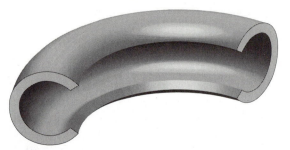

Figure 2.27 Metal C-ring. (*Courtesy of Advanced Products Company*)

pressure side of the system. This shape allows the gasket to be self-energizing. Although more expensive than most gaskets, metal C-ring gaskets are ideal for applications that require low seating loads and high spring-back. Typically they are used for low-vacuum or low-pressure systems.

2.8.6   Metal Spring-Energized Rings

Similar in some respects to metal C-ring gaskets, *metal spring-energized rings* include metal springs inside C-ring gaskets (Fig. 2.28), combining the two elements for a highly energized seal. Such gaskets required higher seating loads but provide a more consistent seal because of the greater load and increased spring-back. Generally expensive, metal spring-energized rings are specified only when the service conditions vary widely. Because critical dimensions, such as those associated with the joint, can change in a varying service, the metal spring-energized

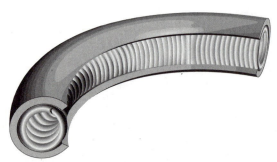

Figure 2.28 Metal spring-energized ring. (*Courtesy of Advanced Products Company*)

ring design allows the gasket to expand or contract during changes in temperature or pressure, while maintaining the seal.

2.8.7 Metal U-Ring Gaskets

Metal U-ring gaskets are designed for high-pressure (up to 12,000 psi or 828 bar working pressure) and high-temperature (up to 1600°F or 871°C) applications where reliability is an important consideration. V-shaped by design (Fig. 2.29) the inside of the U faces the pressure side or faces away when used with a vacuum, using the pressure (or vacuum) to assist with function of the gasket. Because the flared ends of the gasket must keep in constant contact with the top and bottom surfaces, those surfaces must have minimal variation in flatness and must be completely parallel.

2.9 Packing Selection

2.9.1 Introduction to Packing

Any soft material encased in a bonnet (linear and some quarter-turn rotary designs) or in a body (butterfly- and some ball-valve designs) used to seal a valve closure element's stem or shaft is called the *packing*. The packing is normally held in place by a packing follower or by guides, with compression supplied by the gland flange. The *packing follower* is a metal ring used to retain the packing inside the bonnet or bonnet cap, as well as compress the packing in a uniform manner. Packing followers are found in manual on–off or low-performance throttling valve designs. *Guides* are used with throttling valves to keep the stem or shaft of the closure element in correct alignment with the valve body, although the upper guide can also act as a packing follow-

Figure 2.29 Metal U-ring. (*Courtesy of Advanced Products Company*)

er, keeping the packing in place and transferring any force from the gland flange to the packing. The *gland flange* is a thick oblong or rectangular part that is connected to the body with bolting and straddles the guide or packing follower with the stem or shaft extending through a hole in the gland flange. When the bolting is tightened, the gland flange—through the packing follower or upper guide—transfers an axial load to the packing, compressing the packing until a seal is created against the stem or shaft and the inside of the bonnet bore. The *bonnet bore* is a term used to describe the recessed area of the bonnet or body that holds the packing. The configuration of the packing, guides, spacers, etc., is called the *packing box.*

Packing comes in a series of rings: preformed, square, or braided. *Preformed packing* is produced in a particular shape by the packing manufacturer, such as a V-ring configuration. *Square packing,* as the name indicates, is square-shaped and is formed in a solid (unbroken) ring. *Braided packing* is woven strands of a particular elastomeric material, which is manufactured similarly to rope and cut to size.

Individual rings can be grouped together, which is the case with rotary valves (Fig. 2.30), or they can be separated into upper and lower packing sets (Fig. 2.31), which is commonplace with linear valves. The difference between rotary motion—which is circular in nature—and linear motion accounts for the two different designs. Because the linear motion of the plug stem involves pulling some of the medium up into the packing box, a *lower packing set* is necessary to wipe the stem free of the fluid or any particulates in the fluid stream. In other words, the lower packing set is sacrificed to the fluid conditions to allow for proper sealing in the upper portion of the packing box. The *upper packing set* is normally placed far enough away from the contaminated portion of the plug stem to avoid exposure to the fluid medium, allowing

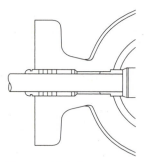

Figure 2.30 Rotary packing box design.
(*Courtesy of Valtek International*)

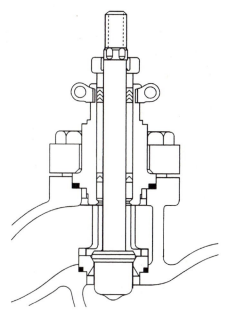

Figure 2.31 Linear packing box design.
(*Courtesy of Valtek International*)

the upper set to seal properly. Because of their circular motion, rotary valves do not require a bottom packing set to wipe the fluid.

In addition, some designs provide an allowance above the packing box to allow the use of live loading. *Live loading* is a mechanical device used to apply constant force to the packing to compensate for *packing consolidation,* which is a reduction in the packing's volume due to wear, cold flow, plastic deformation, or extrusion. In most cases, when consolidation occurs, the packing box will begin to leak and the gland flange bolting must be tightened further to seal the leakage. Using a series of disk springs, live loading avoids the need to constantly retighten the packing when consolidation occurs. With the advent of strict fugitive-emission standards, live loading is becoming a popular option. (Chapter 11 provides a more detailed discussion about live loading and fugitive-emission standards.)

Depending on their material, packings produce a unique deformation when compression is applied. Because all packing materials have some degree of fluid tendencies, the axial load that is applied can result in a wide range of radial loads. Ideally, when axial load is applied, the radial load should be at its greatest in the middle of the packing set where the maximum seal occurs. Of all packing materials, soft packing materials—such as polytetrafluoroethylene packings— provide this ideal situation, as shown in Fig. 2.32.

On the other hand, harder packings—such as graphite packings— are unique in that maximum radial force provides a seal closer to the packing guides rather than in the middle of the packing. This occurs

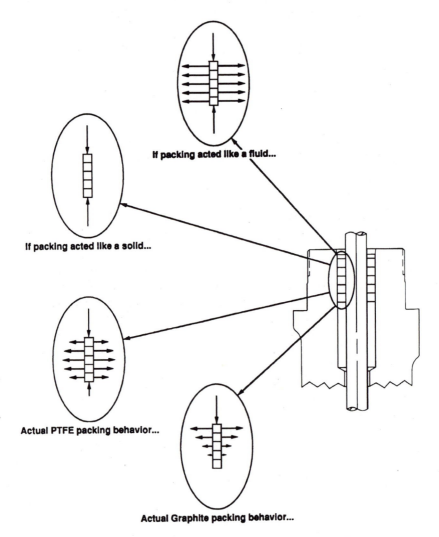

Figure 2.32 Axial pressure effects on packing. (*Courtesy of Fisher Controls International, Inc.*)

because of the high friction between the packing and the stem causes an upward axial force that is inverse to the downward force of the guide. This can be corrected by separating the graphite packing from the guide itself.

Because any variations in the surface of the stem or shaft or the packing box wall can be a potential leak path highly polished surfaces are preferred for nearly all packings. Typically, stems and shafts are polished to between 8 and 4 RMS and bonnet walls between 32 and 16 μin RMS.

Stem and shaft alignment are also critical elements of the packing box's ability to seal. If the stem flexes (inherent to smaller diameter stems) or is not concentric from inadequate guiding, the radial compression of the packing will be unequal, causing a leak path. With rotary valves, often the torque involving certain closure elements (such as a butterfly disk or an eccentric plug) can slightly misalign the shaft, causing a leak path. Obviously, the type and close tolerance of guiding are critical to maintaining the concentricity of both the stem and shaft.

2.9.2 Packing Configurations

The packing box in the bonnet or body should be designed to permit a wide variety of packing configurations. A common configuration is the V-ring design (Fig. 2.33), which uses a series of V-shaped rings designed with "feather" edges and thus provides for an excellent self-adjusting seal with minimal stem or shaft friction. The user should

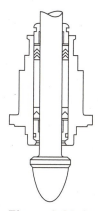

Figure 2.33 Standard V-ring packing configuration. (*Courtesy of Valtek International*)

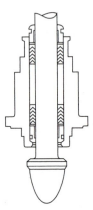

Figure 2.34 Twin V-ring packing configuration.
(*Courtesy of Valtek International*)

note that this design requires the upper packing set to seal and the lower packing set to wipe the stem. The two packing sets are separated by a packing spacer. This design requires an extremely smooth bonnet or body bore—upwards to 4 μin RMS. Leakage can occur if the stem, shaft, or bore is scratched, scored, or otherwise damaged. The twin V-ring configuration is similar to the basic V-ring design, except that the lower packing set has more V-rings (Fig. 2.34), allowing for both the upper and lower packing sets to have equal numbers of rings. In theory, some users prefer twin V-ring configurations with the idea that "if a few are good, then many are better." While this configuration may be right for better wiping of the plug stem (allowing a number of rings to be sacrificed instead of a couple), it is less likely to seal. More axial load from the gland flange must be applied to compress the additional rings, which makes sealing more difficult. In addition, twin V-ring seals are harder to remain leak-free over long periods of time. Other users employ a twin V-ring configuration with a lantern ring, which is a special spacer with holes and an undercut outer diameter in the middle of the spacer. One purpose of the undercut region of the lantern ring is to allow room for a leak to freely circulate. A sniffing device can then be connected to center region of the packing box to warn of lower packing failure and the potential for future upper packing leakage if the leak migrates past the upper set of packing. Lantern rings also permit the circulation of lubrication that may be injected into the packing box. Figure 2.35 shows a typical twin V-ring–lantern-ring configuration.

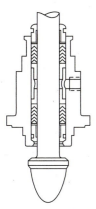

Figure 2.35 Twin V-ring packing or lantern-ring configuration. (*Courtesy of Valtek International*)

Square and braided packing can also be used for standard and twin packing configurations (Fig. 2.36). In the case of the application of square graphite packing, oftentimes a special lubricator is used with twin packing configuration (Fig. 2.37) to allow for the injection of lubrication into the graphite packing. Lubrication keeps the graphite soft and pliable while providing for smooth stem travel. Combinations of square and braided packing are used with a graphite packing configuration, which is normally applied in high-temperature services. Because die-formed solid graphite rings are extremely abrasive and create high stem friction, only one or two are used in the upper pack-

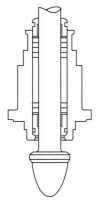

Figure 2.36 Standard square packing configuration. (*Courtesy of Valtek International*)

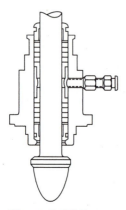

Figure 2.37 Twin square packing or lubricator configuration. (*Courtesy of Valtek International*)

ing set. However, two solid graphite rings will not adequately seal the packing box; therefore, braided graphite rings—which are softer—are used to complete the seal. A braided ring is commonly used for the bottom wiper set. Both standard and twin configurations are possible with square and braided packing (Figs. 2.38 and 2.39).

When the process fluid is at vacuum pressure or below atmospheric pressure, a special packing configuration is required. Because of their superior sealing ability, V-rings are used in a vacuum seal configuration (Fig. 2.40). If the process is always under a vacuum, the V-rings of both the upper and lower set of packing are inverted with the chevron facing away from the closure element. If the process pressure varies

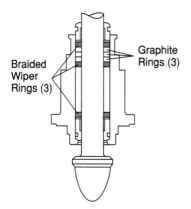

Figure 2.38 Standard graphite packing configuration. (*Courtesy of Valtek International*)

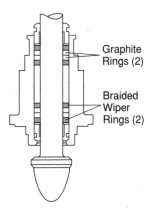

Figure 2.39 Twin graphite packing configuration. (*Courtesy of Valtek International*)

from vacuum to positive pressure at different times, a twin V-ring packing (Fig. 2.41) should be used, with the upper packing set inverted, while the lower set remains in a normal configuration. Occasionally a vacuum seal is necessary inside the packing box, which is independent of the process pressure. In this case, the twin V-ring packing configuration will permit this application. A purge may also be included to create and monitor the vacuum.

With the advent of strict fugitive-emission monitors, several configurations using special packing materials have been designed, which are detailed in Chap. 11.

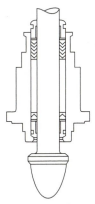

Figure 2.40 Vacuum-seal V-ring packing configuration. (*Courtesy of Valtek International*)

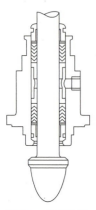

Figure 2.41 Vacuum-seal twin V-ring packing or lantern-ring configuration. (*Courtesy of Valtek International*)

2.9.3   Packing-Material Considerations

Because of the wide variety of valve applications, packing materials must be able to withstand a wide range of temperature changes, as well as withstand contact with the fluid medium, and to generate minimal stem or shaft friction. Packing materials designed for extreme temperatures must sacrifice performance in other ways. For example, graphite is a popular packing for high temperatures, but it is more difficult to achieve a seal without increasing the stem or shaft friction to the point of inhibiting performance.

As a general rule, packing materials are relatively inexpensive for general services and become increasingly more costly for services with higher temperatures and pressures or with corrosive fluids. The ideal packing material is one that operates within the temperature and pressure ranges of the service, creates minimal stem or shaft friction, holds a seal with very little material, and withstands extrusion. *Extrusion* occurs when overcompression of the packing box forces the packing material (especially softer packing materials) to deform and find a path to escape, in most cases up the stem or down into the body (Fig. 2.42).

A *backing ring* (sometimes called an *antiextrusion ring*) is a close tolerance ring made from a harder, less pliable material and is inserted at the top of the packing box to transfer the axial force from the gland flange bolting to the packing. However, the backing ring must also be soft enough to form a seal with the packing. In most cases, backing rings are installed on both sides of the polytetrafluoroethylene packing and provide an exact fit between the ring and the packing box wall as

Figure 2.42 An example of packing extrusion. (*Courtesy of Fisher Controls International, Inc.*)

well as the stem or shaft. This exact fit is critical to preventing the cold flow from extruding past the backing ring. If the ring is too large, it will provide additional friction against the metal surfaces of the valve, as well as prevent the full axial force to be transferred to the packing. If the ring is too small, it will allow extrusion to occur. The backing ring should also be made from a material that allows it to retain its shape even if thermal cycling or high compression rates are required.

2.9.4 Polytetrafluoroethylene Packing

Virgin polytetrafluoroethylene packing (the compound abbreviated as PTFE) is a common and inexpensive packing material and is typically used in the V-ring design. With the combination of PTFE's elasticity and the pressure-energized design of the V-ring, little compression is required to create a long-lasting seal. Its smooth surfaces allow for smooth stroking and minimal break-out force, which is the force necessary to begin the valve lift or stroke. PTFE provides very little friction; therefore, wear or erosion is usually not a concern. Because it is inert to many process fluids, it can be used in a number of general services. PTFE is also available in a braided packing.

One major drawback to PTFE is its limited temperature range.

Because its thermal expansion is 10 times the thermal expansion of steel, PTFE is especially vulnerable to thermal cycling, which can result in packing loss and shorter life. As PTFE is heated by the process, it expands throughout all available space, which may lead to extrusion. As the temperature drops, the packing returns to its original volume, minus the amount lost to extrusion. Because of this loss, less force is exerted against the bonnet wall or the stem or shaft and leakage can occur. Sometimes only one temperature cycle can cause leakage. When thermal cycling is present, live loading (Sec. 11.9.5) is often recommended to allow for continued sealing of the PTFE; however, eventually through a number of thermal cycles, the packing volume will be so reduced that the force provided by live loading will be inadequate to seal the packing.

Because PTFE is very fluid, another disadvantage is its tendency to consolidate over a period of time. This long-term consolidation is called *cold flow,* and occurs when the packing is compressed several times or if live loading is used. This can occur even if minimal compression force is applied. As a result, the packing can eventually extrude out of the packing area and will not have the material volume to respond to further compression. At that point, the only option is to replace the packing. When PTFE cold flows, backing or antiextrusion rings can be installed that will slow the process. Another option is to reduce the packing compression force, which may lead to an increased chance of leakage. Another drawback to PTFE is that it is not suitable for nuclear-certified valves since radiation can quickly deteriorate the material.

Filled polytetrafluoroethylene is similar to virgin PTFE, although it includes some glass or carbon in its content, which provides for a more rigid V-ring that is less likely to consolidate. However, with less elasticity than virgin PTFE, its ability to seal requires greater force and is not as reliable. Occasionally, the user will alternate rings of virgin PTFE and filled PTFE, combining the benefits of both materials.

2.9.5 Asbestos and Asbestos-Free Packings

In the past, asbestos packing has been used as an effective high-temperature packing with good sealability. However, user interest and installation of asbestos packing have waned with recent litigation as well as with health and safety concerns. Asbestos has hook-shaped fibers that can be ingested into the lungs. Some studies indicate that such ingestion can lead to respiratory illnesses. For this reason, asbestos is not specified for use in North America. However, in some industrial areas outside of North America, asbestos is still in common use. In response to the North American process industry's move away from asbestos, a replacement packing called *asbestos-free packing* (abbreviated *AFP*) was developed. AFP uses a number of substitutes for asbestos (such as ceramic fiber paper) and is normally found in high-temperature applications.

2.9.6 Graphite Packing

As a substitute for asbestos, *graphite packing* and other carbon-based packings are specified for high-temperature applications. Generally considered to be one of the more expensive packings, graphite packings can be produced either in die-formed rings or braided rings.

Die-formed rings are produced from graphite ribbon, which is wound and then compressed in a die according to the specified pressure. This pressure to form the rings is less than the force required to compress the rings to seal the packing box. Thus, the graphite rings reach their designed density (approximately 90 to 100 lb/ft^3 or 1440 to 1600 kg/m^3) not at formation but when installed under compression. Figure 2.43 shows the relationship between graphite density and the compression stress. This compression is not permanent, because die-formed rings are resilient to a certain extent, although not even close to the resiliency of PTFE.

Braided graphite is produced by winding small strands of graphite together, which makes it quite pliable as compared to die-formed graphite. When used as a sealing packing, it forms to the stem or shaft so well that the resulting stem friction impedes free movement of the stem or shaft. Because of this problem, braided graphite is not used to seal, but rather to act as an antiextrusion ring on both sides of the die-formed rings. However, this may cause a problem when high compression is needed, since the braided graphite has a tendency to grab the stem or shaft and not transfer the load to the die-formed rings. Hence, higher friction results from the braided graphite, yet leakage may occur because insufficient load is reaching the primary seal, which is the die-formed rings. Another problem associated with braided graphite is that it has a tendency to break down when compression exceeds 4000 psi (275 bar).

Graphite packing comes in high-density and low-density composites. For the most part, high-density graphite is much more durable and holds a seal longer but creates extremely high friction, which can lead to premature wear of the stem or shaft. It may also impede the

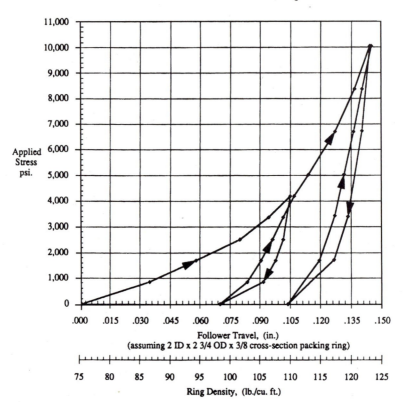

Figure 2.43 Graphite density changes according to compression stress. (*Courtesy of Fisher Controls International, Inc.*)

stem or shaft from stroking freely. On the other hand, low-density graphite is softer and allows for smoother stroking, but must be retightened more often.

Graphite offers a number of advantages. Overall, graphite remains stable through a wide range of thermal cycles. Because its thermal expansion is nearly identical to steel, it does not extrude or lose a seal during thermal cycling. Graphite is fire-safe, which is important to chemical and petroleum refining industries where fire migration is a concern. It is also impervious to radiation and therefore is often recommended for nuclear service. It can be used with a wide range of process fluids without a chemical reaction, with the exception of strong oxidizers. Because graphite is bonded using compression alone, it does not have binding materials that can deteriorate when exposed to extreme temperature or certain chemicals.

The chief drawback to using graphite is that when fully compressed to provide an effective seal, it has a tendency to stick to the stem or shaft—resulting in jerky valve motion and premature wear of moving parts. Because a linear valve's plug stem may not stay in constant contact with the graphite, wear is much slower when compared to rotary valves. With rotary valves, the portion of the stem that makes contact with the packing remains constant, providing no relief to the shaft from the friction and creating faster wear. In some cases, when high compression is required, the graphite can cause the shaft to gall, which leads to packing damage and eventual leakage.

Another major problem with graphite packing is that it is extremely fragile and can be broken easily by mishandling. In addition, if graphite rings are overcompressed, they can be crushed and can lose all ability to seal as the smaller bits of graphite begin to extrude. This is a particular problem, since high compression is required to handle some fluid pressures, as well as to deform the graphite to fill any gaps or voids between the packing and the packing box wall, stem, or shaft. If an accurate torque wrench is not available, the temptation to overcompress the packing exists.

When high compression is required, another problem may occur if the compression is not completely uniform. The stem or shaft may become misaligned and create a new leak path. For this reason, use of larger stem or shaft diameters may avoid any type of flexure or off-center movement that is inherent to smaller diameter stems or shafts.

2.9.7 Perfluoroelastomer Packing

Another packing recently developed for eliminating fugitive emissions is *perfluoroelastomer packing* (compound abbreviated as *PFE*). PFE has a better temperature range than PTFE (Table 2.12) and resists chemical attack. A very versatile packing, PFE's only drawback is its cost, which is very expensive.

2.9.8 Temperature and Pressure Limits for Packing

By virtue of its close proximity to the process, the packing material can be affected by the fluid's temperature and pressure. Obviously, as the

Table 2.12 Temperature Limitations for Common Packing Materials*†

Packing Material	Valve Rating (ANSI/PN)	Standard Length Bonnets	Extended Length Bonnets
PTFE	150 to 600	-20 to 450° F	-150 to 600°F
	16 to 100	-30 to 230°C	-100 to 315°C
	900 to 2500	-20 to 450°F	-150 to 700°F
	160 to 400	-30 to 230°C	-100 to 370°C
Braided PTFE	150 to 600	-20 to 500°F	-150 to 600°F
	16 to 100	-30 to 260°C	-100 to 315°C
	900 to 2500	-20 to 500°F	-150 to 700°F
	160 to 400	-30 to 260°C	-100 to 370°C
Glass-filled PTFE	150 to 600	-20 to 500°F	-150 to 600°F
	16 to 100	-30 to 260°C	-100 to 315°C
	900 to 2500	-20 to 700°F	-150 to 700°F
	160 to 400	-30 to 260°C	-100 to 370°C
Asbestos-free Packing	150 to 600	-20 to 750°F	-20 to 1200°F
	16 to 100	-30 to 400°C	-30 to 650°C
	900 to 2500	-20 to 800°F	-20 to 1200°F
	160 to 400	-30 to 425°C	-30 to 650°C
Graphite	150 to 600	-20 to 750°F	-20 to 1500°F
	16 to 100	-30 to 400°C	-30 to 815°C
	900 to 2500	-20 to 800°F	-20 to 1500°F
	160 to 400	-30 to 425°C	-30 to 815°C
PFE	150 to 600	-20 to 450°F	-20 to 600°F
	16 to 100	-30 to 230°C	-30 to 315°C
	900 to 2500	-20 to 450°F	-20 to 700°F
	160 to 400	-30 to 230°C	-30 to 370°C
PFE (with back-up rings)	150 to 600	-20 to 550°F	-20 to 700°F
	16 to 100	-30 to 290°C	-30 to 370°C
	900 to 2500	-20 to 550°F	-20 to 800°F
	160 to 400	-30 to 290°C	-30 to 425°C

Data Courtesy of Valtek International.

†NOTES:

(1) ANSI B16.34 specifies acceptable pressure/temperature limits for pressure retaining materials.

(2) Appropriate body and bonnet materials must be used.

(3) Graphite packings should not be used above 800°F (424°C) in oxidizing service such as air.

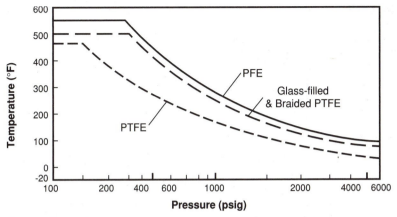

Figure 2.44 Maximum temperature and pressures for packing contained in standard bonnets. (*Courtesy of Valtek International*)

temperature increases, softer packing materials will become more fluid and are more apt to extrude out of the packing box. High pressures also can cause extrusion. Therefore, the combination of high temperatures and high pressures can greatly accelerate extrusion.

Table 2.12 provides a comparison of temperature limits for various packing materials, both standard length and extended length bonnets. The temperature limit for extended bonnets is always higher since they are longer and are designed to place the packing farther away from the temperature of the fluid. Figures 2.44–2.46 provide pressure and temperature limits for common packings.

2.9.9 Packing Lubricants

Packing boxes are often provided with a lantern ring and a tap to allow the injection of lubricant to help minimize stem friction. A number of effective lubricants exist today. The best lubricant is one that reduces stem or shaft friction without increasing the chance of packing box leakage. The lubricant should not react with the process fluid nor attract dirt or other particulate matter, and it must maintain its characteristics during severe temperatures.

The most common stem lubricant is silicone grease, which works well in temperatures up to 500°F (260°C), although it may oxidize at temperatures higher than 500°F and create a leak. In most designs, a lubricator is mounted directly to the bonnet. Turning the screw on the lubricator forces the lubricant into the packing box. An isolating valve

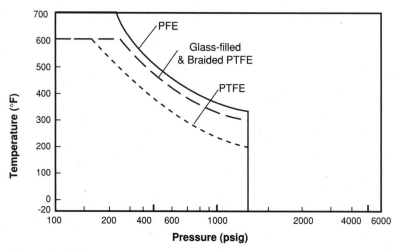

Figure 2.45 Maximum temperature and pressures for packing contained in extended bonnets, ANSI Classes 150, 300, and 600. (*Courtesy of Valtek International*)

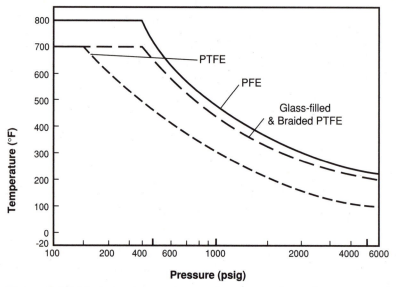

Figure 2.46 Maximum temperature and pressures for packing contained in extended bonnets, ANSI Classes 900, 1500, and 2500. (*Courtesy of Valtek International*)

is required for high-pressure applications to minimize the chance of pressure escaping through the lubricator. With some materials—such as graphite—lubrication is easily absorbed, making the material much more pliable and improving the sealability.

Lubrication has some limitations. Lubrication is not recommended for oxygen service or other flammable services where a petroleum-based lubricant could react with the fluid. When the packing is under high compression, the injection of additional lubricant may be difficult, if not impossible. Therefore, disassembly of the packing box is required. Forcing injection into high-compression packing boxes can sometimes damage the packing and cause leak paths.

3

Manual Valves

3.1 Introduction to Manual Valves

3.1.1 Definition of Manual Valves

By definition, *manual valves* are those valves that operate through a manual operator (such as a handwheel or handlever), which are primarily used to stop and start flow (block or on–off valves), although some designs can be used for basic throttling.

The best manual valves for on–off service are those that allow flow to move straight through the body, with a full-area closure element that presents little or no pressure drop. Usually if a manual valve is used to start and stop flow, as an on–off valve, and the manual operator is placed in a midstroke position, partial flow is possible as a throttling valve. However, some on–off designs in a midstroke position are not conducive to smooth flow conditions and may even cause turbulence and cavitation. Even though a manual on–off valve is being used in a throttling situation, it is not considered a control valve because it is not part of a process loop, which requires some type of self-actuation as well as input from a controlling device to a valve and position feedback. Throttling manual valves used to control flow are those that offer a definite flow characteristic—inherent or otherwise—between the area of the seat opening and the stroke of the closure element.

Besides on–off and throttling functions, manual valves are also used to divert or combine flow through a three- or four-way design configuration.

3.1.2 Classifications of Manual Valves

Manual valves are usually classified into four types, depending on their design and use. The first classification type of manual valves is *rotating valves,* which includes those manual-valve designs that use a quarter-turn rotation of the closure element. Rotating manual valves have a flow path directly through the body and closure element without any right-angle turns. The most common designs in the rotating-manual-valve family are plug, ball, and butterfly valves. They are most commonly used for on–off, full-flow services. In some applications they can be used for throttling control, as well as diversion and combination service. Overall, because rotating valves are inexpensive and versatile, they are the most common type of manual valve used in the process industry today. As a general rule, rotating valves—except butterfly valves—perform well in less-than-clean services, because the rotation of the closure element has a tendency to sever particulates when closing.

The second classification is *stopper valves,* which are defined as those manual-valve designs that use a linear-motion, circular closure element perpendicular to the centerline of the piping. These manual valves use a globe body to direct the flow through a right-angle turn under or above the closure element. If the valve uses an angle body, the flow continues from that right angle. If the valve has a straight-through body design, another right-angle turn is necessary after the closure element for the flow to be redirected in the same direction as the inlet. The two most common designs in the classification are the globe and piston manual valves. Because of the right-angle turns in these valves, stopper valves take more of a pressure drop than other designs. Therefore, among manual valves, they are the most frequently used throttling control and diversion applications, although they are often used for simple on–off service. Because of the stopper design, particulates can trap solids between the closure element and the seat, causing leakage; therefore, stopper valves are preferred for cleaner services.

The third classification is *sliding valves,* which are described as those manual valves that use a flat perpendicular closure element that intersects the flow. Like rotating valves and unlike stopper valves, sliding valves have a body with straight-through flow. Like stopper valves, the closure element—which is a flat element reaching from wall to wall—slides down from its full-open position (which is out of the fluid stream) into the flow stream, acting as a barrier wall. Both gate and piston valves are considered to be sliding valves. The sliding-seal design is best used for on–off service, although it can roughly control flow services where exact positioning is not required. Because the sliding valve seats at the bottom of the valve body, particulates can prevent full seating; therefore, sliding valves are usually used in nonslurry services.

The fourth classification is *flexible valves,* which are defined as valves with an elastomeric closure element and a body that allows straight-through flow. Overall, the design is similar to a sliding-valve design, although the closure element pushes against a highly flexible elastomeric or rubber insert until it meets against the bottom of the body or the other side of an elastomeric inset, literally pinching the flow closed. Both pinch and diaphragm valves are considered to be flexible manual valves. They are typically used in on–off services where tight shutoff (ANSI Class IV) is important or with slurries or other particulate-laden services.

3.2 Manual Plug Valves

3.2.1 Introduction to Manual Plug Valves

By definition, a *plug valve* is a quarter-turn manual valve that uses a cylindrical or tapered plug to permit or prevent straight-through flow through the body (Fig. 3.1). The plug has a straight-through opening. With a full-port design, this opening is the same as the area of the inlet and outlet ports of the valve.

Plug valves can be applied to both on–off and throttling services. Plug valves were initially designed to replace gate valves, since plug valves by virtue of their quarter-turn action can open and close more easily against flow than a comparable gate valve. For this reason, some plug-valve designs are built to face-to-face specifications used for gate valves.

Plug valves are commonly applied to low-pressure–low-temperature services, although some higher-pressure–higher-temperature designs exist. The design also permits for easy lining of the body with such materials as polytetrafluoroethylene (PTFE) for use with corrosive chemical services. They are also ideal for on–off, moderate throttling, and diverting applications. They are applied in liquid and gas, nonabrasive slurry, vacuum, food-processing, and pharmaceutical services. Abrasive and sticky fluids can be handled with special designs.

Depending upon the required end connection, plug valves are commonly found in sizes up to 18 in (DN 450) and in the lower-pressure classes [ANSI Classes 150 and 300 (PN 16 and 40)].

3.2.2 Manual-Plug-Valve Design

The most common plug-valve design allows for straight-through, two-way service (inlet and outlet), with the closure element in the middle

Figure 3.1 Nonlubricated, PTFE-sleeved quarter-turn plug valve. (*Courtesy of The Duriron Company, Valve Division*)

of the body. The closure element, which is a plug and a sleeve, is accessible through top-entry access in the body and is sealed by a *bonnet cap* (sometimes called a *top cap*). Most plug-valve bodies are equipped with integral flanges, but screwed ends are also common. Three-way bodies are also commonplace, with a third port typically at a right angle from the inlet. With the three-way design, the closure element is used to divert or combine the flow, depending on the installation of the valve as well as the position of the plug. Figure 3.2 shows six such three-way flow arrangements.

The face-to-face standard for plug valves is normally associated with ANSI Standard B16.10, with designations for both long and short patterns. However, many manufacturers have elected to use the face-to-face dimensions provided for gate valves. Not only does this standard better fit the design criteria of the plug valve, but it also allows quarter-turn plug valves to replace gate valves in existing process services.

The plug may be cylindrical in shape, which does present some problems in providing a solid seal between the body wall and the plug. The seal is important so that excessive leakage around the out-

Flow Arrangements

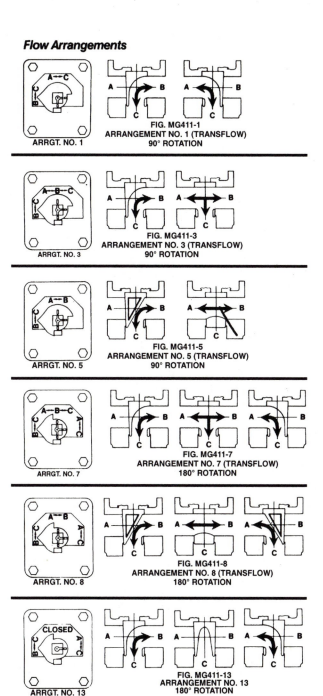

Figure 3.2 Three-way flow arrangements for quarter-turn plug valves. (*Courtesy of The Duriron Company, Valve Division*)

side diameter of the closure element does not occur. It also provides a seal for the top-works of the valve. To provide an adequate seal, three methods are commonly used: a cylindrical sleeve between the plug and the body, a series of O-rings between the plug and the body, and the injection of a malleable sealant. With the cylindrical sleeve, tightening the top-works applies compression to the sleeve against the plug. The force-fit with the O-rings provides an adequate seal also. However, the sealant design poses an inherent maintenance problem with the gradual erosion of the sealant after the valve has been stroked several times. In some high-temperature applications, the sealant may need to be reinjected after each stroke of the valve.

One of the best methods of sealing the plug and the body is to use a tapered plug, which is wedged into the plastic or other nonmetallic sleeve (again refer to Fig. 3.1). As the bonnet cap is tightened, the axial force provided by the tightening of the bonnet cap pushes the tapered plug into the softer sleeve, which provides a tight seal. The sleeve's inside diameter has a smooth surface to help seal the flow against the outside surface of the plug, while the outside surface has a series of ribs to help the sleeve hold its position in the body. The sleeve is typically manufactured from a semirigid elastomer, such as PTFE or other plastic. Because a metal surface slides with minimal friction on a plastic surface, the tapered plug is manufactured from stainless steel or carbon steel with a hard chrome surface.

Plugs can be designed with the flow port in a variety of flow areas, shapes, and functions. A common port design allows for maximum flow area, providing minimal pressure drop. The plug shape can also be characterizable (see Sec. 2.2) for throttling applications. Some cylindrical plugs have full-area ports with the same shape as the flow passages, which allow the passage of a cleaning pig. Self-cleaning ports that prevent particulate clogging or buildup are also available from certain plug-valve manufacturers. Other plugs have multihole designs to prevent or minimize the damage of cavitation (see Sec. 11.2). With cylindrical plugs the shape of the flow port is typically rectangular or round-bored, while tapered plugs are typically triangular. With throttling services, a V-shaped port is used to allow an equal-percentage flow characteristic. The ports of plugs used for three-way services are typically round with vanes contained on the inside diameter of the plug to channel flow, depending on the orientation of the plug in the body.

As previously indicated, a number of sealing designs are used to prevent the fluid from leaking through the closure element: lubricants, O-rings, and sleeves. Overall, the most common method used today is the sleeve and tapered-plug arrangement, which provides not only a good seal through the closure element but also works in conjunction with the top-work's sealing mechanism to prevent atmospheric leakage through the top-works. With most plug valves used in lower-pressure and lower-temperature service, the primary seal to the top-works is the sleeve itself, which seals between the body and the sleeve as well as between the sleeve and the plug. The top-works are further sealed with a metal thrust collar and an elastomeric diaphragm arrangement, which seals around the plug stem. The diaphragm has a spring action that helps provide constant thrust to the plug, keeping it fully seated. The outside portion of the diaphragm also acts as a gasket, sealing the gap between the body and the bonnet cap. Some plug valves—especially those used in higher temperatures and higher pressures—use packing boxes, which effectively seal the stem, but require a gland-flange arrangement to apply compression to the packing. When packing is used, a diaphragm is often not necessary, but a gasket between the body and bonnet cap is required instead.

For some corrosive chemical services (such as hydrochloric acid, sulfuric acid, waste acids, or acid brine), plug-valve bodies are completely lined with PTFE, as well as a similar coating on the plug (Fig. 3.3).

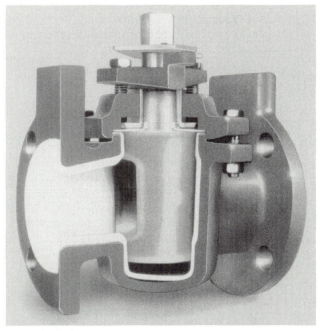

Figure 3.3 Lined quarter-turn plug valve. (*Courtesy of The Duriron Company, Valve Division*)

Other similar linings include PVDF (polyvinylidene fluoride), PVDC (polyvinylidene chloride), polyethylene, and polypropylene. Lined plug valves may have a double seal at the stem to prevent leakage to the atmosphere as well as a corrosion-resistant coating on the exterior surface of the body itself to protect the valve against process drippings. Although lined valves may be more expensive than normal plug valves, they are considerably less expensive than requesting corrosion-resistant metals. As with most corrosion-resistant materials, the lining is completely inert and impermeable. The one disadvantage of a lined valve is that the plastic-on-plastic seal provides a higher break-out torque than a metal-on-plastic seal.

To allow for the correct quarter-turn motion without over- or under-stroking of a plug valve, a stop-collar arrangement is used. The stop collar is designed so that it fits over the flats at the top of the plug and thus turns with the plug stem. A portion of the stop collar is designed with a quarter-turn path, which intersects a fixed key on the bonnet cap, gland flange, etc. As the plug stem is moved, the fixed key keeps the stop collar and the plug from moving outside the quarter-turn range.

3.2.3 Manual-Plug-Valve Operation

When the opening in the plug is in line with the inlet and outlet ports, flow continues uninhibited through the valve, taking a pressure drop through the reduced area of the plug port—although with a full-area cylindrical plug the pressure drop is minimal.

When the hand operator is turned to the full quarter-turn position (90°), the plug's opening is turned perpendicular to the flow stream, with the edges of the plug rotating through the sealing device (sleeve, lubricant, etc.). When the full quarter-turn rotation is reached, the port is completely perpendicular to the flow stream, creating complete shutoff. In throttling situations, where the plug is placed in a midturn position, the plug takes a double pressure drop. The inlet port's flow area is reduced by the turning of the plug away from the full-port position, taking a pressure drop at that point. The flow then moves into the full-port area inside the plug, where a pressure recovery takes place, followed by another restriction at the outlet port. Leakage is prevented through the seat by the compression of the plug against the sleeve or other sealing mechanism, while the packing or the collar–diaphragm assembly prevents leakage through the stem.

With three-way valve arrangements requiring diverting flow, flow enters at the inlet and moves through the plug, which channels the flow to one of the other two outlets. When the plug is moved 90°, the flow is channeled to the other outlet. At a midway position, flow may be equally diverted to both outlets. With combining flow, flow is directed from two inlets to a single outlet. In order for some of these arrangements to occur, the plug must be turned by half-turn (180°) instead of the typical quarter-turn action.

With larger plug-valve sizes [3 in (DN 80) or larger], the torque required for seal breakout may become somewhat excessive. This is caused by the larger contact surface between the plug and sealing device as well as any adverse operating conditions, such as a high process pressure, temperature extreme, corrosion deposits, etc. In this case, handlevers are typically replaced with geared handwheels, which reduce the torque requirement significantly. Table 3.1 shows the turning torque requirements for a typical plug valve for both handlevers and gear-operated handwheels. (The user should note that these numbers are torque values for turning the plug and do not indicate the higher breakout torque.)

3.2.4 Manual-Plug-Valve Installation

Before the valve is installed, the user should check that room exists for the free movement of the wrench, handlever, or gear operator. In some cases where high breakout torque is expected, the wrench or handlever may be of lengths from 24 to 36 in (61 to 91 cm), and room must be available to accommodate the wide quarter-turn arc of the operator. Some plug valves can be universally installed, meaning that either side could be the inlet. If this is the case, the only defining valve orientation would be the location of the operator and the swing of its arc. With other plug valves—such as three-way, anticavitation, or flow characteristic—the flow direction is critical to the operation of the valve. In this case, a flow arrow is attached to the body to ensure that the valve is correctly installed. In this case, the operator may need to be reoriented to ensure free movement by installing the valve upside down or by disassembling the lever–stop-collar mechanism and changing the lever orientation.

Before the plug valve is installed in the process line, the user should take care to remove any foreign material from inside the piping so that it does not become caught between the plug and the body (or sleeve). The gasket surfaces of the end connections of both the valve and the piping should be thoroughly cleaned to ensure an adequate seal. When installing the valve between flanges, new flange gaskets should be used and tightened to the correct torque values.

Table 3.1 Average Run Torques for Manual Plug Valves*

Valve Size	Turning Torque at Plug Stem	Turning Torque with Gear-operator
0.5-inch DN 15	3.0 ft-lbs 4.0 joules	
0.75-inch DN 20	3.0 ft-lbs 4.0 joules	
1.0-inch DN 25	7.0 ft-lbs 9.4 joules	
1.5-inch DN 40	8.0 ft-lbs 10.8 joules	
2.0-inch DN 50	13.0 ft-lbs 17.5 joules	
3.0-inch DN 80	19.0 ft-lbs 25.6 joules	
4.0-inch DN 100	54.0 ft-lbs 72.9 joules	5.0 ft-lbs 6.7 joules
6.0-inch DN 150	140.0 ft-lbs 189.0 joules	8.0 ft-lbs 10.8 joules
8.0-inch DN 200	306.0 ft-lbs 413.0 joules	16.0 ft-lbs 21.6 joules
10-inch DN 250	580.0 ft-lbs 783.0 joules	35.0 ft-lbs 47.3 joules
12-inch DN 300	610.0 ft-lbs 827.0 joules	16.0 ft-lbs 21.6 joules
14-inch DN 350	610.0 ft-lbs 827.0 joules	16.0 ft-lbs 21.6 joules
16-inch DN 400	1170.0 ft-lbs 1587.0 joules	18.0 ft-lbs 24.4 joules
18-inch DN 450	1170.0 ft-lbs 1587.0 joules	18.0 ft-lbs 24.4 joules

Data courtesy of Durco Valve.

If the valve is welded in the line, special precautions should be taken to avoid exceeding the temperature limits of the soft goods inside the valve, such as a sleeve that is made from polyethylene (200°F or 93°C) or polytetrafluoroethylene (400°F or 204°C). To prevent a buildup of heat inside the valve, it should be left in the open position during welding. Any leak-off or buffer connections should be installed after the valve is placed in the line.

During start-up, any potential leak paths should be examined to ensure that no leakage is occurring at the flanges or at the stem. If leakage is occurring, the flange bolting, bonnet-cap bolting, or gland-flange bolting should be tightened until the leakage stops. If the leakage does not stop, the gasket, diaphragm, etc., may have failed and requires disassembly. The valve should be opened and closed several times, ensuring that the manual operator can handle the operating conditions, especially at breakout, and ensuring that the closure element is not overcompressed. Breakout is best checked by allowing the valve to operate for some time before making an attempt to open the valve.

If the valve requires relubrication, it should take place before the valve is operating and the lubrication system should remain in place during the operation of the valve. The lubrication fitting may be located any number of places on the plug valve, but the most common locations are either at the top of the plug stem or at the bottom of the body.

3.2.5 Manual-Plug-Valve Troubleshooting

After the valve has been in service for some time, periodic troubleshooting may be necessary to determine if the valve is operating and holding its seals as expected. First, the valve should be visually checked for signs of process leakage, including the potential leak paths at the flanges, between the bonnet cap and the body, at the plug stem, and at any drains or connection plugs. If leakage is occurring at the flanged end connections, the bolting should be tightened in a criss-cross, even manner—one flat at a time—until the leakage stops. If leakage continues even after tightening, the flange gasket may have failed or the gasket surface may have been damaged or may be unclean, creating a leak path. In this case, the valve will need to be removed from the line, the gasket replaced, and the gasket surfaces examined (and repaired if necessary). If leakage is detected between the body and the bonnet cap, the most common cause is a leak path caused by thermal or pressure cycling. To relieve any pressures or fluids that are trapped in this area, the valve should be depressurized before the bonnet cap bolting is tightened. The user should be careful not to overtighten the bolting beyond industry standard torque values. If leakage continues after retightening the bonnet-cap bolting, the valve will need to be disassembled to replace the diaphragm or gasket and to examine the gasket surfaces of the bonnet cap and body.

If leakage is occurring at the stem, the most likely causes are a poor seal between the body and plug, a worn diaphragm (inside diameter), or consolidated or worn packing (if applicable). If the plug uses lubricant as a seal, additional lubrication should be added to the plug valve. If the plug uses O-rings to provide a seal, the O-rings are most likely worn or damaged and will need replacement. The corrective action for a poor sleeve seal is to tighten the top-works, which pushes the tapered plug farther into the sleeve. This compression provides a better body–sleeve and sleeve–plug seal. Tightening the gland flange will also apply axial force to the diaphragm and packing, providing better contact with the plug stem. If leakage persists, the plug valve will need to be disassembled to replace the soft goods and examine any damage to metal parts that may have led to the creation of the leak path. If leakage is occurring through the closure element, more lubrication is necessary for lubricated plug valves or the O-rings will need to be replaced with concentric plugs. With tapered plugs, the tightening of the gland-flange and/or bonnet-cap bolting may suffice to recreate a good seal. The user should note, when tightening the bolting, that the breakout and turning torque will increase. If the bolting is overtightened, not only will the internal parts wear out faster but the operating torque will dramatically increase. Therefore, after each adjustment, the valve should be operated to ensure that torque is manageable. At some point, however, continued tightening of the plug into the seal will produce a worn sleeve that will no longer respond to continued tightening. At that point, the sleeve will need to be replaced.

3.2.6 Manual-Plug-Valve Servicing

The general servicing guidelines in this section are not intended to supersede any plug-valve manufacturer's specific maintenance and servicing instructions. Instead, they are provided as general guidelines. The manufacturer's instructions should be followed exactly as they are intended.

To disassemble and reassemble the plug valve for servicing or repair, the valve and line should be depressurized and drained of process fluid. If the service is caustic or corrosive, the valve should be decontaminated to prevent harm to personnel or nearby equipment. Figure 3.4 provides an exploded view of a typical tapered-plug valve for reference purposes.

Before the top-works are removed from the plug valve, the manual operator (wrench, handlever, or gear operator) should be removed. To ensure correct orientation of the operator during reassembly, place temporary alignment marks on the plug stem and the operator. This should not be a scratch mark, because that could cause a leak path on the sealing portion of the plug stem.

Once the operator is removed, the stop collar and any associated washers can be removed from the top-works. The gland-flange bolting should be slowly loosened before removing the bonnet-cap bolting. If no gland flange exists, the bonnet-cap bolting should be loosened. With the bolting loosened, the plug should be turned and slightly raised to vent any trapped process fluid in the valve. If a mechanical device is used to lift the plug, the user should make sure the plug stem and end are protected to avoid damaging critical surfaces. As the bolting is removed from the bonnet-cap bolting, the process of lifting and rotating the plug should be continued to carefully vent any trapped fluids or pressure. Once the bolting has been unthreaded, the top-works can be removed from the body. At this point, the gland flange and bolting can be disassembled from the bonnet cap. The plug should be lifted straight up from the body, followed by inspection for any scoring or process damage, such as erosion or cavitation. If the plug uses O-rings as the primary sealing mechanism, the O-rings should be removed. If the plug uses lubricant as a sealant, remove any residue. A number of parts may be attached to the plug stem, such as a diaphragm, spring washers, thrust collars, packing rings, etc. These should be removed by lifting straight off the plug stem, taking care not to damage the plug stem. If the plug valve is equipped with an elastomeric sleeve, the sleeve will need to be cut from the body, taking care not to score the internal bore of the body. The suggested method is to cut the sleeve through the top to the one of the port openings using a screwdriver and mallet. After this cut is made, a pair of pliers can be used to grasp a cut portion and twist the sleeve from the body. This may take some effort since the sleeve has been pressed into the body and the added force of the plug compression reinforces the sleeve-to-body fit.

All disassembled parts should be cleaned thoroughly and then visually inspected for damage or wear. All parts with significant damage should be replaced as well as all soft goods (gaskets, diaphragms, O-rings, sleeve, etc.). The body should be examined carefully for damage, and any external areas of oxidation should be repaired and repainted.

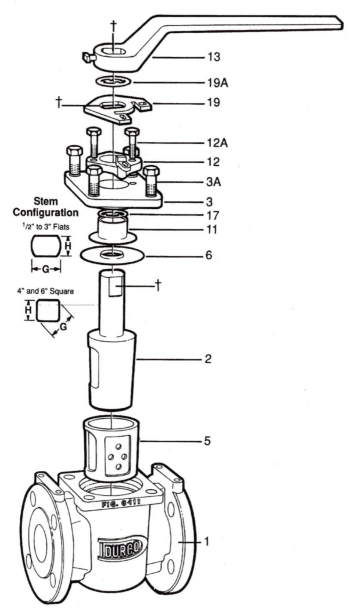

Figure 3.4 Expanded view of a quarter-turn plug valve. Numbered parts are as follows: (1) body, (2) plug, (3) top cap, (3A) top cap bolt, (5) sleeve, (6) diaphragm, (11) thrust collar, (12) adjuster, (12A) adjuster bolt, (13) wrench, (17) grounding spring, (19) stop collar, (19A) stop collar retainer. (*Courtesy of The Duriron Company, Valve Division*)

If the plug valve is equipped with a tapered sleeve, special tools from the manufacturer are required. One tool—a coining die—is needed to center the sleeve over the body bore, and another tool set—a push rod and push guide—is required to drive it into place. A special sealant can be obtained from the manufacturer and applied to the body bore before installation. As the sleeve is placed in the coining die, the ports in the sleeve should match up exactly with the ports of the body. Using the pushing rod, the sleeve is pushed into the body bore until it is firmly seated in the body (Fig. 3.5). At this point, most manufacturers recommend using a special sizing plug to size the inside diameter of the sleeve as well as to provide radial force to anchor the

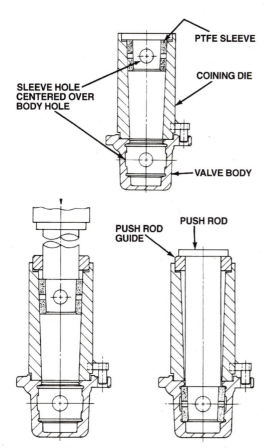

Figure 3.5 Sleeve assembly of a quarter-turn plug valve. (*Courtesy of The Duriron Company, Valve Division*)

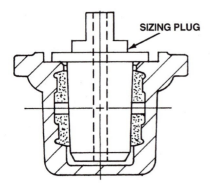

Figure 3.6 Sizing tool to size a quarter-turn plug. (*Courtesy of The Duriron Company, Valve Division*)

sleeve to the body (Fig. 3.6). Lubricant may be spread over the surface of the sizing plug to allow for easy removal of the plug from the sleeve.

Using new soft goods, the O-rings, the diaphragm, and/or packing should be replaced. Before replacing the diaphragm and packing, however, the plug stem must be checked for any surface irregularities that could damage the soft goods during installation. Any minor irregularities in the plug stem should be lightly polished. If major scoring exists, the plug should be replaced. The diaphragm, packing, spring washers, thrust collars, etc., should be reinstalled on the plug stem in the same order as they were removed. The assembled plug is then placed into the body or into the sleeve inside the body. If a sleeve is used, the top of the plug stem should be lightly tapped with a mallet to seat the plug in the sleeve.

The bonnet cap is then replaced, making sure that the diaphragm or gasket is in place to provide a seal between the body and bonnet cap. The bonnet-cap bolting can then be reinstalled, as well as the gland-flange assembly. At this point, the bonnet-cap and gland-flange bolting should be finger-tight. In a criss-cross manner, the bonnet-cap bolting should be evenly tightened to the torque values provided by the manufacturer. After the bonnet-cap bolting has been reinstalled, the plug should be rotated several times to ensure smooth stroking as well as correct orientation of the body, sleeve, and plug ports. The user should note that with the tapered design rotating the plug will cause it to raise slightly, which will cause some misalignment with the ports. This can be corrected by tightening the gland-flange bolting to the torque values provided by the manufacturer. With tapered plugs, this downward force will reposition the plug farther into the sleeve until the ports should be aligned correctly. The stop-collar assembly and the manual operator are now reinstalled. The user should make sure that alignment marks made earlier match up to ensure correct orientation.

If the valve was removed from the line, it should be reinstalled, using new flange gaskets, and retightened to the correct torque values. After the valve is fully assembled, it should be checked using the same procedure outlined in Sec. 3.2.4 before the valve is placed into full-time service.

3.3 Manual Ball Valves

3.3.1 Introduction to Manual Ball Valves

Related in design to the plug valve, the *manual ball valve* is a quarter-turn, straight-through flow valve that uses a round closure element with matching rounded elastomeric seats that permit uniform seating stress. The ball has a flow-through port and is seated on both sides. A common manual-ball-valve design is shown in Fig. 3.7. Because the design of manual ball valves are somewhat different than its automated cousin, the ball control valve, the designs associated with the ball control valve are covered in Chap. 6.

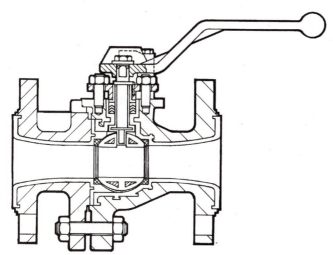

Figure 3.7 Split-body, full-port quarter-turn ball valve. (*Courtesy of Atomac/The Duriron Company*)

Manual ball valves are best used for on–off service, as well as moderate throttling situations that require minimal accuracy. In static high-pressure-drop throttling situations, where the ball's inlet port would be offset from the seal for a long period of time without moving, the velocity may cause the seal to cold flow into the port, creating some interference between the port edge of the ball and the deformed elastomer. This situation can be rectified when the manual ball valve is automated, so that the ball moves more frequently in response to a changing position signal. Ball valves are used in both liquid and gas services, although the service must be nonabrasive in nature. They can also be used in vacuum and cryogenic services.

Because of the wiping rotary motion of ball valves, they are ideal for slurries or processes with particulates, since the ball port has a tendency to separate or shear the particulates upon closing. Occasionally, lengthy thin particulates can foul or wrap around a ball, causing a high-maintenance situation.

When ball valves are applied in highly corrosive chemical services—such as hydrochloric acid, sulfuric acid, waste acid, or acid brine—the wetted surfaces of the body and ball are completely lined with polytetrafluoroethylene, which is inert and impermeable.

Manual ball valves are typically found in sizes up to 12 in (DN 300) and in lower-pressure classes of ANSI Classes 150 through 600.

3.3.2 Manual-Ball-Valve Design

The ball-valve body features a straight-through style, allowing uninhibited flow with minimal pressure drop. A number of body configurations are available, although the most common are the split body (again refer to Fig. 3.7), solid body with side entry (Fig. 3.8), or solid body with top entry (Fig. 3.9). The defining factor for determining the body design is the complexity of installing the ball inside the body. While the split body offers the easiest disassembly and reassembly, it may present problems with an additional joint that can be affected by piping stresses as well as another potential leak path. Face-to-face dimensions for ball valves are established by ANSI Standard B16.10, although with some pressure classifications or special designs manufacturers may use the gate valve face-to-face standard. Face-to-face dimensions are usually specified according to a short pattern (ANSI Class 150) or long pattern for higher-pressure classifications. The most common end connection used with manual ball valves is the integral-flange design.

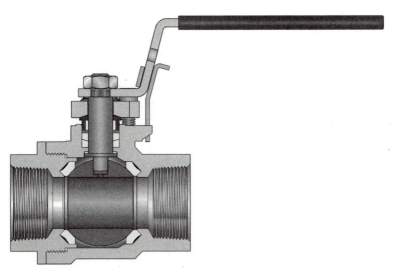

Figure 3.8 Side-entry, full-port quarter-turn ball valve. (*Courtesy of Velan Valve Corporation*)

Figure 3.9 Top-entry, full-port, single-seat with tilt-action quarter-turn ball valve. (*Courtesy of Orbit Valve Company*)

The ball itself can be either round or tapered, depending on the internal seat design. The flow-through port is a reduced area from the body port, approximately 75 percent of the valve's full area. Full-area ports are also available when minimal pressure drop is needed, such as with on–off service, or when a pig is used to scrape the inside diameter of the pipe and a narrow flow restriction in the line would prevent this. Unlike the one-piece plug of plug valves where the stem is an integral part of the plug, the ball is separate from the stem in manual ball valves. A key slot is machined or cast into the top of the ball, into which a key machined into the bottom portion of the stem fits.

Although a ball's port is normally produced in a round flow passage, with either full or reduced area, characterizable balls are also available (Fig. 3.10) with the inlet port of the ball shaped to provide the correct flow-to-position relationship for that flow characteristic. C-shaped balls are also available for eliminating dead spots (Fig. 3.11).

When two round seats are fixed on the upstream and downstream side of the ball, this is commonly called *double seating*. The two seats

Figure 3.10 Characterized ball for throttling applications. (*Courtesy of Atomac/The Duriron Company*)

Figure 3.11 C-ball for eliminating dead spots.
(*Courtesy of Atomac/The Duriron Company*)

are designed to conform with the ball's sealing surface. With moderate pressure drops and elastomeric seating materials, bubble-tight shutoff is possible with double-seated ball valves. Several other seating arrangements are utilized with ball valves. One of the most common arrangements is the *floating ball*, in which the ball is not fixed to the stem and is allowed some freedom of movement through the key slot. With the floating ball, the upstream fluid pressure assists the seal by pushing the ball back against the rear or downstream seat. Another seating arrangement involves a *floating seat*, in which the ball is fixed (called a *trunnion-mounted ball*) at two pivot points, and the process pressure pushes the upstream seat against the ball's sealing surface. The seat can also be prestressed during assembly, using seats that have a spring action. This design applies continuous pressure against a trunnion-mounted ball after the ball is installed, while the top-works apply a load to the entire closure element.

Most seats are made from PTFE, which provides excellent bubble-tight sealing and a temperature range that covers most general services. Buna-N and nylon materials are also specified, but may be limited in pressure ranges and process compatibility. For higher temperatures, metal seats and carbon-based materials are specified, although higher leakage rates are common.

With ball-valve design, the stem is usually sealed by packing rings, with a packing follower and gland flange applying compression. With split bodies and solid bodies with side entry, the stem is installed through the body and the packing installed above the body. Because of the keyed slot, the ball can be turned so that the key and the slot are parallel with the flow passage, allowing the ball to enter from the side and the stem to intersect with the stem key.

With top-entry ball valves that use trunnion-mounted balls and spring-loaded seats, the ball has either an integral or separate lower post that is seated in the bottom of the body. The seats are placed on both sides of the ball and the entire assembly is placed in the body. The top-works—consisting of a bonnet cap, packing box, gland flange, and separate stem—are installed above the ball. When the bonnet-cap bolting is tightened, the resulting compression energizes the seats. The joint between the bonnet cap and the body is sealed using a gasket.

In addition to PTFE, linings can be produced from PVDF, PVDC, polyethylene, and polypropylene. Because of the corrosive nature of the service, lined ball valves are painted with a corrosion-resistant coating on the exterior surface of the body. Although lined valves may be more expensive than normal plug valves, they are considerably less expensive than requesting corrosion-resistant metals. The one disadvantage of lined valves is that the plastic-on-plastic seal provides a higher breakout torque than the metal-on-elastomer seal.

To ensure quarter-turn motion without over- or understroking the valve, a stop-collar arrangement is used. The stop collar is designed to allow only a 90° travel of the wrench or handlever.

3.3.3 Manual-Ball-Valve Operation

With normal service, when the port opening of the ball is in line with the inlet and outlet ports, flow continues uninterrupted through the valve, undergoing a minimal pressure drop if a full-port ball is used. Obviously, the pressure drop increases with the use of a reduced-port ball. When the hand operator is placed parallel to the pipeline, the flow passages of the ball are in-line with the flow passages of the body, allowing for full flow through the closure element. As the hand operator is turned to the closed position, the ball's opening begins to move perpendicular to the flow stream with the edges of the port rotating through the seat. When the full quarter-turn is reached, the port is completely perpendicular to the flow stream, blocking the flow.

In throttling applications, where the ball is placed in a midturn position, the flow experiences a double pressure drop through the valve, similar to a plug valve. The inlet port's flow area is reduced by the turning of the plug away from the full-port position, taking a pressure drop at that point. The flow then moves into the full-port area inside the plug, where a pressure recovery takes place, followed by another restriction at the outlet port.

When a characterizable ball is used to provide specific flow to position, as the ball is rotated from closed to open through the seat, a specific amount of port opening is exposed to the flow at a certain position, until 100 percent flow is reached at the full-open position.

3.3.4 Manual-Ball-Valve Installation

The ball valve should be visually examined before installation to check for any prior damage as well as for the correct specifications (size, materials, operator orientation, etc.) for that particular process. The valve should be fully opened and closed several times to ensure a smooth quarter-turn action. If the valve is damaged, incorrectly built, or not functioning properly, the problem should be corrected before installation. Most manufacturers include installation and maintenance instructions in the shipping box or packaging, which should be kept for reference purposes. To avoid personal injury or damage to the equipment, larger ball valves [3 in (DN 80) or larger] should not be lifted by hand. Instead, lifting straps and a hoist or other mechanical lifting device should be used to lift the valve. Lifting straps should be secured around the valve body.

Before the ball valve is installed, the user should check that space exists for the full quarter-turn range of the wrench or handlever or for the space required to use a gear operator. In some cases where high breakout torque is expected, a longer wrench or handlever may be required and space must be available to accommodate the wider quarter-turn arc of the operator.

As a general rule, manual ball valves can be universally installed, meaning that either side could be the inlet, unless the valves have a special characterizable ball, which would require a specified inlet port. If the ball valve is universal, orientation may be based only upon the location of the operator and the swing of its arc. Most operators can be changed by removing the wrench or handlever and the stop collar and reorienting them to a new turning quadrant. With specialized ball valves, such as those with a specific flow characteristic, the flow direction is determined by a flow arrow that is attached to the body.

After the valve has been prechecked and is ready for installation, the piping should be checked for any foreign material, weld rods, slag, etc. Such material can be swept downstream during startup, which can damage the valve or become caught in the closure element. The gasket surfaces of both the valve and the piping flanges should be clean and free of damage to ensure an adequate seal. When installing the valve between flanges, new flange gaskets should be used and tightened to correct torque values. If the ball valve is equipped with socketweld or buttweld ends so that it can be welded in place, the user should take special precaution to avoid exceeding the temperature limits of the soft goods inside the valve, such as a PTFE seat or packing that would be limited to 400°F (204°C). To prevent excessive buildup of heat inside the valve, the valve's closure element should be left in the open position during the welding procedure.

During the start-up process, all potential leak paths should be carefully examined to ensure that no leakage is occurring at the flanges or at the stem. If leakage is found, the flange bolting, bonnet-cap bolting, or gland-flange bolting should be further tightened until the leakage stops. If the leakage does not stop, even after tightening, the most likely cause is a failed gasket, which will require disassembly and replacement. When possible, open and close the valve several times, ensuring the manual operator can handle the operating conditions—especially the breakout torque. Because breakout torque increases over time in a static situation, the valve should be allowed to remain in one position for some time before attempting to open the valve.

3.3.5 Manual-Ball-Valve Troubleshooting

As part of periodic maintenance, the valve should be checked visually for signs of process leakage, including the potential leak paths at both end connections, at the split-body connection (if applicable), between the bonnet cap and the body, and on the shaft. If process leakage is found at the flanged end connections, the piping bolting should be tightened in a criss-cross, even manner—one flat at a time—until the leakage stops. If leakage continues, even after tightening, the flange gasket may have failed or the gasket surface may be damaged or unclean (creating a leak path). Therefore the valve will need to be removed from the line, the gasket replaced, and the gasket surfaces examined and repaired if necessary.

If leakage is occurring at the joint between body valves of a split-body design, the most common cause is unequal piping forces being applied to the valve, causing unequal compression on the seal. The user should make sure that piping is sufficiently supported in a straight configuration. After this is done, the body bolting should be tightened until the leakage stops. If leakage persists, the gasket or surfaces may have been damaged and replacement or repair will be necessary.

If leakage is detected between the body and the bonnet cap, the leak path was probably caused by thermal or pressure cycling. To relieve any pressures or fluids that are trapped in this area, the valve should be depressurized before the bonnet-cap bolting is tightened. The bolting should not be overtightened beyond the manufacturer's torque values. If leakage continues after retightening the bonnet-cap bolting, the valve will need to be disassembled to replace the gasket and to examine the gasket surfaces of the bonnet cap and body. If leakage is occurring at the stem, the most likely cause is consolidated or worn packing (if applicable). Tightening the gland flange will apply additional axial force to the packing, providing better contact with the plug stem. If leakage persists, the ball valve will need to be disassembled to replace the packing and to examine the plug stem and bonnet bore surfaces. If those metal surfaces have been scored, corroded, or eroded by the process, those affected parts should be replaced.

If leakage is occurring through the ball, any number of problems may have occurred: a damaged or worn seat, foreign material between the seat and the ball, inadequate body (or bonnet) compression of the ball–seat assembly, a scored ball, or a limitation of the full stroke. In any of these cases, the valve will need to be disassembled and examined prior to repair.

3.3.6 Manual-Ball-Valve Servicing

The general servicing guidelines in this section are not intended to supersede any manual-ball-valve manufacturer's specific maintenance and servicing instructions. Instead, they are provided as general guidelines. The manufacturer's instructions should be followed exactly as they are intended.

Before any disassembly of the ball valve takes place, the valve and line should be depressurized and drained of the process fluid. If the service is caustic or corrosive, the valve should be decontaminated to prevent harm to personnel or nearby equipment. Figure 3.12 shows the internal construction of a typical lined ball valve for reference purposes.

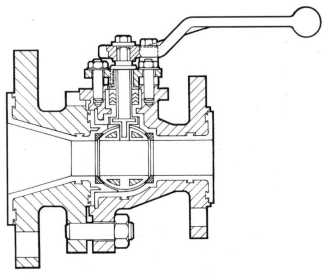

Figure 3.12 Sectioned view of a quarter-turn ball valve.
(*Courtesy of The Duriron Company, Valve Division*)

Because most ball valves have split-body designs, the entire valve will need to be removed from the line for servicing. If the ball has a top-entry design, removal will not be necessary. Prior to disassembling the body, the manual operator (wrench, handlever, or gear operator) should be removed. To ensure the correct orientation of the operator during reassembly, temporary alignment marks on the ball shaft and the operator should be made, although precautions should be taken not to scratch a mark on the sealing portion of the shaft. Once the operator is removed, the stop collar and any associated washers can be removed from the top-works. The packing-box compression should then be relieved by removing the gland flange and bolting.

With the split-body design, the body bolting holding the two halves together should be removed. After separating the half-body, the ball can be removed by turning the ball to the closed position and pushed out—since the key slot and stem key are now parallel with the body flow passage. The seats can now be removed from the two body halves. The stem is removed by pushing it down into the packing box and out through the inside of the half-body. The packing follower, packing, and any washers or antiextrusion rings can now be removed from the packing box. In smaller sizes, removing the rings may not be possible by hand and a wooden dowel should be used. The user should not employ a metal instrument because it may score the smooth walls of the bonnet cap's bore. With the top-entry design, the bonnet-cap bolting should be removed and the top-works lifted off the ball valve. At that point, the stem can be removed and the packing box disassembled. The ball and seats can then be lifted out of the body as an assembly and the seats removed from the ball.

With the side-entry design, the ball is positioned in the closed position to align the ball's key slot with the body's flow passage. The ball retainer can then be removed and the ball slid out of the body. The stem can then be pushed down into the body and then removed through the body port. The ball, stem, and bonnet cap bore should be closely examined for any scoring or process damage. Light damage may possibly be corrected by polishing the affected area. Extensive damage to these three parts will require a new part. Upon reassembly, all soft goods (seats, packing, and gaskets) should be replaced.

The valve may now be reassembled in the reverse order in which the parts were removed. The user should ensure that the body, bonnet-cap, or gland-flange boltings have been tightened to the correct torque values provided by the manufacturer. The stop-collar assembly is then reassembled and the manual operator replaced. The alignment marks made earlier should match up to ensure correct orientation. After the valve has been reassembled, the closure element should be turned to ensure that the ball moves in a smooth manner with the expected torque. If the valve was removed from the line, it should be reinstalled, using new flange gaskets, and retightened to the correct torque values. After the valve is fully assembled, it should be checked using the same procedure outlined in Sec. 3.3.4 before the valve is placed into full-time service.

3.4 Manual Butterfly Valves

3.4.1 Introduction to Manual Butterfly Valves

The manual butterfly valve is a quarter-turn (0° to 90°) rotary-motion valve that uses a round disk as the closure element. When in the full-open position, the disk is parallel to the piping and extends into the pipe itself.

Manual butterfly valves are classified into two groups. *Concentric butterfly valves* are used in on–off block applications, with a simple disk in line with the center of the valve body. Generally, concentric valves are made from cast iron or another inexpensive metal and are lined with rubber or a polymer. For throttling services, *eccentric butterfly valves* are designed with a disk that is offset from the center of the valve body. When butterfly valves are automated, eccentric butterfly valves are preferred since the disk does not make contact with the seat until closing, which prevents premature wear of the seat with the continual positioning associated with automated throttling. In most designs, simple concentric butterfly valves are used for strict on–off service and even when used in throttling applications do not lend themselves as well to automatic control as those butterfly designs designed specifically for throttling control. Because the initial development was for blocking service, concentric butterfly valves have poor rangeability and inadequate control close to the seat, while throttling butterfly valves have design modifications to allow for better flow control through the entire stroke.

Butterfly valves have a naturally high pressure-recovery factor, which is used to predict the pressure recovery occurring between the vena contracta and the outlet of the valve. The butterfly valve's ability to recover from the pressure drop is influenced by the geometry of the wafer-style body, the maximum flow capacity of the valve, and the service's ability to cavitate or choke. Overall, because of the high-pressure recovery, butterfly valves work exceptionally well with low-pressure-drop applications.

The largest drawback to using a butterfly valve is that its service is limited to low pressure drops because of its high-pressure recovery. Although flashing is not associated with butterfly valves, cavitation and choked flow easily occur with high pressure drops. While some special anticavitation devices have been engineered to deal with cavitation, users normally prefer to deal with cavitation with other valve styles that allow the introduction of internal anticavitation devices.

Butterfly valves are used for on–off and flow-control applications. Common service applications include both common liquids and gases, as well as vacuum, granular and powder, slurry, food-processing, and pharmaceutical services.

The sizes of butterfly valves are limited to 2 in (DN 50) and larger because of the limitations of the rotary design. Because of the side loads applied to the disk, the maximum size that a high-performance butterfly can reach is 36 in (DN 900). Manual designs are limited to ANSI Class 150 (PN 16), although some manufacturers offer ANSI Classes 300 and 600.

3.4.2　Manual-Butterfly-Valve Design

When compared to plug and ball valves, butterfly-valve bodies have a very narrow face-to-face. The faces of the butterfly valve body are ser-

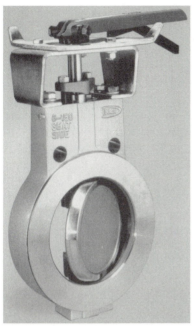

Figure 3.13 Butterfly wafer-style body.
(*Courtesy of The Duriron Company, Valve Division*)

rated to allow the use of flange gaskets for installation in the pipeline. In many cases, this allows the body to be installed between two pipe flanges using a *through-bolt connection.* Through-bolting is only permissible with certain bolt lengths, since thermal expansion of the process itself or an external fire may cause leakage. The butterfly body can be offered in one of two styles. The *wafer body* (Fig. 3.13), sometimes called the *flangeless body,* is a flat body that has a minimal face-to-face, which is equal to twice the required wall thickness plus the width of the packing box. Within this dimension, the disk in the closed position and the seat must fit within the flow portion of the body. Wafer-style bodies are more commonly applied in the smaller sizes, 12 in (DN 300) and less.

The *flanged body* (Fig. 3.14) is used with larger butterfly valves [14 in (DN 350) and larger] that have larger face-to-face dimensions, which are more apt to leak from thermal expansion. Generally, flanged bodies are used with high-temperature or fire-sensitive applications where potential thermal expansion is expected. The flanged style has integral

Figure 3.14 Flanged butterfly body. (*Courtesy of Vanessa/Keystone Valves and Controls, Inc.*)

flanges on the body that match the standard piping flanges with internal room between the flanges for studs and nuts.

As shown in Fig. 3.15, the *lug-body style* has one integral flange with an identical hole pattern to the piping flanges. Each hole is tapped from opposite direction, meeting in the center of the hole. This arrangement allows the body to be placed between two flanges. A stud is then inserted through the piping flange and threaded into the valve's integral flange. After the stud is securely threaded into the integral flange, a nut is used to secure the entire flanged connection. Lug bodies are used for applications in which the risks of straight-through bolting cannot be taken, such as with thermal expansion, when smaller valve size designs cannot permit two integral flanges.

The inside diameter of the butterfly valve is close to the inside diameter of the pipe, which permits higher flow rates, as well as straight-through flow. The closure element of the butterfly valve is called the *disk,* of which the outside diameter fits the inside diameter of the seat. The disk is described as a round, flattened element that is attached to

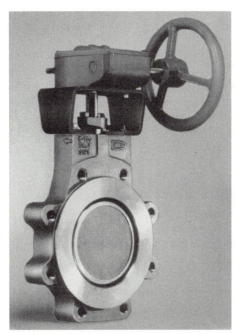

Figure 3.15 Butterfly lug-style body. (*Courtesy of The Duriron Company, Valve Division*)

the rotating shaft with tapered pins or a similar connection. As the shaft rotates, the disk is closed at the 0° position and is wide open at the 90° position. When the shaft is attached to the disk at the exact centerline of the disk, it is known as a *concentric disk* (Fig. 3.16). With a concentric disk, where the middle of the disk and the shaft are exactly centered in the valve, a portion of the disk always remains in contact with the seat regardless of the position. At 0° open, the seating surfaces are in full contact with each other. In any other position, the seating surfaces touch at two points where the edges of the disk touch the seat. Because of this constant contact, the concentric disk–seat design has a greater tendency for wear, especially with any type of throttling application. During throttling, a butterfly valve may be required to handle a small range of motion in midstroke, causing wear at the points of contact. Although the wear will not be evident during throttling, it will allow leakage at those two points when the valve is closed.

To overcome this problem of constant contact between the seating surfaces, butterfly-valve manufacturers developed the *eccentric cammed*

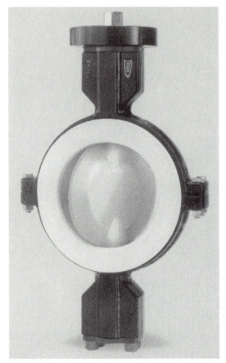

Figure 3.16 Slit-body, lined butterfly body. (*Courtesy of The Duriron Company, Valve Division*)

disk design (Fig. 3.17). This design allows for the disk and seat to be in full contact upon closure, but when the valve opens the disk lifts off the seat, avoiding any unnecessary contact. Such designs allow for the center of the shaft and disk to be slightly offset down and away from the center of the valve, as shown in Fig. 3.18. When the valve opens, the disk lifts out of the seat and away from the seating surfaces, enough to avoid constant contact. If a manual butterfly valve is operated often, the eccentric cammed disk–seat closure element is preferred because of the minimal wear to the seat.

The seat fits around the entire inside diameter of the body's flow area and is installed at one end of the body. If a polymer is used for the seat, it is called a *soft seat*. If a flexible metal is used, it is called a *metal seat*. The seat is installed in the end of the body and is held in place by a *seat retainer,* using screws or a snap-fit to keep the seat and retainer in place. After the seat and seat retainer are in place, the face of the

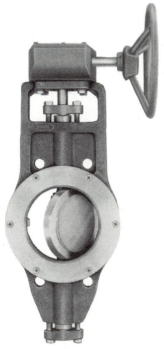

Figure 3.17 Eccentric and cammed butterfly-valve design. (*Courtesy of The Duriron Company, Valve Division*)

retainer lines up with the face of the body—although some seat–retainer designs protrude slightly from the body face, allowing some final gasket compression when the body is installed in the line.

The shaft is supported by close-fitting guides, sometimes called *bearings*, on both sides of the disk, which are installed in the shaft bore, preventing movement of the shaft and disk. Also, thrust washers are often placed on both sides of the disk, between the disk and the body, to keep the disk firmly centered with the seat.

Some concentric valve bodies are lined with rubber or elastomer. This lining has two purposes: First, it protects the metal body from the process, especially if the service is corrosive or has particulates (like sand) that would erode metal surfaces. Second, the lining also acts as the soft seat when the disk is in the closed position.

The rubber or elastomer lining is held in place in one of three ways: First, it can be retained in place by the flanged piping connections. Second, it can also be held in place with a tongue and groove configu-

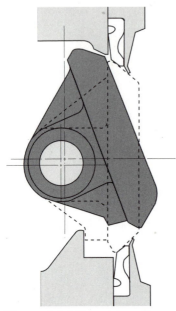

Figure 3.18 Eccentric and cammed disk rotation. (*Courtesy of Valtek International*)

ration, where the rubber lining is U shaped and the body has a T
machined into the inside diameter, allowing the two pieces to inter-
lock. The third arrangement is a split-body design with a liner sandwiched between two body halves and bolted together. All three
designs allow for easy removal of the lining after it becomes worn.
Rubber- or elastomer-lined valves are designed with metal disks that
can also be coated with a similar material. When closed, the rubber-on-
rubber seat makes for a very tight shutoff in low-pressure-drop applications and mild temperatures.

With eccentric butterfly valves, a number of different resilient seat
designs are used to handle higher pressures and temperatures. Some
seat designs use the *Poisson effect*, which refers to a concept that if a
soft metal, O-ring, or elastomer is placed in a sealing situation with a
greater pressure on one side, the softer seat material will deform with
the pressure. When deformation takes place, the pressure pushes the
material against the surface to be sealed (Fig. 3.19). With the Poisson
effect, the greater the pressure, the greater the seal.

Another common resilient seat design utilizes the *mechanical preload*

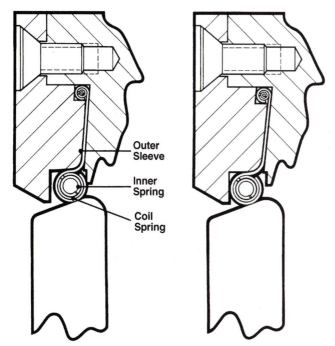

Outer
Sleeve

Inner
Spring

Coil
Spring

Figure 3.19 Butterfly metal seat assisted by process pressure (Poisson effect). (*Courtesy of The Duriron Company, Valve Division*)

effect, which allows the disk's seating surface to slightly interfere with the inside diameter of the seat. As the disk moves into the seat, the seat physically deforms because of the pressure applied by the disk, causing the polymer to seal against the metal surface (Fig. 3.20). When soft seats are used, a gasket is not required to prevent leakage between the body and the retainer because the seat also acts as a gasket.

Metal seats are applied to high temperatures (above 400°F or 205°C). Metal seats can be integral to the seat retainer with a gasket placed in the space where a soft seat is normally inserted. In some designs, both a soft and metal seat can be used in tandem, allowing the metal seat to be a backup in case of the failure of the soft seat (Fig. 3.21). When butterfly valves are specified for fire-safe applications, the tandem seat is preferred.

The body contains the packing box, which is similar to other packing boxes used in plug and ball valves. The packing box features a polished bore and is deep enough to accommodate several packing rings. Normally all that is required is the packing and a packing follower. A

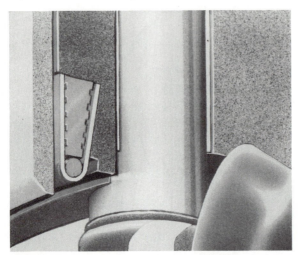

Figure 3.20 Butterfly soft seat assisted by mechanical preloading. (*Courtesy of The Duriron Company, Valve Division*)

gland flange and bolting are used to compress the packing. The shaft bore through the body is usually machined from both ends. A plug or flange cover can be used to cover the bore opening opposite the packing box. On the packing box side of the body, mounting holes are provided allowing the handlever or gear operator to be mounted.

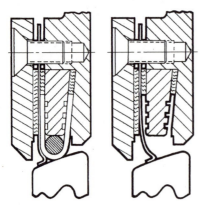

Figure 3.21 Combined metal and soft seat used for fire-safe applications. (*Courtesy of The Duriron Company, Valve Division*)

The designs of common rotary handlevers, gear operators, and actuation systems are detailed in Chap. 7.

3.4.3 Manual-Butterfly-Valve Operation

In butterfly valves, the fluid moves from the inlet to the outlet, with the only obstruction to the flow being the disk itself. Unlike gate- or globe-valve designs, where the closure element moves out of the flow stream, the butterfly disk is located in the middle of the flow stream, creating some turbulence to the flow, even in the open position. This turbulence occurs when the flow reaches the disk and is temporarily divided into two flow streams. As the flow rejoins after the disk, turbulent eddies are created. To offset a potential problem, the disk is designed with gradual angles, as well as smooth and rounded surfaces. These design modifications allow the flow to move past the disk without creating substantial turbulence.

In closing the valve, as the manual operator is turned in a rotary motion, the shaft can turn anywhere between 0° (full-closed) and 90° (full-open). In throttling situations, as the disk closes by approaching the seat, the full fluid pressure and velocity act upon the full area of the face or back side of the disk, depending on the flow direction. Generally, the major drawback of butterfly valves is that control stability is difficult when the disk is nearing the seat. Because the rangeability of butterfly valves is quite low (20:1), the final 5 percent of the stroke (to closure) is not available because of this instability.

As the disk makes contact with the seat, some deformation is intended to take place. Such deformation allows the resilience of the elastomer or the flexible metal strip with metal seats to mold against the seating surface of the disk and create a seal.

As the valve opens, the rotary motion of the shaft causes the disk to move away from the seating surfaces. Because of the mechanical and pressure forces acting on the disk in the closed position, a certain amount of breakout torque must be generated by the manual operator to force the disk to open. The butterfly valves with the greatest requirement for breakout torque are those designs that require a great deal of operator thrust to close and seal the valve. This is why some manufacturers utilize fluid pressure to assist with the seal—in effect, less breakout torque is required.

As the valve continues to open, the disk is in a near-balanced state. As one side resists the fluid forces, the other side is assisted by the fluid forces. If both sides of the disk were identical, the disk could achieve a balanced state. However, both sides of the disk are not identical—usually the shaft is located on one side, while the other side is more flat. This creates a slight off-balance situation. Therefore, the flow direction has a tendency to either push a disk open or pull it closed. When the shaft portion of the disk is facing the outlet side, the process flow tends to open the valve. When the shaft portion is facing the inlet side, the flow tends to close the valve.

Because of the design limitations of the butterfly disk, a particular flow characteristic cannot be easily designed into a butterfly valve, unlike the trim of a globe valve. Therefore, the user must use the inherent flow characteristic of the butterfly valve, which is parabolic in nature.

3.4.4 Manual-Butterfly-Valve Installation

After the butterfly valve has been received, it should be carefully inspected to ensure that it meets the application's requirements. Correcting problems with incorrect or damaged seating surfaces, materials, packing boxes, etc., should be done prior to installation, since most problems cannot be remedied with the valve in-line.

Before installation takes place, the entire length of the upstream pipe should be cleaned of any foreign materials such as dirt, welding chips or rods, or scale. If this is not done, debris may become caught between the disk and seat, causing leakage and/or damage to seating surfaces. To avoid valve damage or possible injuries to personnel, larger valves [4 in (DN 100) or larger] should be lifted using lifting straps and a mechanical lifting device.

The user should make sure that side and/or top clearance is available for the valve to be removed from the line. Also, room should exist for full quarter-turn movement of the valve's operator. Most applications allow for a butterfly valve to be installed in either a vertical or horizontal position. A flow plate or cast arrow on the body indicates the flow direction for the butterfly valve.

When installing the butterfly valve between piping flanges, special care should be taken to ensure that enough room exists between the pipe flanges to allow for the face-to-face of the wafer body, the width of the gaskets, and clearance to allow for free movement of the valve into place. If the piping is fixed rigidly in place, too much space between flanges may make it difficult to tighten the flange bolting to prevent gasket leakage. As shown in Fig. 3.22, installation involves loosely placing and threading the bottom two studs and nuts between the piping flanges, which creates a cradle to support the valve body. With the flange gaskets in their correct positions against the serrated

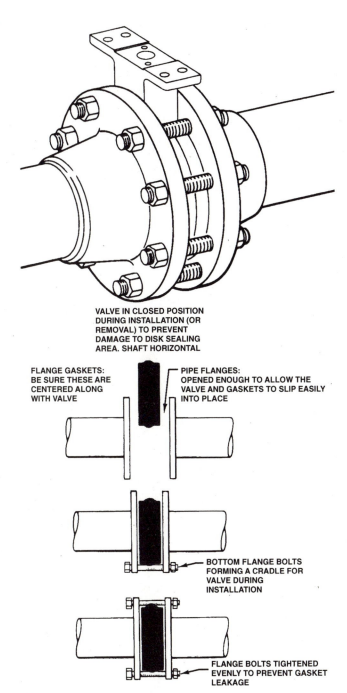

VALVE IN CLOSED POSITION
DURING INSTALLATION (OR
REMOVAL) TO PREVENT
DAMAGE TO DISK SEALING
AREA. SHAFT HORIZONTAL

FLANGE GASKETS:
BE SURE THESE ARE
CENTERED ALONG
WITH VALVE

PIPE FLANGES:
OPENED ENOUGH TO ALLOW THE
VALVE AND GASKETS TO SLIP EASILY
INTO PLACE

BOTTOM FLANGE BOLTS
FORMING A CRADLE FOR
VALVE DURING
INSTALLATION

FLANGE BOLTS TIGHTENED
EVENLY TO PREVENT GASKET
LEAKAGE

Figure 3.22 Butterfly body installation procedure. (*Courtesy of Fisher Controls International, Inc.*)

valve body faces, the body subassembly is slipped into place, sitting on the two piping studs and nuts. With the valve in position, the remainder of the bolting is installed. Some butterfly-valve bodies have fins with holes to act as guides for the bolting to ensure position stability. To ensure correct and equal gasket compression, the bolting should be tightened in a criss-cross pattern by tightening one bolt and then the one opposite. This will allow the flanges to stay square with the valve body and gaskets. Manufacturer or industry torque values should be met when tightening the bolting. Low-strength bolting is used for flanged connections, while intermediate- or high-strength bolting is used for through-bolted joints. Typical ANSI flange bolting specifications are found in Table 3.2.

Because the disk moves into the piping itself, especially when full-open, some heavy scheduling piping or piping with a cement lining may interfere with the movement of the disk. To remedy this situation the piping must be modified to allow for free movement of the disk, or another type of valve must be used. If the shaft moves with the signal, but the valve does not respond in kind, the disk–shaft connection (usually shaft pins) has failed.

Because of consolidation, packing can lose some compression between the factory and the final installation. The packing should be retightened according to the manufacturer's written procedures, although the user should not overtighten the packing, which may increase stem friction. It may also cause erratic shaft motion and premature wear of the shaft and packing. If the application has wide temperature swings, the packing tightness should be checked after the temperature excursion has taken place. Also, with wide temperature excursions, the pipe flange bolting should be checked for leakage, especially if straight-through bolting is used with a wafer-style body.

With manual handwheels, the lubrication around the handwheel stem should be checked. If a lubrication fitting exists, lubrication should be added.

3.4.5 Manual-Butterfly-Valve Troubleshooting

Butterfly-valve life can be extended by periodically troubleshooting the valve for proper operation, as well as performing preventative maintenance. In nearly all cases, manual butterfly valves can be checked for proper operation while in-line but must be removed from the line if servicing is necessary. Servicing should be done according to the methods and procedures outlined by the valve manufacturer, although a number of general procedures are outlined in this section.

Table 3.2 Flange Bolting Specifications*†

Valve Size	ANSI/PN Class Rating	Bolt Length (inches/cm)	Torque for Low Strength (ft. lbs./joules)	Torque for Intermediate Strength (ft. lbs./joules)
1-inch DN 25	150/PN 16	2.50/6.35	23/31	61/83
	300/PN 40	3.00/7.62	46/62	122/165
	600/PN 100	3.50/8.89	46/62	122/165
1.5-inch DN 40	150/PN 16	2.75/6.99	23/31	61/83
	300/PN 40	3.50/8.89	82/111	218/296
	600/PN 100	4.25/10.80	82/111	218/296
2-inch DN 50	150/PN 16	3.25/8.26	46/62	122/165
	300/PN 40	3.50/8.89	46/62	122/165
	600/PN 100	4.25/10.80	46/62	122/165
3-inch DN 80	150/PN 16	3.50/8.89	46/62	122/165
	300/PN 40	4.25/10.80	82/111	218/296
	600/PN 100	5.00/12.70	82/111	218/296
4-inch DN 100	150/PN 16	3.50/8.89	46/62	122/165
	300/PN 40	4.50/11.43	82/111	218/296
	600/PN 100	5.75/14.61	132/179	353/479
6-inch DN 150	150/PN 16	4.00/10.16	82/111	218/296
	300/PN 40	4.75/12.07	82/111	218/296
	600/PN 100	6.75/17.15	199/270	531/720
8-inch DN 200	150/PN 16	4.25/10.80	82/111	218/296
	300/PN 40	5.50/13.97	132/179	353/479
	600/PN 100	7.50/19.05	296/401	789/1070
10-inch DN 250	150/PN 16	4.50/11.43	132/179	353/479
	300/PN 40	6.25/15.88	199/270	531/720
	600/PN 100	8.50/21.59	420/570	1119/1517
12-inch DN 300	150/PN 16	4.75/12.07	132/179	353/479
	300/PN 40	6.75/17.15	296/401	789/1070
	600/PN 100	8.75/22.23	420/570	1119/1517

Courtesy of Valtek International.
†Note: Lengths are based upon ANSI Standard B16.5 studs used with raised face bodies.

The piping end connections should be carefully inspected for signs of process leakage to ensure that thermal expansion has not weakened the integrity of the seal or that the flange gasket has not failed. The top of the packing box, body end plate or plug, and any other pressure-retaining connections should also be closely examined. Leakage from the piping flanges may indicate a misalignment of the upstream and downstream piping or can be caused by failure of the flange gaskets. If the gaskets or the gasket surfaces are dirty, leakage may occur through a large particulate. Rubber- or elastomer-lined bodies may not seal well when gaskets are used in conjunction with the lining. This is because more compression is needed than usual to seal the rubber soft surfaces.

If process leakage is found between the piping flanges and the body, the piping flange bolting should be tightened in a criss-cross manner until the leak is eliminated. If the leakage continues, the gasket has failed or is contaminated, and disassembly and possible replacement of the gasket are necessary. If leakage is occurring through the packing, the gland-flange bolting should be tightened, being careful to follow the manufacturer's recommended torque value or procedure until the leakage stops. As noted earlier, overtightening the packing may crush the packing rings or cause consolidation and extrusion.

Because prior cleaning may mask signs of leakage, the valve should be only cleaned after a check for leakage has been completed. Any process buildup from the shaft, packing follower, gland flange, or gland-flange bolting should be thoroughly removed. If painted metal surfaces show signs of severe oxidation, they should be cleaned with a wire brush and repainted using a rust inhibitor, especially if the outside environment is corrosive, such as the atmospheric salt that is prevalent near sea water. Nearby piping lines should be checked to ensure that process drippings are not falling on the butterfly valve. If so, those piping leaks should be corrected, or the valve must be shielded from the process drips.

If possible, the valve should be stroked to ensure that it is capable of a full and smooth stroke. If the shaft binds or moves in an erratic or jerky manner, it may indicate an internal galling or packing tightness. In high-temperature services, some graphite packings are known to cause jerky motion when properly compressed and, in moderate amounts, this erratic shaft motion is considered normal.

A common malfunction of the butterfly valve is a seat that leaks beyond the expected leakage rate. Some leakage through the seat may be expected, unless the valve has a elastomeric seat and is classified as having bubble-tight shutoff. In this case, the term *leakage* is used when the measured leakage is beyond what is permitted by the user. If the valve has been operating satisfactorily for a reasonable period before seat leakage occurs, the likely cause is a worn or damaged seat and/or disk. Probable causes are process erosion, mechanical failure of the seat, frictional wear between the two mating seating surfaces of the seat and the disk (especially if the valve closes often), damage from a foreign object caught between the seat and disk, or cavitation damage to the disk. The soft seat or metal seat gasket may also have failed. This will cause leakage through the joint between the body and the seat retainer, even when the disk and seat are successfully sealing the flow. The disk and seat may be misaligned if the shaft guides or bearings are worn, resulting in a damaged seat and leakage. Seat leakage may be caused by a galled shaft, which may be preventing full motion of the disk, not allowing the disk to reach the seat.

Leakage commonly occurs through the packing box. If retightening the packing fails to stop the leakage, the packing has most likely consolidated or extruded, and the packing box must be rebuilt using new packing. As a general rule, graphite packings are more abrasive than others, and the quarter-turn movement has a tendency to wear the shaft where it makes continual contact—as opposed to linear-motion valves, where friction can be spread over the entire length of the linear stroke. If the packing is highly abrasive, these frictional losses can eventually cause leakage unless the packing is retightened. Eventually, the continual wear will lead to eventual replacement of the shaft.

In a few applications, disk movement is impeded by high temperatures outside the design parameter of the valve, which can generate thermal expansion of parts. Some parts, such as shafts and guides or bearings, have exact tolerances, and the thermal expansion can result in sticking parts.

3.4.6 Manual-Butterfly-Valve Servicing

The general servicing guidelines in this section are not intended to supersede any manual-butterfly-valve manufacturer's specific maintenance and servicing instructions. Rather, they are provided as general guidelines. The manufacturer's instructions should be followed exactly as they are intended.

When a manual butterfly valve is serviced to remedy a problem, it should be removed from the line. If the process must stay in operation during servicing, any existing bypass block valves should be used to channel the flow around the affected valve. That portion of the line or the entire pipeline (if not bypassed) should be completely depressurized and decontaminated prior to removing the piping flange bolting.

Before the line flange bolting is removed, the disk should be placed in the closed (or seated) position so that the valve-body subassembly will clear the piping. The valve should be fully supported with a hoist or other means before loosening the piping-flange bolting. In some cases, a tight face-to-face fit coupled with pressure adhesion of the gasket causes difficulty in freeing the valve from the piping. If this is the case, the manual operator should not be used as a convenient tool for leverage because this can damage the shaft. Also, the user should not use a screwdriver, crowbar, or wedge to work the surfaces loose as this can damage gasket surfaces. The best method is to place mechanical spreaders between the flanges and to spread the piping slightly to release the valve. With the valve closed and completely perpendicular to the piping, the valve should be carefully removed from the piping without scraping or damaging the gasket surfaces.

Before the valve is disassembled, the user may need to determine if disassembly is easiest with the operator removed from the shaft. If this is the case, the bolting holding the operator to the valve should be loosened and the operator removed from the valve. To ensure correct orientation of the operator during reassembly, alignment marks should be placed on the shaft and the operator. However, the user should not scratch a mark on the sealing portion of the shaft since this may create leakage problems with the packing box. Once the operator is removed, the stop collar and any associated washers can be removed from the top-works.

Unless the valve body is rubber- or elastomer-lined (which does not require a separate seal), the seat should be removed from the body for inspection or possible replacement. To do this, the seat retainer should be removed. Some retainers are bolted into place, while others use a snap ring or clip arrangement. These fasteners may become corroded over time and may require some effort to remove them. Once removed, the seat should be carefully checked for any unusual damage or wear. If the seat is worn or damaged, it should be replaced unless seat leakage is not a major concern. As a general rule, the seat of a concentric valve (or an eccentric valve that closes often) is routinely replaced during periodic maintenance.

To remove the shaft and disk, the user should check to see if the body has a plug or flange that seals a blind end of the body (the end opposite from the operator). If so, the plug or flange and the gasket should be disassembled. Referring to Figs. 3.23 and 3.24, the gland flange and bolting should be loosened or removed to decompress the packing box. At this point, the shaft should turn freely; therefore, the user should be careful to prevent the disk from slipping, which may cause damage or injury. With the disk in the open position, the pins or keys used to hold the disk to the shaft should be removed without damaging or deforming the shaft, pins, or guides or bearings. Once the packing is decompressed and the disk is disengaged from the shaft,

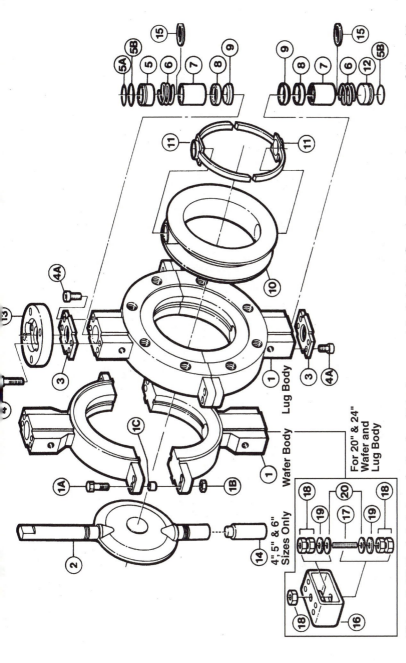

Figure 3.23 Exploded view of a lined, split-body, concentric disk butterfly valve. Numbered parts are as follows: (1) body, (1A) hexagonal head cap screw, (1B) hexagonal nut, (1C) bushing, (2) disk, (3) retainer plate, (4 and 4A) socket-head cap screw, (5) gland at top, (5A) O-ring inboard, (5B) O-ring outboard, (6) spring, (7) bearing, (8) ring-stem wedge, (9) ring-stem compression, (10) liner, (11) seat energizer, (12) gland at bottom, (13) plate mounting, (14) stem extension, (15) spacer, (16) bracket, (17) stud, (18) hex nut, (19 and 20) washer. (*Courtesy of The Duriron Company, Valve Division*)

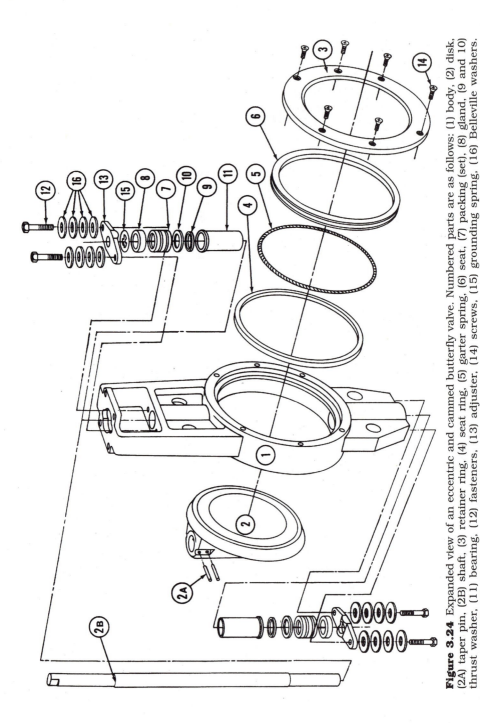

Figure 3.24 Expanded view of an eccentric and cammed butterfly valve. Numbered parts are as follows: (1) body, (2) disk, (2A) taper pin, (2B) shaft, (3) retainer ring, (4) seat ring, (5) garter spring, (6) seat, (7) packing (set), (8) gland, (9 and 10) thrust washer, (11) bearing, (12) fasteners, (13) adjuster, (14) screws, (15) grounding spring, (16) Belleville washers.

the shaft should be carefully slid out the body. The user should take special care to prevent scoring or scratching of shaft surfaces that touch the packing and guides or bearings. Depending on the design, the shaft may be pulled out through either the packing end or blind end of the body. While pulling the shaft out, the disk should be fully supported to prevent it from falling loose and damaging critical surfaces or personnel. The user should be careful not to twist the shaft so that it binds on the guides or bearings.

Once the shaft and the disk have been removed from the body, the bearings or guides can be removed. In some designs, the bearings or guides are pressed into the body, requiring some mechanical force to remove them. If mechanical force is used, care should be taken not to damage the guides unless they will be replaced. Once the disk is removed, the disk seating surface should be checked for any signs of wear or damage. A wooden dowel may be used to push the packing out of the packing box from inside the body. Once removed, the packing should be inspected for signs of consolidation or extrusion.

Before the butterfly valve is reassembled, all parts should be thoroughly cleaned. Any areas of severe oxidation should be removed and painted with an antioxidation paint. All damaged or worn parts should be replaced with new or reconditioned parts.

To reassemble the valve, the guides or bearings should be reinstalled into the body. Some guides and bearings can be placed by hand, while others have a force-fit and will need to be pressed into the body bore. Once the guides or bearings are placed, the disk should be repositioned inside the body's flow area, aligning the disk correctly in relation to the shaft and disk stop. To avoid injury, large disks should be supported rather than held by hand. The shaft should be reinserted through either the packing end or the blind end of the body and disk until the shaft is correctly in place. The user should take special care to prevent scratching or scoring the sealing or guiding surfaces of the shaft. Using the pins, keys, or other fasteners, the disk should be attached to the shaft. The fasteners should be firmly secure so that they will not work free during normal service. The shaft should now be manually turned through the entire quarter-turn motion to ensure that the valve functions in a smooth stroke without binding of the guiding surfaces or misalignment of the disk with the body. The plug end or plate should be replaced, if necessary, using a new gasket or O-ring.

The packing box should be replaced, using new packing rings. As the rings are placed over the sharp edges at the end of the shaft and slid into the body bore, the user should be careful not to damage the individual rings. The correct number and order of packing rings, spacers, antiextrusion rings, and packing should be verified for the required packing-box design. After the packing box is rebuilt, the gland flange and bolting can then be replaced. If the gland-flange bolting is corroded and cannot be adequately cleaned, it should be replaced because corroded bolting may provide a false torque reading (if torque values are required for correct packing compression). Using the gland-flange bolting, the packing should be correctly compressed according to the torque value or procedure provided by the valve manufacturer. The user should be careful not to overcompress the packing, which can lead to excessive stem friction and wear.

If a separable seat is used with that particular design, it should now be reinstalled, making sure that the seat is correctly positioned. Some seat designs can only be installed one way, while others are universal. After the seat is placed correctly in the body, the retainer or metal seal should be replaced, using the fasteners needed to secure the seal assembly. Once again, corroded fasteners should be replaced. The disk should be rotated into the seat to ensure a proper sealing function. With spring-loaded seats, some slight interference is expected. If the seat and disk do not fit together or if significant interference is experienced, the disk should not be forced. The user should examine the cause of the mismatch, such as a seat misalignment or incorrect disk reassembly, and correct it.

Before remounting the operator, the user should verify that the disk is in the correct position (full-open or full-closed) by using the previous alignment marks made on the stem. After the operator has been reinstalled, the butterfly valve should be operated to ensure smooth quarter-turn travel, after which it may be returned to service.

3.5 Manual Globe Valves

3.5.1 Introduction to Manual Globe Valves

As shown in Fig. 3.25, a *manual globe valve* is a linear-motion valve characterized by a body with a longer face-to-face that accommodates flow passages sufficiently long enough to ensure smooth flow through the valve without any sharp turns. It is used for both on–off and throttling applications. The most common closure element is the *single-seat design*, which operates in linear fashion and is found in the middle of the body. The single-seat design uses the plug–seat arrangement: a linear-motion plug moves into a seat to permit low flows or closure, or moves away from the seat to permit higher flows. By virtue of its

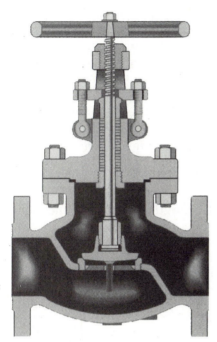

Figure 3.25 Manually operated globe valve. (*Courtesy of Pacific Valves, a unit of the Crane Valve Group*)

design, a globe valve is not limited to an inherent flow characteristic like some quarter-turn valves. A particular flow characteristic can be designed into the shape of the plug.

Manually operated globe valves are somewhat more versatile in application than other manual valves, although the overall cost and size factors are higher. Manual globe valves can be applied in both gas and liquid services, although the service should be relatively clean to avoid particulates from being caught in the seat and creating unwanted leakage. Common manual-globe-valve applications include on–off and flow control, frequent stroking, vacuum, and wide temperature extremes. Although the globe body design can handle high-pressure classes (up to ANSI Class 2500 or PN 400), manual globe valves are usually applied to lower-pressure applications because of the thrust limitations of the hand operator. High-pressure applications will require the use of a gear operator. Using the largest available hand operators, a manual handwheel is limited to 9,000 to 13,000 lb (40 to 60 kN), although some designs—such as a nonrotating plug and a stem nut supported by a roller bearing—may surpass this limit. Globe valves can be designed to handle higher-pressure classes by increasing the wall thickness of the body and using heavier-duty flanges, bolting, and internal parts. Manual globe valves are found in sizes from 0.5 to 48 in (DN 6 to 1200).

The majority of globe-valve designs feature top-entry to the trim (the plug and seat). This design permits easier servicing of the internal parts by disassembling the bonnet flange and bonnet-flange bolting and removing the top-works, bonnet, and plug as one assembly. Unlike rotary-motion manual valves, globe-valve bodies with top-entry access can remain in-line while internal maintenance takes place. Because of top-entry access, globe valves are preferred in the power industry where steam applications require the welding of the valve into the pipeline.

The largest drawbacks to the globe valve are that it can weigh considerably more than a comparable rotary valve and is much more costly. Sizewise, it is not as compact as a rotary valve.

3.5.2 Manual-Globe-Valve Design

The globe-style body is the main pressure-retaining portion of the valve and houses the closure element. The flow passages in a globe valve are designed with smooth, rounded walls with no sharp corners or edges, thus providing a smooth process flow without creating unusual turbulence or noise. The flow passages themselves must be of constant area to avoid creating any additional pressure losses and higher velocities. With two widely spaced end connections, globe-valve bodies are adaptable to nearly every type of end connection, although the face-to-face is to too long to accommodate a flangeless design (bolting the body between two pipe flanges, which is commonplace with a rotary valve). With globe valves, mismatched end connections are also acceptable.

The globe valve's trim is more than just a closure element (because a throttling valve does more than just open or close), but rather it is a regulating element that allows the valve to vary the flow rate against the position of the valve according to the flow characteristic, which may be equal percentage, linear, or quick-open (see Sec. 2.2). Typically, this trim consists two key parts: the *plug,* which is the male portion of the regulating element, and the *seat ring,* which is the female portion. The portion of the plug that seats into the seat ring is called the *plug head,* and the portion that extends up through the top of the globe valve is called the *plug stem.* The plug stem may be threaded at the top of the stem to allow for interaction with the handwheel mechanism. The chief advantage of the single-seated trim design is its tight shutoff possibilities—in some cases better than 0.01 percent of the maximum flow of the valve. This occurs because the force of the manual operator is applied directly to the seating surface.

Two sizes of trim can be used in manual globe valves. *Full trim* is the most common and refers to the area of the seat ring that can pass the maximum amount of flow in that particular size of globe valve. On the other hand, *reduced trim* is used when the valve is expected to throttle a smaller amount of flow than that size is rated for. If full trim is used, the valve must throttle close to the seat, as well as in small increments—which is difficult to achieve with a hand operator. The preferred method, then, is to use a smaller seat diameter with a matching plug, which is called reduced trim.

The *bonnet* is a major element of the valve's top-works and acts as a pressure-retaining part, providing a cap or cover for the body. Once mounted on the body, it is sealed by bonnet or body gaskets. It also seals the plug stem with a packing box—a series of packing rings, followers or guides, packing spacers, and antiextrusion rings that prevent or minimize process leakage to atmosphere. Mounted above the packing box is the gland flange, which is bolted to the top of the bonnet. When the gland-flange bolting is tightened, the packing is compressed and seals the stem as well as the bonnet bore.

Keeping the plug head in alignment with the seat ring is important for tight shutoff. To maintain this alignment, one of two types of guiding mechanisms is used: double-top stem guiding or seat guiding. *Double-top stem guiding* uses two close-fitting guides at both ends of the packing box to keep the plug concentric with the seat ring (Fig. 3.26). These guides can be made entirely from a metal compatible with the plug to avoid galling and can include a hard elastomer or graphite liner. The ideal arrangement is for the two guides to be located as far apart as possible to avoid any lateral movement caused by the process fluid acting on the plug head. The guides, bonnet bore, and actuator stem must all be held to close tolerances to maintain a fit that will allow smooth linear motion without binding or slop.

The other common type of guiding in manual globe valves is the *seat-guiding* design, where the plug stem is supported by one upper guide (which also acts as a packing follower). As a second guiding surface, the outer diameter of an extension of the plug head guides inside the seat (Fig. 3.27). This means that the lower guiding surface remains inside the flow stream, so therefore the process must be relatively clean. The lower portion of the plug head has openings that allow the

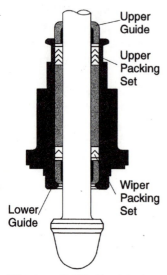

Figure 3.26 Double-top stem guiding. (*Courtesy of Valtek International*)

flow to move through the plug head to the seat during opening. By varying the size and shape of these openings, reduced flow and flow characteristics can be introduced. Because the length between the upper guide and the lower guide are at a maximum length, lateral plug movement due to process flow is not an issue and the tolerances

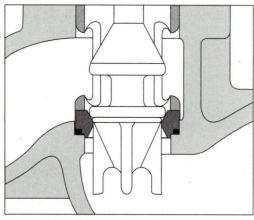

Figure 3.27 Seat-guiding design. (*Courtesy of Valtek International*)

required for this type of guiding are not required to be as close as double top-stem guiding. This design minimizes any chance of vibration of the plug in service. When the plug and seat are made from identical materials, galling may occur during long-term or frequent operation. High temperatures may also lead to thermal expansion and binding.

The metal seat surface of the plug is designed to mate with the metal seating surface seat ring, using angles that slightly differ. Normally the plug has a steeper seating angle than the seat ring. This angular mismatch assures a narrow point of contact, allowing the full axial force of the operator to be transferred to a small portion of the seat only, assuring the tightest shutoff possible for metal-to-metal contact. In most designs, the seat ring for manual globe valves is threaded into the body. This sometimes requires a tool to turn the seat ring into a body with limited space. With threaded seat rings, exact alignment between the seating surfaces of the plug head and seat ring must require *lapping*—a process where an abrasive compound is placed on the seat surface. The plug is then seated and turned until a full contact is achieved. Although simple in concept, threaded seats have some disadvantages. First, in corrosive or severe services the threads can become corroded, making disassembly difficult. Second, alignment between the plug and seat ring require the additional step of lapping to achieve the required shutoff. And third, in situations where vibration is present and the seat ring is not held in place by the plug in the closed position, the seat ring may eventually loosen, allowing leakage through the seat gasket and/or misalignment of the seating surfaces.

Some globe-valve applications require bubble-tight shutoff, which cannot be attained with a metal-to-metal seal. To accomplish this, a soft elastomer can be inserted in the seat ring. In this case, the seat ring is a two-part design with the elastomer sandwiched between the two halves (Fig. 3.28). The metal plug surface pressing against the seat ring's soft seat surface can achieve bubble-tight shutoff if the plug and

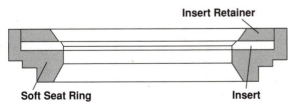

Insert Retainer

Soft Seat Ring **Insert**

Figure 3.28 Soft-seat design. (*Courtesy of Valtek International*)

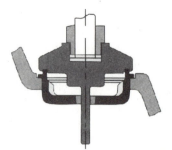

Figure 3.29 Soft-plug design. (*Courtesy of Pacific Valve Group, a unit of the Crane Valve Group*)

seat-ring surfaces are concentric. Some manufacturers also insert the elastomer into the plug, which achieves the same effect (Fig. 3.29).

3.5.3 Manual-Globe-Valve Operation

Most manual globe valves use a T-style body, allowing the valve to be installed in a straight pipe. Flow enters through the inlet port to the center of the valve where the trim is located. At this point, the flow makes a 90° turn to flow through the seat, followed by another 90° turn to exit the valve.

The flow direction of globe valves is defined by the manufacturer or the application, although in most manual applications, flow direction is almost always under the plug. Seating the plug against the flow provides constant resistance but not enough to be insurmountable. With under-the-plug flow, the valve is relatively easy to close as long as the fluid pressure and flow rate are low to moderate. In addition, under-the-plug flow provides easy opening by the flow pushing against the bottom of the plug.

Manual-globe-valve trim can be modified to allow for equal-percentage, linear, or quick-open flow characteristics. As explained in detail in Sec. 2.2, flow characteristics determine the flow rate (expressed in flow coefficient or C_v) expected at a certain valve position. Therefore, with a certain flow characteristic, the user can roughly determine the flow rate by the linear position of the manual handwheel. If the plug head is in a throttling position (between full-open or full-closed), because of the pressure drop the flow moves toward the flow opening in the seat. In a throttling position, the plug head extends somewhat into the seat ring, providing only so much flow in that particular position for a given flow characteristic. As the plug retracts further away from the seat, more flow is provided. If the plug extends further into the seat, less flow is allowed. As the flow moves through the seat, fluid pressure decreases as velocity increases. After the fluid enters the lower portion of the globe body, the flow area expands again, the pressure recovers, and the velocity decreases.

As the flow enters the seat or plug area of the valve, an important design consideration is the gallery area of the body. In ideal situations the flow should freely circulate around the plug and seat, allowing flow to enter the seat from every possible direction. If the gallery is narrow in any one area (for example, in the back side of the plug), velocities can increase, causing noise, erosion, or downstream turbulence. In addition, unequal forces acting on the plug head can cause slight flexing of the plug head if it is not guided by the seat.

When the globe valve closes, the axial force of the manual handwheel is transferred to the plug. The plug's seating surface is forced against the slightly mismatched angle of the seat ring, not allowing any flow to pass through the closure element. In the full-open position, the entire seating area is open to the flow.

Process flow is retained inside the body and bonnet by the static seals of the gaskets in the end connections [if flanges or ring-type joint (RTJ) end connections are used]. Flow seeking to escape through the sliding stem of the plug is prevented by the packing's dynamic seal in the bonnet's packing box. Depending on the shutoff requirements of the user, flow may or may not be leaking through the regulating element itself.

3.5.4 Manual-Globe-Valve Installation

Before installation, the globe valve should be inspected carefully against the packing list to ensure that the valve is built according to specification. Many manufacturers provide installation and maintenance instructions inside the shipping carton, which should be collected and kept for reference. Larger globe valves [2 in (DN 50) or larger] should be lifted, using lifting straps and a mechanical lifting device with the lifting straps placed around sturdy members of the valve. If an operator uses a yoke to connect the valve and handwheel, the lifting straps can be placed around the yoke legs. Some manufacturers provide metal lifting rings as part of the bonnet-flange bolting, which may be used for lifting. A sticker or tag advises the user of these lifting points. Extreme caution should be taken whenever lifting a valve with straps and a mechanical lifting device, because failures can happen and several hundred pounds of valve can be quite unforgiving. Many certified drawings show where a valve's center of gravity is on the valve. Knowing the location of the center of gravity can help in lifting the valve, as well as warning the user if the valve will be off-balance when lifted.

Before the globe valve is installed, the line should be cleaned of any foreign material, such as dirt, welding chips or rods, or scale. A pig cannot be used with globe valves because it cannot navigate the right angles of the body. Because of their linear motion, globe valves have a greater height than other manual valves, which must be taken into consideration during installation. Access to the internal parts of the valve involves top-entry in nearly all globe valves, which requires removing the bonnet-flange bolting and lifting the entire top-works (manual handwheel, bonnet, and plug head) out of the body. The *disassembly clearance* is the amount of space needed between the top of the manual operator to lift the top-works completely out of the body and is provided by the manufacturer.

Globe valves generally should be installed in the horizontal position (body parallel to the ground), which provides for easy maintenance, calibration, and operation. In some cases, the valve must be installed in the vertical position (body perpendicular to the ground) to function correctly. Maintenance with globe valves is easier when the top-works can be lifted straight out of the body. To remove the valve's top-works from a vertical or 45° installed position requires constant support to prevent damage to parts or injury to workers. Y-body valves are ideal for piping placed at a 45° angle, since the top-works remain vertical to the ground.

Maintaining the correct flow direction is very important, since flow direction plays an important role in the proper operation of the valve. Flow direction is almost always indicated on the side of the globe valve.

If a valve is welded into the line, requiring buttweld or socketweld end connections, excessive heat should not be allowed to build up in the valve body. Thermal expansion may cause seizing or galling between parts if the valve is operated while still hot. Extreme heat may also melt polymer gaskets or packing, causing leak paths or extrusion of the material. In this situation, keeping the valve open during welding will allow for any thermal expansion to dissipate.

If the valve is equipped with integral-flange end connections, correct alignment of the matching ends on the piping flanges is important. Because the flanges on the body are fixed, they are carefully machined so that the hole pattern lines up exactly with the vertical axis of the valve. If the piping flanges are installed so that they are not lined up exactly with the vertical axis, the valve will have a tendency to lean, making disassembly more difficult. Some play exists between the size of the flange hole and the required stud, which may help minimize the misalignment. If only one piping flange is misaligned with the vertical axis, installation may be difficult, even impossible, since the holes on the two mating flanges may match up on one connection but be misaligned on the other connection. Some users correct this misalignment by widening the holes on the piping flange, but this weakens the integrity and strength of the flange. Another option is to cut the flange off the pipe and to reweld it to the correct position. Specifying separable flanges on the valve can solve the misalignment problem. As outlined in Sec. 2.4.3, separable flanges are not integral to the valve body; rather individual flanges are held in place by half-rings. Because the separable flanges can spin in either direction, misalignment problems are quickly remedied. This also allows for the globe valve to be installed in the vertical position. Separable flange bolting must be correctly tightened to ensure enough friction at the body–half-ring–flange connection to prevent the entire valve from accidentally rotating—which may happen with tall valves or if line vibration is prevalent. In this case, some users add a number of tack welds to maintain the position of the flange with the valve following installation.

Once the globe valve is installed in the line, is should be stroked, observing the motion of the plug stem. The plug stem should move in a smooth linear motion. If the plug seems to stick or move in an erratic manner, something is internally wrong with the valve or the manual operator, and this should be thoroughly investigated and corrected before further operation takes place.

Over time, the packing may lose some compression through consolidation between the factory and the final installation. A safe precaution is to retighten the packing according to the manufacturer's written procedures after installation, but before start-up. The user should be careful not to overtighten the packing since this will increase stem friction, causing erratic stroking as well as premature stem and packing wear. If temperature swings are expected as part of the service, the temperature excursion should be allowed to take place during start-up. The packing box should then be rechecked and retightened, if necessary. Following the temperature excursion, checking the body and bonnet gaskets for leakage is also recommended.

3.5.5 Manual-Globe-Valve Troubleshooting

The user should periodically check and troubleshoot the manually operated globe valve for proper operation. In most cases, this can be accomplished while the valve remains in the line. If any portion of the internal parts of the globe valve must be checked or replaced, the process service will need to be interrupted or blocked around the valve before being serviced.

First, the entire valve should be examined for signs of process leakage. The following areas should be examined: the end connections, especially if a flanged connection is used and a body face gasket may fail; the gap between the bonnet flange and body where body or bonnet gaskets may fail; a body plug; at the top of the packing box where the plug exits the bonnet; or any other pressure-retaining connection. If leakage occurs with a static seal, the leak will appear immediately after full process pressure is applied to the valve. If a gasket maintains its seal at start-up, it should last for the duration of the maintenance cycle. Eventual failure will also occur if a slight flaw in the gasket grows under pressure or if the gasket material is attacked by the process fluid. In the latter case, the gasket material should be changed. With liquid services, this examination can be performed visually. In gaseous service, evidence of leakage may not be as obvious; therefore, a "sniffer" sensing device may be necessary. If leakage is found, tightening the necessary compression bolting to the pipe or body flange, bonnet flange, or gland flange will most likely eliminate the leak. If tightening the bolting does not stop the leak, the gasket or packing has probably failed, and disassembly and replacement of the soft goods will be necessary to correct the problem.

In some rare cases, the body or bonnet will fail because of undetected porosity, fracture, erosion, or cavitation. Sometimes if the failure is minor, the affected area can be ground out and then filled with a weld. This procedure should be performed using established methods to ensure the integrity of the pressure retaining vessel.

If possible, the valve should be stroked to verify a full and smooth stroke. Erratic or jerky stem movement may indicate an internal galling, packing tightness, or manual-operator problem. Some packings, especially graphite packing, are known to cause erratic (or jerky) stroking when properly compressed. With manual handwheels, the lubrication should be checked around the handwheel stem. If a lubrication fitting exists, proper lubrication should be added.

Excessive leakage occurs through the valve seat when the leakage exceeds that amount permitted by the user. If the valve has been providing adequate service for some time before seat leakage occurs, the most probable cause is a worn or damaged seat ring–plug combination. Seat leakage may be caused by process erosion, by wear between the two mating seating surfaces of the seat ring and the plug if the valve is closed often, or if a corrosive, cavitating, or flashing process has deteriorated metal parts. Also, the seat-ring gasket or split-body gaskets may have failed, allowing leakage between the seat ring and the body. When a process line is started up for the first time, sometimes damage occurs to the seating surfaces when a foreign object— such as a weld rod left in a pipeline—becomes caught in the seat. If the valve has recently been disassembled for servicing and then reassembled, the most common cause of leakage is misalignment of the plug to the seat. Because threaded seat rings are in a fixed position, they must be lapped in relationship with the plug every time the seat ring or plug has been removed and replaced. If lapping is not performed or not performed correctly, small gaps may exist between the seating surfaces, which will cause seat leakage.

Occasionally, plug movement is impeded by a process service that is outside the range of the valve. For example, high temperatures can generate thermal expansion of parts with tight tolerances, causing additional friction. Also, higher-than-expected process pressure may create a situation where the operator does not have the necessary leverage or thrust to overcome the line pressure to close the valve.

3.5.6 Manual-Globe-Valve Servicing

The general servicing guidelines in this section are not intended to supersede any manual-globe-valve manufacturer's specific maintenance and servicing instructions. Rather, they are provided as general guidelines. The manufacturer's instructions should be followed exactly as they are intended.

If the valve must be disassembled to inspect or replace trim parts or to replace leaking gaskets, the process line should be depressurized and drained. If the process fluid itself is corrosive or dangerous to human exposure, it should be completely decontaminated. Since most manual globe valves have top-entry access, the trim can be reached by removing the bonnet flange bolting and lifting the entire top-works (plug, bonnet, bonnet flange, and manual operator) straight out of the body. This must be done carefully to avoid damaging the critical surfaces of the plug and seat.

To disassemble the packing box, the manual operator will need to be removed from the top-works. To do this, the tension on the packing should be released by loosening or removing the gland-flange bolting. The plug can now be unthreaded from the handwheel stem. In sizes 6 in (DN 150) or smaller, this can be done by turning the plug head. In larger sizes, the plug can be removed by using a wrench on the flats on the plug stem. After the plug is released from the actuator stem, it can be pulled out of the bottom of the bonnet. Because of packing consolidation, some resistance may be felt as the plug is pulled through the packing box. If the packing material appears capable of possible reuse and is not consolidated or damaged, it can be used again, and some care should be taken that the plug stem threads do not score the inside diameter of the packing. Following the removal of the plug, the manual operator can be removed from the top-works assembly.

After the plug has been removed from the bonnet, the seating surface should be inspected for deep scratches, erosion, or pitting from cavitation. Minor damage can sometimes be corrected by turning and polishing the seat surface, taking care not to affect too large of an area that may affect critical tolerances. Some users attempted to remove large imperfections by machining the seating surface. In this case, the user should make sure that the manual operator can compensate for this slightly longer stroke. The plug stem should also be examined for any scratches, scoring, or galling, which may impede or damage the sealing ability of the packing. Minor scratches can oftentimes be polished using a very fine abrasive compound substance. More substantial scoring requires a new plug.

If the packing is being replaced, prior packing compression or extrusion makes removal difficult. A wooden dowel is recommended to push the packing out from the bottom of the bonnet, since metal tools may slip out of position and score the sealing surface of the bonnet bore. As the guides (or packing follower), packing rings, extrusion rings, lantern rings, and/or spacers are taken from the packing box, their order should be noted for reassembly. The packing should be examined for evidence of consolidation or extrusion, so that preventative measures may be taken in the future.

After the bonnet bore is cleaned, it should be examined for any pitting, corrosion, or erosion. If the bonnet bore has been significantly damaged, the entire bonnet will need to be replaced, especially if fugitive emissions are a concern. The inside diameters of the upper and lower guides (if applicable) should also be examined for any abnormal wear or scoring.

At this point, the seat should be removed from the body. Since most seats in manual globe valves are threaded, this procedure will most likely require a special tool from the manufacturer, as well as a thread-loosening compound, especially if the surfaces are corroded. After the seat is removed, the seat-ring gasket should be removed for future replacement.

The entire body should now be checked for signs of process damage, such as process erosion, cavitation, or pitting. Special care should be taken to visually inspect the bonnet, body, and seat-ring gasket surfaces to ensure they are not damaged.

Prior to reassembly, all parts should be thoroughly cleaned, using cleaning agents that are approved for that particular process. If the pressure-retaining parts have experienced minor oxidation, corrosion, or cavitation from the process or environment—meaning that the destruction did not penetrate minimal wall thicknesses—those areas may be repaired and repainted.

The packing box should now be rebuilt by placing the plug stem into the bonnet bore. The guides, packing, lantern rings, extrusion rings, and/or spacers should be replaced in the same order they were removed (unless a new packing-box configuration is being installed). Each piece should be slipped over the top of the plug stem—being careful not to damage the inside diameters of these parts on the plug-stem threads—and down the plug stem into the packing box. Depending on the type of packing used, the packing may need to be tapped into place, using a spare packing spacer or two.

If a different packing material or arrangement is being installed, such as a twin seal or a vacuum-pressure packing arrangement, some caution should be taken. For example, the packing box may vary in height as the number of packing rings or type of spacers changes. Sometimes a small variance will be permissible with the depth of the bonnet bore. On the other hand, if the height variance is significant, a packing spacer of a different length may be needed. When changing the packing arrangement, the user should be especially careful to ensure that the packing is installed correctly. With vacuum-pressure packing arrangements, for example, the upper seal portion has its chevron packing placed upside down (chevron facing downward) for correct function. Also, the user should be careful not to add extra rings of packing in addition to the number recommended by the manufacturer. Sometimes the user believes that "more is better," but actually too many rings may amplify the effects of thermal expansion of the packing, increase stem friction, and require high compression levels to compress additional rings. Sometimes with extra rings, the gland-flange bolting may be incapable of providing the correct compression for that number of rings. The manual operator or the top-works assembly should not be replaced at this time, since the plug will need to be free for lapping the seat.

As the body is reassembled, new bonnet, body, and seat-ring gaskets should be used—especially if spiral-wound gaskets are used. Since the height of a spiral-wound gasket is dependent on the enclosed metal strips, which are crushed during assembly, the gasket does not recover any height following decompression. Therefore, a reused gasket will nearly always cause leakage. In an emergency situation, polymer gaskets could be reused, since they do recover their height somewhat following decompression. However, because of consolidation, some height is lost by prior compression, and using old gaskets is not worth the risk of eventual leakage—especially since the cost of gaskets is minimal as compared to the cost of unplanned maintenance.

As the seat ring is rethreaded into the body, the user should make sure the seat ring is fully seated in its bore, engaging the gasket. Some manufacturers provide a torque value for seat-ring installation to ensure proper gasket compression. To lap the seat, lapping compound is first placed on the seat-ring seating surface. Without using body or bonnet gaskets, the bonnet should be replaced on the body and the plug rotated in the seat to provide an even fit between the plug and seat. The bonnet can then be removed, the lapping compound wiped from the seating surfaces, and the bonnet replaced with the proper gaskets for the final assembly. Some technicians prefer to build the final assembly and then lap the seat and plug, not bothering to clean up the lapping compound. This latter procedure is quite common if the process allows some impurities from the used lapping compound.

The bonnet flange bolting can be tightened in one of two ways: First, the manufacturer may list torque values for the bolting in the maintenance instructions, which requires the use of a torque wrench. Using torque values ensures that the correct amount of compression of the gaskets is achieved as well as providing enough force to hold the trim in place without unnecessary stress to the valve's internal parts. With this type of design, the user must use a properly calibrated torque wrench as well as the right torque value. Guessing the correct torque value may result in either a leaking, misaligned body subassembly if undertightened or overcompressed gaskets and stressed parts if overtightened. The second method uses an exact-tolerance step: When fully tightened, the body and bonnet reach a metal-to-metal state with each other, ensuring correct gasket compression as well as providing trim stability. The metal-to-metal step can be felt through the wrench.

At this point, the manual operator can be remounted onto the body subassembly, although the user should take special precautions not to turn the plug in the seat to avoid galling at the seating surfaces. Following reassembly and before the valve is placed in service, the valve should be stroked several times to ensure smooth stem travel. If hydrotest facilities are available (and the valve has been removed from the line), the valve can be tested under pressure to ensure the integrity of the seat, gasket, and packing.

3.6 Manual Gate Valves

3.6.1 Introduction to Manual Gate Valves

A *gate valve* is a linear-motion manual valve that uses a typically flat closure element perpendicular to the process flow, which slides into the flow stream to provide shutoff. Overall, the simplicity of the gate-valve design and its application to a large number of general, low-pressure-drop services makes it one of the most common valves in use today. It can be applied to both liquid and gas services, although it is mostly used in liquid services. The gate valve was designed primarily for on–off service, where the valve is operated infrequently. For the most part, it can be used in either liquid or gas services. It is especially designed for slurries with entrained solids, granules, and powders and cryogenic and vacuum services. As an on–off block valve, it can be designed for full-area flow to minimize the pressure drop and allow the passage of a pipe-cleaning pig. When compared to other types of manual valves, the gate valve is relatively inexpensive as well as easy to maintain and disassemble. When used with a metal seat, a gate valve is inherently fire-safe and is often specified for fire-safe service.

Gate valves do have some limitations. Gate valves do not handle throttling applications well because they provide inadequate control characteristics. Therefore, they are most commonly applied in simple on–off services as a block valve. They also have difficulty opening or closing against extremely high pressure drops. Tight shutoff is not easily attained in some applications. In addition, they can become fouled with those processes that have entrained solids. Because they are known for lengthy strokes, they take longer to open than other manual valves.

As a general rule, gate valves are divided into one of two designs: parallel and wedge-shaped. The *parallel-gate valve* (Fig. 3.30) uses a flat disk gate as the closure element that fits between two parallel seats—an upstream seat and a downstream seat. To achieve the required shutoff, either the seats or the disk gate are free-floating, allowing the upstream pressure to seal the seat and disk against any unwanted seat leakage. In some designs, the seat is spring-energized by an elastomer that applies constant pressure to the disk gate's seating surface. For the most part, the application of parallel-gate valves is limited to low pressure drops and low pressures, and where tight shutoff is not an important prerequisite.

Some variations of the parallel-gate valve have been designed for specific applications. The *knife-gate valve* (Fig. 3.31) has a sharp edge on

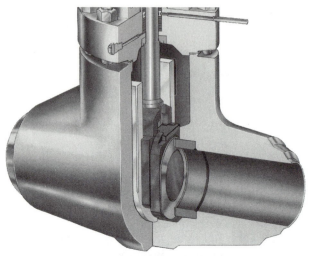

Figure 3.30 Parallel-gate valve. (*Courtesy of Velan Valve Corporation*)

Figure 3.31 Bidirectional knife-gate valve. (*Courtesy of DeZURIK, a unit of General Signal*)

Figure 3.32 Through-conduit valve. (*Courtesy of Daniel Valve Company, a division of Daniel Industries, Inc.*)

the bottom of the gate to shear particulates or other entrained solids as well as to separate slurries. The *through-conduit gate valve* (Fig. 3.32) has a rectangular closure element with a circular opening equal to the full-area flow passageway of the gate valve. By lowering or raising the element, the opening is exposed to the flow or the barrier shuts off the flow, respectively. The through-conduit gate-valve design allows the seating surfaces of the gate to be in contact with the gate at all times. With a full-area opening, it also allows the use of a pig to scour the inside diameter of the line.

The second classification of gate valves, the *wedge-shaped gate valve* (Fig. 3.33), uses two inclined seats and a slightly mismatched inclined gate that allows for tight shutoff, even against higher pressures. The inclined seats are designed 5° to 10° from the vertical plane, while the inclined gate can be designed with a close, but not exact angle. When the seat and gate angles are slightly mismatched, either the seat or gate is designed with some free movement to allow the seating surfaces to

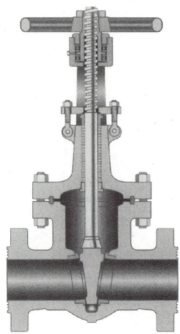

Figure 3.33 Wedge gate valve. (*Courtesy of Pacific Valves, a unit of the Crane Valve Group*)

conform with each other as the manual actuator force is applied. This can be accomplished through either a floating seat and a solid gate or by a *flexible* or a *split-wedge gate* that provides flexure (or "give") of the gate seating surfaces (Fig. 3.34). Also, pressure-energized elastomer inserts can be installed on a solid gate to provide a tight seal (Fig. 3.35).

Flex Wedge   Split Wedge

Figure 3.34 Flexible and two-piece split wedges. (*Courtesy of Pacific Valves, a unit of the Crane Valve Group*)

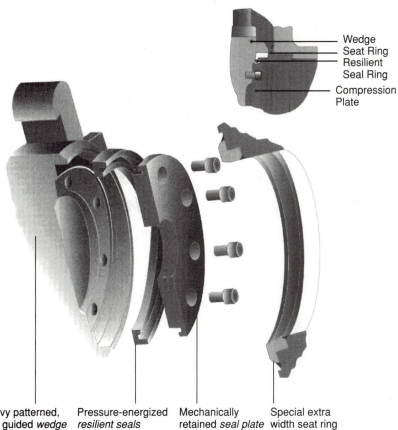

Wedge
Seat Ring
Resilient
Seal Ring
Compression
Plate

| Heavy patterned, fully guided *wedge* is precision fitted before resilient seals are installed. | Pressure-energized *resilient seals* available in PTFE(BTT) for maximum versatility. | Mechanically retained *seal plate* allows fast, easy resilient seal replacement. | Special extra width seat ring allows for wear without loss of seal. |

Figure 3.35 Pressure-energized wedge design (with soft seats). (*Courtesy of Pacific Valves, a unit of the Crane Valve Group*)

Gate valves are commonly found in sizes of 2 through 12 in (DN 50 through DN 300) in ANSI Class 150 (PN 16), although larger sizes are sometimes custom designed.

3.6.2 Manual-Gate-Valve Design

The gate is attached to the manual operator through the *gate stem,* which may be either fixed (rising stem) to the gate or threaded (nonrising stem) to the gate. The fixed-gate stem does not turn with the man-

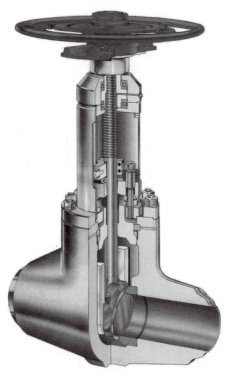

Figure 3.36 Rising-stem gate-valve design.
(*Courtesy of Velan Valve Corporation*)

ual operator but stays stationary with the gate (Fig. 3.36). As the handwheel is turned, the threads (which are located above the packing box)
retract the gate from the flow stream, causing the threaded portion of
the stem to rise above the handwheel. With a threaded gate stem, the
stem is threaded to the gate itself. Turning the handwheel threads the
stem into the gate, causing the gate to lift out of the flow stream (Fig.
3.37). The gate stem is not integral to the gate but rather uses a T-
shaped collar that fits into a T-shaped slot in the gate. The T-slot is parallel to the flow stream, but may also be perpendicular to the flow
stream in certain designs.

With both parallel- and wedge-shaped gate valves, the screw-driven
manual operator lowers or raises the gate either into the flow stream
or out of the flow stream. Most designs call for the gate to rise above
the flow stream into a cavity created by the bonnet cap, although some
designs allow the gate to be extended into a lower body cavity.

The body itself is a straight-through design with a special face-to-

Figure 3.37 Nonrising-stem gate-valve design.
(*Courtesy of American Flow Control, a division of
American Cast Iron Pipe*)

face for gate valves (ANSI Standard B16.34). In most cases, the body is designed with flanged end connection, although buttweld, socketweld, and screwed ends are sometimes offered. With simple wedge-gate valves, an integral seating surface may be machined directly into the body. Separable wedge seats are installed through the valve's top-entry opening. The upstream and downstream parallel-gate seats are a floating design and are held in place between the body and gate, which also acts as guides for the gate.

With wedge gates, guiding takes place with slot–rib combinations between the body wall and the gate. In some cases, a rib fits a matching slot in the gate, or a slot in the body fits a matching slot in the gate. Although the slots are machined, the ribs are of a cast finish, providing only simple positioning (enough to place the gate and seats into position) after which the force of the operator seals the seating surfaces together and prevents any leakage between the body and gate during guiding.

The bonnet cap not only provides top-entry to the gate, but also encloses the packing box, which seals the gate stem to prevent process leakage. A gland flange is used to apply compression to the packing.

3.6.3 Manual-Gate-Valve Operation

In the open position, as flow moves into the inlet of the valve, it continues through the flow-through globe body with minimal, if any, pressure drop occurring. This happens because most gate valves have full-area seats and are used for simple on–off blocking applications. Any pressure drop that occurs is due to the geometry of the seats, body guides, or cavities. In the open position, wedge gates and parallel gates are normally located above the seat in the upper body cavity, away from the flow stream. With conduit parallel gates, when the gate is in the open position, the flow opening of the gate is exposed to the full flow.

When the valve begins to close, the rotation of the manual operator turns the threads of the gate stem against either the operator itself (rising stem) or into the gate (nonrising stem). In either case, the gate begins its downward travel into the flow stream. Because gate valves operate in low-pressure or low-pressure-drop applications, the introduction of the gate into the flow stream is met with only moderate resistance.

As the gate valve closes, a parallel gate begins to seal the flow as the upstream pressure builds. In the parallel-gate design, the upstream pressure acts upon the floating seat, pushing the seat against the seating surface of the gate and providing the necessary seal. In the wedge-gate design, when the wedge gate reaches the seat, the thrust applied by the manual operator pushes the gate into the seat. As noted in Sec. 3.6.2, the wedge gate and the seats have some resilience as well as mismatched angles between seating surfaces. As additional thrust is applied, the wedge gate is pushed harder into the seats, providing tighter shutoff.

While the parallel valve requires minimum thrust to close, upon opening it must overcome a greater breakout force because of the upstream pressure pushing against the floating seat, especially if the valve has been in the closed position for some time. Once the flow begins to move through the seat and velocity builds, the upstream pressure is reduced and the gate slides easily to the full-open position without much resistance from the flow.

On the other hand, with a wedge valve, less breakout force is required due to the mismatched seating surfaces and wedge gate, which have a tendency to repel each other upon opening. This action is also enhanced by the natural resilience of the wedge gate. As the operator thrust is reversed and the valve begins to open, the gate and seats separate easily without hindrance or assistance by the flow.

With most applications, unless the flow rate is minimal, keeping the gate in a throttling position (midstroke) results in flutter of the gate as well as vibration and unnecessary wear. Because gate valves create additional flow turbulence in a midstroke position, they are not normally specified for throttling applications.

3.6.4 Manual-Gate-Valve Installation

When unpacking the gate valve, it should be inspected carefully against the packing list to ensure that the valve meets the requirements of the application. Installation and maintenance instructions included with the valve should be collected and kept for reference. Larger gate valves [3 in (DN 80) or larger] should be lifted using lifting straps and a mechanical lifting device, placing the lifting straps around sturdy members of the valve or manual operator. Extreme caution should be used whenever lifting a valve, because the equipment may drop and injure personnel or the equipment itself. Knowing the location of the center of gravity can help in lifting the valve as well as warning the user should the valve be off-balanced when lifted. Before the gate valve is installed, the line should be cleaned of any foreign material, such as dirt, welding chips or rods, or scale. If a pig is used to clean the line of debris, the user should ensure that the gate valve has a full-area seat to ensure passage of the pig.

Because of their linear motion, larger gate valves have a greater height than other manual valves. This must be taken into consideration during installation, especially if the design has a rising-stem action. Access to the trim of the valve involves top-entry in nearly all gate valves, which requires removing the bonnet or gland-flange bolting and lifting the entire top-works (manual actuator, bonnet, and gate) out of the body. Gate-valve bodies, for the most part, should be installed in the horizontal position (perpendicular to the ground) with the manual operator in the top position. This position provides for the easiest access to the body internal parts through the top-entry bonnet. Also, maintenance on gate-valve body subassemblies is easier when the top-works can be lifted straight out of the body.

Although gate valves are usually installed in horizontal lines, smaller-sized gate valves also can be installed in vertical lines. The user should remember, however, that gravity has a tendency to pull the gate out of alignment with the seats in larger gate-valve sizes, requiring some additional mechanical support. Some parallel-gate-valve designs actually seal better in a vertical line carrying liquids, since the weight of the liquid, along with the pressure, can be applied against the floating seat (when the upstream portion is above the valve).

With most gate valves, flow direction is not a consideration since the valve can be universally installed from either direction. Because of the design of the floating seats, some parallel-gate valves may require a specific flow direction. If the body does not show flow direction, the gate valve is considered to be universal and can be mounted in either direction.

If the gate valve comes with integral-flanged end connections, the user should ensure that the matching end piping flanges are aligned. Because the flanges on the body are in a fixed position, the hole patterns are designed to line up with the vertical axis of the valve. If the piping flanges are installed off the vertical axis, the valve will lean slightly. The user should see some minor play between the size of the piping-flange hole and the required stud. This may be used to minimize the misalignment with the vertical axis. However, if one piping flange is misaligned with the vertical axis, overall installation may be impossible since the holes on the two mating flanges may match up on one connection but be misaligned on the other connection. Some users correct this misalignment by widening the holes on the piping flange. Because this weakens the integrity of the flange, it is a poor option. The best solution is to cut the flange off the pipe and to reweld it in the correct position.

After the gate valve is installed in the line, the manual operator should be used to fully stroke the valve, observing the motion of the gate stem. The stem should move in a smooth linear motion. If the plug seems to stick or moves in an erratic manner, something may be wrong with the valve internally or the manual operator, and it should be investigated thoroughly. Before start-up, the packing compression should be checked as some packing decompression occurs between the factory and the final installation. The user should be careful to not overtighten the packing since this will increase gate or stem friction, causing erratic stroking and premature stem and packing wear. If temperature swings are normally expected as part of the service, the temperature excursion should take place during start-up and then the packing box rechecked (and retightened if necessary). Following the temperature excursion, checking the body and bonnet gaskets for leakage is also recommended.

In gate valves used in applications in which adhesive solids in the process fluid may settle during closing (causing difficulties in opening the valve), purged lines are installed below the seat to push the material out away from the seat.

3.6.5 Manual-Gate-Valve Troubleshooting

Once a gate valve is in operation, periodic troubleshooting and servicing should take place to ensure longer life. In most cases, troubleshooting can take place while the valve remains in the line with minimal interruption to the service. If the gate, seats, or packing box must be visually checked or repaired, the process service will need to be interrupted or blocked around the valve being serviced.

The valve should be carefully examined for any signs of process leakage. The end connections between the pipe and the valve should be checked first, especially if a flanged connection is used where a body face gasket may fail. The gap between the bonnet cap and the body should be examined in case the body or bonnet gaskets have failed. The top of the packing box should be checked, especially where the gate stem exits the bonnet. Any other pressure-retaining connection, such as a body plug, should be examined. Normally, if a gasket maintains its seal at start-up, it should last for the duration of the maintenance cycle. Eventual failure may occur, however, if a slight flaw in the gasket grows under pressure or if the gasket material is attacked by the process fluid. In the latter case, the gasket material is misapplied and should be changed to a compatible material. With liquid services, leakage examination can be performed visually. In gaseous service, evidence of leakage may not be as obvious; therefore, a "sniffer" sensing device may be necessary. If leakage is found, tightening the necessary compression bolting to the pipe or body flanges, bonnet cap, or gland flange may eliminate the leak. If that does not eliminate the leak, the gasket or packing has probably failed, and disassembly and replacement of the soft goods will be necessary.

In some rare cases, the body or bonnet will fail because of undetected porosity, fracture, erosion, or cavitation. If the failure is minor, the affected area can be ground out and then filled with a weld, as long as it is done using established methods to ensure the integrity of the pressure-retaining vessel.

If possible, the manual operator should be used to fully stroke the valve, from full-open to full-closed. Erratic or jerky stem movement may indicate an internal galling, a packing tightness, or a manual operator problem. With manual handwheels, the user should regularly check the lubrication around the handwheel stem. If a lubrication fitting exists on the operator housing, a lubricant should be injected on a regular basis.

Excessive leakage may occur through the valve seat. For manual valves, excessive leakage is usually defined as leakage beyond what is permitted by the system. If the valve has been providing adequate service for some time before the seat leakage occurs, the most probable cause is a worn or damaged seating surface of the seat or the gate. Probable causes are process erosion, normal day-to-day wear when the valve is opened and closed often, or a corrosive process that has corroded metal parts. When a process line is started up for the first time, damage sometimes occurs to the seating surfaces when a foreign object (such as a weld rod left in a pipeline) becomes caught between the gate and the seat.

With some applications in which a metal seat is used with a solid wedge gate, galling can occur when temperature excursions occur. If the gate valve is seated at the low end of the temperature range, as the temperature increases thermal expansion can lock the seating surfaces of the metal seat to the solid wedge gate. This situation could result in a gate that is stuck in place until the temperature cools. It could also result in galling of the seating surfaces when the gate breaks free from the seat. In this case, the gate should be changed to a split gate or similar design where some resilience would avoid this problem.

3.6.6 Manual-Gate-Valve Servicing

The general servicing guidelines in this section are not intended to supersede any manual-gate-valve manufacturer's specific maintenance and servicing instructions. Rather, they are provided as general guidelines to the user. The manufacturer's instructions should be followed exactly as they are intended.

If the gate valve must be disassembled to inspect or replace trim parts or to replace leaking gaskets, the process line should be depressurized and drained. If the process fluid itself is corrosive or dangerous to human exposure, it should be completely decontaminated.

Most gate valves have top-entry access through the bonnet cap. By removing the body bolting, the manual operator, bonnet cap or gland flange, and gate can be lifted out of the valve body. In some designs, this allows access to the separable seats, which are threaded into place and can be removed with common tools. In other designs, special tools must be obtained from the manufacturer to remove the seats. In some highly corrosive situations, seat removal may be difficult or even impossible. If the seats cannot be removed, the entire body will need replacement. Body replacement is also necessary with smaller-sized gate valves that use integral seats. After disassembly, the entire body should be examined for signs or evidence of process damage, such as erosion, cavitation, or pitting.

The seating surfaces of the gate should be examined to ensure that they are not galled or scored. If some difficulty has arisen while stroking the valve, such as tightness or jerky stem travel, the threaded stem-to-gate connection (nonrising-stem design only) should be checked to ensure that the threads are undamaged. If the gate valve has a rising-stem design, the stem-to-gate connection should be checked for smooth rotation.

If the packing needs to be replaced, the manual operator will need to be removed from the top-works. To do this, first the tension on the packing should be released by loosening or removing the gland-flange bolting. The gate stem should then be removed by unthreading or unbolting it from the handwheel stem. After the gate stem is separated from the manual operator, it can be pulled out of the bottom of the bonnet cap. Because of packing consolidation, some resistance may be felt as the gate stem is pulled through the packing box. If the packing material is capable of reuse and is not consolidated or damaged, it can be reinstalled again. Caution should be taken so that the gate-stem threads (rising-stem design only) do not damage the inside diameter of the packing.

After the gate is removed from the bonnet cap, the gate stem should be examined for any scratches, scoring, or galling, which may impede or damage the sealing ability of the packing. Minor scratches can oftentimes be polished using a very fine abrasive compound substance. More substantial scoring requires a new gate stem or gate if both pieces are integral.

If the packing has been compressed over time or extrusion has occurred, the packing should be removed and replaced. To remove the packing, use a wooden dowel to push the packing out from the bottom of the bonnet cap, as metal tools may slip from their position and score the sealing surface of the bonnet bore. As the guide and packing follower and packing rings are taken from the packing box, their order or orientation should be noted for reassembly. Once the packing has been removed, the extent of damage caused by packing consolidation or extrusion should be noted so that preventative measures may be taken in the future. After cleaning the bonnet bore, the bore should be examined for any pitting, corrosion, or erosion—especially if packing-box leakage has occurred. If the bonnet bore has been significantly damaged, the entire bonnet cap will need to be replaced. The inside diameters of the upper guide (or packing follower) should also be examined for any abnormal wear or scoring.

Prior to reassembly of the gate valve, all parts should be thoroughly cleaned, using cleaning agents that are approved for use in a particular process. If the pressure-retaining parts have experienced minor oxidation, the user should check to see that the oxidation did not penetrate the minimum wall thickness of the pressure-retaining parts. The user should also look for ways to prevent the valve from being exposed to process drippings or atmospheric conditions. Areas of concern should be repaired and repainted.

The packing box can be rebuilt by placing the gate stem into the bonnet bore. The guide and packing should be replaced in the same order they were removed unless a new packing-box configuration is being installed. Each piece should be slipped over the top of the gate stem and down the plug stem into the packing box. (The user should take care not to damage the inside diameter of the packing rings as they are placed over the stem threads.) Depending on the type of packing used, the packing may need to be tapped into place, using a spare packing spacer or two. If a different packing material or arrangement is being installed, the packing box may vary in height, as the number of packing rings or type of spacers changes. Sometimes a small variance will be permissible with the depth of the bonnet bore. On the other hand, if the height variance is significant, a different-length packing spacer may be needed or fewer rings may be used. When changing the packing arrangement, the user should be especially careful to ensure that the packing is installed correctly and should not add extra rings of packing beyond the number recommended by the manufacturer. Too many rings may amplify the effects of thermal expansion of the packing, increase stem friction, and require increased compression to seal the additional rings. Sometimes when extra rings are added, the gland-flange bolting is incapable of providing enough axial force to seal the packing box. Finally, the gland flange should be replaced and the gland-flange bolting tightened to the recommended torque value.

Once the packing box is rebuilt, the manual operator should be reinstalled to the bonnet cap, making sure the operator threads are thoroughly lubricated (rising-stem design only). If the gate was disassembled from the gate stem during service, it should be reattached. If the gate is threaded to the gate stem (nonrising design) and the service allows, an appropriate thread lubricant should be used to ensure a smooth stroking action. If the parts are separate, the seats should be reinstalled. In threaded seats with no gasket, they should be fully seated in the body. If a gasket and threaded seat are used to seal the joint between the seat and the body, the manufacturer may provide a torque value to ensure proper gasket compression. In parallel-gate-valve designs that feature floating-seat designs, O-rings or special gaskets are used to ensure the floating design. These soft goods must be installed correctly to ensure the proper seat compression.

The bonnet cap and gate subassembly should be reinstalled into the body, using new body or bonnet gaskets. During this process, the user should take special care not to force the gate into the seats, especially with a parallel-seat design. After the body and bonnet cap are reinstalled, the bonnet-cap bolting should be threaded into the body. While the bolting is finger-tight, the valve should be fully stroked to ensure smooth operation and the bolting tightened to the manufacturer's torque values. Following reassembly and before the valve is returned to service, the valve should again be stroked several times to ensure that the gate and seats remain in alignment during the bolt tightening. If jerky stem travel is noticed, a problem may exist in the alignment of the seats with the gate, the threading to the gate or manual operator, or the packing.

3.7 Manual Pinch Valves

3.7.1 Introduction to Manual Pinch Valves

A *pinch valve* is any valve with a flexible elastomer body that can be pushed together—or "pinched"—through a mechanism or through fluid pressure (Fig. 3.38). In most cases, the elastomeric body is simply a complete liner that lines the entire inside flow passage as well as the flanges. The liner keeps all moving parts outside of the flow stream; therefore these nonwetted parts can be made from less expensive materials, such as carbon steel. Because the fluid is completely contained inside the liner, the valve has the added benefit of not requiring a packing box or gaskets.

In pinch valves, when the liner seals, the sealing area is large as opposed to a single sealing point with most valves. Because of this characteristic, large objects or particulates can be trapped in the sealed area of the valve, yet the seal can be maintained. For this reason, pinch valves are ideal for particle-entrained fluids or slurries, such as processed food, sand-entrained water systems, sewage treatment, unprocessed water, granular flows, etc. Because of the resilience associated with elastomeric liners, the liner wall effectively resists abrasion

Figure 3.38 Pinch valve closing against entrapped solids. (*Courtesy of Red Valve Company, Inc.*)

damage that results from the passage of solid matter. Also, depending upon the material selection of the liner, pinch valves do exceptionally well with corrosive fluids that may attack metal surfaces.

The main limitation of pinch valves is that they are used in lower-pressure applications, because of the pressure and temperature limits of the elastomer liner. Since common liner materials (polytetrafluoroethylene, Neoprene, Buna-N, and Viton) are also associated with rubber hoses, rubber-hose pressure ratings are used for pinch valves in lieu of common valve pressure ratings. Although elastomeric pressure ratings are typically low, these limits can be increased by using specialized liners or body designs. For example, the pressure limits can be increased by using a rubber liner that has a metal mesh woven into the rubber or by injecting an outside fluid (under pressure) around the liner to offset the fluid pressure.

Another limitation is that if the pressures inside the process system move toward vacuum or if a high pressure drop is experienced, the liner can collapse with a valve in the open position unless the liner is attached physically to the closure mechanism. Pinch valves also work poorly in pulsating flows, where the liner expands and contracts constantly, causing premature failure. When these valves are used in liquid service, the liquid must have some fluid movement to allow for the displacement of fluid by the large sealing area associated with the liner. Otherwise, the incompressible nature of liquids can place additional strain on the liner and cause it to burst.

With straight-through or uninhibited flow, pinch valves have little or no pressure drop and are ideal for on–off service. Because they are commonly used in lower-pressure services, they can be throttled quite easily and provide good flow control at the last 50 percent of the stroke. This is because the smooth walls and resilience of the liner do not provide a significant pressure drop until at least 50 percent of the stroke has been achieved. Therefore, some pinch valves made for throttling service are designed for maximum opening at 50 percent to avoid using the ineffective half of the full stroke.

With services that are extremely erosive (especially with sharp particulates), the recommended practice is not to throttle the valve close to shutoff since the particulates can etch the liner, causing grooves that can potentially tear. Another positive aspect of the liner is that the smooth walls and gentle turns of the fluid produce minimal turbulence and line vibration. The resilient liner also achieves bubble-tight shutoff easily.

Most pinch valves are operated through the injection of air pressure or another fluid or by manual operators. They can also be automated and used as a control valve. Pinch valves are commonly found in sizes of 2 to 12 in (DN 50 to DN 300) in ANSI Class 150 (PN 16).

3.7.2 Manual-Pinch-Valve Design

Two designs are prevalent in pinch valves: the open body and enclosed body. The *open-body pinch valve* has no metal body casing and relies upon a skeletal metal structure. This skeletal structure consists of two cross-bars fastened to metal flange supports. The metal flange supports are designed in halves, allowing the rubber liner to be placed between the halves during assembly (Fig. 3.39). Top and bottom supports are used to connect the cross-bars or flange support halves into one structure unit. The top support is also threaded to accept the threaded handwheel stem. This stem has a free-moving connection to a moving closure bar, called the *compressor*, which is located directly above the liner. When the handwheel is turned, the compressor is lowered, squeezing the liner against the bottom support.

The open-body design is fairly simple, does not require expensive metal castings, and allows for easy inspection of the liner for bulges, leaks, tears, or other failures. A primary disadvantage of this design is

Figure 3.39 Open-body pinch valve. (*Courtesy of Red Valve Company, Inc.*)

that the liner is exposed to the adverse effects of the outside environment, which may shorten the life of the liner.

The *enclosed-body pinch valve* has the appearance of most flow-through globe valves (Figs. 3.40 and 3.41), although the body is not actually a body but rather a protective casing for the liner. The closure mechanism is similar in design to the open-body pinch valve, except that the compressor is totally enclosed inside the body above the liner. The body can be designed with an integral bar cast into the bottom of the casing, perpendicular to the flow stream, which acts as the static closure bar. Other designs do not have this integral cast bar, called a *weir*, using the full compression of the liner against the bottom of the casing to shut off the valve. To allow for each assembly of the liner, the casing is split along the axis of the flow passage and bolted together. A drain can be included in the bottom half of the body as a tell-tale indicator that the liner has failed.

The advantage of using the enclosed-body pinch valve is that an outside fluid or pressure can be introduced through a tapped connection into the casing, assisting the liner in staying open or closed. For

Figure 3.40 Enclosed-body pinch valve. (*Courtesy of Red Valve Company, Inc.*)

example, if the process involves a vacuum, the internal casing area outside the liner in the casing can be depressurized to vacuum. This prevents the liner from collapsing when open. In some applications, additional air pressure is introduced into the casing to assist closing.

Manual handwheel operators are simplified in pinch valves because packing boxes are not required. A threaded bonnet and threaded stem (connected to the handwheel) are used to adjust the height of the compressor when operating the valve.

Another common design of pinch valves is the *pressure-assisted pinch valve,* which uses an outside fluid pressure only to close the valve (instead of a manual operator). This design (Fig. 3.42) uses a casing similar to an enclosed-body pinch valve, except the closure mechanism and operator are missing. Fluid is introduced to the inside of the casing (but outside the liner) through tapped connections. When the pres-

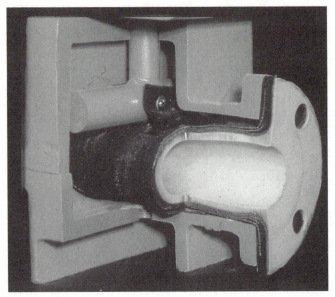

Figure 3.41 Internal view of enclosed-body pinch valve. (*Courtesy of Red Valve Company, Inc.*)

Figure 3.42 Pressure-assisted pinch valve. (*Courtesy of Red Valve Company, Inc.*)

sure of the introduced fluid overcomes the process-fluid pressure, the liner closes and remains closed until either the system pressure increases or the introduced fluid pressure decreases. This design is very inexpensive, although it is limited to on–off service only. Throttling service is difficult because changes to the downstream pressure will automatically change the position of the valve, requiring the introduced fluid pressure to be reset.

3.7.3 Manual-Pinch-Valve Operation

Generally, pinch-valve operation is quite simple. Turning the handwheel lowers the compressor and moves the upper wall of the liner toward the static lower wall, which is supported by the bottom of the casing or the bottom bracket. In throttling situations, the manual operator is turned until the required flow is achieved and is then left in that position. In on–off situations, the manual operator is turned until the closure mechanism presses the upper wall of the liner against the lower wall, which is supported by either a static lower bar or the bottom of the casing. As more thrust is applied by the manual operator, the two surfaces seal more tightly. When the pinch valve opens, the turning action of the manual operator is reversed, raising the compressor and allowing the liner to open as it moves toward its natural relaxed position. As the opening increases, the pressure of the process pushes the liner against the closure mechanism, which widens the flow area more as the closure mechanism is raised. Eventually, at the full-open position, the liner will have reached its full area capacity.

With pressure-assisted pinch valves, fluid pressure is introduced above and below the body liner. When the introduced pressure is greater than the pressure of the process fluid, the liner begins to collapse. As the introduced pressure builds, the liner begins to collapse, restricting the flow until the liner totally collapses and forms a seal between the upper and lower walls. When the introduced pressure is relieved or if the process pressure builds, the forces reverse and the liner walls separate, opening the pinch valve.

3.7.4 Manual-Pinch-Valve Installation

Before the pinch valve is installed, it should be carefully inspected against the packing list to ensure that the valve is built according to specification. Special care should be taken to ensure that the liner is made from the specified material. Many manufacturers provide installation and maintenance instructions inside the shipping carton, which should be collected and kept for reference. Larger pinch valves [3 in (DN 80) or larger] should be lifted, using lifting straps and a mechanical lifting device. The valve should not be lifted by the handwheel, since it can turn the valve unexpectedly during lifting. Before the pinch valve is installed, the line should be cleaned of any large foreign material, such as welding chips or rods. However, most manufacturers do not recommend the driving of a cleaning pig through the pinch valve because it can damage the softer liner material.

Pinch valves generally can be installed in either the horizontal or vertical position, although the horizontal position provides for easy operation of the manual operator. Pinch valves can be universally installed without regard to flow direction.

Since nearly all pinch valves are equipped with integral-flange end connections, correct alignment of the body with the piping flanges is important during installation. Because the flanges on the body are fixed, they are carefully machined so that the hole pattern lines up with the vertical axis of the valve. If the piping flanges are installed so that they are not lined up exactly with the vertical axis, the valve will have a tendency to lean, although some play exists between the size of the flange hole and the required stud, which may help minimize the misalignment. If only one piping flange is misaligned with the vertical axis, installation may be difficult. The holes on the two mating flanges may match up on one connection but be misaligned on the other connection. Some users correct this misalignment by widening the holes on the piping flange, but this weakens the integrity of the flange. The best option, but which is more expensive in terms of time and effort, is to cut the flange off the pipe and reweld it in the correct position. Since the lining extends out to the flanges, the lining is also used to seal the piping connection. Because the lining also acts as a flange gasket, using a separate line gasket is not needed. The use of gaskets may actually increase the chance of leakage through the end connection.

When the pinch valve is installed in the line, it should be fully stroked, ensuring that it moves in a smooth, linear motion. If the closure mechanism seems to stick or move in an erratic manner, a problem may exist with the stem threads. The closure element may also be binding on the casing or with its stem connection. If a pressure-assisted pinch valve is used, the external fluid pressure supply should be connected and the valve operated. If installed correctly, the valve should show no indication of leakage from the end connections.

If the pinch valve is an open-body design, it should be examined during use for any bulges, indicating a weakened portion of the liner.

3.7.5 Manual-Pinch-Valve Troubleshooting

The most common problem with pinch valves is the eventual failure of the liner, especially in those services in which the valve is closed and opened often or in which the flow pulsates. Leakage is evident immediately with an open-body design since the liner is exposed. With enclosed-body designs, a leak may not be noticed for some time until leakage is visually detected coming from a drain or through the joint between the lower and upper halves of the body. When the liner fails, the only option is to replace it.

Another common problem is poor shutoff, which can occur for several reasons. A large or long object, such as a weld rod, may be caught in the liner so that the liner cannot make full contact at one portion of the seal, creating a leak path. The closure mechanism may not have its full stroke available because of binding or damaged stem threads. The compressor may also be binding on the sides of the casing or side support brackets. If erosive, the process may have damaged the internal surface of the liner, creating additional leak paths. This problem also occurs when the pinch valve is used for throttling service near the closed position. The higher velocities caused by the restriction can also cut grooves in the liner, especially if particulates are present.

3.7.6 Manual-Pinch-Valve Servicing

The general servicing guidelines in this section are not intended to supersede any manual-pinch-valve manufacturer's specific maintenance and servicing instructions. Instead, they are provided as general guidelines. The manufacturer's instructions should be followed exactly as they are intended.

Most problems requiring servicing deal with the closure mechanism or failure of the liner. Otherwise, the simple design associated with pinch valves (no gaskets, O-rings, or wetted metal parts) requires very little, if no, servicing.

If the valve must be disassembled to inspect or replace the liner or to repair the closure element, the process line should be depressurized and drained. If the process fluid itself is corrosive or dangerous to human exposure, it should be completely decontaminated. With enclosed-body designs, the user should make sure that no pressure or fluid is trapped between the liner and the casing. With pressure-assisted pinch valves, the introduced fluid should be completely depressurized.

With open-body designs, the liner is replaced by disassembling the upper and lower closure bars. This allows the skeleton structure to split into two halves, each half with a cross-bar attached to two flange support halves. This allows for removal and replacement of the liner, after which the structure can be reassembled and the valve operated.

With enclosed-body and pressure-assisted designs, the split-body casing must be disassembled to remove the liner. This is done by removing the casing bolting, removing the lower half of the casing, and exposing the liner. The liner can then be removed and replaced, followed by reassembly of the split casing.

Whenever the closure mechanism is not operating, the problem is most likely caused by binding or galling associated with a lack of lubrication on the threads or in the guiding portion of the compressor. Lubrication is possible without disassembly.

If the closure mechanism is not operating correctly on an open-body design, it can be disassembled by simply removing the compressor and examining the threads of the handwheel stem and the intersecting threads of the top support. The user should look for signs of galling or thread damage. Also, the connection between the compressor and the handwheel stem should move freely. The closure bar should be checked to see if it is hanging on its guides. During reassembly, all sliding surfaces should be lubricated, if necessary, to ensure smooth operation.

To service the closure mechanism with an enclosed-body design, the split body will need to be disassembled and the liner removed. This will expose the closure element for inspection and lubrication, if necessary.

When the valve has been serviced and reassembled, it should be stroked to ensure smooth operation. Following installation, it should again be stroked to ensure that the liner is providing a strong seal.

3.8   Manual Diaphragm Valves

3.8.1   Introduction to Manual Diaphragm Valves

Related to the pinch valve, the *diaphragm valve* uses an elastomeric diaphragm instead of a liner in the body to separate the flow stream from the closure element (Fig. 3.43). When compressed, the diaphragm is pushed against the bottom of the body to provide bubble-tight shutoff.

The advantage of a diaphragm valve is similar to a pinch valve. The closure element is not wetted by the process and therefore can be made from less expensive materials in corrosive processes. The flow stream is straight-through or nearly straight-through, providing a minimal pressure drop, which makes it ideal for on–off service, as well as avoiding the creation of turbulent flow. Diaphragm valves can also be

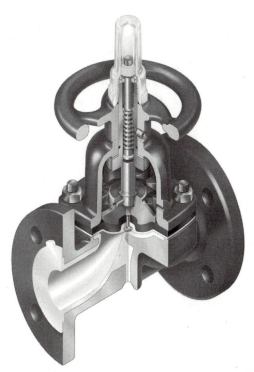

Figure 3.43 Diaphragm valve. (*Courtesy of ITT Engineered Valves*)

used for throttling service. However, maintaining a throttling position close to the bottom of the valve body can sometimes result in erosion as the particulates can cut grooves into the diaphragm and the bottom of the body. Because the diaphragm is contained in a pressure-retaining body, a diaphragm valve is able to handle somewhat higher pressures than a pinch valve, although the overall pressure and temperature ratings are dependent upon the flexibility of the material or reinforcement of the diaphragm. The design of the body flow passageway (such as the addition of a weir) has a bearing on the amount of flexibility of the diaphragm. Another advantage of the diaphragm valve is that if the diaphragm fails, the body can contain the fluid leak better than a pinch valve casing.

Diaphragm valves have an application similar to pinch valves. The resilience of the diaphragm allows it to seal around particulates in the fluid, making it ideal for service with slurries, processed food, or solid-entrained fluids.

When compared to the pinch valve, the primary disadvantage of the diaphragm valve is that the body can cost more than a pinch-valve casing because the body material must be compatible with the process fluid. Also, while the resilience of the diaphragm has a tendency to resist erosion damage from the process, the body can erode, making shutoff more difficult.

Depending on the design, diaphragm valves are available in larger sizes than pinch valves, typically up to 14 in (DN 350), although some special designs can reach up to 20 in (DN 500). Because of the pressure limitations of the liner, diaphragm valves are nearly always rated at ANSI Class 150 (PN 16).

3.8.2 Manual-Diaphragm-Valve Design

Two designs are typically associated with diaphragm valves: the straight-through design and the weir-type design. The *weir-type diaphragm valve* has the same construction as the straight-through design except for the body and diaphragm. As shown in Fig. 3.44, the body has a raised lip that raises up to meet the diaphragm, allowing the use of a smaller diaphragm. This body design is self-draining,

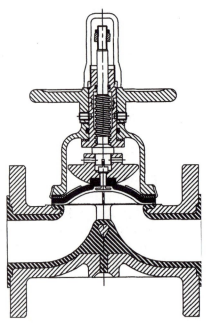

Figure 3.44 Weir-style diaphragm valve. (*Courtesy of ITT Engineered Valves*)

which makes it ideal for food-processing applications. Since the diaphragm can be made from heavier materials, the body can also be used with high-pressure services, which are not as flexible and do not allow for a long stroke. Heavier, reinforced diaphragms also allow the weir-style design to be used for vacuum services. On the other hand, the *straight-through diaphragm valve* has a body in which the bottom wall is nearly parallel with the fluid stream, allowing the flow to move uninhibited through the valve with no major obstructions (Fig. 3.45). The flexibility of the diaphragm allows it to reach the bottom of the valve body. Above the diaphragm is the compressor, a round part shaped much like the body's flow passage, which is connected to the handwheel stem. The diaphragm is attached to the bottom of the compressor to ensure that the diaphragm is lifted out of the flow stream during full-open. The compressor, the nonwetted portion of the valve, and the handwheel mechanism are contained by the bonnet cap, which is bolted to the body. The diaphragm itself is used as the gasket between the body and bonnet cap and prevents leakage to the atmosphere.

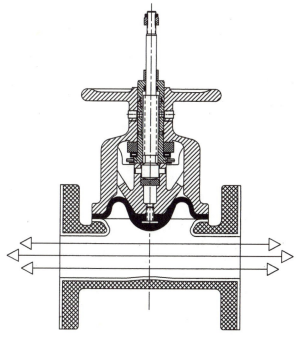

Figure 3.45 Straight-through diaphragm valve. (*Courtesy of ITT Engineered Valves*)

3.8.3 Manual-Diaphragm-Valve Operation

Manual-diaphragm-valve operation is very similar to the operation of a pinch valve. Turning the handwheel lowers the compressor, which begins to move the diaphragm toward the bottom wall of the body. In throttling situations, the manual operator is turned until the required flow is achieved and is then left in that position. In on–off situations, the manual operator is turned until the compressor pushes the diaphragm against the bottom wall of the body. As more thrust is applied by the manual operator, the two surfaces seal tighter until maximum compression is achieved. When the diaphragm valve opens, the turning action of the manual operator is reversed, raising the compressor and allowing the diaphragm to separate from the bottom body wall. As the opening increases, the pressure of the process keeps the liner pushed against the compressor, widening the flow area as the closure mechanism is raised. Eventually, at the full-open position, the compressor is fully retracted inside the bonnet cap and the diaphragm is out of the flow stream. At this point, the valve is at its full-area capacity.

Generally, diaphragm valves offer an inherent equal-percentage flow characteristic, which tends to move toward linear when installed (Sec. 2.2.5).

3.8.4 Manual-Diaphragm-Valve Installation

When the diaphragm valve is received from the manufacturer, it should be carefully inspected against the packing list to ensure that the valve is built according to specification. Special care should be taken to ensure that the diaphragm is made from the specified material. Many manufacturers provide installation and maintenance instructions inside the shipping carton, which should be collected and kept for reference. Large diaphragm valves [3 in (DN 75) or larger] should be lifted, using lifting straps and a mechanical lifting device. The valve should not be lifted by the handwheel, since it can turn the valve unexpectedly during lifting.

Before installation takes place, the line should be cleaned of any large foreign material such as welding chips or rods. Exceptionally large foreign materials can become caught in the valve and inhibit the operation of the valve, requiring immediate disassembly.

Straight-through diaphragm valves can be installed in either the horizontal, vertical, or 45° position, although the horizontal position provides for the easiest operation of the manual operator. Weir-type diaphragm valves should be installed horizontally if the self-draining feature is important to the process.

Although diaphragm valves can be built with most common end connections, the most common is the integral flange. When using flanges, correct alignment between the piping and valve flanges is important during installation. Because the integral flanges on the body are fixed, the hole pattern lines up with the vertical axis of the valve. If the piping flanges are installed so that they are not lined up exactly with the vertical axis, the valve will have a tendency to lean. Some play exists between the size of the flange hole and the required stud, which may help minimize the misalignment. If only one piping flange is misaligned with the vertical axis, installation may be difficult. This is because the holes on the two mating flanges may match up on one connection, but be misaligned on the other connection. Some users correct misalignment by widening the holes on the piping flange, but this weakens the flange's integrity. Another option is to cut the flange off the pipe and to reweld it to the correct position. Because the lining extends out to the flanges, the lining is also used to seal the piping connection.

After the valve has been installed, it should be stroked fully to make sure it operates in a smooth, linear motion. If the compressor sticks or moves erratically, a problem may exist with the handwheel stem threads or the compressor may be binding with its stem connection.

3.8.5 Manual-Diaphragm-Valve Troubleshooting

Overall, diaphragm valves are relatively trouble-free, due to their design simplicity. As with most soft goods, the diaphragm will eventually fail, especially with those applications in which the valve's full stroke is operated often. Diaphragm leakage is detected through the top of the bonnet cap, where the handwheel stem is found. If the leak is small, it may go undetected for some time until the fluid buildup exits the valve or the tear in the diaphragm widens and releases more fluid.

A malfunctioning diaphragm valve can also have poor shutoff. In some applications, large objects can become caught in the valve body and prevent the diaphragm from making full contact with the bottom of the valve body. The compressor may also be binding on the handwheel stem, generating unnecessary friction and preventing full compression. An erosive process may have damaged the internal surface of the diaphragm, creating additional leak paths. This problem may occur when the diaphragm valve is used for throttling service near closing. The erosive nature of the service under higher velocities can cut grooves in the diaphragm and/or valve body.

3.8.6 Manual-Diaphragm-Valve Servicing

The general servicing guidelines in this section are not intended to supersede any manual-diaphragm-valve manufacturer's specific maintenance and servicing instructions and are provided as general guidelines. The manufacturer's instructions should be followed exactly as they are intended.

To replace the diaphragm, the valve body must be disassembled. The top-entry design of the diaphragm valve allows for the body to remain in line while the top-works (including the diaphragm) can be removed. Before disassembly begins, the process line should be depressurized and drained. If the process fluid itself is corrosive or dangerous to human exposure, it should be completely decontaminated. The user should make sure that no pressure or fluid is trapped inside the bonnet cap due to process leakage into that area.

By removing the body bolting between the body and bonnet cap, the top-works can be lifted off, including the diaphragm. The diaphragm will need to be removed from the compressor if replacement is necessary. While the compressor is exposed, the compressor–handwheel stem connection should rotate freely without any binding or galling. If resistance is felt, the connection should be lubricated. If galling has taken place, one or both parts will need to be replaced. A new diaphragm should be replaced under the compressor, making sure the holes for the bolting line up with the holes in the bonnet cap. The top-works on the body should be reinstalled and the body bolting reinstalled. The body bolting should be tightened in a criss-cross manner using the correct torque values provided by the manufacturer.

When the compressor is not operating smoothly, most problems involve a lack of lubrication between handwheel and bonnet-cap threads or in the connection between the compressor and the handwheel stem. Lubrication is sometimes possible if the threads are exposed when the handwheel is retracted. In other cases, the handwheel mechanism will need to be disassembled to expose the threads.

When the valve has been serviced and reassembled, it should be stroked to ensure smooth operation. Once installed in the process system, it should again be stroked to ensure that the liner is providing a strong seal.

4

Check Valves

4.1 Introduction to Check Valves

4.1.1 Definition of Check Valves

Check valves (also known as *nonreturn valves*) are automatic valves that prevent a return or reverse flow of the process. Check valves are unique in that they do not require an outside power supply or a signal to operate. Rather, the check valve's operation is dependent upon the flow direction of the process, which may be created by a pump or a pressure drop. If the flow stops or if pressure conditions change so that flow begins to move backward, the check valve's closure element moves with the reverse flow until it is seated, preventing any backward flow. The check valve remains closed until positive flow direction is again achieved, at which time it opens with the flow direction and remains open as long as the flow continues.

Although the primary purpose of check valves is to prevent fluid backflow, its shutoff tightness is dependent upon the reverse pressure drop, where the downstream pressure exceeds the upstream pressure. The larger this reverse pressure drop, the greater the downstream pressure will push against the closure element providing better shutoff. When the backflow is slight, the minimal pressure drop provides poorer shutoff. Without a high downstream pressure, the closure element closes but not tightly, allowing some fluid to escape into the upstream line. Because of the limitation of this self-actuation principle, most check valves really slow down the backflow—they cannot totally prevent fluid and pressure from escaping. Because check valves can be used in a wide variety of applications, with a wide range of requirements, shutoff may be either fast or gradual, depending on the specific design of the valve.

Although the principle of operation is very similar, a wide variety of check valve designs exist. The *lift check valve* uses a free-moving closure element (similar to a globe valve's plug head without a stem) that is placed above the seat. The *swing check valve* uses a hinged closure element that is similar to a common door arrangement (where the hinged area is located outside the seating area). The *tilting disk check valve* has a closure element much like a butterfly disk that has two pivot points located on each side of the seat. The *split disk check valve* uses two half-circle disks hinged together that fold upon positive flow and retract to a full circle to stop reverse flow. The *diaphragm check valve* uses a preformed elastomeric closure element that opens with upstream flow and returns to its preformed closed shape with reverse flow. Diaphragm check valves can also use an elastomeric diaphragm that forms against a seat.

4.1.2 Common Check-Valve Applications

Check valves are used for two main purposes: first, to prevent backflow, and second, to maintain pressure. Check valves are oftentimes installed in a process system to maintain pressure in a line after a system pump has stopped or failed. Check valves are also important in the correct function of compressors and reciprocating pumps. Check valves may be a requirement for secondary systems where pressures can exceed that of a primary system. They also prevent any damage associated with reversal of rotary pumps and compressors.

If a system is apt to pulsate or create some fluctuations, the check valve is commonly placed as far away as possible from the portion of the system causing the fluctuations. If the check valve is installed close to the source of the fluctuations, it could further aggravate the situation by continually opening and closing, increasing the magnitude of the fluctuation, and causing wear on the moving parts and seating surfaces of the valve. A nonslamming check valve has a tendency to dampen pulsating flow and can sometimes be placed close to the source for that purpose.

Nonslam designs aside, the speed by which the closure element closes is often a measurement of the distance between the full-open and full-closed position of the closure element. In most cases, this is a function of the size of the check valve: the smaller the closure element, the smaller the travel from open to closed. Therefore, in smaller-sized applications, the check valve would be expected to close much faster than in larger-sized applications.

Generally, check valves can be used in both horizontal and vertical pipelines, although the use of a typical check valve in a vertical pipeline requires a spring to close the closure element (since gravity has a minimal effect on the piston, poppet, ball, disk, etc.). In vertical pipelines, flow must always rise up under the closure element. Otherwise, the check valve would have to have enough pressure reversal to lift the closure element into the seated position against its weight, which is unlikely in most cases.

4.2   Lift Check Valves

4.2.1   Introduction to Lift Check Valves

A common lift-check-valve design is the *piston check valve* (Fig. 4.1), which uses a globe body design with either a piston or a ball installed above the seat. The piston is pushed up by the flow (flow-under-the-seat) until the flow reverses when gravity and downstream pressure close the closure element against the seat. A *nonslamming piston* is a special design that is vented to allow it to move slowly to the closed position. Another design variation of the lift check valve is the *ball type*, which uses a spherical ball instead of a piston or poppet

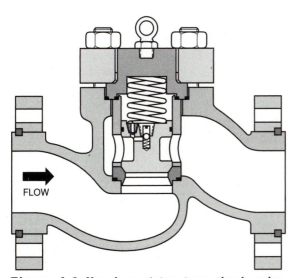

Figure 4.1 Nonslam piston-type check valve. (*Courtesy of Valtek International*)

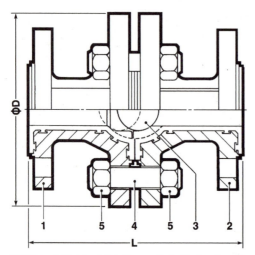

Figure 4.2 Straight-through ball check valve. Numbered parts are as follows: (1 and 2) body piece, (3) ball, (4) stud fastener, (5) hexagon nut. (*Courtesy of Atomac/The Duriron Company*)

(Fig. 4.2). The linear design includes an integral seat and a cage (which is sometimes an integral part of the bonnet cap) to keep the ball in place during flow. The advantage of using a ball as the closing device is that it can rotate with the flow, utilizing the entire spherical surface as a seating surface to minimize wear, especially in high-cycle applications. Also, when it is applied in viscous services or services where particulates are present, the ball is less likely to become dirty and stick. The ball does not require venting because the chamber above the ball is never completely blocked off. Another advantage of using the ball as the closure element is that it can be designed exclusively for use in vertical pipelines, using a straight-path flow body.

Another common lift check valve is the *plug-type* design (Fig. 4.3), which uses a simple plug to lift when flow is under the seat and to allow the downstream pressure and gravity to seat the plug quickly. Because the plug does not require the sealing of a chamber above the plug, the tolerances between the cage or seat retainer and the plug are not as close as the nonslamming design, allowing for smooth movement without the increased possibility of galling or sticking. A spring can be placed above the plug to assist with closure when installed in a vertical line.

Another variation of the plug-type check valve is the *poppet-type*

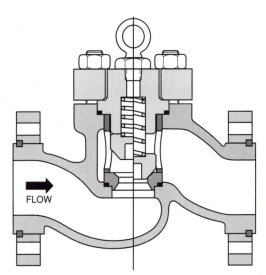

Figure 4.3 Plug-type check valve. (*Courtesy of Valtek International*)

spring check valve (or *center-spring check valve*), which uses a special plug with loose tolerances (called a *poppet*) as the closure element. A spring is placed behind the poppet to keep the closure element closed unless positive line pressure is applied. Unlike other lift check valves, flow is straight through the body without any right turns. Poppet-type spring check valves can be used in either horizontal or vertical lines, and because of their smaller sizes [2 in (DN 50) and smaller] are available with threaded or coupled connections (Fig. 4.4). Some poppet-type spring check valves are available in larger sizes above 2 in and have flanged connections (Fig. 4.5).

Because of the close tolerances generally associated with lift check valves, the service should be free of foreign materials and large particulates to avoid galling or sticking. Although some seat leakage is expected with lift check valves, the presence of large particulates may prevent adequate seating and increase the chance of leakage in the closed position.

Lift check valves are found in nearly all sizes and pressure classifications, from 1 to 36 in (DN 25 to DN 900) in pressure classes between ANSI Classes 150 and 300 (PN 16 to PN 40), and in sizes from 1 to 12 in (DN 25 to DN 300) in pressure classes above ANSI Class 900 (PN 160).

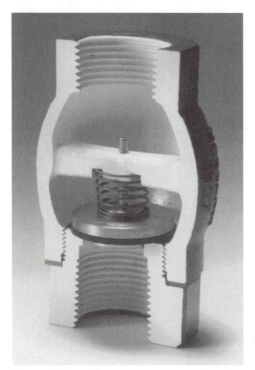

Figure 4.4 Poppet-type spring check valve with screwed ends. (*Courtesy of Mueller Steam Specialty, a Division of Core Industries, Inc.*)

4.2.2 Lift-Check-Valve Design

In most cases, the lift check valve uses a standard globe body with top-entry to the closure element, although some designs use a Y-body design (Fig. 4.6). The primary advantage of using a Y-body design is that the check valve can be used in either horizontal or vertical pipelines with the same 45° orientation, although it requires the use of a spring above the piston or plug for proper operation. As with any Y-body design, the check valve has better straight-through flow, a larger flow rate, and a lower pressure drop than the straight-through globe valve body design.

Depending on the design, the body may have an integral seat to allow for a maximum flow area. The trim involves a full-area integral seat or a separate seat ring as well as a matching closure element. Three closure-element designs are typically available: the nonslam piston, the quick-closure plug (or poppet), and the ball. The nonslam piston (again refer to Fig. 4.1) is shaped much like a pressure-balanced linear globe plug head.

Figure 4.5 Poppet-type spring check valve with flanged ends. (*Courtesy of Mueller Steam Specialty, a Division of Core Industries, Inc.*)

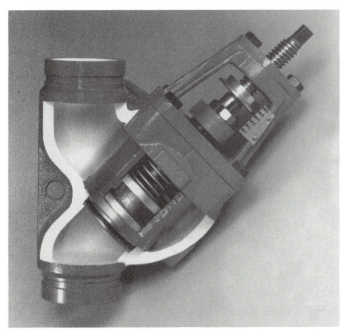

Figure 4.6 Y-body check valve. (*Courtesy of Mueller Steam Specialty, a Division of Core Industries, Inc.*)

The piston is guided on a cage or seat retainer. The tolerances between the cage and the piston are reasonably close—similar to a pressure-balanced trim design in linear globe valves—to allow for the use of piston rings. Obviously, in larger sizes, the piston can become quite heavy and the weight alone may resist the opening of the check valve. In this case, the weight of the piston can be reduced by machining away the middle portion of the piston, leaving the two ends as guiding surfaces. The key element to designing any closure element of a check valve is to balance the need for low inertia against the need for closing forces. To achieve movement with positive flow, low inertia is permitted through lightweight construction of the closure element. However, it cannot be so lightweight that it does not respond well to lessening, stagnant, or reversed flow.

The design includes a fluid chamber placed above the piston to assist with the nonslamming mechanism, which is sealed by the piston rings between the piston and the cage. Upon opening, flow is removed from this chamber by a small ball check valve inside the piston. The nonslamming principle is provided by an orifice drilled into the piston or by using a removable orifice. The size of this orifice determines the speed of closure for the piston, which allows flow to move into the chamber at a particular flow rate. Therefore, this principle applies: the smaller the orifice, the slower the closure. The trim is held in place and sealed by a bonnet or bonnet cap, which may be bolted or screwed into place. Gaskets are provided between the body (or cage) and cage (or bonnet), as well as a gasket between the seat (or body).

When the check valve is placed in a horizontal line, the weight of the piston assists with the closure. However, when the check valve is placed in a vertical line, the piston's closure cannot be assisted by gravity. In this case, a light-duty spring is placed in the chamber. This spring is intended to generate only enough force to overcome the cracking pressure and the friction between the piston and the cage. The spring also serves to prevent the piston from sticking at the top of the stroke—sometimes the free-floating piston can become slightly cocked and stick to the cage, preventing closure during flow reversal. The spring is also designed to help provide better shutoff, especially when the pressure differential during reversal is slight. One drawback with a spring is that the cracking pressure may be slightly higher because the reversal flow must also overcome the spring force.

4.2.3 Lift-Check-Valve Operation

Nonslamming piston-type lift check valves are always installed in the process line with flow-under-the-piston. When the upstream pressure

(P_1) is at a higher pressure than the downstream pressure (P_2) the flow is through the valve, pushing the piston away from the seat and allowing full flow. As described in Sec. 4.2.2, some pistons have a ball-type check valve in the piston to relieve the fluid above the piston. The opening speed of the piston is dependent on the orifice size of the piston's check valve: the larger the orifice, the faster the opening of the piston. In most piston-type lift check valves, the orifice of the ball check valve is as large as possible for quick opening.

If the flow ceases movement or if it reverses direction (the downstream pressure exceeds the upstream pressure), the weight of the piston—or spring force if the valve is installed vertically—causes it to move toward the seat. As the piston begins to drop, the ball check valve in the piston closes with the flow reversal. As the piston drops, the pressure in the chamber above the piston decreases. Without an outlet the chamber would achieve vacuum, which would prevent the piston from moving at all. The higher fluid pressure on the opposite side of the piston causes flow to move into the chamber above the piston, although its flow rate is limited by the size of the limiting orifice. As the flow moves into the chamber, the piston drops slowly until it finally seats. The seat tightness blocks the flow, but provides only limited shutoff unless the downstream pressure is significantly higher than the upstream pressure.

When the flow reversal ends, or if the upstream pressure exceeds the downstream pressure, the piston opens at a predetermined cracking pressure between the two sides of pressure. With pulsating or fluctuating flow, a nonslam piston check valve is continually moving toward or away from the seat with pulses of flow. However, because of the limiting orifice, the piston cannot move as quickly with the pulses, which allows the valve to modify the flow.

As with nonslam piston check valves, the plug-type check valve is installed with the flow under the plug and in principal works with the same pressure differentials. The only difference is that no chamber exists above the plug to prevent the plug's weight (or spring) from quickly closing the valve. For this reason, the plug-type check valve is used for services where closure must take place immediately.

With ball-type check valves, the flow direction is flow-under-the-ball, using the same flow principle as plug-type check valves.

4.2.4 Lift-Check-Valve Installation

Larger lift check valves, 3 in (DN 80) and larger, should be lifted into position using a mechanical hoist and straps to avoid personal injury or equipment damage. Ideally, a lift check valve should be installed five pipe diameters away from any source of process turbulence to avoid flutter of the closure element, which can dramatically increase valve wear. If five diameters are not possible, the user should place the valve as far away as possible from the source. If the valve is installed in a vertical line, the user should ensure that the design includes a spring behind the piston or plug to ensure proper closing operation. Special precautions should be taken to ensure that the lift check valve is installed in the correct direction. A flow tag or cast arrow is indicated on the side of the check valve. If the valve is installed in the wrong direction, positive flow will keep the valve shut.

Because of the close tolerances associated with both piston and plug styles, the user should ensure that no foreign material is included in the valve or on either side of the piping that may lead to sticking, galling, or increased leakage during operation.

Once the check valve is installed in the process line, the valve should be soon checked for proper operation. First, the entire line should be pressurized. The upstream pressure is then drained off until the downstream pressure is greater than the upstream pressure and the check valve closes. Typically, horizontally mounted lift check valves close when the upstream pressure is 1 psi (0.1 bar) less than the downstream pressure. If a nonslamming piston is used, the valve should close a few seconds after the pressure reversal has occurred. If a new valve closes too soon or too late, a different limiting orifice may be used to speed up or slow down the valve's closing.

After the valve has closed, the user should increase the upstream pressure until it is higher than the downstream pressure. When the cracking pressure is reached, the valve should open.

4.2.5 Lift-Check-Valve Troubleshooting

If possible, the upstream pipeline should be depressurized until the valve closes. The system should then be repressurized until the valve opens at the correct cracking pressure. If the check valve fails to close or open or if notable leakage is occurring, the piston, plug, or poppet may be sticking, the seat ring may be out of alignment, or the seating surfaces may be dirty or damaged. In this case, the lift check valve will need to be disassembled and repaired.

The user should visually check the end connections and around the bonnet or bonnet cap for signs of process leakage. If leakage is occurring, the check valve needs to be disassembled to replace the gaskets.

If a nonslam piston design is being used, the closure speed should be checked to make sure it has not changed drastically. An increased closing speed may indicate either worn piston rings (allowing flow to escape through the rings) or a stuck-open ball check valve in the piston itself. If the valve is closing too slowly, the most likely cause is contamination buildup inside or around the limiting orifice.

4.2.6 Lift-Check-Valve Servicing

Before disassembling the lift check valve, the process fluid should be drained from the valve and the pipeline depressurized to atmospheric pressure. If the valve was used in a caustic fluid, it may need to be decontaminated before servicing proceeds.

After removing the bonnet-flange bolting, the bonnet cap and gasket should be lifted off the top of the body. If included, the spring should be removed, followed by the cage or seat retainer, piston or poppet, and seat ring. The user should examine the cage or seat retainer and piston or poppet, looking for indications of galling or scoring between the two parts. Unless the damage is very minor, worn or scratched sliding surfaces usually require new parts.

If a nonslamming piston is used, the ball check valve should be disassembled. The limiting orifice should be cleaned thoroughly of any process or dirt buildup. The piston rings should be checked for damage or unusual wear.

The seating surfaces of the seat ring and the piston or poppet should be checked to ensure an adequate seal. If the seating surface is damaged or worn, remachining the seat angles may help avoid replacement for the time being. If the seat angles are remachined, the original angle should be maintained. If an integral seat is used in lieu of a seat ring, machining that surface may be extremely difficult and may best be left to a valve repair facility.

The overall body should also be inspected for process damage. Areas of severe corrosion should be blasted and repainted. After all parts have been thoroughly inspected and cleaned, as well as repaired or replaced if necessary, the lift check valve should be reassembled in the reverse order as the parts were removed during disassembly. All gaskets should be replaced during this process. The user should make sure that the piston or plug moves freely inside the cage or seat retainer and that the piston or plug makes full contact with the seating surface.

After the internal parts have been replaced, the bonnet cap should be reinstalled and the body bolting tightened, using the correct tightening method (and torque values if applicable). Insufficient bolting tightness will result in inadequate gasket compression and will lead to process leakage. Once the check valve is reassembled, the user should check for correct operation by proceeding through the process described in Sec. 4.2.4.

4.3   Swing Check Valves

4.3.1   Introduction to Swing Check Valves

One of the more simple check valve designs is the *swing check valve,* which uses a hinged door to open during positive flow and to close against a pressure reversal (Fig. 4.7).

From outside appearances, the body used with the swing check valve appears similar to a globe-style body, although it actually has a straight-through flow path rather than flow under (or over) the closure element. The body is usually top-entry and uses standard end connections. Although the face-to-face is typically longer than a gate valve, it is shorter than a globe valve's face-to-face.

The basic advantage of using the swing check valve is its design simplicity. Other than an occasional problem with binding between the pin and the disk arm, little else can go wrong. Even if particulates are

Figure 4.7 Swing check valve. (*Courtesy of Velan Valve Corporation*)

present in the flow stream and become caught between the disk and the seat, when the valve opens the offending matter flows downstream. The primary disadvantage is that swing check valves are best utilized in lower-pressure classes (ANSI 150 to 600 or PN 16 to PN 100). In higher-pressure classes, the disk must be made with a greater wall thickness. This greater mass and weight make it harder for the disk to swing open as well as to produce some slamming when flow reverses. Generally, this pressure weight issue precludes the use of swing check valves in pressure classes higher than ANSI Class 900 or PN 160. Swing check valves are normally found in sizes from 1 to 36 in (DN 25 to DN 900), although most are less than 12 in (DN 300) in size.

4.3.2 Swing-Check-Valve Design

The closure element is normally a round disk and a seat, which can be a separate part or integral to the body. The disk has an integral arm that is attached to a pivoting hinge pin. To avoid the design and manufacturing difficulties of attaching the hinge pin directly to the body, an internal hanger is used to attach the hinge pin to the disk arm. The hanger can be attached to the body directly or to the bottom of the bonnet cap. Attaching the hanger to the bonnet cap has the added advantage of being able to adjust the alignment of the disk and seat. Some manufacturers avoid internal arrangements by inserting the hinge pin through a bore in the body wall. Once the hinge pin is installed, the bore is sealed using a retainer plug.

When a separable seat is used in the design, it is usually screwed into place. Because the disk may have some difficulty seating during stagnant flow or with a minimal flow reversal, the seat can be slightly inclined (similar to a wedge-style gate valve) to allow for better seat contact at the closed position. The seating surfaces of both the seat and the disk are flat and make full contact when seated. The area of contact can be narrow or broad, depending on the manufacturer's design.

The bonnet cap, which is used to seal the top-entry portion of the body, can include an integral disk-stop. This disk stop is used to limit the swing of the disk at a certain angle (normally 60° to 85° from the centerline of the pipe). With the disk-stop, the disk remains in the flow stream and is subject to instant closing when the flow stops or reverses.

4.3.3 Swing-Check-Valve Operation

When the upstream pressure (P_1) exceeds the downstream pressure (P_2) the flow moves through the seat and against the disk, causing it to move in the same direction as the flow. As the positive flow increases, the flow pushes the disk farther away from the flow path until the disk is nearly out of the flow stream. At this point the disk-stop, if used in the design, stops the swing travel of the disk.

As the flow lessens, the weight of the disk allows it to drop until the drag of the flow prevents it from traveling any further. If the flow stops, the weight of the disk allows it to drop down until its seating surface makes contact with the seat. If the flow reverses, the greater downstream pressure, P_2, pushes the disk toward the seat, improving the shutoff. The leakage during shutoff depends on the reverse pressure differential: the greater the reverse pressure differential, the better the shutoff. The disk remains in the closed position until a positive flow is again achieved and the cracking pressure is reached, allowing for the disk to again begin swinging open.

4.3.4 Swing-Check-Valve Installation

To avoid personal injury or damaging the equipment, swing check valves larger than 2 in (DN 50) should be lifted into position using a mechanical hoist and straps. Ideally, a swing check valve should be installed five pipe diameters away from any source of process turbulence to avoid check valve flutter, which can dramatically increase valve wear. If five diameters are not possible, the user should place the valve as far away as possible from the turbulent source. For correct operation, the user should make sure that the swing check valve is installed in the correct direction by visually checking the flow tag or cast arrow on the side of the check valve. If the valve is installed incorrectly, positive flow will keep the valve closed. To avoid any damage to the seating surfaces or binding of the hinge pin, foreign material (such as welding rods, scale, etc.) should be cleaned from both sides of the piping. Before installing the valve in the line, the disk should be checked to make sure it moves freely without binding or sticking. The user should not use his or her hands to do this, since some disks are heavy with sharp edges, which may pinch or cut fingers.

Once the swing check valve is installed, the entire line should be pressurized to operating conditions. Proper function of the check valve can be quickly checked by draining off the upstream pressure until the downstream pressure is greater than the upstream pressure and the valve shuts off with the reserve flow. Typically, horizontally mounted lift check valves close when the upstream pressure is 1 psi (0.1 bar) less than the downstream pressure. After the valve has closed, the upstream pressure should be increased until it is higher than the downstream pressure. When the cracking pressure is reached, the valve should open.

4.3.5 Swing-Check-Valve Troubleshooting

If the system permits, the upstream pipeline should be depressurized until the valve closes, followed by repressurization until the valve opens at the cracking pressure. If the check valve fails to open or close, or if notable leakage is occurring through the closure element, a number of causes may be possible: the hinge pin may be sticking or galling, or the seating surfaces may be dirty or damaged. In these cases, the swinging check valve will need to be disassembled and repaired.

The end connections and the gap between the body and the bonnet cap should be inspected for signs of process leakage. If leakage is occurring, the valve will need to be disassembled and the gaskets replaced. If the swing check valve has a hinge pin that is installed through a bore in the body wall, the seal of the retainer plug should be checked.

If the valve is closing too slowly or experiencing erratic operation, the most likely cause is process buildup around the hinge pivot point or galling between moving parts.

4.3.6 Swing-Check-Valve Servicing

If the swing check valve is not shutting off or opening properly, it should be disassembled and repaired. Before disassembly takes place, the pipeline should be depressurized to atmospheric pressure and the process fluid should be drained from the valve—especially if the process is caustic.

After removing the body bolting, the bonnet cap should be lifted off the top the body. The gaskets may then be removed. The gasket surfaces should be inspected for any signs or erosion or process leakage. The user should test the travel of the disk to check for a smooth action. If the pivot point between the pin and the disk handle shows signs of process buildup, the buildup should be removed and cleaned until the disk moves freely. If the pivot point continues to bind, the pivot point assembly should be disassembled and examined for signs of galling. If substantial galling has occurred, the damaged parts should be replaced. If minor galling has occurred, it may be repaired using an abrasive cloth. If a retainer plug is used to seal a hinge pin bore, it should be removed prior to removing the pin.

The seating surfaces of the seat and the disk should be checked to ensure an adequate seal. The separable seat is usually threaded and may require a special tool provided by the manufacturer. With all internal threading arrangements, corrosion can pose difficulties in removing the seat without applying significant force that could further damage the part. If the seating surfaces appear to be slightly worn or damaged, remachining the flat surfaces of the seat may keep the valve functioning without having to replace the seat and disk.

The overall body should be also inspected for process damage. Areas of severe corrosion should be blasted and repainted (if applicable). All other parts should be thoroughly inspected and cleaned. If possible, minor damage should be repaired. Severely damaged parts, as well as all gaskets, should be replaced as standard practice. The lift check valve should be reassembled in the reverse order as the parts were removed during disassembly. After the pivot-pin assembly has been reassembled, the user should make sure that the disk moves freely without too much play. Too much play in the hinge could cause chatter and premature wear; therefore the hinge should be snug, but not too tight. If the hinge is too tight, sticking and premature wear will result. Some manufacturers recommend lubrication of the hinge pin in certain applications. If this is the case, the lubrication must be compatible with the process. After the hinge-pin assembly is complete, the disk should make full contact with the seat's seating surface.

After the internal parts have been replaced, the bonnet cap and the body bolting should be reinstalled, using the correct tightening method. Insufficient bolting tightness will result in inadequate gasket compression and will likely lead to process leakage. If the bonnet cap has an internal disk-stop, the user should make sure that the orientation is correct so that the stop and disk make full contact. Once the check valve is reassembled, the user should check for correct operation by proceeding through the check process (Sec. 4.3.4).

4.4   Tilting-Disk Check Valves

4.4.1   Introduction to Tilting-Disk Check Valves

Similar in some respects to a swing check valve, the *tilting-disk check valve* uses a pivoting disk as the closure element, except the pivot point is roughly through the center of the disk instead of above the disk. In many respects, it follows the design criteria of an eccentric disk in a high-performance butterfly valve. This design allows the disk to rotate rather than swing outward from the seat. The pivot point is slightly offset from the true center of the disk so that gravity can act on the disk, allowing it to move to the closed position when neutral flow exists. However, the offset is only slight, so that a low cracking pressure will open the disk and allow flow through the valve. At its full-open position, when the seat is perpendicular to the pipe centerline, the full rotation of the disk is parallel to the centerline and 90° from the seat. Some tilting-disk check valves are designed with the seat at a 45° angle so that the disk is only 45° from the seat at the open position, which allows for much faster operation.

The chief advantage of a tilting-disk check valve is its speed of closure, since the disk requires very little travel to close. Because the pivot point is close to the center of the mass of the disk, the disk has a tendency to avoid slamming shut (as opposed to the swing check valve), and hence is popular for use in water services where water-hammer effects occasionally occur. The chief disadvantage is that—similar to a butterfly valve—the closure element is in the flow stream and can generate turbulence, vibration, and noise in systems with high velocities. Another disadvantage is maintenance, since disassembling the rotating disk is somewhat difficult as opposed to a swinging disk. Another disadvantage with tilting-disk check valves with a split-body design is that the check valve must be removed from the pipeline for maintenance.

Tilting-disk check valves are normally used in large water lines—up to 48 in (DN 1200)—and are rarely seen in lines below 4 in (DN 100). They are rated to the lower-pressure classifications of ANSI Classes 125 to 150 (PN 6 to PN 16). Some higher-pressure designs are available but usually require some type of internal hanger to attach the pivot pin to so as to avoid using a retaining plug in high-pressure service.

4.4.2 Tilting-Disk-Check-Valve Design

The tilting-disk-check-valve design includes an integral seat inside the body, although separable seats are available that are screwed or welded into place. The seating surfaces for the seat and the disk are beveled and make full contact in the closed position.

The tilting-disk-check-valve design can be achieved with either a Y-body design (Fig. 4.8) or a conventional straight-through body design (Fig. 4.9). Normally, the pivot point is achieved using pins, which are

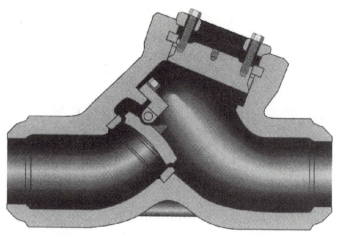

Figure 4.8 Y-body tilting check valve. (*Courtesy of Pacific Valves, a unit of the Crane Valve Group*)

installed through the body wall, necessitating the use of retainer plugs to seal the body. When the straight-through or Y-body designs are utilized, the top-entry portion of the body is sealed using a bonnet cap. The split-body accesses all moving parts through the middle connection and does not require any top-entry access.

Figure 4.9 Straight-through split-body tilting check valve. (*Courtesy of ITT Engineered Valves*)

4.4.3 Tilting-Disk-Check-Valve Operation

When the flow is in a neutral state (upstream pressure equals the downstream pressure), the disk has a tendency to stay closed because of the offset pivot point. However, when the upstream pressure increases (or downstream pressure decreases), the positive flow moves through the seat and pushes against the disk, causing it to rotate to its open position. Because the center of the disk mass is below the pivot point, the larger half of the disk rotates away from the flow, while the shorter half rotates into the flow. The disk continues to travel until it is stopped by the physical limitations of the body—ideally the plane of the disk is parallel to the flow.

As the flow slows or ceases, the dominant weight of the lower half of the disk allows it to begin dropping toward the bottom of the seat. As the flow stops, the disk's inertia position is achieved, making seat contact. If the flow reverses, the greater downstream pressure P_2 applies pressure to the disk in the seat, improving the shutoff. As the reverse-pressure differential increases, shutoff improves significantly. The tilting check valve remains in the closed position until positive flow occurs and the cracking pressure is reached or surpassed, allowing for the disk to again begin rotating open.

4.4.4 Tilting-Disk-Check-Valve Installation

Because tilting-disk check valves are 4 in (DN 100) and larger, their weight requires the use of a mechanical hoist and straps to avoid personal injury. The best method to avoid unnecessary valve flutter is to install the tilting-disk check valve five pipe diameters away from any source of process turbulence. If five diameters are not possible, the valve should be placed as far away as possible from the turbulent source. As with all check valves, correct operation is dependent upon flow orientation of the valve, which is determined by a flow tag or cast arrow on the side of the body. If the valve is installed incorrectly, positive flow will keep the valve closed, while reverse flow will open the valve.

Any type of foreign material, such as welding rods, scale, or trash, should be removed from both sides of the piping. The inclusion of such material could damage the seating surfaces or bind the motion of the disk. Before installing the valve in the line, the disk should be checked to see that it moves freely without binding or sticking. A safe procedure is to use a wooden tool to move the disk, instead of one's hand, since the quick-closing motion could catch or pinch fingers. If the disk does not move freely, the problem should be corrected before the valve is placed in line.

Although some tilting check valves work well in either horizontal or vertical lines, the best option for a vertical line is a Y-body design where the disk remains in the same position (45°) despite the line orientation.

Once the tilting-disk check valve is installed, the entire process line should be pressurized to operating conditions and the valve should be checked. The check valve can then be tested for correct function by draining off the upstream pressure until the downstream pressure is greater than the upstream pressure. At this point, the valve should shut off quickly without slamming. Typically, horizontally mounted lift check valves close when the upstream pressure is 1 psi (0.1 bar) less than the downstream pressure. After the tilting-disk check valve has closed against the reverse flow, the upstream pressure should be increased until it exceeds the downstream pressure. When the cracking pressure is reached, the valve should open.

4.4.5   Tilting-Disk-Check-Valve Troubleshooting

Periodically, the tilting-disk check valve should be checked for correct operation, similar to the procedure outlined in Sec. 4.4.4. If possible, the upstream pipeline should be depressurized until the valve closes, followed by repressurization until the valve opens at the cracking pressure. If flow seems restricted when positive flow is achieved, most likely the disk is not rotating freely, but is restricted in its movement. If excessive reversal flow is noticed even after the valve should have shutoff, the disk may not be seating due to a pin that is binding or galling, or the seating surfaces may be severely damaged. In this case, the swinging check valve will need to be disassembled and repaired. If the valve seems to respond sluggishly or the cracking pressure has increased significantly, the most likely cause is a pivot pin that is bound by process buildup or galling.

The end connections and all joints in the body, especially the retaining plugs, should be inspected for signs of process leakage. If leakage is occurring, the valve should be disassembled and the gaskets replaced.

4.4.6   Tilting-Disk-Check-Valve Servicing

When the tilting-disk check valve requires service, the symptoms are failure to close against reverse flow, sluggish response, or an increased cracking pressure. In all of these cases, the valve will need to be disassembled and serviced. Before any service on the valve takes place, the line and valve should be depressurized to atmospheric pressure. If the process fluid is caustic or corrosive, it should be drained from the valve and the valve decontaminated before the valve is worked on.

If the check valve has a split-body design, the entire valve will need to be removed from the line for disassembly. If a straight-through globe or Y-style body is used, the valve can stay in the line while the service is being completed through the valve's top-entry access. With split-body designs, the body bolting should be removed and the two halves separated, exposing the disk and seat. With straight-through globe or Y-style bodies, the top-entry access can be exposed by removing the body bolting and the bonnet or bonnet cap.

The gaskets should be removed and all gasket surfaces should be inspected for any signs or erosion or process leakage. The movement of the disk should be checked for ease of travel. If the pivot point shows any signs of process buildup, it should be removed and cleaned until the disk moves freely. If the pivot point still continues to bind, the retaining plugs (if applicable) should be removed and the pivot-point assembly disassembled. The bores or pin should be examined for any signs of galling. If substantial galling has occurred, the damaged parts should be replaced, while minor galling may be repaired using an abrasive cloth.

The seating surfaces of the seat and the disk should be thoroughly examined to ensure an adequate seal. If the seat is threaded, it should be removed using a special tool provided by the manufacturer. With all internal threading arrangements, corrosion can pose difficulties in removing the seat without applying significant force that could further damage the part. If the seat is welded into place, the anchor bead weld between the body and the seat ring will need to be ground out, being careful not to remove more material than necessary. If the seating surfaces appear slightly worn or damaged, replacement of the parts may or may not be necessary. If the user attempts to remachine the seating surfaces, the beveled angles of the original design should be maintained. The overall body should also be inspected for process damage. Areas of severe corrosion should be sandblasted and repainted.

Before reassembly takes place, all parts should be thoroughly inspected for wear or damage and cleaned using an appropriate cleaning solution or solvent. Any minor damage should be repaired; severely damaged parts and all gaskets should be replaced. The check valve should be reassembled in the reverse order in which the parts were removed during disassembly. If a separate seat is used, it should be securely tightened (if threaded) or welded. When welding the seat into place, the welder should be careful to dissipate the heat, so that heat transfer does not warp the seating surface of the seat ring. After the pivot-pin assembly has been reassembled, the disk should move freely without too much play. Too much play at the pivot point can cause chatter and premature wear. If the pin assembly is too tight, however, sticking and premature wear will result. After the assembly is complete, the disk should make full contact with the seat's seating surface. Some manufacturers recommend lubrication of the pin in certain applications. If lubrication is required, the lubricant should be compatible with the process.

After the closure element has been reassembled, the bonnet cap and the body bolting should be replaced, using the correct tightening method (if the body is top-entry). Insufficient bolting tightness will result in inadequate gasket compression and will likely lead to process leakage. If a split body is used, the two halves should be bolted together, ensuring even tightness and correct gasket compression. Once the check valve is reassembled, correct operation can be checked by proceeding through the process described in Sec. 4.4.4.

4.5   Double-Disk Check Valves

4.5.1   Introduction to Double-Disk Check Valves

The *double-disk check valve* uses two half-sphere disks and a connecting hinge to create a closure element that collapses when positive flow exists, and folds back to a full circular disk when reverse flow exists (Fig. 4.10). It is also known as a *split-disk* or *wafer check valve.*

Generally, double-check valves are popular because of their low cost and minimal size. Their chief disadvantage is poor sealability in low-back-pressure situations. Also, the unsealed joint between the two half-disks allows for some backflow to move through the closure element when it is in the closed position. The inclusion of the spring in the flow stream may present a problem if the spring fails and metal parts are washed downstream, which may foul or damage downstream equipment. The introduction of the closure element into the middle of the flow stream may also present some turbulence problems with high flow rates and velocities.

Double-disk check valves are found in sizes 2 through 36 in (DN 50 through DN 900) in ANSI Pressure Classes 150 through 600 (PN 16 through PN 100). Larger sizes—12 in (DN 300) and higher—are available only in ANSI Class 150 (PN 16).

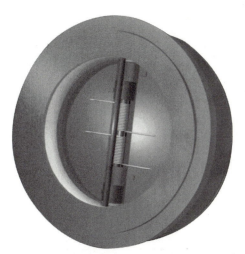

Figure 4.10 Double-disk check valve. (*Courtesy of Gulf Valve Company*)

4.5.2 Double-Disk-Check-Valve Design

With a hinge similar to a door hinge, the two half-disks are connected using a straight pin that is inserted through a bore in the body wall with the bore sealed by a retaining plug. At least one spring is installed in the hinge and around the pin itself to provide a spring action toward closing, although some designs use two springs—one for each half-disk. In some designs, a second pin is inserted just downstream from the half-disks, which is used as a disk-stop for the two half-disks when they are open, as well as an antirotation device for the spring end(s). The other spring end is straight and lays flat against the back side of the half-disk. The springs are made from materials that provide good spring action, as well as resistance to corrosion, such as Inconel. The spring rate can be varied by the type and design of the spring, allowing for different cracking pressure and closing speeds. In some applications, the user may not want the additional potential leak paths presented by the retaining plugs. In this situation, a special assembly uses a self-contained retainer to hold the pins, half-disks, and springs, which is then inserted into the body (Fig. 4.11).

In the closed position, the two half-disks' seating surfaces rest against the flat surface of an integral seat, which is machined into the body, providing a full contact seal. The seat is perpendicular to the centerline of the pipeline. This 90° seat is not apt to provide a strong seal with the half-disks when stagnant flow or minimal reverse flow

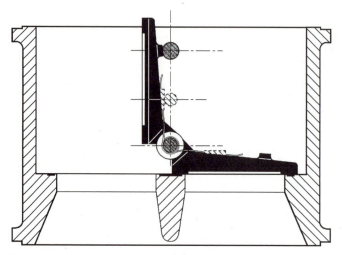

Figure 4.11 Internal view of double-disk check valve.
(*Courtesy of Gulf Valve Company*)

exists, as opposed to other check-valve designs that have an inclined
seat and can use the weight of the disk against the incline to provide a
tight shutoff at a low pressure differential. To compensate for this
problem, a soft seat can be specified to help achieve better shutoff at
low back pressures.

The body used for the double-disk check valve is a wafer-style body,
which is a flat body with a minimal face-to-face. The wafer-style body
allows the check valve to be installed between two flanged piping connections with straight-through bolting. In sizes below 12 in (DN 300)
and in applications where high temperatures are not a concern, the
wafer-style body is typically used. In larger sizes, or where high temperature may cause potential leakage through thermal expansion, a
lug-body design is used. A lug body utilizes flanges with drilled and
tapped holes for direct connection to flanged piping using studs and
nuts. If room exists between the body's integral flanges for nuts,
tapped holes are not necessary.

4.5.3 Double-Disk-Check-Valve Operation

When the upstream pressure (P_1) is higher than the downstream pressure (P_2) the positive flow pushes against both half doors. When the
pressure differential exceeds the spring force, the doors open with the

flow and continue to open wider as the flow increases. Finally, at full flow the half-disks reach their full motion when the disk-stops (located on the back side of the half-disks) meet the second pin or the opposite disk-stop. As the flow lessens, the spring action begins to force the half-disks back toward their closed position. As flow stagnates, the half-disks are in their full-closed position, with their flat seating surfaces meeting up with the flat seating surface of the body. As reverse flow continues to build in strength, the shutoff improves.

4.5.4 Double-Disk-Check-Valve Installation

After the double-disk check valve has been received from the manufacturer, the user should carefully inspect the materials and spring rate to ensure its proper function in the system. Larger check valves—3 in (DN 80) and larger—should be installed using a mechanical hoist and straps. With any large and heavy equipment, caution should be taken to avoid dropping the valve, which may hurt personnel or damage the equipment.

Ideally, the double-disk check valve should be installed five pipe diameters away from any source of process turbulence to avoid check valve flutter, which can dramatically increase valve wear. If five diameters are not possible, the user should place the valve as far away as possible from the source of turbulence. Before the check valve is installed in the pipeline, the entire length of the upstream and downstream piping should be checked for any foreign materials, such as welding rods, and cleaned of scale or any other foreign material.

Double-check valves can be installed in either the horizontal or vertical position. During installation, the user should make sure that the valve is installed with the flow in the correct direction. A flow tag or a cast arrow on the external wall of the body should indicate this. (If flow is not indicated, the open action of the door should move with the positive flow direction.)

When installing the double-disk check valve between piping flanges, care should be taken to allow enough room for the face-to-face of the wafer body, the width of the gaskets, and some clearance for the free movement of the valve to slide into place. If the pipeline is rigidly fixed and too much space exists between flanges, tightening the bolting between the piping flanges may make it hard to achieve correct gasket compression. The best installation method is to loosely install the bottom two studs and nuts, creating a cradle for the check valve to sit. With the gaskets in the correct position, the remaining bolting can be installed. If the check valve has integral flanges, they may be installed using a stud and nut for a tapped hole or a stud and two nuts for an untapped hole.

To determine if the double-disk check valve is installed correctly and operational, both the upstream and downstream lines must be pressurized to operating conditions. By draining off the upstream pressure until the downstream pressure is higher than the upstream pressure, the valve should shut off quickly without a slamming. Most double-disk check valves should close when the upstream pressure is 1 psi (0.1 bar) less than the downstream pressure. After the double-disk check valve has closed, the upstream pressure should be increased until it again exceeds the downstream pressure. The valve should open when the cracking pressure is reached.

4.5.5 Double-Disk-Check-Valve Troubleshooting

As part of periodic maintenance, the double-disk check valve should be checked for correct operation. To do this, the process system will need to be deviated from normal service by depressurizing the upstream line until the valve closes. Following closure, the upstream line should again be pressurized until the valve opens at the correct cracking pressure. If the valve opens before the cracking pressure is reached or seems to flutter at a low pressure differential, the disk spring may have failed. If flow is restricted when positive flow is achieved or if the valve seems to be sluggish upon opening, most likely one or both double disks are not rotating freely or are binding on the pin. If excessive reversal flow is noticed even after the valve should have shutoff, one (or both) of the half-disks may have remained opened (especially if the spring has failed). Also, the half-disk may not be seating due to process binding or extensive galling between the pin and half-disk. In this case, the double-disk check valve will need to be disassembled and repaired.

All potential leak paths in the body (especially the retaining plugs) should be inspected for signs of process leakage. If leakage is occurring, the valve should be disassembled and the internal gaskets replaced. If the service has temperature swings, the user should check for adequate tightness of the line's through bolting.

4.5.6 Double-Disk-Check-Valve Servicing

When the double-disk check valve fails to close against reverse flow, is sluggish in response, or has a higher cracking pressure, the valve will need to be disassembled and serviced. However, before any service on the valve takes place, the line and valve should be depressurized to atmospheric pressure. If the process fluid is caustic, corrosive, or dangerous to workers, it should be drained from the valve and the valve completely decontaminated.

Because of the wafer-body design, the piping flange bolting will need to be disassembled. With a horizontal line, a safe procedure is to leave two bottom studs and nuts (loosened) in place so that valve does not fall during line disassembly. Because process residue and line pressures can cause the valve to stick between the flanges, the flanges may need to be worked loose to remove the valve. If the valve is stuck between the flanges, a crowbar or other lever should not be used between the gasket surfaces of the flanges or valve since this may scar or indent the serrations and make a future seal difficult.

After the valve is taken out of the line, the gaskets should be removed from the gasket surfaces, and those surfaces should be inspected for any signs of erosion or process leakage. The half-disks should be pushed open using a wooden dowel. If the springs are intact, the disks should return smoothly to the closed position without hanging up. A visual check of the springs is not enough, since a spring may be broken but may appear to be intact. If the hinge pin shows any signs of process buildup, it should be removed and cleaned until the half-disks move freely. If the hinge pin continues to bind, the retaining plugs should be removed and the hinge pin pushed out. The pin, half-disk hinges, or body wall bores should be examined for any signs of galling. If substantial galling has occurred, the affected parts (or entire valve, if necessary) should be replaced, while minor galling may be repaired using an abrasive cloth.

The seating surfaces of the body and the half-disks should be thoroughly examined to ensure an adequate seal. If the seating surfaces appear slightly worn or damaged, replacement of the parts may not be necessary. The seating surfaces might be capable of light machining or grinding, although the user must maintain the perpendicular flat seat surfaces of the original body and half-disks.

The body should also be inspected for process or atmospheric damage. Areas of severe corrosion should be sandblasted and repainted, if necessary. After all parts have been inspected for damage or wear, they should be thoroughly cleaned using an appropriate cleaning solution or solvent. Any minor damage should be repaired, while severely damaged parts and all line gaskets should be replaced. The check valve should be reassembled in the reverse order in which the parts were removed during disassembly. The spring(s) should be correctly repositioned so that the spring ends press against the back of the half-disks. The pin is then reinserted through the body bore and loosely retained by the body plugs. Some manufacturers recommend lubrication of the pin in certain applications. If lubrication is required, the lubricant must be compatible with the process. After the hinge pin has been reassembled, the disk should move freely without too much play or tightness. Too much play can cause chatter and premature wear. If the hinge is too tight, sticking and premature wear may occur. The half-disks should be visually checked to ensure that they make full contact with the body's seating surface. After the half-disks are assembled and working smoothly, the retaining plugs should then be tightened, using the correct torque and sealant to ensure a tight pressure-retaining seal.

Using new line gaskets and two piping flange studs and nuts as a cradle, the valve should be replaced between the piping flanges. Using studs and nuts, the flanges should be tightened with the appropriate torque value. Insufficient bolting tightness will result in inadequate gasket compression and will likely lead to process leakage. After or when the double-disk check valve is reassembled, correct operation should be checked by proceeding through the process mentioned in Sec. 4.5.4.

4.6 Diaphragm Check Valves

4.6.1 Introduction to Diaphragm Check Valves

The diaphragm check valve is one of the more simple check-valve designs, using a preformed elastomeric sleeve that opens to allow positive flow, yet returns to its preformed shape when reverse flow occurs (Fig. 4.12). Diaphragm check valves are ideal for processes with low-pressure drops (a ΔP less than 150 psi or 10.3 bar) and for moderate temperatures (less than 160°F or 70°C). It works better than most check valves in slurries or liquids that contain particulate matter.

Diaphragm check valves are available in nearly all pipe sizes and in low-pressure classes (ANSI Class 150 or PN 16 and smaller).

4.6.2 Diaphragm-Check-Valve Design

The diaphragm sleeve is made of rubber for water services or a special elastomer if applied in a chemical process. The elastomer is reinforced with layers or plies of nylon. When closed, its opening has the appear-

Figure 4.12 Diaphragm check valves. (*Courtesy of Red Valve Company, Inc.*

ance of a tight oval, with the lips of the diaphragm sleeve touching. When the diaphragm sleeve opens, the lips part and the distance widens as the flow increases. Eventually, at maximum flow, the shape of the opening is closer to a broad oval.

The diaphragm check valve has a straight-through body with the diaphragm sleeve retained one of two ways: First, it can be bolted to one of the body flanges, allowing the sleeve to fit inside the body. Or, second, it can use a split-body design, retaining the diaphragm sleeve between the joint where the body valves are bolted together. Most common end connections are available, although integral flanges are the most commonly specified.

4.6.3 Diaphragm-Check-Valve Operation

In a static flow condition, the diaphragm sleeve remains in its preformed, static state with the lips touching and no gaps in the sleeve.

As the upstream pressure increases, causing a pressure differential, the flow moves through the sleeve by separating the pliable lips. As the flow continues to increase, the gap opens until it eventually reaches its maximum opening. As the upstream pressure decreases or the downstream pressure builds until a reserve flow situation occurs, the diaphragm sleeve moves back to its preformed state because of the greater downstream pressure. Because the preformed state is lip-to-lip, the sleeve is closed against the reverse flow and remains in that position until positive flow occurs again. Closure is very quick, although water hammer is usually avoided because the diaphragm sleeve has a tendency to absorb any shock associated with a sudden closure.

4.6.4 Diaphragm-Check-Valve Installation

To avoid injury or damage to the valve, larger diaphragm valves [3 in (DN 80) and larger] should be moved using a mechanical hoist and straps. The rubber sleeve should be inspected from inside the valve; in the preformed state, the sleeve's lips should be in full contact. The entire length of the upstream and downstream piping should be checked for any foreign materials, such as welding rods, and cleaned of scale or any other foreign material *before* the valve is installed.

Diaphragm check valves work best in horizontal lines, but can be placed in vertical lines in certain applications. During installation, the user should make sure that the diaphragm valve is installed with the flow in the correct direction. A flow tag or a cast arrow on the external wall of the body will indicate this. If flow is not indicated, the pleated portion of the sleeve should be on the downstream side of the valve.

Before the diaphragm check valve is placed into service permanently, it should be tested for correct operation by pressurizing both the upstream and downstream lines to operating conditions. When the upstream pressure is drained until the downstream pressure is higher than the upstream pressure, the valve should shut off quickly. After the diaphragm check valve has closed, the upstream pressure should be increased until it again exceeds the downstream pressure. The valve should open when the cracking pressure is reached.

4.6.5 Diaphragm-Check-Valve Troubleshooting

Periodically, the diaphragm check valve should be checked for correct operation and the causes of any failures should be examined. To troubleshoot the valve, the upstream line should be depressurized, if possible, until the valve closes. Following closure, the upstream line should again be pressurized until the valve opens at the correct cracking pressure.

If excessive reversal flow is noticed even after the valve should have shutoff, most likely the diaphragm sleeve has failed. Over time, the sleeve may age and develop cracks or split at the pleat. If a foreign object becomes wedged between the lips, leakage can also occur. In addition, if the object is trapped in the sleeve for any period of time, the sleeve may deform, which will continue to produce excessive leakage even after the object is washed downstream.

If the body is a split-body design, the joint between the body halves should be inspected for signs of process leakage. The end connections should also be checked. If leakage is occurring, the valve should be disassembled and the leaking gasket replaced. If the body is supplied with a drainage plug, the user should check that seal for leakage also.

4.6.6 Diaphragm-Check-Valve Servicing

When the diaphragm check valve fails to completely close against reverse flow, most likely a problem exists with the diaphragm sleeve and the valve will need to be disassembled and serviced. Both the line and valve should be depressurized to atmospheric pressure before any service on the valve takes place. If the process fluid is caustic, corrosive, or otherwise dangerous to workers, it should be drained completely from the valve and the valve decontaminated.

Because most diaphragm check valves use a split-body design and feature no top-entry, the entire valve will need to be removed from the process line and the piping flange bolting will need to be disassembled. Because process residue and line pressures can cause the valve's end connections to stick between the flanges, the flanges may need to be worked loose to remove the valve. The temptation to wedge a crowbar or other lever between the gasket surfaces of the flanges or valve should be avoided, since this may scar or indent the serrations—making a future seal difficult.

Once the valve is taken out of the line, the line gaskets should be removed from the flanges' gasket surfaces. Those surfaces should be thoroughly inspected for any signs of erosion or process leakage. From one side of the valve, the diaphragm sleeve should be checked for any cracks, slits, indentations, or other damage that would preclude the diaphragm from providing a tight shutoff. In some cases, debris may have become caught in the diaphragm sleeve. If this is the case, the debris should be removed and the sleeve checked for damage. If the diaphragm is distorted toward the upstream side of the valve, the pressure drop is greater than the application parameters of the diaphragm sleeve, and the process will need to be adjusted to avoid such a high-pressure drop.

If the diaphragm is damaged or worn, requiring replacement, it should be removed from the body. Some split-body designs retain the diaphragm sleeve between body halves, while others bolt the retainer to one end of the body. The body should also be inspected for process or atmospheric damage. Areas of severe corrosion should be sandblasted and repainted (if applicable). All metal parts of the diaphragm check valve should be inspected for damage or wear. Following inspection, they should be thoroughly cleaned using an appropriate cleaning solution or solvent. Any minor damage should be repaired; severely damaged parts and all line gaskets should be replaced. If the diaphragm sleeve is replaced, it should be carefully inspected to ensure that the lips make full contact in the closed, preformed position.

The diaphragm check valve should be reassembled in the reverse order as the parts were removed during disassembly. The diaphragm sleeve should be correctly repositioned with the folded lips facing the downstream side of the valve. If the diaphragm sleeve is retained by the two halves of the split body, the user should make sure that the sleeve fully seats between the two body valves. If the diaphragm sleeve is attached to the upstream end of the valve, the retaining bolting should be sufficiently tightened to provide a good seal.

Once the diaphragm check valve has been reassembled, the line gaskets should be replaced prior to reinstalling the valve in the line. With line flanges, the bolting should be tightened in an even manner and to the torque values suggested by the manufacturer. Once the double-disk check valve is reassembled, correct operation should be assured by pressurizing the line and draining off the upstream pressure until the flow reverses and the valve shuts off.

5

Pressure-Relief Valves

5.1 Introduction to Pressure-Relief Valves

5.1.1 Definition of Pressure-Relief Valves

A *pressure-relief valve* is a self-operating valve that is installed in a process system to protect against overpressurization of the system. When excess line pressure is detected, the pressure-relief valve automatically opens and the excess pressure is relieved. A pressure-relief valve is installed in a process system where excess pressure constitutes a safety concern of the piping or equipment bursting, or where an abnormal operating condition could damage the process product. Following the depressurization of the process line to safe or normal limits, the pressure-relief valve automatically closes again to allow for normal system operation. Pressure-relief valves can be used for both gas and liquid services, although the design varies with each.

Pressure-relief valves are actuated (opened and closed) by one of two common methods. The first is through *system actuation* (also commonly called a *direct-acting pressure-relief valve*), where the process pressure acts on one side of the closure element (commonly a disk and a nozzle), while a predetermined spring provides a mechanical load on the other side of the closure element (Fig. 5.1). The correct spring rate is critical to the function of the pressure-relief valve, as well as the nozzle opening. When the line pressure reaches its maximum limit, the spring rate is such that the pressure force overcomes the spring force,

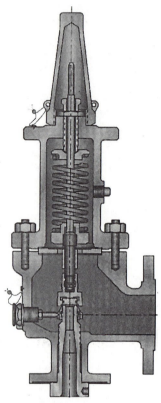

Figure 5.1 Direct-acting spring-loaded pressure-relief valve. (*Courtesy of Anderson, Greenwood & Co., a Keystone International company*)

causing the spring and disk to retract, allowing for pressure to escape through the nozzle at a predetermined flow rate.

The second actuation method is *pilot actuation* (also called a *pilot pressure-relief valve*) in which a pilot-valve mechanism monitors the system pressure and triggers the main valve to open when the pressure exceeds the limit (Fig. 5.2). Although some pilot actuation designs place the pilot mechanism inside the valve, most have the mechanism attached to an outside surface (such as to the bonnet cap or to a body wall using a bracket). The pilot mechanism also determines the operating characteristics of the valve. With pilot pressure-relief valves, the closure element is held in place by the process fluid or an external power supply (or a combination of both).

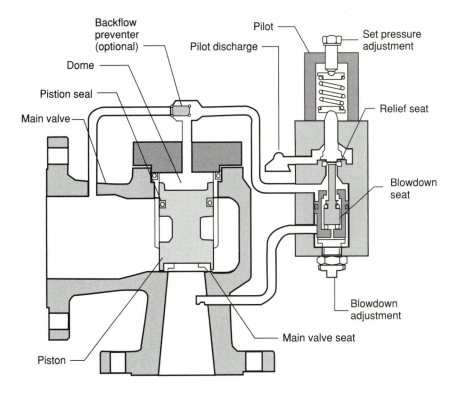

Figure 5.2 Pilot-actuated pressure-relief valve. (*Courtesy of Anderson, Greenwood & Co., a Keystone International company*)

Pressure-relief valves open one of two ways. The first method is called *full lift:* When the overpressurization causes the disk to open slightly, the valve opens to the full-open position immediately and allows for immediate depressurization of the line. The second method is called *modulating lift,* in which the valve only partially opens—just enough to relieve the overpressurization but not enough to depressurize the line entirely.

Pressure-relief valves are divided into two families according to application. *Relief valves* are used in liquid services and can be either full-open or modulating, depending on the application. One major characteristic with relief valves is that because of the incompressibility of liquids, any overpressurization of the service automatically opens the valve. On the other hand, *safety valve* is a term used for pressure-relief valves used in gas or vapor services, although the large majority of applications for such valves is steam service. Because of the dangers associated with high pressure and superheated steam, most safety valves are used for the protection of personnel and associated systems (and hence the name *safety* valves). Generally, gas services are designed to allow some overpressurization—because of the inherent compressibility of gases—before the pressure-relief valve opens.

Pressure-relief valves are commonly found in inlet sizes of 0.25 in (DN 6) to 20 in (DN 500).

5.1.2 Pressure-Relief-Valve Design

Because of the critical nature of relief and safety valves, their design is strictly regulated by local, regional, and national codes. The code with the largest influence over relief-valve design is the ASME Boiler and Pressure Vessel Code, which provides guidelines for overpressurization. This code calls for the accreditation of pressure-relief-valve manufacturers and repair organizations to certain manufacturing and testing methods. It also sets standards for laboratory testing of pressure-relief valves. Following compliance with this code, an ASME code symbol is used to show compliance by the manufacturer or service provider. Internationally, ISO Standard 4126 has been adopted to monitor design and performance standards for pressure-relief valves (ISO denotes the International Standards Organization). Other groups, such as HPGCL in Japan and Technicscher Ueberwachungs Verein (TUV) in Germany, provide performance and specification regulations. For applications, the American Petroleum Institute (API) has a wide selection of standards for the design and sizing of pressure-relief valves as well as for their proper testing, storage, and handling.

As shown again in Fig. 5.1, the main closure element of the pressure-relief valve consists of the *disk* (sometimes called the *pallet*) and the *nozzle*. As the name implies, the disk is a round, flat part that covers the area of the nozzle. The disk itself is guided by the stem, or it can be guided directly by a disk guide. The nozzle can be an integral part of the body or a separate part that is threaded into the body from the bottom port. The inside diameter of the nozzle determines the flow area of the valve. A number of disk and nozzle seating arrangements are found, although the most common metal-seated design is one in which the disk has an inverted groove for the nozzle. The seating angles are slightly mismatched to allow for full contact. In addition, a soft seat or O-ring can be attached to the bottom of the disk to provide a bubble-tight seal. The disk is attached to the spring assembly and adjusted through a separate stem that is bolted to the disk.

A pressure-relief valve has two ports: an inlet, which is defined as the pressure-bearing side of the closure element, and an outlet, which is the downstream pressure (or atmospheric pressure, if vented). The outlet is always larger by one or two line sizes than the inlet to accommodate the large flow associated with a blowdown. The outlet is typically discharged into a vessel or protected area for safety reasons, or to recover lost product. Pressure-relief valves use typically a 90° angle body design. One-piece or two-piece bodies are common, with two-piece designs having the versatility of changing the direction of the outlet in any one of four quadrants.

Above the body is the bonnet, which supports and encloses the spring and spring compression adjustment. Two types of bonnets are common: the open bonnet and the pressurized bonnet. The *open bonnet* exposes the spring to the atmosphere (Fig. 5.3) and is used in high-temperature applications. A problem with springs is that they lose

Figure 5.3 Safety valve with open bonnet. (*Courtesy of Valve & Controls Division, Dresser Industries, Inc.*)

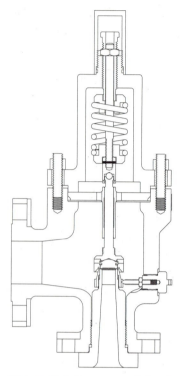

Figure 5.4 Pressure-relief valve with pressurized (closed) bonnet. (*Courtesy of Hydroseal Valve Company, Inc.*)

their strength when heated. The opening allows the spring to remain cool, thereby maintaining the correct spring rate. With an open-bonnet design, because the bonnet area is not sealed completely from the body area, some gas or liquid can leak out of the bonnet openings, although the majority of the fluid will escape through the body outlet. The *pressurized bonnet* (sometimes called a *closed bonnet*) is sealed with gaskets and has few or no vent openings (Fig. 5.4). The no-vent design is used to prevent the leakage of any dangerous gases to atmosphere. Because of the loss of spring strength at high temperatures, a pressurized bonnet must be used in mild temperature applications. For exceptionally high-temperature applications, such as some steam services, some safety-valve manufacturers may use a special spool inserted between the body and the bonnet to distance the spring from the body.

The spring is placed between two spring buttons. The lower spring button is attached to the disk stem, which is the dynamic end of the spring, although its position on the stem is fixed. The upper spring button is in a fixed position and does not move during operation of the pressure-relief valve. However, although the upper spring button is fixed for operation, its position can be changed by the adjusting screw, which sets the spring compression.

In some smaller sizes or lower pressures, calibrated weights may be used instead of springs to provide the closing force on the closure element (although this practice is not as prevalent as it once was). Weights are heavier and require large bonnet areas, making their use especially impractical in larger sizes and pressures. Overall, springs have the greater advantage in that the spring rate can be changed with larger and longer springs. Springs are not sensitive to line vibration. Overall, the force characteristic of the spring is similar to the requirements of the flow characteristic of a typical full-lift pressure-relief valve. For these reasons, a wide majority of pressure-relief valves use springs for closure and actuation.

A lifting lever is provided on safety valves to allow for periodic testing of the valve to ensure correct operation or to manually depressurize the line. The lever is cammed so that lifting the handle raises the spindle, causing the disk to lift off the nozzle, thus allowing the pressure-relief valve to actuate. Generally, lifting levers are used to unseat the valve when the system pressure is at least 75 percent of the set pressure. Using the lifting lever at pressures less than 75 percent can result in misalignment of internal parts.

5.1.3 Pressure-Relief-Valve Operation

Pressure-relief valves are designed to stay closed during the *operating pressure*, which is defined as the service pressure for which the process system was designed. At this point, the downward force on the disk produced by the spring rate and compression is greater than the upward force produced by the pressure against the disk. The downward force is also increased by any significant *back pressure*, which is the pressure on the outlet side of the valve. It is also affected by the temperature of the service, which may have an effect on the spring force.

Each pressure-relief valve has a defined *set pressure*, which is the point where the overpressurization of the system overcomes the spring force holding the disk to the nozzle. When the set pressure is reached, the pressure-relief valve begins to open. The *seating-pressure differential* is a term used to express the difference between the operating and set pressures. For sensitive or critical applications, the seating-pressure differential may be quite small, while for broader applications (where minor overpressurization is acceptable) the setting may be higher so that the valve does not open at a slight increase and upset the process. For gas services, the seating-pressure differential is usually set between 1 and 5 percent, while in liquid services it is set between 5 and 20 percent. Because the actuation of a pressure-relief valve can upset a process and cause problems with production, the seating-pressure differential must be set to balance the needs of the process and the safety management requirements of the system.

With full-lift valves, the valve does not immediately go to the full-open position at the set pressure. Instead, further overpressurization is required before the valve fully opens. This point is called the *opening pressure*. With modulating valves, the valve typically opens in a linear characteristic, opening only enough to accommodate the overpressurization.

As the overpressurization decreases, the pressure will reach a point where the downward force of the spring overcomes the upward force of the fluid pressure, causing the disk to lower and finally seat. The point where the disk reseats is called the *reseating pressure*. This action in which the overpressurization causes the valve to move to an open position and the pressure is relieved until the valve again seats is called the *blowdown*, which is defined as the difference in pressure between the set and reseating pressures. Some users prefer the blowdown to be of a short duration for prevention of fluid loss and for more efficient production. However, the valve can become somewhat unstable if the blowdown is too short. Ideally, the best solution is determined by the application itself. If a number of blowdowns are expected—resulting in lost production—shorter blowdowns are best. If blowdowns are infrequent, a longer blowdown is best since valve stability is ensured. Blowdown may be adjusted by a number of means. For example, an upper blowdown ring can be installed, which is used to shorten or lengthen the disk guide. Raising the blowdown ring decreases the blowdown, while lowering the ring increases the blowdown. If a separate nozzle is threaded into the body, it may be adjusted up and down (similar to the blowdown ring). However, this can adversely affect the opening pressure differential. A throttling regulator can also be added to regulate the forces acting on the back side of the disk.

Spring-operated pressure-relief valves fall under Section I of the ASME code, which establishes that full rated discharge must take place at a maximum pressure of 3 percent above the valve's set pressure. On the other hand, the blowdown is required to be between 2 and 4 percent below the set pressure. If the relief valve is set below 100 psi (6.9 bar), the reseating pressure is required to be between 2 and 4 psi (between 0.1 and 0.3 bar).

5.1.4 Pressure-Relief-Valve Installation

As the pressure-relief valve is removed from its packaging, the user should make sure that all flange protectors and packing materials are discarded. All gasket surfaces should be checked to make sure that they are clean and free of any scratches that may lead to gasket failure. Before installation takes place, the valve should be inspected to ensure that it has been certified by an approved accreditation program (ASME or other comparable program). The valve should have a seal attached with the proper approval stamps or other identification marks.

If the valve is provided with a lifting lever, the user should never use it as a convenient device to lift the valve, since this action may cause the seating surfaces to rub against each other and cause a potential leak path. Lifting by the lever can also cause the valve to rotate, causing a safety hazard to nearby workers and equipment. A good practice to follow with pressure-relief valves it to use a light-gauge wire or tape to attach the handle to the bonnet in order to prevent a future accidental discharge.

Before the valve is installed, the line should be cleaned of dirt, welding chips, scale, or other matter. A mechanical lifting device and secure straps should be used to lift the valve into place, making sure that no gasket surfaces are damaged when placing the valve between piping flanges. Pressure-relief valves are normally installed with the bonnet up, although other orientations are sometimes possible. Valves that use weights must be installed with the bonnet upright. Tags, plates, or embossed letters in the casting indicate the inlet, which must be connected to the upstream piping. The outlet is connected to the tank or other receiving vessel. Discharge piping is short and pointed away from walkways or other regions where workers may be. In most cases, the discharge piping is upward. In all cases, the discharge piping should be anchored to a nearby structure to prevent excessive vibration during blowdown.

Room should exist to allow for access to the pressure-relief valve and to remove it, if necessary. If the valve is equipped with a lifting lever, the user should make sure the lever action is not restricted by other piping or equipment. All piping connections should be properly tightened and sealed, using the correct size of studs and nuts, as well as the correct torque values.

After the valve has been installed, the system should be started up and tested to the set pressure point to see if overpressurization will cause the valve to open and blowdown to properly occur. If the set opening or reseating pressure is incorrect, the manufacturer's instructions should be checked for possible calibration procedures. Many settings on a pressure-relief valve are very sensitive, and large movements of the setting can upset the operation of the valve, causing further difficulties and lost start-up time. If the user is not familiar with the valve or its calibration procedures, the calibration should be completed by the manufacturer's service technicians or by an authorized service operation.

5.1.5 Pressure-Relief-Valve Troubleshooting

The pressure-relief valve should be inspected periodically to ensure proper operation. As a matter of procedure, the seals should be checked to ensure that they are intact, as broken seals may indicate that unauthorized adjustments have taken place. As part of monthly troubleshooting, the valve should be opened with the lifting lever or through planned system overpressurization. Those valves in services that are extremely corrosive may require constant checking to avoid sticking problems.

The most common problem with pressure-relief valves is a tendency for the closure element to pulsate when it is opening or in the open position. This causes chatter when the disk is barely off the nozzle or fluttering of the disk when the valve is open and away from the nozzle. The most common cause of chatter or flutter is an oversized valve, which happens all too frequently when the user and manufacturer add safety factors to the information used to size the valve, causing an inflation of the sizing factors. The only remedy for the problem of oversizing is to restrict the lift of the pressure-relief valve. However, if that fails to correct the problem, a smaller-sized valve will need to be installed.

Another frequent cause of chatter or flutter is the close proximity of a piece of process equipment that causes pressure fluctuations or process turbulence. Reciprocating pumps, compressors, orifice plates, or pressure-reducing valves can all cause such fluctuations. To deal with the effects of other equipment, the pressure-relief valve will need to be installed further down the line, putting some distance between the valve and the problem equipment. Such fluctuations have a tendency to dissipate over distance but not completely. Another option is to adjust the pressure-relief valve for a higher blowdown setting or a higher set pressure.

Chatter and flutter can be caused by other factors. The back pressure may be higher than the valve is set for. The opening pressure differential may be set at such a low setting that standard leakage flow is enough to open the valve. A pressure loss may be taking place between the compressor or pump and the pressure-relief valve. And finally, the valve may be designed for maximum flow, but pressure variations are only minor, not allowing the valve to move to the full-open position. In this case, another pressure-relief valve can be installed before the existing maximum flow valve to handle the minor process upsets.

Valve leakage in a pressure-relief valve is evident by the presence of liquid or fluid in the outlet piping. This is caused by a worn disk or nozzle (especially in high-use situations), a damaged disk or nozzle, a valve that chatters because it operates too close to the set pressure, or piping stresses that may affect the alignment of the valve body and closure element.

High blowdown is a problem with adjustments of the pressure-relief valve itself, although it can be increased by the presence of drops of liquid in the gas or vapor or the buildup of back pressure over time.

5.1.6 Pressure-Relief-Valve Servicing

Because of the critical nature of pressure-relief valves, their servicing and adjustments should be made by a trained and certified technician. Generally, pressure-relief-valve adjustments are extremely sensitive and even minor misadjustments can upset the operation of the valve. The best procedure to follow is to refer to the manufacturer's printed maintenance literature before making any adjustments. Because of the dangerous applications associated with safety valves, such as those in steam service, servicing and calibration of the valve must be done by trained technicians who are certified to disassemble and reassemble such valves.

To disassemble a standard pressure-relief valve, the valve should be removed from the process line. Before that takes place, the line should be completely depressurized and decontaminated, if necessary. All process fluids should be completely drained. For most servicing, the valve body can remain bolted to the process line. If the body or nozzle needs servicing or replacement, the entire valve should be removed from the line.

The first step is to break the seals and remove the cap. Because the spring can cause injury or damage if removed under compression, the bonnet bolting (holding the bonnet to the body) should not be removed until the spring compression has been relieved. This varies according to design; however, the most common method is to remove the adjustment screw cap or adjusting bolt, which is found above the bonnet. At this point, a good idea is to measure the distance of the top of the adjustment screw to a fixed surface for easy reassembly. Turning the adjustment screw itself relieves the spring compression. The screw may need to be turned out altogether, although when the spring is decompressed, it can be felt easily.

After the spring is decompressed, the bonnet bolting should be removed and the bonnet assembly lifted out of the body. At this point, the disk and the nozzle seating surfaces can be inspected for damage or wear. If the disk is being removed, it must be unbolted from the stem. Once the disk is removed, guides, spring buttons, and the spring can be removed.

All moving and pressure-retaining parts should be checked for signs of wear, galling, or damage. Process buildup should be removed and all parts cleaned. All gaskets, O-rings, or other elastomer sealing materials should be replaced. If the seating surfaces of the disk and nozzle are lightly worn or damaged, they may be repaired using emery cloth or similar materials. A good practice is to use a flat grinding surface to ensure parallel surfaces of the two parts. To avoid galling the spring adjustment screw, thread lubricant should be applied to those threads prior to reassembly.

After cleaning and lubrication, the user should carefully reassemble the parts in the reverse order in which they were removed and return the spring adjustment screw to its original setting so as to maintain the proper spring compression. The valve should then be returned to service in the process line and the start-up procedure performed, including overpressurization, to ensure that the valve operates at the correct set pressure and blowdown (Sec. 5.1.4).

If a valve has been completely rebuilt, the best procedure to follow is to perform a pressure test with the valve. The ASME code requires that each valve be tested for correct operation by use of a steam test (see Fig. 1.7 in Chap. 1). This test should be conducted by ASME-approved facilities and technicians and should mirror the basic operating conditions of the actual service.

6
Control Valves

6.1 Introduction to Control Valves

6.1.1 Definition of Control Valves

Over the years, some confusion has existed between the definitions of a throttling valve and a control valve. Some use the words interchangeably because they both have a similar purpose: to regulate the flow anywhere from full-open to full-closed. For the most part, a *throttling valve* is any valve whose closure element has a dual purpose of not only opening or blocking the flow but also moving to any position along the stroke of the valve, thus regulating the process flow, temperature, or pressure. Using the term *closure element* is not adequate in describing this portion of the throttling valve; thus, for purposes of differentiation, the term *regulating element* is used to describe any portion of the valve that allows for throttling control. A throttling valve is designed to take a pressure drop in order to reduce line pressure, flow, or temperature. The interior passageways of a throttling valve are designed to handle pressure differential, while on–off valves are designed to allow straight-through flow without allowing a significant pressure drop. Because the purpose of the throttling valve is to provide reduced flow to the process, rangeability is a critical issue. The valve's trim size is almost always smaller than the size of the pipeline or flow passages of the valve. Using a full-size valve in a similarly sized pipe will provide poor controllability by not utilizing the entire stroke of the valve. Throttling valves must have some type of mechanical device that uses power supplied by a human being, spring, air pressure, or hydraulic fluid to assist with this positioning. Some manually operated on–off valves can be used or adapted for throttling service. Pressure regulators are also considered throttling valves, since they vary in the position of the regulating element to maintain a constant pressure downstream.

By definition, a *control valve* (also known as an *automatic control valve*) is a throttling valve, but is almost always equipped with some sort of actuator or actuation system that is designed to work within a control loop. As discussed in Sec. 1.3.5, the control valve is the final control element of a process loop (consisting of a sensing device, controller, and final control element). This involvement with the control loop is what distinguishes control valves from other throttling valves. Manually operated valves and pressure regulators can stand alone in a throttling application, while a control valve cannot, hence the difference: a control valve is a throttling valve, but not all throttling valves are control valves. In some cases, a manually operated valve can be converted to a control valve with the addition of an actuation system and can be installed in a control loop—thus in the pure sense of the definition it becomes a control valve.

Control valves are seen as two main subassemblies: the body subassembly and the actuator (or actuation system). This chapter will concentrate on the operation, design, installation, and maintenance of body subassemblies, while Chap. 7 will detail actuators and actuation systems.

Generally, control valves are divided into four types: globe, butterfly, ball, and eccentric plug valves. Variations of these four types have resulted in dozens of different available designs, the most common of which will be covered in this chapter. Each design has specific applications, features, advantages, and disadvantages. Although some control valves have a wider application than others, no control valve is perfect for all services, and each design should be examined to find the best solution at minimal cost.

6.2 Globe Control Valves

6.2.1 Introduction to Globe Control Valves

Of all control valves, the linear-motion (also called rising-stem) globe valve is the most common, due in part to its design simplicity, versatility of application, ease of maintenance, and ability to handle a wide range of pressures and temperatures. The globe valve is the most commonly found control valve in the process industry, although demand is not as great with the advent of high-performance rotary valves, which offer lower cost and smaller packages, size for size. Sizes range from 0.5 to 42 in (DN 12 through DN 1000) in lower-pressure classes (up through ANSI Class 600 or PN 100); from 1 to 24 in in ANSI Classes 900 to 2500 (PN 160 through PN 400); and from 1 to 12 in in ANSI Class 4500 (PN 700).

By definition, a *globe valve* is a linear-motion valve characterized by a globe-style body with a long face-to-face dimension that accommodates smooth, rounded flow passages. The most common regulating element is the *single-seat design*, which operates in linear fashion and is found in the middle of the body. The single-seat design uses the plug–seat arrangement, where a plug moves into a seat to permit low flows or away from the seat to permit higher flows. The alternative to the single-seat arrangement is the double seat, which will be discussed in detail in Sec. 6.2.4.

The advantages of globe control valves are many—hence their overall popularity. Generally, globe valves are quite versatile and can be used in a wide variety of services. The same valve can be used in dozens of different applications as long as the pressure and temperature limits are not exceeded, and the process does not require special alloys to combat corrosion. This versatility allows for reduction in spare parts inventory and maintenance training. Their simple linear-motion design permits a wider range of modifications than other valve styles. Because of the linear motion, the force generated by the actuator or actuation system is transferred directly to the regulating element; therefore, a minimal amount of the energy to the regulating element is lost. On the other hand, rotary valves lose some transfer energy and accuracy because of the dead band (amount of input change needed to observe shaft movement) associated with linear- to rotary-motion linkage. For this reason, globe valves are capable of high performance and are used in applications where such performance is mandatory.

A major advantage to using globe control valves is their ability to withstand process extremes. They are designed to work in extremely high pressure drops, handling pressure differentials of thousands of pounds of pressure (or hundreds of kilograms per centimeter squared). Globe valves can be designed to handle higher pressure classes by increasing the wall thickness of the body and using heavier-duty flanges, bolting, and internal parts. Severe temperatures can be handled with extension modifications to the bonnet or the body, keeping the top-works (actuator, positioner, supply lines or tubing, and accessories) away from the process temperature.

An important advantage of a globe control valve is that it can have a flow characteristic designed into the trim or the regulating element itself—unlike butterfly valves whose design only allows for an inherent characteristic.

Most globe control valves with single seats have top-entry to the trim (plug, seat, and cage or retainer). This allows easy entry into the valve to service the trim by removing the bonnet flange and bonnet-flange bolting and removing the top-works, bonnet, and plug as one assembly. Unlike rotary valves, globe valves can remain in the line during internal maintenance. For this reason, globe valves are preferred in the power industry where steam applications require the welding of the valve into the pipeline.

As mentioned earlier, the main disadvantages of globe valves are that, size for size, they are larger, heavier, and more expensive than rotary valves. They present seismic problems because of their greater height—a problem where an earthquake or process vibration could cause the top-works to place stress on the body subassembly or line.

Another disadvantage is that globe valves are restricted by the significant stem forces required by the throttling process. Globe valves with pneumatic actuators are restricted to sizes smaller than 24 in (DN 600), or 36 in (DN 900) with a hydraulic or electrohydraulic actuator. With higher-pressure classes, the bulk of the globe-valve body assembly, as well as the stem forces, decreases the size availability even more. When large flows must be regulated beyond the size capabilities of a globe valve, users sometimes divide the flow between two smaller pipelines, preferring smaller valves. In some cases, butterfly or eccentric disk rotary valves are used instead.

6.2.2 Globe-Control-Valve Design

In describing the design elements of a globe valve, the *globe body* is the main pressure-retaining portion of the globe valve, which has matching end connections to the piping and also encloses the trim (Fig. 6.1). The flow passages in a globe valve are designed with smooth, rounded walls without any sharp corners or edges, thus providing a smooth process flow without creating unusual turbulence or noise. The flow passages themselves must be of constant area to avoid creating any additional pressure losses and higher velocities. Globe-valve bodies are adaptable to nearly every type of end connection, except the flangeless design. Obviously with a long face-to-face dimension, the long bolting required between two pipe flanges would be susceptible to thermal expansion during temperature cycles.

The *single-seated trim* is more than just a closure element, because a throttling valve does more than just open or close; rather, it is a regulat-

Figure 6.1 Globe-style control valve. (*Courtesy of Valtek International*)

ing element that allows the valve to vary the flow rate with respect to the position of the valve according to the flow characteristic, which may be equal percentage, linear, or quick open (Sec. 2.2). Typically, the trim consists of three parts: the *plug*, which is the dynamic portion of the regulating element; the *seat ring*, which is the static portion; and the *seat retainer* or *cage*. The portion of the plug that seats into the seat ring is called the *plug head*, and the portion that extends up through the top of the globe body subassembly is called the *plug stem*. The plug stem is threaded to the *actuator stem*, allowing a solid connection without any play or movement. The actuator stem is assembled to an actuator piston or diaphragm plate, which transfers pneumatic or hydraulic force to the regulating element. The basic advantage of the single-seated trim design is that it allows the tightest shutoff possible, usually better than 0.01 percent of the maximum flow or C_v of the valve. This is because the actuation force can be applied directly to one seating surface. The greater the actuation force, the greater the shutoff of the valve.

Two sizes of trim can be used in globe valves. *Full trim* refers to the area of the seat ring that can pass the maximum amount of flow in that particular size of globe valve. On the other hand, *reduced trim* is used when the globe valve is expected to throttle a smaller amount of flow than that size is rated for. If full trim is used, the valve would have to throttle close to the seat as well as in small increments—which is difficult for some actuators. The preferred method, then, is to use a seat ring with a smaller seat area—with a matching plug—which is defined as *reduced trim*. Most manufacturers offer four or five sizes of reduced trim for each size of valve.

The *bonnet* is an important pressure-retaining part that has two purposes. First, it provides a static cap or cover for the body, sealed by bonnet or body gaskets. Second, it seals the plug stem with a packing box—a series of packing rings, followers or guides, packing spacers, and antiextrusion rings that prevent or minimize process leakage to atmosphere. Mounted above the packing box is the gland flange, which is bolted to the top of the bonnet. When the gland-flange bolting is tightened, the packing is compressed and seals the stem as well as the bonnet bore.

Guiding the plug head in relation to the seat ring is accomplished by two types of guiding: double-top stem guiding or caged guiding. *Double-top stem guiding* uses two close-fitting guides at both ends of the packing box to keep the plug concentric with the seat ring (see Fig. 6.2). These guides can be made entirely from a compatible, dissimilar

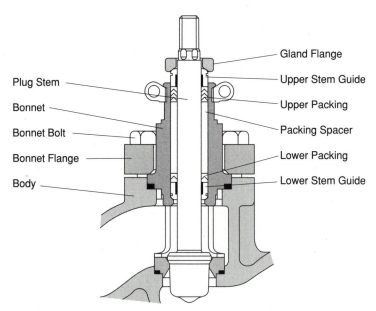

Figure 6.2 Double-top stem guiding in a globe valve. (*Courtesy of Valtek International*)

metal with the plug to avoid galling or can include a hard elastomer or graphite liner. The key element of double-top stem guiding is that the guides must be widely separated to avoid any lateral movement from the process fluid acting on the plug head, which is exposed to the forces of the process stream. The guides—as well as the bonnet bore and the actuator stem—must be held to close tolerances to maintain a fit that will allow a smooth linear motion without binding or slop. To avoid lateral movement as the process impinges on the plug head, some plugs have large-diameter stems to resist flexing. However, when compared to smaller-diameter stems, larger plug stems do have an increased circumference, which increases the sealing surface and the possibility of seal leakage as well as packing friction. However, the stem-friction problem is easily rectified by using higher thrust actuators, such as piston cylinder actuators, which can easily handle the increased stem friction.

The second type of guiding configuration is *caged guiding*. With the cage-guided design (Fig. 6.3), the upper guide is placed at the top of the packing box and the lower guiding surface is placed inside the flow stream, using the outside diameter of the plug head to guide within the inside diameter of the cage. Because the distance between

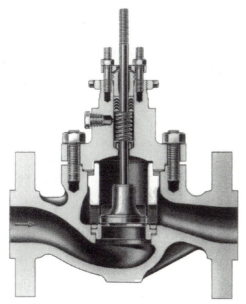

Figure 6.3 Caged-guided trim in a globe valve.
(*Courtesy of Fisher Controls International, Inc.*)

the upper guide and the lower guide is at a maximum length, lateral plug movement due to process flow is not an issue and the tolerances required for this type of guiding are not required to be as close as double top-stem guiding. This also permits the use of smaller-diameter plug stems, providing a smaller sealing surface and decreased stem friction (which is necessary when lower-thrust diaphragm actuators are used). Caged guiding also minimizes any change of vibration of the plug in service and helps support the weight of the plug head. Because this guiding surface is in the flow stream, the process must be relatively free from particulates, or binding or scoring may occur. In some situations, identical or similar materials between the plug head and the cage may gall during prolonged operation. High temperatures may also lead to thermal expansion and binding. Galling and temperature problems can be remedied using guiding rings made from an elastomer or nongalling metal, which are installed in grooves machined into the plug head.

Cages are designed with large flow holes (anywhere from two to eight) that allow passage of the flow into or from the seat, depending on the flow direction. They can also be modified to allow a staged pressure drop—reducing the pressure drop and velocities inside the valve to avoid cavitation, flashing, erosion, vibration, or high noise levels. To ensure the alignment of the plug seating surface with the seat-ring seating surface, some designs combine the cage and the seat ring into one part. This one-piece design maintains the concentricity between the inside diameter of the cage and the inside diameter of the seat.

The cage is also used to determine the flow characteristic. The flow holes in the cage are sometimes shaped such that the plug lifts from the seat ring. In this way a certain percentage of the flow hole is opened up, allowing only so much flow at that portion of the stroke. By varying the size and shape of the hole, certain flow characteristics can be generated. Figure 2.2 in Chap. 2 shows a variety of shapes available according to the flow characteristic.

In trim designs that do not feature cages (such as those that use a seat-ring retainer or screwed-in seats, which is discussed later), the plug head can be machined to a particular shape that provides an inherent flow characteristic. Figure 2.3 in Chap. 2 shows how the contour of a plug head can be turned to provide the flow characteristic. In contrast, Fig. 6.4 shows a V-port plug head, which is cylinder shaped with V-shaped grooves machined into the cylinder for a linear characteristic.

With globe valves, the seating surface of the plug is designed to make full contact with the seating surface of the seat ring at the point

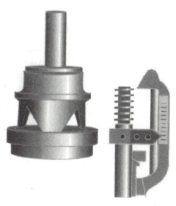

Figure 6.4 V-ported characterized plug.
(*Courtesy of Pacific Valves, a unit of the Crane Valve Group*)

of closure. Although some early valve designs used identical angles, current designs use angles that slightly differ, with the plug at a steeper angle than the seat ring. This slight mismatch ensures a narrow point of contact, allowing the full axial force of the plug to be transferred to the seat, ensuring the tightest shutoff possible for metal-to-metal contact (normally ANSI Class II shutoff is standard, although Class IV shutoff can be achieved with high-thrust cylinder actuators). Even with ANSI Class IV shutoff, metal-to-metal seats can never completely shut off the flow, as the classification allows a small amount of process leakage.

The seat ring is fixed in the body, while the gap between the seat ring and the body is sealed by a gasket. The seat ring can be fixed in the body by one of two arrangements. First, a common method of fixing the seat ring is through a retained arrangement. The seat ring is inserted into a slightly larger diameter machined into the body and held in place by a part between the bonnet and the seat ring, called the *seat retainer.* If the retainer is used to guide the plug head, it is called a cage, but it can serve the dual purpose of retaining the seat ring. If the diameter machined into the body is wide enough, the seat ring will have some play, allowing lateral movement, which can lead to a quick, easy method of correct plug and seat-ring alignment. During assembly, and before the bonnet-flange bolting is completely tightened, a signal can be sent to the actuator to seat the plug in the seat, providing the correct alignment between the matching seat surfaces of the two parts. After the plug and seat ring are aligned, the bonnet-flange bolting is tightened and the subsequent force is transferred through the retainer or cage to secure the location of the seat ring with the plug head. If the seat ring does not have this self-adjustment feature, its seating surface must be lapped with the seating surface of the plug head. *Lapping* is the process in which an abrasive compound is placed on the seat-ring seat surface and the plug is seated and turned until a full contact is achieved. The retained seat ring is also known for easy disassembly, especially in corrosion-prone applications, since it just lifts out of the body once the bonnet and seat retainer or cage are removed. The only disadvantage to retained seat rings is that they work best when a high-thrust actuator is used, since high seating force is needed to ensure a good seat-ring gasket seal.

The second method of securing the seat ring is the threaded arrangement in which the seat ring is threaded into the body. This process normally requires a special tool from the manufacturer to turn the seat ring into the body. The major advantage of this design is that no other part is needed to retain the seat ring, providing a simplified trim arrangement, as well as no cage or seat retainer to restrict the flow. With three-way or double-seated valves, the use of seat retainers or cages is not possible from a design standpoint, and the only alternative is to use threaded seats. Threaded seat rings are widely used with cryogenic applications in which the top of the body must be elongated to provide a fluid barrier between the process and the packing box and top-works.

The disadvantages of threaded seats are threefold. First, and most evident, the threads can become corroded, making disassembly difficult, if not impossible in some long-term situations. Second, alignment between the plug and seat ring will require the extra step of lapping to achieve the required shutoff. And third, in situations in which vibration is present and the seat ring is not held in place by the plug in the closed position, the seat ring may eventually loosen and allow leakage and misalignment. Overall, the disadvantages of the threaded seat ring far outweigh the advantages; therefore many newer single-seat designs use the retained arrangement. When a seat retainer or cage is not possible or preferred and the application is too corrosive to allow a threaded seat ring, a split-body arrangement is a practical substitute.

Some globe-valve applications require bubble-tight shutoff (ANSI Class VI), which cannot be attained with a metal-to-metal seal. To accomplish this, a soft elastomer can be inserted in the seat ring. In most designs, the seat ring is made from two parts with the elastomer sandwiched between the two, as shown in Fig. 6.5. The combination of the metal plug surface pressing against the seat ring's soft seat surface can achieve bubble-tight shutoff if the plug and seat-ring surfaces are

Figure 6.5 Exploded view of soft-seat design.
(*Courtesy of Valtek International*)

concentric. Some manufacturers also insert the elastomer in the plug, which achieves the same effect (Fig. 3.29, Chap. 3).

6.2.3 Globe-Control-Valve Operation

The most common globe valve uses a T-style body, which allows the valve to be installed in a straight pipe with the top-works or actuator perpendicular to the line and will be used to explain the basic operation of a globe valve. Flow enters through the inlet port to the center of the valve where the trim is located. At this point, the flow must make a 90° turn to flow through the seat, followed by another 90° turn before exiting the valve through the outlet port.

The flow direction of globe valves is defined by the manufacturer and in many applications is critical to the valve's operation. With standard single-seated globe valves using inlet and outlet ports, the two choices are flow-under-the-plug and flow-over-the-plug. With manually operated globe valves, flow is almost always under the plug. The plug closing against the flow provides constant resistance, but not enough to be insurmountable, and is relatively easy to close as long as the fluid pressure and flow rate are low to moderate. Flow-under-the-plug provides for easy opening, as the fluid pushes against the bottom of the plug. However, flow direction is an important consideration with control valves equipped with diaphragm actuators, which are not capable of high thrusts. If the flow is over the plug and the process involves high pressures, the diaphragm actuator is not usually stiff enough to prevent the plug from slamming into the seat ring when throttling is close to the seat. Also, the actuator must pull the plug out of the seat against the full upstream pressure, which may be difficult in a high-pressure application. Therefore, lower-thrust actuators demand flow-under-the-plug, allowing the full thrust to close against the upward force of the fluid pressure. Another situation in which flow-under-the-plug is an issue is with fail-open applications, where the service requires the valve to remain open during a signal or power failure. Even if an actuator with a fail-safe spring is rendered inoperable during a fire, the flow-under-the-plug design will ensure continued flow as the flow pushes the plug away from the seat.

Inversely, flow-over-the-plug is important in fail-closed situations, where the service requires the valve to shut during a loss of signal or power. If the actuator fails and the fail-safe spring also fails, the flow acts on the top of the plug to push it into the seat. Obviously, with flow-over-the-plug situations, throttling close to the seat presents a problem if the actuator does not have sufficient stiffness (the ability to hold a position despite process forces). The actuator must have enough thrust to pull the plug out of the seat against the fluid's upstream pressure—which increases to its maximum value in a nonflow state. As the issues of stiffness and thrust are considered, in a majority of situations where the flow must be over the plug, piston cylinder actuators are preferred over diaphragm actuators.

As alluded to earlier in Sec. 6.2.2, the globe-valve trim can be modified to allow for equal-percentage, linear, or quick-open flow characteristics. As explained in detail in Sec. 2.2, flow characteristics determine the expected flow rate (expressed in flow coefficient or C_v) at a certain valve position. Therefore, with a particular flow characteristic, the user can determine the flow rate at a given instrument signal. As the flow reaches the trim, and if the trim is in a throttling position, the flow is directed to a restriction. This restriction may be created by the exposed portion of a hole in a cage, which is based upon the linear position of the plug. It may also be created by the portion of the V-shaped slot of a V-port plug that is exposed above the seat ring. Also, the restriction may be created by the amount of the seat that is open to the flow when the area of a contoured plug is filling a portion of the seat area. When a pressure-drop situation occurs (the downstream pressure is lower than the upstream pressure), the flow moves from the inlet through the seat to the outlet. As the flow moves through the seat, line pressure decreases as velocity increases. After the fluid enters the lower portion of the globe body, the area expands, the pressure recovers to a certain extent, the velocity decreases, and flow continues through the outlet port and downstream from the valve. As the flow enters the trim area of the valve, an important consideration is the gallery area of the body surrounding the trim. In ideal situations the flow should freely circulate around the trim, allowing flow to enter the trim from every possible direction. With cages and retainers, flow should enter equally from every hole to provide equal forces to act on the plug head. If the gallery is narrow in any one area (for example, in the back side of the cage), velocities can increase, causing noise, erosion, or downstream turbulence. In addition, unequal forces acting on the plug head can cause slight flexing of the plug head if it is not supported by a cage.

When the globe control valve closes, the axial force from the actuator is transferred to the plug and its seating surface makes contact against the slightly mismatched angle of the seat ring. As full contact is made, the valve is closed, allowing minimal or no flow to pass through the trim according to the ANSI leakage classification. If the axial force is applied in the opposite direction, the plug lifts and, in the full-open position, the entire seating area is open to the flow as well as the holes of the cage or retainer.

Because the process flow is under pressure and the environment outside the valve is at atmospheric pressure, the flow seeks to escape through the gaps in the valve. This leakage is prevented by the static seal of the gaskets in the end connections (if flanges or RTJ end connections are used) and the bonnet gaskets. Flow seeking to escape through the sliding stem of the plug is prevented by the packing's dynamic seal in the bonnet's packing box. In closed positions, flow may escape through the seat but is prevented by the static seal between the seat ring and the body.

6.2.4 Globe-Control-Valve Trim Variations

With special service requirements, globe control valves can use a number of specialized trims for unique flow requirements. Some applications require extremely low flow coefficients, with C_vs anywhere down to 0.000001. Because of these extremely low flows, these designs are found only in smaller valve sizes (less than 2 in or DN 50). The

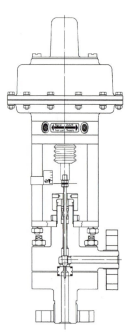

Figure 6.6 Low-flow control valve with needle
trim. (*Courtesy of Kammer Valves*)

plug head is shaped very narrowly, earning the designation *needle-
valve trim* because of its needlelike appearance (Fig. 6.6). Because even
the smallest variations in diameter can have a wide impact on the
overall flow coefficient and flow rate, needle plugs are machined using
special micromachining procedures (using technologies developed by
the watchmaking industry). These precise trims require the flow characteristic to be machined into the plug head contour. Needle-valve
trim requires a very precise method of adjustment of the distance
between the seat and plug-seating surfaces. A very fine thread (twice
the magnitude of a normal plug thread) is normally required, allowing
a very minute amount of linear adjustment per turn.

Pressure-balanced trim is defined as a special trim modification that
allows the upstream pressure to act on both sides of the plug head, significantly reducing the off-balance forces and operator thrust needed
to close the valve. It is sometimes used to replace normal trim arrangements when the valve must close against a large seat diameter coupled
with high-pressure process forces or high-pressure drops. Because the
regulating element must overcome these forces, exceptional actuator

force from a high-thrust actuator or a larger lower-thrust actuator must be used to close the valve. In other applications, a standard valve may need a smaller actuator size to fit into a tight space. In this case, pressure-balanced trim reduces the valve's need for a larger standard actuator by reducing the off-balanced area of the trim. Pressure-balanced trim is common with valves in larger sizes [size 12 in (DN 300) and higher] in which a large amount of flow is passing through a large seat and where the cost of a larger actuator would be greater than the cost of the pressure-balanced trim.

Pressure-balanced trim requires a special plug and *sleeve,* which is similar in many respects to a cage. These parts allow the upstream pressure to act on both sides of the plug, as shown in Fig. 6.7. The sleeve's inside diameter is slightly larger than the inside diameter of the seat ring. The plug requires a smaller plug stem to minimize the off-balance area, and is equipped with metal piston rings, O-rings, or polymer rings that, when installed inside the sleeve, create a pressure chamber above the plug. One or two holes are machined through the plug head, allowing the fluid pressure to act on both sides of the plug. In effect, this results in a net force equal to the pressure multiplied by the off-balance area.

With high inlet pressures and a large seat area, a high actuator force is required to close the valve. With standard trim (unbalanced plug),

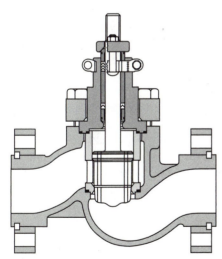

Figure 6.7 Globe-body subassembly with pressure-balanced trim. (*Courtesy of Valtek International*)

the force necessary to close the valve is the total *off-balance area*, which is written as

$$F_{OBA} = P_1(A_S - A_{stem}) - P_2(A_S)$$

where F_{OBA} = actuator force required to overcome the off-balance area
P_1 = upstream pressure
P_2 = downstream pressure
A_S = area of the inside diameter of the seat
A_{stem} = area of the outside diameter of the plug stem

However, with pressure-balanced trim and its counter-balanced design, the off-balance area is far less, which requires less actuator force, as written in the following equation:

$$F_{OBA} = P_1(A_{sleeve} - A_{stem}) - P_2(A_S)$$

where A_{sleeve} = area of the inside diameter of the sleeve

With pressure-balanced trim, the larger the off-balance area (slight as it may be), the greater the shutoff. For example, in smaller globe-valve sizes (0.5 through 3 in or DM 12 through DN 80), the off-balance area is slight and an ANSI Class II shutoff is usually the standard—ANSI Class II calls for a maximum leakage rate of 0.5 percent of rated valve capacity. On the other hand, for sizes of 4 in (DM 100) and larger, the off-balance area of the trim increases and ANSI Class III shutoff is possible—ANSI Class III calls for a maximum leakage rate of 0.1 percent of rated valve capacity.

With standard unbalanced trim, the direction of the flow assists with the motion of failure (flow-over-the-plug is used for fail-closed and flow-under-the-plug is used for fail-open cases). With pressure-balanced trim, however, the opposite occurs. Flow direction is under the plug for fail-closed situations and over the plug for fail-open situations. The actuator force required to fail-open or fail-closed is related to the off-balance area. Hence, for flow-over-the-plug and fail-closed situations, this off-balance area is equal to the sleeve area minus the seat-ring area. The spring must be able to overcome this off-balance area, which can be written as

$$F_{open} = P_1(A_{sleeve} - A_{seat})$$

where F_{open} = spring force required to fail-open

With flow-over-the-plug and fail-closed applications, the off-balance area is equal to the sleeve area minus the plug stem area, as indicated in the following equation:

$$F_{closed} = P_1(A_{sleeve} - A_{stem} - A_{seat})$$

where F_{closed} = spring force required to fail-closed

In standard services, the major advantage of using pressure-balanced trim is that smaller or less powerful actuators can be used. Another advantage is that high-pressure drops or higher process pressures can be handled without resorting to expensive, large nonstandard actuators. In some instances, use of pressure-balanced trim is the only method by which some applications can be handled because an actuator with extremely high thrust may not be available for the required valve size or may not fit in the available space.

On the other hand, pressure-balanced trim has four major disadvantages: First, because pressure-balance trim only works with a sliding seal between the plug and the sleeve, the fluid must be relatively clean and free from particulates; otherwise, the seals can be damaged and cause leakage or galling between the plug and sleeve. Second, because of the balanced nature of the plug, coupled with the lower thrust of a smaller actuator, leakage rates through the seat are not as good as with unbalanced trim—ANSI Class II is normal. Third, pressure-balanced trim is more costly initially than standard trim, although the use of a smaller actuator may offset that cost or even make the overall cost more attractive. And fourth, because of the seal within the process flow, the trim may require a shorter servicing cycle, especially if the process has entrained particulates.

Double-ported trim is a special trim design used to fill the same purpose as pressure-balanced trim: to reduce the effect of the process forces on the plug, thereby lowering the thrust requirement and allowing the use of smaller actuators. Flow is directed by the inlet port to the body gallery and the trim, which features two seats and a single plug that features two plug heads, one above the other (Fig. 6.8). In air-to-open (fail-closed) applications, the plug–seat combination at the top of the gallery is a flow-under-the-plug design, while the plug–seat combination at the bottom is a flow-over-the-plug design. In air-to-close (fail-open) applications, the opposite design is used. The plug–seat arrangement at the top is flow over the plug and that at the bottom is flow-under-the-plug.

Upon opening, the net forces working on these two seats nearly cancel each other out. The fluid pressure is pushing the upper plug head

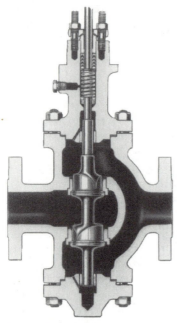

Figure 6.8 Double-ported globe-body assembly.
(*Courtesy of Fisher Controls International, Inc.*)

out of the seat, while the lower plug head is pulling out against the fluid pressure. Upon closing the opposite occurs. The upper plug head pushes against the flow, while the lower plug head is assisted by the flow. Although in principle double-seated valves are close to pressure-balanced valves, in reality they are somewhere between pressure balanced and unbalanced. This is because the fluid is acting against the plug contour with one seat and the top of a plug head (usually a flat surface) with the other seat, creating a dynamic imbalance. With double-seated valves, flow characteristics are nearly always determined by the contour of the plug head. Guiding is accomplished with upper and lower guides. The upper guide is placed above the upper seat, while the lower guide is located in the lower body region with a lower body cap for access and assembly. This arrangement also allows for easy reversal of the stroke direction (air-to-open to air-to-close, or vice versa). The body can be inverted, with the bonnet and the lower body cap retaining their previous positions.

Double-ported trim can also be used with three-way valves for diverting, combining, or dividing flows. In the case of diverting flow, the plugs are offset, meaning that one of the two plug heads is always seated, while the other is in the full-open position. As the valve moves from one end of the stroke to the other, the opposite occurs: the previously closed plug head moves to the full-open position and the previously open plug head moves to full-closed. To divide flow between the two outlets, this same arrangement can be used, except that the stroke remains in the middle as if throttling, allowing both seats to be open to some extent and flow to move down both outlets. For combining flows, the flow direction of the valve is reversed, allowing for two inlet ports and a single outlet port. Using a double-seated valve for three-way service means that a lower guide surface as part of the body is not possible, since that area is used as a port. In these cases, the plug head is designed to guide in the seats, using notches in the plug head to achieve flow control.

Double-ported trim does have drawbacks: First, the alignment of the plug and the seat is critical in T-line valve styles (one inlet and one outlet), and if one plug head is out of alignment, one may fully seat, while the other will be slightly off the seat, allowing leakage through that seat. Because of the extreme difficulty of aligning the two seats to provide equal shutoff, allowable leakage is 0.5 percent of the rated flow of the valve. Thermal expansions can also cause the distance between the seats to widen, leading to increased leakage. The second drawback is that the design requires screwed-in seat rings, which are prone to corrosion and must be lapped to ensure tight shutoff.

Another trim variation is *sanitary trim*, which is required for those valves used in the food and beverage industry. Such valves require stainless-steel construction of all wetted parts and are specified with angle-style bodies, which allow the downstream port to be 90° from the inlet port. In other words, the flow is directed straight down from the seat ring. With sanitary applications, pockets of fluid cannot be allowed to stand or pool; otherwise contamination or bacterial growth can result. When the system is flushed by water or steam, the self-draining allows for the system to quickly dry and be readied for another type of process fluid or for the system to remain dormant.

Sanitary-trim design (Fig. 6.9) allows the valve to self-drain when the system is depressurized or if the valve is closed, allowing the outlet side to drain. To avoid pockets of trapped fluid, sanitary trim has very few flat areas and no walled pockets. In some designs, the seating surface is machined into the body to avoid a gap between a seat ring and the body. The plug head is tapered on its top side until it reaches the plug stem. Because sanitary services must have tight shutoff, the plug head is fitted with an elastomeric insert to provide bubble-

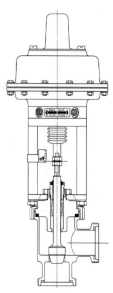

Figure 6.9 Sanitary-trim control valve.
(*Courtesy of Kammer Valves*)

tightness. Because of possible pooling areas, pressure-balanced trim is never an option with sanitary services. Most sanitary valves also require stainless-steel actuators to avoid any sort of oxidation in the clean environment.

6.2.5 Globe-Control-Valve Body Variations

Globe valves are considered to be one of the most versatile valve designs because the body can be varied in numerous ways to allow for different piping configurations or functions. The most common single-seated globe body style is the flow-through design (or sometimes called the *T-style body*), which is shown in Fig. 6.10. Basically, this body style allows the valve to be installed in a straight piping configuration, with the rising-stem action perpendicular to the centerline of the piping. Unlike most quarter-turn valves or gate valves where the flow moves straight through the body relatively unimpeded, the flow-through design brings the flow through two right-angle turns, allowing for a significant pressure drop, which is essential for some applications. As the flow moves through the inlet port, the flow passage shifts up (or down, depending on the flow direction) approximately 30° until

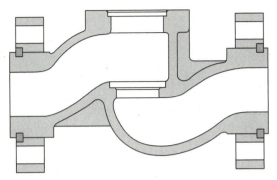

Figure 6.10 Globe body with top-entry to the trim and separable flanges. (*Courtesy of Valtek International*)

the flow reaches the gallery of the body, bringing the flow above (or below) the seat, which is usually on the piping centerline. At that point, in flow-over-the-plug situations, the flow enters the gallery area that surrounds the trim. The flow then turns 120° to flow through the seat. At this point, the flow is perpendicular to the piping centerline. As the flow exits the seat, it turns 120° again by the flow passage, shifting up (or down) until the flow meets the outlet port and moves out into the downstream piping.

Globe flow-through bodies can be modified with a elongated body chamber above the regulating element (Fig. 6.11) for cryogenic applications. The upper chamber of this body style allows for a small amount of liquefied gas to vaporize between the process and the packing, acting as a vapor barrier—the pressure from the vaporization actually prevents any further liquid from entering the chamber.

An alternative single-seated body style, somewhat related to the flow-through style, is the *angle-body* style (Fig. 6.12). Instead of the two ports being in-line with the straight piping configuration, one port is turned 90° from the other port (or at a right angle) to match piping that requires such a turn. The port that is perpendicular to the rising stem is called the *side port,* and the port that is in-line with the rising stem is called the *bottom port.* Valves with an angle-style body are used in a number of applications. First, angle valves are sometimes used in cavitating services where the imploding bubbles are channeled directly into the center of the downstream piping. Depending upon the severity of the cavitation, the bubbles may not directly impact a metal wall (such is the case with the bottom of the globe straight-through body). Rather, they implode harmlessly in the middle of the pipe. If

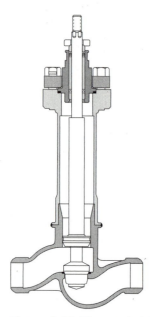

Figure 6.11 Elongated globe body for cryogenic service. (*Courtesy of Valtek International*)

the control valve is part of a piping system that discharges into a tank, an angle valve can be used so that any cavitating liquid can flow into the large vessel, where it will not affect any nearby metal surfaces. An angle valve also allows the use of a *Venturi seat ring* (Fig. 6.13), which is an extended seat ring that can protect the sides of the bottom port and downstream piping from adverse process effects, such as abrasion or erosion. Also, because of the right-angle turn in the body design,

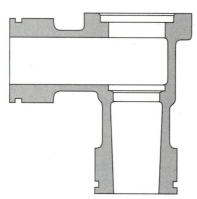

Figure 6.12 Angle body with top-entry to the trim and separable flange hubs. (*Courtesy of Valtek International*)

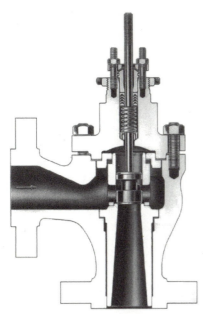

Figure 6.13 Venturi seat ring design. (*Courtesy of Fisher Controls International, Inc.*)

angle valves can be installed in services that have a natural upward flow, such as in crude oil or natural gas applications or boiler services. A special kind of angle valve, called a *choke valve*, is used for most wellhead applications. Many mining applications involve gas services that have particulate matter such as sand or dirt, which have a tendency to erode—a process similar to sandblasting. Modified-sweep-style angle valves (Fig. 6.14), with trim made from ceramic for durability, allow the particulates to be channeled down a pipe without directly impinging on any body walls. Also, angle valves allow for easy draining, since no pockets exist that allow the fluid to pool.

One disadvantage of using an angle valve is that turbulent flow created by the regulating element can channel the turbulence directly into the downstream piping, creating more vibration and noise than would be created using a flow-through body. The downstream side of the flow-through body is quite stiff, handling some of the flow's energy conversion in an unyielding vessel before the flow proceeds into downstream piping. Angle valves also have a higher pressure recovery than other types of globe valves, resulting in a lower σ value (the cavitation index, Sec. 11.2), which means an increased chance of cavitation.

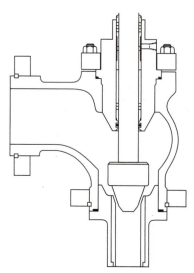

Figure 6.14 Sweep-angle body subassembly.
(*Courtesy of Valtek International*)

A variation of the globe straight-through style is the *expanded-outlet* style, which is basically a straight-through design except that the end connections are a larger pipe size than the trim is designed for. For example, a 4 × 2-in expanded outlet valve would have 4-in end connections (for mounting to a 4-in pipe), but would have the full-area trim for a 2-in valve. Expanded-outlet valves are used to lower the cost of welding or installing piping increasers to the valve body. The expanded-outlet body's face-to-face is also shorter than a normal globe straight-through valve with increasers, which may be important in piping systems with limited space. This style is also a cost-saving measure when a larger valve size is required with reduced trim. The smaller trim size may also act as a reduced trim—although technically it is considered a full-area trim for the smaller valve size.

Another variation of the globe straight-through style is the *offset body* style, which provides for straight-through flow except that the inlet and outlet ports are parallel and not in-line with each other (Fig. 6.15). The seat is placed in a center position between the two piping centerlines. Offset valves are used for unique piping configurations because the flow passages do not shift up or down to bring the flow above and below the seat. Unlike the T-style globe body, less pressure drop occurs with the offset body.

The *split-body* style involves a body made of two separate parts: the

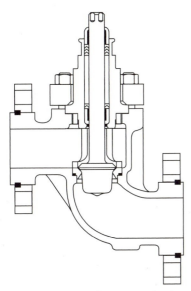

Figure 6.15 Offset globe-body subassembly.
(*Courtesy of Valtek International*)

upper body half and the lower body half (Fig. 6.16). These two body parts connect at the center of the valve body with the seat ring sandwiched between the two body parts. Body bolting is used to secure the two body halves together. Two gaskets are used on both sides of the seat ring to ensure pressure retention. The bonnet can be integrally connected to the upper body half. This is preferred, since a good design should minimize potential leak paths—having a separate bonnet would add another potential leak path. Using a split-body design offers several advantages. First, the seat ring is retained in place without a seat retainer or cage to center or hold the seat ring in place, in effect, combining the advantages of both retained and threaded seat rings. If the application is such that the plug and seat ring must be inspected or replaced often, such as in chemical services that are highly corrosive, the simplicity of construction and disassembly permits frequent inspections. The split-body design also reduces the trim by one part, which may be a factor if the valve body is made from an exotic alloy. It also avoids any flow difficulties associated with a cage or retainer, such as galling or noise. Second, the seat ring can be removed with minimal disassembly, although the lower body half would need to be removed entirely from the line. And third, in some designs, the two body halves can be disassembled and turned 90° in

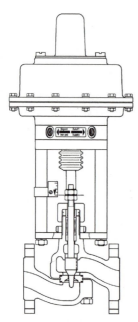

Figure 6.16 Split-body control valve. (*Courtesy of Kammer Valves*)

either direction to provide a right-angle valve, perpendicular to the rising stem, as opposed to a true angle valve where the lower port is in-line with the rising stem. With a split body, the actuator or manual handwheel could remain upright. With a true angle valve, the actuator would be on its side. The split-body valve has some limitations. For example, it is usually only specified with flanged end connections. It cannot be used in steam or other high-temperature services where buttweld or socketweld end connections are required for welding the valve into the line, since the body could not be disassembled to access the seat ring. If process leakage occurs at the body connection, the body bolting is located where fluid could cause corrosion, making disassembly difficult.

Another unique body style is the *Y-body* style, which is a body where the rising stem is inclined 45° (or sometimes 60°) from the axis of the inlet and outlet ports, which are in-line with the piping (Fig. 6.17). Y-body valves are the best type of globe control valve for passing the largest C_v possible with minimal pressure drop—short of using a globe body with an integral seat and an oversized plug. Also, because the body avoids the right-angle turns and the plug pulls nearly out of the

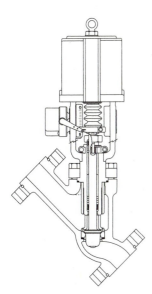

Figure 6.17 Y-body control valve. (*Courtesy of Valtek International*)

flow stream, less turbulence is generated through the body, which may reduce noise. Y-body valves are also commonly applied in piping systems with piping set at 45°, allowing the valve body to be in-line with the piping, while the top-works is vertical to the ground. This allows easier maintenance and better operation. Because the body, when placed at a 45° angle, has little if no pockets for a fluid pool, the Y body is often applied in self-draining applications.

A *three-way body* style has three ports: two ports in-line with the piping centerline and one port in-line with the rising stem. This design uses a plug head featuring an upper and lower seating surface and two matching seats (Fig. 6.18). Depending on the position of the plug or the orientation of the piping, the process flow can be diverting, splitting, or mixing. With diverting flow, the flow enters a side port and, if the plug is fully extended into the lower seat, the flow is diverted out the opposite side port. If the plug is fully retracted into the upper seat, the flow is diverted through the bottom port. When the plug remains in a throttling position between the two seats, flow is diverted to both the side and bottom ports for when the flow needs to be split. Combining two separate flows can be accomplished with the same body style, except that the opposite side port and the bottom port both receive the upstream process flow. When the plug is placed

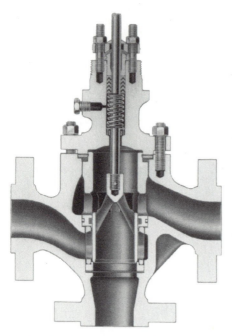

Figure 6.18 Three-way body subassembly with integral three-port body and pressure-balanced trim. (*Courtesy of Fisher Controls International, Inc.*)

in midposition, both processes flow together and combine before exiting the side port.

Another optional design with three-way valves involves the use of a *three-way adapter* with a conventional globe straight-through body (Fig. 6.19). The adapter consists of an upper-body extension that is mounted above the body where the bonnet normally sits. An upper seat ring is sandwiched between the body and the adapter. The adapter is equipped with a side port, which can be mounted in any one of four quadrants if the end connection can be used without interfering with another port. One exception is flanged end connections, which can only be possible at right angles since the flanges would interfere with the in-line piping or other flanged connections. The bonnet sits above the adapter and a special three-way, dual-seating plug is used to divert, mix, or separate process flow. The obvious advantage to this type of design is that a valve can be converted to three-way service without a new body—only a new adapter, upper seat ring, and plug are required. The disadvantage is that an additional possible leak path is added to the body subassembly.

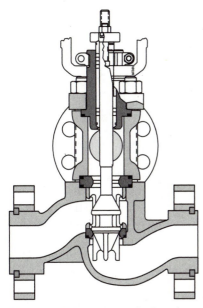

Figure 6.19 Three-way body subassembly with three-way adapter. (*Courtesy of Valtek International*)

6.2.6 Globe-Control-Valve Installation

Following receipt of the valve from the manufacturer, the globe valve should be carefully inspected against the packing list to ensure that the valve is built according to specification. Correcting problems with actuator accessories, tubing, end connections, packing boxes, etc., is almost always more easily accomplished before the valve is installed in the pipeline, where space or access to the valve may be limited. Many manufacturers also provide installation and maintenance instructions inside the shipping carton, which should be collected and kept for reference.

Larger globe valves [2 in (DN 50) or larger] should be lifted using lifting straps and a mechanical lifting device, placing the lifting straps around sturdy members of the valve or actuator. When an actuator uses a yoke, the lifting straps can be placed around the yoke legs. The lifting straps should not make contact with tubing or actuator accessories in order to avoid any damage to less sturdy parts. Some manufacturers provide metal lifting rings as part of the bonnet-flange bolting or actuator, which may be used for lifting. Usually a sticker or tag advises the user of these lifting points. Extreme caution should be used whenever lifting a valve with straps and a mechanical lifting device because failures can happen. Many certified drawings show where a valve's approximate center of gravity is located on the valve. Knowing the location of the center of gravity can help in lifting the valve as well as warn the user should the valve be off-balance when lifted. When lifting, the user should not allow any portion of the valve, such as the actuator, to rotate while lifting, since this may cause damage to the internal or external portions of the valve.

Before the globe valve is installed, the line should be cleaned of any foreign material, such as dirt, welding chips or rods, or scale. This is especially critical if a severe service trim is used, where a tortuous or multihole flow path is required. These trims can act as sieves by not allowing larger particulates or debris to flow through the valve. A blockage in the flow passages results in limited flow and marginal performance. Such a situation will require disassembly of the valve to remove the debris. Even if special trim is not used, foreign materials in the pipeline can pick up speed and damage seating surfaces or plug stems, which can damage packing and/or cause leakage through the packing box.

Because of their linear-motion design and the addition of an actuator, globe control valves have a greater height than other styles of valves, which must be taken into consideration during installation. Access to the trim in nearly all globe valves, except in split-body valves, involves top-entry, which requires removing the bonnet-flange bolting and lifting the entire top-works (actuator, bonnet, and plug head) out of the body. The *disassembly clearance* is the amount of space needed between the top of the actuator or handwheel to lift the top-works completely out of the body. It is often provided in a certified dimensional drawing for that valve. In some designs, valves can be disassembled with less than required disassembly clearance, but the procedure involves angling the entire top-works to one side while lifting up—a difficult maneuver without damaging valve parts or creating safety hazards for personnel. This should only be used as a last resort and even then with extreme caution.

Globe valves generally should be installed in the vertical position, which provides for the easiest maintenance, calibration, and operation. In some cases, the valve must be installed in a horizontal line to function correctly. For example, valves used for cryogenic services use a special chamber between the process flow and the bonnet to allow some liquid vaporization to take place. This action can only take place if the valve is installed in a horizontal line. Valves that have great heights—for example, a valve with an extended bonnet, actuator, and top-mounted handwheel—may place exceptional radial stresses on the end connections if placed horizontally with no secondary support provided. In addition to stresses on the piping connections, gravitational forces may cause some abnormal wear on moving parts or may cause erratic operation. As explained earlier, maintenance with any linear-motion valves, such as globe valves, involves the top-works being lifted straight out of the body. To remove the valve's top-works from a horizontal or 45° installed position requires constant support to prevent damage to parts or injury to workers. Y-body valves are ideal for piping placed at a 45° angle, since the top-works remain vertical, perpendicular to the ground.

Another consideration during installation is seismic requirements. With dangers associated with a process under pressure, extreme temperatures, or other dangerous fluid properties, valves with unusually great heights are subject to severe stresses during earthquakes or line vibration. During a seismic disturbance, pressure-retaining joints may fail with tall globe valves. In applications where seismic requirements exist, attaching the valve's top-works to a nearby fixed support may be necessary.

Maintaining the correct flow direction is very important, since flow direction may play an important role in the functions of the globe control valve—such as the failure mode, cavitation or noise control, or pressure balancing. Flow direction is indicated on the side of the globe valve.

If a valve is welded into the line requiring buttweld or socketweld end connections, special care should be taken to avoid excessive heat buildup in the valve body. Thermal expansion may cause seizing or galling between parts if the valve is operated while still hot. The extreme heat may also melt polymer gaskets or packing, causing leak paths or extrusion of the material.

If the valve is equipped with integral-flange end connections, correct alignment of the matching end flanges on the piping flanges is important so that the valve can easily be installed. Because the flanges on the body are fixed, they are carefully machined so that the hole pattern lines up with the vertical axis of the valve. If the piping flanges are installed so that they are not lined up exactly with the vertical axis, the valve will have a tendency to lean, making disassembly a bit more difficult. Some play exists between the size of the flange hole and the required stud, which may help minimize the misalignment. If only one piping flange is misaligned with the vertical axis, installation may be impossible, since the holes on the two matching flanges may match up on one connection but be misaligned on the other connection. Some users correct this misalignment by widening the holes on the piping flange, but this weakens the integrity of the flange. Another option is to cut the flange off the pipe and to reweld it in the correct position.

Specifying separable flanges on the valve can solve the misalignment problem. As detailed in Sec. 2.4, separable flanges are not integral to the valve body, rather individual flanges are held in place by half-rings. Because the separable flanges can spin in either direction, misalignment problems are quickly remedied. Caution should be used with separable flanges, because the bolting must be correctly tightened to ensure adequate friction at the body–half-ring–flange connection. Otherwise, if the flanges are insufficiently tightened, the entire valve may accidentally rotate, especially if it is extremely tall or if line vibration exists. As a precaution, a number of tack welds should be used to fix the position of the flange to the valve.

After the globe control valve is installed in the line, and if automatic control is required, the actuator, actuation system, or positioner (if applicable) should be connected to the instrumentation signal and power supply. A tag or sticker on the actuator indicates the maximum power supply (pneumatic, electric, or hydraulic) that the actuator can handle. If pneumatic, a pressure regulator may be necessary to lower the plant's air supply to the required operational levels, especially with diaphragm actuators. With some actuators, such as spring cylinder actuators, high air-supply pressure up to 150 psi (10.3 bar) actually enhances the thrust capabilities, and a pressure regulator is not necessary. Unless the plant air supply is exceptionally clean, an air filter should be installed before the actuator (or positioner, if necessary) to prevent water, oil, dirt, or other debris from entering and fouling the pressure chambers. Most air filters are designed with the filter cartridge below the air connections. This portion should always point downward to allow gravity to assist with the entrapment of any foreign materials.

Once the globe control valve is installed in the line and the instrumentation signal and power supply are connected, the valve should be stroked fully and the stroke indicator observed on the actuator or handwheel. The plug should move in a smooth linear motion. If the plug seems to stick or moves in an erratic manner, something may be internally wrong with the valve or actuator, or the instrument signal or power supply may be intermittent. Regardless, the cause should be thoroughly investigated before further operation takes place.

The entire instrument signal change should be made to ensure correct calibration. For example, a common range is 3 to 15 psi (0.2 to 1.0 bar) for pneumatic signals and 4 to 20 mA for electrical signals. All air and electrical connections should be inspected to ensure no leakage or loose connections exist, respectively.

Because the packing may lose some compression through consolidation between the factory and the final installation, a wise procedure is to retighten the packing to the manufacturer's written procedures. The user should be careful not to overtighten the packing since this will increase stem friction, causing erratic stroking and premature packing box wear. If wide temperature swings are normally expected as part of the service, the temperature excursion should be allowed to take place, and then the packing box rechecked and retightened if necessary. Following the temperature excursion, checking the body and bonnet gaskets for leakage is also recommended. If the packing box is supplied with a lubricator or lubricator fitting, the user should ensure that the supply is adequate and add lubrication if low. A fluorinated general-purpose grease handles most common liquid and gas services at temperatures from −5 to 550°F (−20 to 285°C). It should be nonflammable and chemically inert and should maintain its lubrication in harsh processes. For less-harsh services—such as water, steam, or mild chemicals—a molybdenum disulfide lubricant can be used in temperatures from 32 to 500°F (0 to 260°C). Graphite petrolatum lubrication can be used for high-temperature applications that range up to 1000°F (540°C), while synthetic oil-based lubricants can be used for low temperatures down to −100°F (−75°C).

If the valve is configured for a failure action, the actuator power supply should be disconnected to verify that the valve fails in the correct position (open, closed, or at last position).

6.2.7 Globe-Control-Valve Preventative Maintenance

Valve manufacturers recommend that the user periodically check the globe control valve for proper operation and perform preventative maintenance. In most cases, globe valves can be serviced while remaining in the line. If any portion of the internal parts of the globe valve must be checked or replaced, the process service will need to be interrupted or blocked around the valve being serviced.

Although most manufacturers provide recommended servicing instructions with their globe valves, the following general guidelines are provided to assist with periodic preventative maintenance. The entire body subassembly should be examined for signs of process leakage. Areas of concern include the end connection, especially if a flanged connection is used where a body face gasket may fail; the joint between the bonnet flange and body where gaskets may fail; a body plug; the top of the packing box where the plug exits the bonnet; or any other pressure-retaining connection. If the valve includes a metal bellows seal (Sec. 11.9), the telltale tap and/or sensor should be checked for possible bellows failure. With liquid services, this check can be performed visually, while gaseous services can be examined using a "sniffer" sensing device. If leakage is evident, the necessary compression bolting to the pipe or body flanges, bonnet flange, or gland flange should first be tightened in an attempt to eliminate the leak. If that does not totally eliminate the leak, the gasket or packing have most likely failed, and disassembly and replacement of the soft goods will be necessary.

Following the initial troubleshooting inspection, the valve should be thoroughly cleaned. If severe oxidation is occurring in metal surfaces, those areas should be repainted using a rust inhibitor. This is especially important if the atmosphere is highly corrosive, such as near seawater. If process drippings are falling on the valve from nearby piping, steps should be taken to remedy those leaks or to shield the valve from those leaks. If the valve is equipped with exterior soft goods, such as rubber bellows to protect a stem surface, they should be checked to ensure that they are not worn or cracked from exposure to the elements.

Occasionally, the pressure-retaining vessel will begin to fail because of undetected porosity or fracture or through natural wear from erosion or cavitation. Sometimes if the failure is minor, the affected area can be ground out and then filled with a weld, using established methods to ensure the integrity of the pressure-retaining vessel.

If the bonnet's packing box is equipped with a lubricator or lubricator fitting, it should be examined to ensure that adequate lubrication exists. If not, additional lubrication should be added. If the process allows or if the valve is blocked from the process, the valve should be stroked to verify a full and smooth stroke. Erratic or jerky stem movement may indicate internal galling, packing tightness, or an actuator or positioner problem. Some packings, especially graphite packing, are known to cause jerky stroking even when properly compressed. If possible, the power supply should be removed from the actuator to see if the valve fails in the correct position. When the plug stem is fully retracted (open), the exposed stem should be checked for any unnatural wear or galling. If process buildup is evident, it should be removed using established safety standards.

All linkage and bolting should be checked to ensure that they are properly fastened. Sometimes line vibration can loosen bolting, which can lead to misalignment or malfunction of any number of areas of the valve, such as the regulating element, positioner linkage, packing box, etc. The actuator should be examined for any leaks of the power supply. With pneumatic-powered actuators, a soap solution can be applied to joints to detect air leaks. If the pneumatic actuator has an air filter, the cartridge should be replaced if blocked. With hydraulic or electro-hydraulic actuators, the actuator should be visually checked for any hydraulic fluid leaks. With manual handwheels, the lubrication should be checked around the handwheel stem.

6.2.8 Globe-Control-Valve Troubleshooting

The design complexity of globe valves, as well as their actuators or actuation systems, makes determining the causes of a failure or malfunction difficult. This section is intended to acquaint the user with common failures and malfunctions of globe valves but is not intended to replace the manufacturer's specific instructions for repairing the problem.

The most common malfunction is when excessive leakage occurs through the valve seat. This is defined as leakage beyond what is permitted by the ANSI seat-leakage classification. With some unbalanced and pressure-balanced valve designs, flow direction is critical to the correct shutoff function of the valve. This is especially true with flow-to-close applications, where flow is needed to help the actuator close the valve and maintain closure. If the valve is installed incorrectly, allowing the flow to flow under the plug, the thrust capability of the actuator will not be enough to allow complete shutoff.

If the valve has been in satisfactory service for some time before seat leakage occurs, the probable cause is a worn or damaged seat-ring–plug combination. This may occur because of process erosion, wear between the two matching seating surfaces of the seat ring and the plug, or a cavitating process that has worn away metal parts. Also, the seat-ring gasket or gaskets (split-body design) may have failed, allowing leakage between the seat ring and the body. When a process line is started up for the first time, damage sometimes occurs to the seating surfaces when a foreign object, such as a weld rod left in a pipeline, becomes caught in the seat. Misalignment of the plug and the seat may also cause leakage because the plug is not able to fully engage the seat. This happens when the plug is threaded too far into the actuator or handwheel stem, not allowing the plug to fully reach the seat upon full extension of the actuator. In some cases, an auxiliary handwheel or limit-stop is incorrectly positioned, which may also inadvertently restrict the travel of the plug.

If the valve has been disassembled recently for servicing and then reassembled, the most common cause of leakage is misalignment of the plug and the seat. If a retained seat ring is used, misalignment may be caused by improperly tightened bonnet-flange bolting (with top-entry globe valves) or body bolting (with split-body valves). If the bolting is not tightened equally, the mating parts (body, bonnet, cage or seat retainer, etc.) may not make proper contact with each other, being cocked at slight angles or providing a slight misalignment between the plug and seat ring.

Because threaded seat rings are in a fixed position, they must be lapped in relationship with the plug every time the seat ring or plug is removed and replaced. If lapping is not completed, small misalignments are likely between the seating surfaces, which will cause excessive seat leakage. Using a lower-thrust actuator, such as a diaphragm actuator, can lead to seat leakage, especially if retained seat rings are used. With inadequate thrust, the actuator cannot push the plug hard enough into the seat to achieve the necessary shutoff.

Leakage can also occur to atmosphere through the static seals, such as the bonnet gaskets (top-entry globe valves), flange gaskets, or seat-ring gaskets (split-entry valves). Usually, if leakage occurs with a static seal it will happen immediately after full process pressure is applied to the valve. If a gasket maintains its seal at start-up, it will likely last for the duration of the maintenance cycle. Eventual failure sometimes occurs if a slight flaw in the gasket grows under pressure or if the gasket material is attacked by the process fluid, in which case the gasket material should be changed. Leakage can also occur through the packing box, which is a dynamic seal with the sliding plug stem moving against stationary packing. If the packing fails to stop the leakage even after tightening the gland-flange bolting, the packing has completely consolidated or extruded and is incapable of providing a full-contact seal around the plug stem or bonnet bore. At this point, the user has no choice but to rebuild the packing box with new packing.

Another common problem associated with globe valves is erratic or impeded stem movement. One of the most common causes is over-tightened packing, which causes the packing to grip the stem. As the actuator attempts to position the plug, the packing resists any movement until enough thrust is generated to cause the plug stem to break loose and jump in a sudden motion. With some packings, such as braided graphite, the packing actually becomes more resistant as more thrust is applied, creating a very difficult control situation. One problem with graphite packing, which is used for high-temperature applications, is that the graphite tends to nearly form a perfect seal with the plug stem. This creates greater friction than softer, more pliant packings, such as PTFE. Occasionally, plug movement is impeded by a process service that is outside the range of the valve. For example, high temperatures can generate thermal expansion of those parts designed with tight tolerances, causing unplanned friction. In other applications, higher-than-expected process pressure can create a situation where the actuator cannot overcome the pressure to close the valve.

Poor performance of the regulating element can be linked to problems with the actuator, such as a low air supply to a pneumatic actuator or a malfunctioning positioner. If the valve is providing inadequate flow, the most likely causes are a malfunctioning positioner or actuator, improper plug adjustment, or an incorrect flow characteristic, especially if piping and pump effects have changed the inherent flow characteristic to an installed flow characteristic (Sec. 2.2).

Instability can also be caused by a valve that is equipped with reduced trim in an application that requires full-area trim. Typically, the valve moves quickly to the full-open position and locks in position, as the starved system seeks more flow, pressure, or temperature. Instability is also a characteristic of a valve experiencing choked flow (see Sec. 11.4).

Another common occurrence is when the plug slams into the seat ring, which causes water hammer (Sec. 11.6). Water hammer takes place when the actuator is underpowered or is too small, providing insufficient stiffness to hold a low-flow position close to the seat ring when flow is over the plug. As the actuator moves the plug toward the seat, the process pressure will overcome the actuator force and slam the plug into the seat (sometimes referred to as the "bathtub stopper effect"). This can also be caused by trim that is too large for the flow, which is remedied by installing reduced trim.

If the valve fails in the wrong direction, the most common cause is that the valve is installed in the wrong direction, allowing the flow to counteract the failure mode of the actuator.

6.2.9 Globe-Control-Valve Servicing

This section is intended to acquaint the user with general rules and procedures associated with disassembling and reassembling globe valves. In no way is it intended to replace the manufacturer's installation, operation, and maintenance (IOM) instructions for a specific valve model. If the valve must be disassembled to inspect or replace trim parts or to replace leaking gaskets, the process line should be depressurized and drained. If the process fluid itself is corrosive or dangerous to human exposure, it should be decontaminated.

Since most globe valves offer a top-entry design, access to the body internal parts, including the trim, is made by removing the bonnet-flange bolting and lifting the entire top-works (plug, bonnet, bonnet flange, and actuator) straight out of the body. This must be done carefully to avoid damaging critical surfaces of the plug and seat. With split-body designs, the top body half plus the top-works must be removed.

If the plug is guided on a cage or shares a seal with a pressure-balanced sleeve, sometimes the cage or sleeve is lifted out with the plug. If this happens, the cage or sleeve may slide off and damage the valve or injure workers. To prevent this from happening, with the plug fully extended, several blocks of wood can be placed between the cage or sleeve and bonnet. When the plug is slowly retracted using the actuator, the cage or sleeve can slide off the plug head. If extensive galling has occurred between the cage or sleeve and plug, separating the two parts may be very difficult, if not impossible, where extensive bonding has taken place.

To disassemble the packing box, removing the top-works from the valve body works best. Depending on the design of the valve, the actuator may or may not need to be removed, which is done by removing the actuator nut or slip clamps. The top-works should be placed onto its side during disassembly to prevent the plug from falling free during disassembly, which could damage valve parts or hurt personnel. After releasing the tension on the packing by loosening or removing the gland-flange bolting, the plug can be removed by unthreading it from the actuator stem or handwheel stem. In sizes less than 6 in (DN 150), this can be done by turning the plug head. In larger sizes, the plug can be removed by using a wrench on the flats of the plug stem. When the plug is released from the actuator stem, it can be pulled out of the bottom of the bonnet. Because the packing may be consolidated, some resistance may be felt as the plug is pulled through the packing box. If the packing is a type that can be used again, some care should be taken that the plug stem threads do not score the inside diameter of the packing.

After the plug is removed from the bonnet, it should be carefully examined for damage to the seating surface, such as gouges, erosion, or pitting from cavitation. If the plug head is used as the lower guide (with cage guiding), it should also be examined for signs of galling. If major galling has occurred between the plug head and cage, it will be very evident during earlier disassembly, since separating the two parts will be very difficult. In this case, both parts will need to be replaced. Minor galling can sometimes be corrected by lightly grinding or polishing the affected area, taking care not to work on too large an area, which may affect critical tolerances.

The plug stem should be examined for any scratches, scoring, or galling, which may impede or damage the sealing ability of the packing. Minor scratches can oftentimes be polished using a very fine abrasive compound substance. More substantial scoring usually requires a new plug. Some light-duty designs feature a threaded or pinned plug stem to a plug head, allowing the plug stem to be replaced separately from the plug head.

If the packing needs to be replaced, prior compression may make removing the packing somewhat hard. A wooden dowel may be used to push the packing out from the bottom of the bonnet. As the guides or packing follower, packing rings, extrusion rings, lantern rings, and/or spacers are taken from the packing box, their order should be noted for reassembly. The packing should be examined for evidence of consolidation or extrusion, so that preventative measures may be taken in the future. If a lubricator is used with the packing box, old lubrication should be cleaned from the lantern ring, spacers, and guides, as well as the bonnet bore. Using a flashlight, the bore should be examined for any pitting, corrosion, or erosion. If substantial damage has occurred to the bonnet bore, which is a sealing surface, the entire bonnet will need to be replaced, especially if fugitive emissions is a concern. The inside diameter of the guide(s) should also be examined for any abnormal wear or scoring. If the guide has an elastomer or graphite liner, it should be removed for inspection. If wear is minimal, the liner can be reused, except for a graphite liner, which should always be replaced.

Prior to reassembly, all parts should be thoroughly cleaned with cleaning agents that are approved to be used for a particular process. If the pressure-retaining parts have experienced minor oxidation, corrosion, or cavitation from the process or environment (meaning that the destruction did not penetrate minimal wall thickness), those areas may be repaired and repainted if necessary.

Reassembly begins with rebuilding the packing box. After placing the plug stem into the bonnet bore, the guides, packing, lantern rings, extrusion rings, and/or spacers should be replaced in their correct order. Each piece is slipped over the top of the plug stem (being careful of the plug-stem threads), down the plug stem into the packing box. Depending on the type of packing used, it may need to be tapped into place. Polymer V-rings are the easiest to repack, while rope packings are the most difficult. Using an old packing spacer that has the correct outside and inside diameters is the ideal procedure for this task.

In some cases, a different packing material or arrangement may be required, such as a twin seal or a vacuum-pressure packing arrangement. In this case, the packing box may vary in height, as the number of packing rings or type of spacers changes. Sometimes a small variance will be permissible with the depth of the bonnet bore. On the other hand, if the height variance is significant, the packing space length might need to be modified or additional spacers added. When changing the packing arrangement, the user should be especially careful to ensure that packing is installed correctly. For example, with vacuum-pressure packing arrangements, the upper seal portion has its chevron packing placed upside down for correct function. Also, extra rings of packing should not be added to the manufacturer's packing box design. Sometimes the user believes that "more is better," but actually too many rings may amplify the effects of thermal expansion of the packing, increase stem friction, and require high compression levels to compress additional rings—sometimes the gland-flange bolting is incapable of providing the right compression.

As the bonnet and trim are reinstalled in the body, new gaskets should always be used, especially if spiral-wound gaskets are used. Since the height of a spiral-wound gasket is dependent on the metal strips, which are crushed during assembly, the gasket does not recover any height following decompression. Therefore, a reused gasket would nearly always leak. In an emergency situation, polymer gaskets could be reused since they do recover somewhat following decompression. However, because of consolidation, some height is lost by prior compression, and using old gaskets is not worth the risk of eventual leakage, especially since the cost of gaskets is minimal when compared to the cost of unplanned maintenance.

As the seat ring is replaced in the body, it should be carefully seated in its bore, fully engaging its gasket(s). As the body, bonnet, cage or seat retainer, etc., are replaced, all parts should be square with each other. If the seat ring is retained, the plug should be fully seated in the seat ring (to ensure concentricity) before tightening the body or bonnet-flange bolting.

Body or bonnet-flange bolting can be tightened in one of two ways. First, the manufacturer may list torque values for the bolting in the maintenance instructions, which requires the use of a torque wrench.

Using torque values ensures that the correct amount of compression of the gaskets is achieved as well as enough force is provided to hold the trim in place without unnecessary stress to the valve's internal parts. With this type of design, the user should always use the manufacturer's recommended torque value, since guessing the torque may either result in a leaking, misaligned body subassembly (if undertightened) or overcompressed gaskets and stressed parts (if overtightened). The second method takes the guesswork out of bolting tightening. Some designs have an exact-tolerance step such that when the bolting is tightened, it reaches a metal-to-metal state with another part, ensuring correct gasket compression as well as providing trim stability.

With globe valves using a retained seat ring, the actuator must be reinstalled before final tightening of the bonnet-flange or body bolting takes place. As the bolting is tightened, a retained seat ring is fixed by the axial force applied by the bonnet via the cage or seat retainer (top-entry valves) or by the two body halves (split-body valves). With pneumatic actuators, the plug can be left fully extended in the seat during this final step, ensuring full contact of the seat surfaces. However, if hydraulic or electrohydraulic actuators are used, the plug should be retracted from the seat before final tightening of the bolting. This is because these actuators have no allowance for a stem back-drive occurring during final bolt tightening, as opposed to a pneumatic design that can tolerate some back-drive.

With threaded seat rings, the actuator is not installed until after lapping takes place, because the plug must be free so as to allow it be turned in the seat. After lapping, the entire bolting can be tightened and the actuator reinstalled.

Following reassembly and before the valve is placed in service, the valve should be stroked several times to ensure smooth stem travel. If hydrotest facilities are available (and the valve has been removed from the line), the valve can be tested under pressure to ensure the integrity of the seat, gasket, and packing.

6.3   Butterfly Control Valves

6.3.1   Introduction to Butterfly Control Valves

Although the butterfly valve has been in existence since the 1930s, it was used mainly as an on–off block valve until the past two decades, when it began to be used for throttling services. In the late 1970s, design advancements were made to the butterfly valve that not only made it more applicable for throttling service, but also made it preferred over globe valves in some applications. Such butterfly control valves are differentiated from their on–off block cousins by the name *high-performance butterfly valves*. In simple terms, the high-performance butterfly control valve is a quarter-turn (0° to 90°) rotary-motion valve that uses a rotating round disk as a regulating element. Typically, butterfly control valves are available in sizes 2 through 8 in (DN 50 through DN 200) from ANSI Classes 150 to 600 (PN 16 through PN 100); 10 and 12 in (DN 250 and DN 300) in ANSI Classes 150 and 300 (PN 16 and PN 40); and 14 through 36 in (DN 350–900) in ANSI Class 150 (PN 16).

When fully open, the disk actually extends into the pipe itself, which makes butterfly valves distinct from other valve designs. Butterfly-valve bodies have very narrow face-to-face dimensions compared to other types of valves, allowing the body to be installed between two pipe flanges without any special end connections. This type of arrangement is called a *through-bolt connection* and is only permissible with certain bolt lengths. If the bolt length is too long, the bolting may be subject to thermal expansion of the process or during an external fire, causing leakage.

Initially, butterfly control valves were designed as automatic on–off block valves. However, with recent improvements to rotary-valve actuators and body subassemblies, they can now be used in throttling services with the addition of an actuator or an actuation system. As detailed in Sec. 3.4, the family of butterfly valves is classified into two groups. *Concentric butterfly valves* are normally used in on–off block applications, with a simple disk in-line with the center of the valve body. Generally, concentric valves are made from cast iron or another inexpensive metal and are lined with rubber or polymer. Because of their lower performance, they are normally equipped with manual operators. In some applications, the manual operators are replaced with an actuation system for throttling service. In most applications, however, simple concentric butterfly valves are used strictly for on–off service. Even when used in ·throttling applications, they do not lend themselves as well to automatic control as other butterfly designs specifically designed for throttling control. This is because the initial development was for blocking service. Concentric butterfly valves have poor rangeability, while throttling-specific butterfly valves have design modifications to allow for better flow control through the entire stroke.

Eccentric butterfly valves are valves designed specifically for high-performance throttling services, using a disk that is offset from the center of the valve body. The majority of butterfly valves used as control valves feature the eccentric design. For the most part, eccentric butterfly valves are specified in common valve materials, such as carbon, stainless, or alloy steels. When equipped with actuators and positioners, they are much more precise than concentric butterfly valves that have been automated.

Compared to other types of throttling valves, eccentric butterfly valves are one of the fastest growing types of control valves today for a number of reasons. Because of the increased dead band associated with the mechanical conversion of linear motion to rotary motion, globe valves are more precise in high-pressure-drop applications than butterfly valves. However, the control provided by today's butterfly valves is more than adequate for many low-pressure-drop applications and other standard services.

When compared to globe control valves, butterfly control valves are much smaller and lighter in weight because the butterfly valve's body subassembly weight can be anywhere from 40 to 80 percent of a comparable valve and less than half the mass of the globe body subassembly. In addition, smaller actuators can often be used with butterfly valves since the weight of the regulating element is not a critical factor in factoring the necessary actuator force. The difference in regulating-element weight between butterfly and globe control valves becomes much more evident as sizes become larger, as shown in Table 6.1. This means that butterfly valves are preferred in applications where limited space or weight is a consideration.

Another major benefit of using a butterfly control valve is that, size for size, it has a larger flow coefficient, producing a greater flow than comparable globe valves. Because the shaft of the butterfly valve moves in a rotary motion instead of a linear motion, the frictional forces are far less than a linear-motion valve, requiring less thrust and permitting a smaller actuator. A butterfly valve has a naturally high pressure-recovery factor (Sec. 9.2.9). This factor is used to predict the pressure recovery occurring between the vena contracta and the outlet of the valve. The butterfly valve's ability to recover from the pressure drop is influenced by the geometry of the wafer-style body, the maximum flow capacity of the valve, and the service's ability to cavitate or choke. Overall, because of the high-pressure recovery, a butterfly valve works exceptionally well with low-pressure-drop applications.

The largest drawback to using a butterfly valve is that its service is usually limited to low-pressure drops because of its high pressure recovery. Although flashing is normally not associated with a butterfly-valve design, cavitation and choked flow occur easily with a but-

Table 6.1 Weight Comparisons between Globe and Butterfly Valves*

Valve Size	Flanged Globe Valve Standard Valve with Actuator, ANSI Class 150	Flangeless Butterfly Valve Standard Valve with Actuator, ANSI Class 150	Percent Reduction
2-inch	75 pounds	40 pounds	47%
DN 50	34 kilograms	18 kilograms	
3-inch	160 pounds	46 pounds	71%
DN 80	73 kilograms	21 kilograms	
4-inch	240 pounds	52 pounds	78%
DN 100	109 kilograms	24 kilograms	
6-inch	360 pounds	96 pounds	73%
DN 150	163 kilograms	44 kilograms	
8-inch	590 pounds	110 pounds	81%
DN 200	268 kilograms	50 kilograms	
10-inch	1050 pounds	267 pounds	75%
DN 250	477 kilograms	121 kilograms	

*Data courtesy of Valtek International.

terfly valve installed in an application with a high-pressure drop. Although some special anticavitation devices have been engineered to deal with cavitation, users prefer to deal with cavitation in a globe valve because of its design versatility in allowing the inclusion of an anticavitation device. Another disadvantage is that a butterfly valve has a poor-to-fair rangeability of 20 to 1 because of the difficulty the disk has in holding a position close to the seat. The process pressure applied to the butterfly disk creates a significant side load, which can only be remedied by using a larger-diameter shaft. Another drawback to the butterfly control valve is the increased hysteresis and dead band associated with the mechanical transfer of linear action from the actuator to the rotary motion needed for the regulating element. Valve manufacturers have utilized splined shafts or other secure linkages to minimize this problem, although a globe valve avoids this problem altogether with its direct linear motion. The sizes of butterfly valves are also limited to 2 in (DN 50) and larger because of the limitations of the rotary regulating element. Because of the side loads applied to the disk, the maximum size that a high-performance butterfly can reach is 36 in (DN 900).

6.3.2 Butterfly-Control-Valve Design

The butterfly body typically involves one of two styles. The *wafer body* (sometimes called the *flangeless body*) is a flat body that has a minimal face-to-face, which is equal to double the required wall thickness plus the width of the packing box (Fig. 6.20). Within this dimension, the disk in the closed position and the seat must fit within the flow portion of the body. Because the wafer-style body has a minimal face-to-face, straight-through bolting using the two flanged piping connections is possible without fear of thermal expansion causing leakage. Wafer-style bodies are more commonly applied in the smaller sizes, 12 in (DN 300) and less. The other body style is the *flanged body*, which is

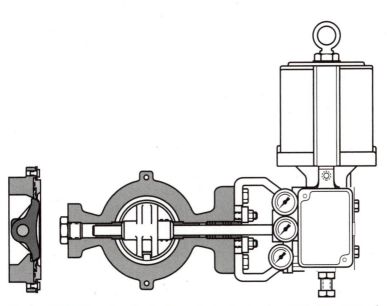

Figure 6.20 Flangeless butterfly control valve (wafer style). (*Courtesy of Valtek International*)

Figure 6.21 Flanged butterfly control valve. (*Courtesy of Valtek International*)

used with larger butterfly valves [14 in (DN 350) and larger] that require a longer face-to-face (Fig. 6.21) when a higher degree of thermal expansion is expected or when the regulating element cannot fit within the wafer-style body. The flanged style has integral flanges on the body that match the standard piping flanges.

As shown in Fig. 6.22, another body style is the *lug-style body*, in which the butterfly body has one integral flange that has an identical hole pattern to the piping flanges. Each hole is tapped from each direction, meeting in the center of the hole. This arrangement allows the body to be placed between two flanges. Studs are then inserted through the piping flange and threaded into the valve's integral flange. After the stud is securely threaded into the integral body flange, a nut is threaded to the stud to secure the piping flange to the body. Lug bodies are used in applications in which the risks of straight-through bolting cannot be taken—such as with thermal expansion—in smaller valve sizes that do not permit the use of two integral flanges.

The faces of the butterfly-valve body are often serrated to fix and secure the location of the flange gaskets between the pipeline and the valve. The inside diameter of the butterfly valve is close in size to the inside diameter of the pipe, which permits higher flow rates as well as

Figure 6.22 Lug-style butterfly control valve.
(*Courtesy of Automax, Inc. and The Duriron
Company, Valve Division*)

straight-through flow. Perpendicular to the flow area of the valve is the shaft bore, which is drilled from both sides. Drilling from one side through the entire body is extremely difficult without the wandering associated with using a long drill bit.

The regulating element of the butterfly valve is the called the *disk,* which rotates into the *seat.* The disk is described as a round, flattened element that is attached (usually by tapered pins) to the rotating shaft. As the shaft rotates, the disk is closed at the 0° position and wide open at the 90° position. As explained earlier in Chap. 3, if the shaft is attached to the disk at the exact centerline of the disk, it is known as a *concentric disk.* When the disk is offset both vertically and horizontally (refer to Fig. 3.18), it is referred to as an *eccentric cammed disk.*

The disk is designed to minimize interruption of the flow as the process fluid moves through the valve. Slight angles and rounded surfaces are characteristic of a common disk design. When closed, the flat side (facing the seat) is called the *face*, while the opposite side is called the *back side*. The face is often designed slightly concave so that maximum flow can be achieved in the open-flow position. On the backside, sometimes a *disk-stop* is provided that matches up with a similar stop inside the body's flow area. This stop prevents the valve from overstroking. Overstroking can cause the disk to drive through the seat, irreparably damaging the seat. The circumference of the seat wraps around the entire inside diameter of the body's flow area and is installed at one end of the body. If a polymer is used for the seat, it is called a *soft seat*. When a flexible metal is used as the seating surface, it is called a *metal seat*. The seat is installed in the end of the body and is held in place by a *seat retainer*, using screws or a snap-fit to keep the seat and retainer in place. After the seat and seat retainer are in place, the face of the retainer usually lines up with the face of the body. In some designs, the seat–retainer design protrudes slightly from the body face, allowing some gasket compression when the body is installed in the line.

The disk is attached to the shaft with the use of one or more tapered pins. The shaft is supported by close-fitting *guides* (sometimes called *bearings*) on both sides of the disk, which are installed in the shaft bore to prevent lateral movement of the shaft and disk that can cause misalignment. Thrust washers may also be placed on both sides of the disk, between the disk and the body, to keep the disk firmly centered with the seat.

A number of different resilient seat designs exist for eccentric butterfly control valves, which are designed to handle higher pressures and temperatures—most of which operate by similar principles. One of the most common soft-seat designs is the seat that utilizes the *Poisson effect*, which states that if an O-ring or an elastomer is placed in a seating situation with a greater pressure on one side, the soft material will deform away from the pressure. In other words, deformation takes place when the pressure pushes the softer material against the surfaces to be seated (Fig. 6.23). With the Poisson effect, the greater the upstream pressure compared to the downstream pressure, the greater the seal. Because of their flexibility, O-rings encased in a polymer work exceptionally well with the Poisson effect. Related to the Poisson effect is the *jam-lever* or *toggle effect*, which uses a hinged elastomer that is designed to be thinner in the midsection than at the outside or inside diameter. This design permits the outside diameter of seat to flex and seal against metal surfaces when process pressure is applied (Fig. 6.24). A third resilient seat design uses the *mechanical preload effect*, which calls for the inside diam-

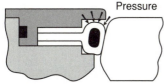

Pressure

Seat is forced into gap between body
and disc causing valve to seal.

Poisson Effect with Pressure Upstream

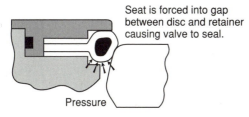

Seat is forced into gap
between disc and retainer
causing valve to seal.

Pressure

Poisson Effect with Pressure Downstream

Figure 6.23 Poisson effect on a butterfly seal
for both upstream and downstream pressures.
(*Courtesy of Valtek International*)

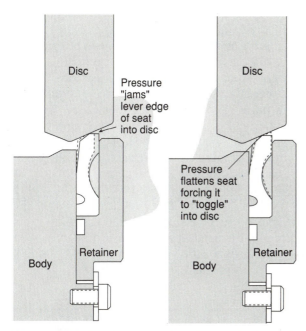

Disc

Pressure
"jams"
lever edge
of seat
into disc

Disc

Pressure
flattens seat
forcing it
to "toggle"
into disc

Retainer

Body

Retainer

Body

Figure 6.24 Jam lever or toggle effect on the butterfly
seal. (*Courtesy of Valtek International*)

eter of the seat to slightly interfere with the outside diameter of the disk. As the disk approaches the seat to close, it makes contact with the seat. As the disk moves further into the seat, the seat physically deforms because of the pressure applied by the disk, causing the polymer to seat against metal surfaces. In some cases, a manufacturer may use both the mechanical preload and Poisson effects to achieve the correct shutoff (Fig. 6.25). When a soft seat is used, it also has a secondary purpose, acting as a gasket between the body and the retainer. Metal seats are typically applied to high temperatures (above 400°F or 205°C). Metal seats are integral to the seat retainer—with a gasket placed where a soft seat is normally inserted (Fig. 6.26). In some designs, both a soft and metal seat can be used in tandem, allowing the metal seat to be a backup in case of failure of the soft seat (Fig. 6.27). When butterfly valves are specified for fire-safe applications, the tandem seat is installed. In pure throttling applications, where the valve is intended to remain in midstroke at all times and never close, the valve can be built without a seat as a cost-saving measure.

A butterfly valve's packing box is similar in some regards to the globe valve's packing box. The packing box has characteristics similar to all packing boxes: a polished bore and a depth to accommodate various packing designs. One major difference, however, is that a butterfly valve does not require a lower set of packing. Because of the rotary-motion design, the stem rotates and never changes linear position. In other words, the packing always remains in contact with the same region of the stem. Since the stem never moves its linear position, a "wiper" packing set is not necessary. All that is required is an optional spacer, the packing, and a packing follower. An upper guide or bearing is not needed at the open end of a butterfly-valve packing box as the shaft has its own guides on each side of the disk. The shaft can also be guided by a bearing in the actuator's transfer case. A gland flange and packing follower are used to compress the packing.

Because the shaft bore is normally machined from both ends, a plug or flange cover can be used to cover the bore opening opposite the packing box. To retain the body pressure, a gasket or O-ring is required. If a threaded plug end is used, it should not come in contact with the shaft, since the quarter-turn action of the shaft could possibly rotate the end plug, causing process leakage to atmosphere.

On the packing box side of the body, mounting holes are provided allowing the transfer case to be mounted. The *transfer case* contains the linear-motion to rotary-motion mechanism that allows a linear-motion actuator to be used with a quarter-turn valve. The end of the shaft that fits into the transfer case is either splined or milled with several flats

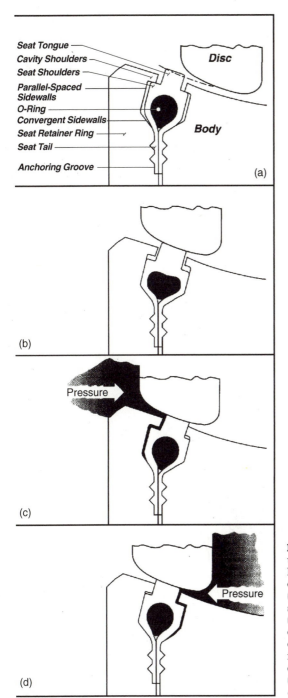

Figure 6.25 Butterfly seal using both mechanical pre-loading and the Poisson effect. (*a*) Basic seal design, (*b*) preloading effect on the seat caused by disk seating (with minimal pressure effects), (*c* and *d*) Poisson effect on the seat caused by increased upstream or downstream pressures. (*Courtesy of Flowseal, a unit of the Crane Valve Group*)

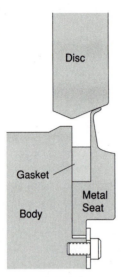

Figure 6.26 Butterfly metal seat design.
(*Courtesy of Valtek International*)

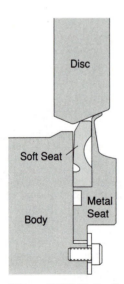

Figure 6.27 Butterfly dual soft- and metal-seat
design. (*Courtesy of Valtek International*)

to allow for attachment of the linkage. The designs of common rotary actuators, actuation systems, and handwheels are detailed in Chap. 7.

6.3.3 Butterfly-Control-Valve Operation

As the process fluid enters the butterfly body, it moves in a straight direction through the flow passage. The only obstruction to the flow is the disk itself. In the open position, the gradual angles and smooth, rounded surfaces of the disk allow the flow to continue past the regulating element without creating substantial turbulence. However, some turbulence should always be expected because the disk is located in the middle of the flow stream. In closing the valve, as the signal is received by the actuator or actuation system, the force is transferred to rotary motion, turning the shaft in a *quarter-turn motion,* which is defined anywhere between 0° (full-closed) and 90° (full-open). As the disk approaches the seat, the full pressure and velocity of the process fluid are acting on the full area of the face or back side of the disk (depending on the flow direction), which makes stability difficult. This instability may be compounded when diaphragm actuators are used, since they do not generate high thrust to begin with. Because the rangeability of butterfly valves is so poor (20 to 1), the final 5 percent of the stroke (to closure) is not available to the user. As the disk makes contact with the seat, some deformation takes place, allowing the resilient elastomer or flexible metal strip to mold against the seating surface of the disk.

To open the valve, the signal causes the disk to move away from the seating surfaces. Because of the mechanical and pressure forces acting on the disk in the closed position, a certain amount of rotary-motion force, called *breakout torque,* must be generated by the actuator or handwheel to allow the disk to open. The designs with the greatest requirement for breakout torque are those designs that require a great deal of actuator thrust to close and seat the valve. Therefore the greater the actuator force for closure, the greater the breakout torque. When fluid pressure is utilized to assist with the seat, less actuator force is required and thus less breakout torque.

In principle, the opening disk is nearly in a balanced state, since one side is pushing against the fluid forces, while the other side is pulling with the fluid forces. However, because both sides of the disk are not identical—the shaft is connected on one side, while the opposite side is more flat—flow direction has a tendency to either push a disk open or pull it closed. In most cases, when the shaft portion of the disk is facing the outlet (downstream), the process flow tends to open the valve. On the other hand, when the shaft portion is facing the inlet side (upstream), the flow tends to close the valve. The failure mechanism of the actuator must complement the flow direction, so that the proper failure mode will occur.

With concentric disk–seat arrangements (the center of the disk and the shaft are exactly centered in the valve), a portion of the disk always remains in contact with the seat in any position. At 0° open, the seating surfaces are in full contact with each other. In any other position, the seating surfaces touch at two points where the edges of the disk touch the seat. Because of this constant contact, the concentric disk–seat design has a greater tendency for wear, especially with automated control applications. During throttling, a butterfly valve may be required to handle a small range of motion in midstroke, causing wear at those two points of contact. Although the wear will not be evident during throttling, it will eventually allow leakage at those two points when the valve is closed. To overcome this problem of constant contact between the seating surfaces, some butterfly-valve manufactures prefer to use the eccentric cammed disk–seat configuration, which allows for the disk and seat to be in full contact upon closure, but when the valve is open the disk and seat are no longer in contact. Such designs allow for the center of the shaft (and disk) to be slightly offset down and away from the center of the valve. When the valve opens, the disk lifts out of the seat and slightly away from the seating surfaces—enough to avoid constant contact.

Because of the design limitations of the disk and seat arrangement, a flow characteristic is not easily designed into the body subassembly, unlike the trim of a globe valve. Thus, a butterfly valve must use its inherent flow characteristic, which is parabolic in nature. To achieve a flow characteristic, an actuator with a cammed positioner must be used to provide a modified flow characteristic.

A feature unique to high-performance valves is the ability to mount the valve on either side of the pipeline so that the shaft orientation (shaft upstream or shaft downstream) and the failure mode (fail-open and fail-closed) can operate in tandem with the air-failure action of the actuator. Figure 6.28 shows the four common orientations [(1) fail-closed, shaft upstream, air-to-open; (2) fail-open, shaft upstream, air-to-close; (3) fail-open, shaft downstream, air-to-close; and (4) fail-closed, shaft downstream, air-to-close].

6.3.4 Butterfly-Control-Valve Installation

After the butterfly valve has been received from the manufacturer, it should be carefully inspected to ensure that it was specified according

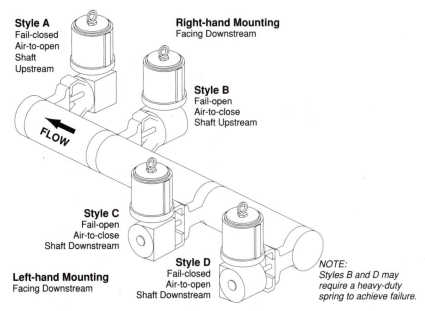

Figure 6.28 Rotary actuator mounting orientations. (*Courtesy of Valtek International*)

to the user's requirements. Correcting problems with incorrect or damaged actuator accessories, tubing, packing boxes, etc., are best corrected prior to installation. The user should check the shipping box or packaging for installation and maintenance instructions, which should be kept for reference. To avoid back injuries, large butterfly valves [4 in (DN 100) or larger] should not be lifted by hand. Rather, lifting straps and a hoist or other mechanical lifting device should be used to handle the valve. If used, lifting straps should be secured around the valve-body subassembly as well as the actuator or handwheel, avoiding contact with the tubing or actuator accessories. Before the butterfly valve is installed in the pipeline, the entire length of the upstream pipe should be cleaned of any foreign material, such as dirt, welding chips, spent rods, or scale. If this is not done, debris may collect around the disk and seat, and the seating surfaces may be damaged. With butterfly valves, the disassembly clearance may differ significantly from that of globe valves (which require top-entry to the trim). Side and/or top clearance must be available for the body to be removed from the line. In addition, side and/or top clearance must be available to allow for disassembly of the actuator. The valve's certified dimensional drawing often provides the necessary clearances for that particular valve.

Sometimes butterfly valves can be disassembled with less than required disassembly clearance, but the procedure involves removing the actuator or handwheel assembly first before removing the body subassembly—a time-consuming but possible solution. A majority of applications allow a butterfly valve to be installed with the body in a vertical position and the pipe in the horizontal position. However, some applications—such as untreated water that may contain particulates—require vertical piping so that any buildup of particulates can fall free after the valve is opened. If an actuator is included in the valve, it should be installed right-side up (unless space restrictions are present), making maintenance and calibration easy. If a valve with an actuator and positioner must be installed horizontally (vertical piping), the best orientation is to install the valve so that the positioner is facing up, which allows for easier calibration.

A flow plate or cast arrow on the body indicates flow direction for the butterfly valve. Maintaining the correct flow direction is critical to the failure mode for the butterfly valve, since the flow direction must coincide with the correct disk orientation as well as the failure mode of the actuator. With many throttling butterfly valves, the actuator's failure mode is determined by how the actuator is mounted to the transfer case. Mounting the actuator on one side of the transfer case may produce a downward motion toward closure, while mounting on the opposite side will produce a downward motion toward opening. Refer to Fig. 6.28 for typical actuator and body orientations for both flow directions and failure modes.

When installing the butterfly valve between piping flanges, care should be taken to ensure that enough room exists between the pipe flanges to allow for the face-to-face of the wafer body and the width of the gaskets, as well as a working clearance to slide the valve into place. If the piping is fixed rigidly in place, too much space between flanges may make it difficult to tighten the flange bolting to prevent gasket leakage. An insufficient space may prevent the installation of the valve or make it extremely difficult. Installation of a butterfly valve involves loosely installing the bottom two studs and nuts between the piping flanges, which creates a place for the valve body to sit on. With the flange gaskets in their correct positions against the serrated valve body faces, the body subassembly is slipped into place on the two previously installed studs. With the valve correctly positioned, the remainder of the bolting can be installed. To ensure position stability, some butterfly-valve bodies have fins with holes to act as guides for the bolting. The user should also take care to apply equal gasket compression. Ideally, the bolting should be tightened in a criss-cross pattern by tightening one bolt and then the one opposite, allowing the flanges to stay square with the valve body and gaskets. The manufacturer's recommended torque values should be known and used when tightening the bolting. Low-strength bolting is used for flanged connections, while intermediate- or high-strength bolting is used for through-bolted joints. Typical ANSI flange bolting specifications are found in Sec. 3.4.

If a larger butterfly valve is designed with integral flanges, the matching end flanges on the piping flanges should be correctly aligned, avoiding any offset or misaligned hole patterns. Because the flanges on the body are fixed, the hole pattern should match up with the vertical axis of the valve. If the piping flanges are true to each other yet not square with the vertical axis, the butterfly valve will lean. Some play exists between the size of the flange hole and the required stud, which may help minimize this misalignment. If one of the two piping flanges is misaligned with the vertical axis, installation may be difficult unless the misaligned flange is corrected. Rather than cut out a flange and weld it correctly in place, some users widen the holes on either the piping flange or valve flange. This practice weakens the integrity of the flange and is not recommended.

With automatic control, the actuator, actuation system, and/or positioner should be connected to the instrumentation signal and power supply. The actuation system typically displays a tag or sticker that indicates the maximum power supply. If a pneumatic actuator is used, a pressure regulator may be necessary to lower the plant air supply to the required levels, especially with diaphragm actuators. With spring cylinder actuators, high air-supply pressure up to 150 psi (10.3 bar) can be handled without a pressure regulator. Unless the plant air supply is exceptionally clean, an air filter should be installed before the actuator or positioner to prevent water, oil, dirt, or other debris from entering and fouling the actuator. Most air filters are designed with a filter cartridge, which should always point downward to allow moisture or any other materials to be collected.

Once the high-performance butterfly valve is installed and the instrumentation signal and power supply are connected, the valve should be fully stroked. By visually observing the stroke indicator mechanism on the transfer case, the shaft should turn in a smooth, rotary fashion. If the shaft seems to catch or turn erratically, a problem may exist with the disk–seat alignment, shaft guides or bearings, packing tightness, linkage, or actuator, which should be investigated and corrected immediately before further operation. The entire instrument signal change should be made, from full-open to full-close, to ensure correct calibration of the range: for example, 3 to 15 psi (0.2 to 1.0 bar) for pneumatic signals or 4 to 20 mA for electrical signals. All connections (either electrical or pneumatic) should be inspected to ensure that no leakage or loose connections exist.

Packing compression can occur through consolidation between the valve's assembly at the factory and the final installation. Thus, retightening the packing should be done according to the manufacturer's written procedures, taking care not to overtighten the packing. Overtightened packing will cause increased shaft friction, erratic shaft motion, and premature wear of the shaft and packing. If the service has wide temperature swings, the packing tightness should be checked after the temperature excursion has taken place. With temperature excursions, the pipe-flange bolting should be checked for possible leakage, especially if straight-through bolting is used with a wafer-style body.

To check the failure action of the butterfly control valve, the power supply should be shut off or disconnected from the actuator, allowing the valve to fail in the correct position (open, closed, or at last position).

6.3.5 Butterfly-Control-Valve Preventative Maintenance

Butterfly-valve life can be extended by periodically checking the valve for proper operation as well as performing preventative maintenance. In most cases, butterfly valves are serviced with the valve removed from the line. If any portion of the internal parts of the valve must be checked or replaced, the process service will need to be interrupted or blocked around the valve being serviced. Servicing should be done according to the methods and procedures outlined by the valve manufacturer, although a number of general procedures are outlined in this section.

The first step should be to inspect the piping end connections for signs of process leakage to ensure that thermal expansion has not weakened the integrity of the seal or that the flange gasket has not failed. The packing box, body end plate or plug, and any other pressure-retaining connection should be closely examined. With liquid processes, this check can be done visually, while gas processes should be examined using a "sniffer" atmospheric sensing device. If leakage is evident between the piping flanges and the body, the piping-flange bolting should be tightened until the leak is eliminated. If that does not totally eliminate the leak, the gasket has most likely failed, and disassembly and replacement of the gasket are necessary. If leakage is occurring through the packing, the gland-flange bolting should be tightened, being careful to follow the manufacturer's recommended torque value or procedure until the leakage stops. The user should be careful not to overtighten the packing as this may crush it or cause consolidation and extrusion.

Only after the valve is inspected should it be thoroughly cleaned— since cleaning prior to inspection may mask signs of leakage. If leakage has occurred from the packing box, any process buildup should be removed from the shaft, packing follower, gland flange, or gland-flange bolting. If metal surfaces show signs of severe oxidation, they should be cleaned with a wire brush and repainted using a rust inhibitor, especially if the outside environment is corrosive, such as the atmospheric salt prevalent near seawater. The piping above the valve should be inspected carefully to ensure that process drippings are not falling on the butterfly valve. If so, those piping leaks should be corrected or the valve shielded from the process drips.

If the butterfly-valve body is removed from service, it should be inspected for any porosity or fracture that may have occurred during service that may lead to more serious problems. The inner flow area of the body subassembly should also be inspected for signs of erosion or cavitation. Sometimes if the crack or porosity is minor, that area can be ground out and filled with a weld. If this is done, the welding should be done according to established methods so that the integrity of the pressure-retaining vessel will not be jeopardized.

If the valve is installed in a noncritical process or if the valve is blocked from the process, it should be stroked so that the user can be assured that the valve is capable of a full and smooth stroke. If the shaft moves in an erratic or jerky manner, it may indicate an internal galling, packing tightness, or actuator or positioner problem. Some packings, especially graphite packing, are known to cause jerky motion when properly compressed.

If the process allows, the power supply should be removed or shut off from the actuator or actuation system, allowing verification that the butterfly valve fails to the correct position. Most transfer cases have a cover that can be removed so that all linkage and bolting can be checked to see if they are properly fastened. Sometimes line vibration can loosen bolting in the linkage, which can lead to increased mechanical hysteresis or malfunction of any number of areas of the valve, such as the regulating element, positioner linkage, etc. If the transfer-case cover is used to guide the end of the shaft, the user should be careful not to stroke the butterfly valve when the cover is off, since the shaft is unsupported and may cause internal damage.

If the valve is used for automatic control, the actuator should be checked for any leaks of the power supply. With pneumatic actuators, a soap solution can be used with the tubing or actuator joints to detect air leaks. If the pneumatic actuator is equipped with an air filter, the cartridge should be checked and replaced, if necessary. If the valve uses a hydraulic or electrohydraulic actuator, it should be visually inspected for any hydraulic fluid leaks. With auxiliary handwheels, the user should check the lubrication around the handwheel stem. If a lubrication fitting exists, lubrication should be added.

6.3.6 Butterfly-Control-Valve Troubleshooting

Because eccentric butterfly valves are much more complex than concentric butterfly valves, especially when actuators or actuation systems are involved, the user should be acquainted with common failures and malfunctions. However, these troubleshooting tips are only intended to provide ideas about a valve's malfunction and should not replace the manufacturer's specific instructions for troubleshooting or repairing the problem.

The most common malfunction of a butterfly valve is leakage through the valve seat. Unless the control valve is classified as having ANSI Class VI (bubble-tight) shutoff, some minor leakage through the seat is expected. Therefore, *leakage* is a term used to indicate that the measured leakage is beyond that which is permitted by the classification. If the valve has been operating satisfactorily for a reasonable period before seat leakage occurs, the likely cause is a worn or damaged seat and/or disk. Probable causes of a worn seat are process erosion, mechanical failure of the seat, frictional wear between the two mating seating surfaces of the seat and the disk (especially if the valve closes often), damage from a foreign object caught between the seat and disk, or cavitation damage to the disk. The soft seat or seat gasket (metal seat designs) may also have failed, causing leakage in the gap between the body and the seat retainer—even when the disk and seat are successfully shutting off the flow.

If the shaft guides or bearings are worn, the disk and seat may become misaligned. Ultimately this can result in a damaged seat and increased leakage. Leakage can also be caused by a mechanical problem, such as galling of the shaft, which may be preventing full motion of the disk and not allowing the disk to reach the seat. In other cases, an auxiliary handwheel or limit-stop may be incorrectly positioned, which may inadvertently limit the travel of the disk.

Leakage commonly occurs through the packing box, which is a dynamic seal with the rotating shaft moving against stationary packing. If tightening the packing fails to stop the leakage, in most cases the packing has completely consolidated or extruded. When this happens, the packing is incapable of providing a full-contact seat around the shaft or body bore, and the packing box must be rebuilt using new packing. As a preventative measure, antiextrusion rings can be added to the packing-box configuration.

Leakage from the piping flanges may indicate a misalignment of the upstream and downstream piping or could be caused by failure of the flange gaskets. If the gaskets and valve and piping gasket surfaces are dirty or are not cleaned of particulate matter during installation, leakage may occur.

Another common problem associated with butterfly valves is erratic or impeded shaft movement. One of the most common causes is over-tightened packing, which results in the packing gripping the shaft. As an attempt is made to position the disk, the overtightened packing resists any rotational movement until a buildup of thrust causes the shaft to break loose, causing a sudden, jerky motion. Other probable causes of erratic shaft travel can be tight or misadjusted linkage between the shaft and actuator, actuator failure (such as internal galling), or worn or damaged shaft bearings.

Because the disk moves into the piping itself, especially when full-open, some heavy scheduling piping or piping with a cement lining may interfere with the movement of the disk. To remedy this situation the piping must be modified to allow for free movement of the disk. If the shaft moves with the signal, but the valve does not respond in kind, the disk–shaft connection has failed.

Because some packing materials are more abrasive than others—such as graphite—the quarter-turn rotation of the valve has a tendency to wear the shaft where it makes continual contact. If the packing is highly abrasive, these frictional losses can eventually cause leakage through the packing box unless the packing is tightened. Eventually, the continual wear will require replacement of the shaft.

Some packings, such as braided graphite rings, are known to become more resistant as more thrust is applied, creating a very erratic control situation. A problem with high-temperature applications is that the recommended packing material, graphite, forms a nearly perfect seal with the shaft, which creates greater friction than softer, more pliant packings, such as PTFE.

In a few applications, disk movement may be impeded by high temperatures that are outside the temperature limits of that particular valve design. During a temperature extreme, thermal expansion of those moving parts with tight tolerances (such as shafts and guides or bearings) can result in sticking or galling. If the process pressure is higher than expected, the actuator designed for that application may not be able to overcome the process forces to correctly operate the valve.

In other cases, poor throttling function or limited closure of the body may be caused by the pneumatic actuator, which may have a low air supply or a malfunctioning positioner. If an actuator fails, such as when a diaphragm bursts or a stem seal fails, the actuator will not be able to hold a position and the valve will move to the failure position. This can also occur when the linkage between the actuator and shaft fails.

If the valve is providing poor flow control, the most likely cause is a malfunctioning positioner or actuator, or an incorrect flow characteristic cam in the positioner. As discussed in Sec. 2.2, the valve's flow characteristic can be affected by piping and pump effects that will change the inherent flow characteristic to an installed flow characteristic.

Another common problem occurs when the disk slams into the body seat, causing a water-hammer effect (Sec. 11.6). When water hammer occurs, the likely cause is an undersized or underpowered actuator that has insufficient stiffness to hold a low-flow position close to the seat ring when flow is shaft upstream. As the actuator moves the disk toward the seat, the process pressure may overcome the actuator and the disk is sucked into the seat.

If the valve fails in the wrong direction, the most common cause is that the valve is installed in the wrong direction, allowing the flow to counteract the failure mode of the actuator.

6.3.7 Butterfly-Control-Valve Servicing

The general servicing guidelines in this section are not intended to supersede any butterfly-control-valve manufacturer's specific maintenance and servicing instructions—they are provided as general guidelines. The manufacturer's instructions should be followed exactly as they are intended.

If a valve must be serviced or repaired to remedy an internal problem, it must first be removed from the line. If the process is expected to stay in operation during this repair, bypass block valves should be used to channel the flow around the control valve in question. That portion of the line (or the entire pipeline if not bypassed) should be completely depressurized and decontaminated prior to removing the piping-flange bolting.

Before the piping-flange bolting is removed, the user should make sure that the disk is in the closed position. Otherwise, the valve-body subassembly will not be able to clear the piping. In fail-open configurations, the actuator, actuation system, or handwheel should be used to manually place the disk in the closed position during removal from the line. The valve should be fully supported with a hoist or other means before loosening the piping-flange bolting. Sometimes a tight space, along with adhesion of the gasket, may make it difficult to break the valve free from the piping. If this is the case, the user should not use the actuator or auxiliary handwheel as leverage, as this may damage the shaft. Also, a large screwdriver, crowbar, or wedge should not be used to work the surfaces loose, which may damage gasket surfaces. Instead, mechanical spreaders placed between the flanges may be necessary to release the valve-body subassembly. With the valve closed and completely perpendicular to the piping, the body subassembly should be carefully removed from the piping, being careful not to scrape or damage the gasket surfaces.

In most cases, the body subassembly is easier to disassemble when the actuator or actuation system is removed. To do this, the actuator must be completely depressurized. If a failure spring is involved, it must be decompressed for safety reasons. Any linkage attached to the shaft must be disconnected and the transfer-case connection removed from the body. Sometimes the clamping device used to hold on to the splined or flattened end of the shaft continues to hold the shaft firmly even after the bolting has been removed. In this case, the user should carefully wedge the halves of the clamp apart until the shaft is freed.

Unless the valve body is rubber- or elastomer-lined (which does not require a separate seat), the seat should be removed from the body for inspection or possible replacement. To do this, the seat retainer must be removed first. Some retainers are bolted into place, while others use a snap-ring or clip arrangement. Fasteners can become corroded over time and may require some effort to remove them. The seat should be inspected carefully for unusual damage or wear. Any worn or cut portion of the seat should be replaced, unless seat leakage is not a major concern or the valve is normally throttling in a midstroke position. During periodic maintenance of a concentric valve or an eccentric valve that closes often, the seats are often replaced.

If the valve body has a plug or flange and gasket to cover the blind end of the body (opposite the actuator), it should be removed before disassembling the shaft–disk connection. The packing should be decompressed by removing the gland flange and bolting, allowing the shaft to turn freely. With the shaft in the open position, the pins or keys used to hold the disk to the shaft should be removed. During this procedure, care should be taken not to damage or deform the shaft, pins, guides, or bearings. With the packing decompressed and the disk loose, the shaft can now be slid out the body, being cautious of the shaft surfaces that interface with the packing and guides or bearings. Depending on the design, this may be through the packing end or blind end of the body. While pulling the shaft out, the disk should be adequately supported to prevent it from falling loose and damaging critical surfaces. The user should be careful not to twist the shaft so that it binds on the guides or bearings. With the shaft and disk removed from the body, the bearings or guides can then be removed. Some bearings or guides are pressed into the body and thus will require mechanical force to remove them. If mechanical force is used, care should be taken not to damage the guides, unless they will be replaced. The disk seating surfaces should be checked for any signs of wear or damage.

At this point, the packing box should be disassembled so that the packing can be replaced. A wooden dowel is the best tool to push the packing out of the packing box from inside the body. Each ring of packing should be examined for signs of consolidation or extrusion.

Before reassembling the body subassembly, all parts should be thoroughly cleaned. Areas of severe oxidation should be removed and painted with an antioxidation paint. Any damaged or worn parts should be replaced by new or properly reconditioned parts.

The guides or bearings should be reinstalled in the body first. Some guides or bearings can be placed by hand, while others have a force-fit and will need to be pressed into place. Once the guides or bearings are placed, the disk should be positioned inside the body's flow area, being careful to align the disk correctly in relation to the shaft or diskstop. Large disks may need to be supported rather than held by hand. The shaft should then be inserted through either the packing end or the blind end of the body (depending on the design) and through the disk and the remainder of the body, being careful not to scratch or score the seating or guiding surfaces of the shaft. Using the pins, key, or other fasteners, the disk should be securely attached to the shaft. The fasteners should be firmly in place so that they will not work free during normal service. The shaft should be turned through the entire quarter-turn motion to ensure that no binding of guiding surfaces or misalignment of the disk with the body exists.

The packing box should be rebuilt using new packing rings. As each ring is placed on the shaft, care should be taken not to scar any of the soft sealing surfaces as the ring is placed over the end of the shaft. Some shafts may have sharp edges from the splines or flats. Once in place, each ring can be slid down the shaft into the body bore of the packing box. The user should ensure that the correct number and order of packing rings, spacers, antiextrusion rings, and packing are used for the required packing-box design. After the packing box is rebuilt, the gland flange and bolting can be reinstalled. If the bolting is corroded and cannot be adequately cleaned, it should be replaced, because corroded bolting may provide a false torque reading and subsequent incorrect packing compression. The gland-flange bolting should be tightened according to the torque value or procedure provided by the valve manufacturer. The user should be careful not to overcompress the packing, which can lead to excessive stem friction and wear. The end plug or flange can then be reinstalled on the blind end of the body, ensuring that new gaskets or O-rings are used.

The seat (or gasket if a metal seat is used) can now be reinstalled. The user should make sure it is correctly positioned, since some seat designs can only be installed one way, while others are universal. After the soft seat or gasket is in place, the retainer or metal seat should be installed using the fasteners needed to secure the seat assembly. If the fasteners are corroded, they should be replaced. The disk can now be turned into the seat to test for proper sealing function. Before remounting the actuator or actuator system onto the body subassembly, the user should verify that the disk is in the correct position (full-open or full-closed) according to the failure mode of the actuator. The actuator's transfer case can now be slid onto the shaft with the connection end of the shaft fully engaging the linkage connection. The actuator transfer case can then be bolted to the body subassembly and the linkage clamp tightened to secure the shaft. If the end of the shaft is supported by a guide in the transfer case, the user should make sure the shaft is supported by the gasket before operation. If necessary, any joints of the linkage should be lubricated. If the actuator is provided with limit-stops, the valve should be slowly stroked to ensure that disk travel is limited to the parameters of the application. In some cases, if the limit-stops are not properly set, the seat can be destroyed by the overtravel of the disk, or when a disk-stop is present, it can twist or shear the shaft. After the butterfly valve has been tested for a smooth quarter-turn travel, it may be reinstalled in the line and returned to service.

6.4 Ball Control Valves

6.4.1 Introduction to Ball Control Valves

Similar in many respects to the butterfly control valve, ball valves have been used for throttling service for the past two decades. As control valves, they have been adapted from the automation of simple on–off valves to automatic control valves designed specifically to accurately control the process. Improved sealing devices and highly accurate machining of the balls have provided tight shutoff as well as characterizable control. For the most part, they are used in services that require high rangeability. Ball control valves typically handle a rangeabilty of 300 to 1, notably higher than butterfly control valves that offer 20 to 1. Such high rangeability is permitted by the basic design of the regulating element, which allows the ball to turn into the flow without any significant side loads that are typical of a butterfly disk or a globe-valve plug.

Ball control valves are also well suited for slurry applications or those processes with fibrous content (such as wood pulp). The rotary action of the ball provides a shearing action against the seal, which allows for clean separation of the process during closure. The same process would clog or bind in a butterfly or globe control valve (which uses a regulating element or trim directly in the path of the process flow). Similar to the butterfly-valve design that features straight-through flow, a ball valve can be installed in a vertical pipeline (Fig. 6.29) to avoid the settling or straining of fibrous or particulate matter. A globe valve, on the other hand, allows heavier portions of the process to settle at the bottom of the globe body (horizontal line installations) or in the body gallery (vertical line installations).

Figure 6.29 Ball control valve mounted in a vertical line. (*Courtesy of Valtek International*)

Tight shutoff is a characteristic of ball control valves, since the ball remains in continual contact with its seal. With soft seals, ball control valves can achieve ANSI Class VI shutoff (bubble-tight) but have a limited temperature range. For higher-temperature ranges, metal seals are used although they permit greater leakage rates (ANSI Class IV). Ball valves are also capable of higher flow capacity than globe valves, and even butterfly valves where the presence of the disk in the flow stream can restrict the flow capacity. Because the flow capacity of a typical ball valve can be two to three times greater than that offered by a comparably sized globe valve, a smaller-sized ball valve can be used, which may be a significant economic consideration. Table 6.2 shows a

Table 6.2 C_v Comparisons Globe vs. Ball Valves*

Valve Size	Globe Valve (T-body style, flow-over-the-plug, full area trim, 100 percent open)	Ball Valve (Wafer-style, shaft downstream)	Percentage Increase
2-inch DN 50	46	104	126%
3-inch DN 80	104	275	164%
4-inch DN 100	179	445	149%
6-inch DN 150	355	844	138%
8-inch DN 200	606	1338	121%
10-inch DN 250	897	3180	255%
12-inch DN 300	1310	4150	217%

Data courtesy of Valtek International.

comparison of flow capacity between globe (both T and Y styles), butterfly and ball valves.

One major disadvantage of ball control valves is that as the valve throttles the geometry changes dramatically, providing lower pressure differentials, higher pressure drops, and an increasing chance of cavitation, although the straight-through flow style of ball valves provides a minimal pressure drop. Therefore if the service conditions are likely to result in cavitation, larger-sized ball valves may be required to provide higher differentials and to prevent a high-pressure drop from developing—defeating one of the purposes of ball valves, which is to use a smaller-sized valve with a large C_v. Using a larger ball valve also means that a good portion of the valve stroke will not be available for control purpose, utilizing the portion of the stroke closest to the closed position.

Two basic ball-valve designs are used today: the *full-port ball valve* and *characterizable-ball valve*. Similar in design to a manually operated on–off block ball valve, a full-port ball valve uses a spherical ball as the regulating element, characterized by a hole that is bored to the same inside diameter as the pipeline (Fig. 6.30). When the full-port ball valve is wide open, the flow continues unimpeded through this hole.

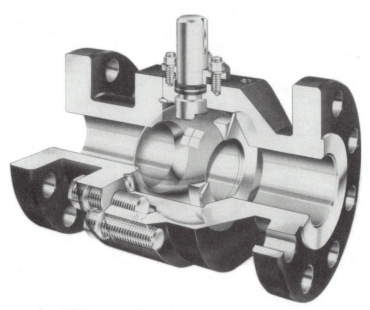

Figure 6.30 Full-port ball valve with floating seal. (*Courtesy of Vanessa/Keystone Valves and Controls, Inc.*)

Therefore, the flow does not impinge on a regulating element or trim, creating little (if any) pressure drop as well as minimal process turbulence. Although best utilized for on–off services, a full-port valve is rarely used for a pure throttling service because the sharp edges associated with the ball's bore may create noise, cavitation, erosion, and an increased pressure drop. Although a full-bore ball valve is often associated with on–off services, it is also applied where a pig or cleaning rod is used to clean out the interior of the pipeline. (This requires using a valve with straight-through flow that does not have a regulating element in the flow stream.) Because of the design limitations of full-port ball, a flow characteristic cannot be designed into the ball. The machining of orifice shapes other than circular is exceptionally difficult and expensive. The inherent flow characteristic associated with full-port valves is close to the equal-percentage characteristic, and any flow characteristic modifications must be made with a positioner cam.

The characterizable-ball valve (Fig. 6.31) does not use a spherical ball. Instead, it uses a hollow segment of a sphere that, when full-open, is turned out of the path of the process flow. This allows reasonably smooth flow through the valve body, although the contours of the

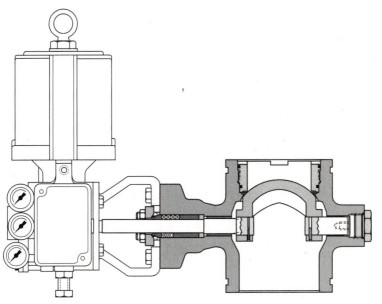

Figure 6.31 Characterizable-ball control valve. (*Courtesy of Valtek International*)

body and geometry of the characterized ball will take a small pressure drop and may create some turbulence. However, as the valve moves to a midstroke throttling position, the characterized ball moves into the flow path. The flow characteristic is cut into the ball with either a V-notch or a parabolic curve to provide the necessary flow per position. As the valve continues through the quarter-turn motion, this notch or curve becomes progressively smaller until the entire surface of the ball is exposed to the flow area, providing a full-closed position. The V-notch provides an inherent linear flow characteristic, which can become close to the equal-percentage characteristic when installed. The parabolic notch can be modified to meet specific flow requirements.

Ball control valves are typically found in sizes 1 through 12 in (DN 25 through DN 300) in pressure classes up through ANSI Class 600 (PN 100).

6.4.2 Ball-Control-Valve Design

Outside of the regulating element, ball control valves are similar in many regards to butterfly control valves: quarter-turn motion, rotary-action actuators, and packing boxes without wiper (lower) packing.

As described in Sec. 6.4.1, two basic ball-valve styles exist: the full-port ball valve and characterizable-ball valve. The regulating element of the full-port body subassembly features a spherical ball that is supported by one of two methods. The first is a *floating-seal* design (Fig. 6.30), similar to most manual ball-valve designs, where two full contact seals are placed on both the inlet and outlet ports, in which the ball is fully supported by these two seals without coming in direct contact with the body. The ball is connected to the shaft using a slip fit or other comparable connection. This connection must be extremely tight to avoid any mechanical hysteresis, especially in light of the continuous seal friction evident in this design. The basic advantage of this design is that a blind end bore is not required to support the nonshaft end of the ball. The disadvantage is that the sphere must have extremely tight tolerances to ensure constant contact at both seals. These seals are designed for more rigorous, heavy-duty service since they must both seal the flow and support the ball. Because this design is dependent upon the support of the seals, it is specified for general services featuring moderate pressures and temperatures.

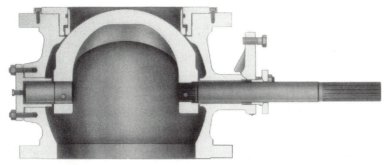

Figure 6.32 Trunnion-mounted segmented-ball valve. (*Courtesy of Fisher Controls International, Inc.*)

The characterizable ball is typically *segmented,* meaning that only a portion of the sphere is used instead of an entire sphere. The segmented ball includes only enough of the sphere to entirely close off the flow area plus enough ball surface to provide a seal. A segmented ball is normally *trunnion-mounted* (Fig. 6.32). With trunnion mounting, the ball is supported by both the shaft and the side opposite the shaft using another shaft or post, which can be separate or integral to the ball. Because support is not handled by a seal, trunnion-mounted balls are normally designed with one seal (although two-seal designs are available), which provides less friction between the ball and seal. Trunnion-mounted designs are best for more severe services where higher pressures and temperatures are involved.

Ball valves can be provided with either soft or metal seals. With soft seals, the elastomer seal is provided with a metal or hard-elastomer backup ring to apply continual pressure to the sealing surface, act as a backup in case the elastomer fails, and to provide additional wiping of sealing surfaces. With highly corrosive or nonsparking services—such as an oxygen application—metal backup rings are prohibited in favor of hard elastomers. If a metal seal is required because of temperature extremes, care must be taken to provide complementary metals so that galling or scoring does not take place. Metal seals require heat treatment and/or coating of the ball.

The style of the body determines how the seals are held in place in relation to the ball. With one-piece bodies, the ball is installed followed by the seal, which is held in place by a retainer. Most retainers are threaded into the body, allowing for minute adjustments of the retainer to increase or decrease the compression of the seal against the ball. This design balances the integrity of the seal versus increased ball–seal friction. Ideally, the retainer should not encompass the entire gasket region surface of the body face but should share it with the body. If the retainer does handle the entire seal, its compression of the seal will be affected by the piping forces. With uneven piping forces, they can create an uneven seal. To ensure uniform seal tightness, shims of varying width are often used between the retainer and the seal.

A few ball-valve bodies use two-piece designs in which the body is divided in half (much like a split-body globe valve), allowing for easier assembly and the use of a floating ball. The major drawback to using the two-piece design is that piping forces or process temperature can alter the seal tightness. As with all split-body designs another potential leak path is created at the joint between the two body halves.

Because the body's face-to-face is dependent upon the design of the body subassembly, that dimension varies from manufacturer to manufacturer. No overall standards have been established that all manufacturers adhere to, as opposed to ANSI/ISA Standard S75.15 or ANSI/ISA Standard S75.16 for globe-style valves. Because the ball-valve face-to-face is larger than the thin wafer-style body of the butterfly valve, yet smaller than the globe body, its body can be installed between piping flanges in some applications. When high temperatures or thermal cycling are present, the longer bolting between the piping flanges can result in lost compression through thermal expansion and cause leakage. Also, even if temperatures are moderate, the bolting associated with larger valves [8 in (DN 200) or larger] can stretch over time and cause leakage. For those applications in which a flangeless design is not practical, ball valves are also available with integral flanges or separable flanges. Integral flanges offer solid, one-piece structure integrity, while separate flanges offer lower cost (with alloy bodies) as well as easier installation when piping does not match up with the valve flanges.

The packing box is nearly identical to that found in butterfly control valves. Similar to other packing boxes, the bore is polished and deep enough to accommodate a wide variety of packing designs. As is the case with butterfly valves, the rotary quarter-turn action of the ball valve does not require a lower set of packing to wipe the shaft of any process. A typical packing box will include the packing set, an optional spacer and a packing follower (which is used to transfer the force of the gland flange to the packing). Unlike globe valves, an upper guide or bearing is not needed at the open end of a ball-valve packing box as the shaft is normally guided on each side of the ball. In some automated rotary-motion valves, the shaft is also guided by a bearing in the actuator's transfer case.

For machining simplicity of the trunnion-mounted design, the shaft bore is machined from both ends of the body, and a plug or flange cover (plus a gasket or O-ring) can be used to cover the bore opening opposite the packing box. If a threaded plug is used, it should not come in contact with the shaft, since the quarter-turn action of the shaft could unthread the plug, causing process leakage to atmosphere. Mounting holes are provided on the packing-box side of the body, allowing the transfer case of the actuator to be mounted. As with all automated rotary valves, the transfer case contains the linear-motion to rotary-motion mechanism that allows a linear-motion actuator to be used with a quarter-turn valve. The end of the shaft that fits into the transfer case is either splined or milled with several flats to allow for attachment of the linkage. The designs of common rotary actuators, actuation systems, and handwheels are detailed in Chap. 7.

6.4.3 Ball-Control-Valve Operation

As with all rotary-action valves, the ball valve strokes through a quarter-turn motion, with 0° as full-closed and 90° as full-open. The actuator can be built to provide this rotary motion, as is the case with a manual handlever, or can transfer linear motion to rotary action using a linear actuator design with a transfer case.

When full-open, a full-port valve has minimal pressure loss and recovery as the flow moves through the valve. This is because the flow passageway is essentially the same diameter as the pipe inside diameter, and no restrictions, other than some geometrical variations at the orifices, are present to restrict the flow. The operation of throttling full-port valves should be understood as a two-stage pressure drop process. Because of the length of the bore through the ball, full-port valves have two orifices, one on the upstream side and the other on the downstream side. As the valve moves to a midstroke position, the flow moves through the first narrowed orifice, creating a pressure drop, and moves into the larger flow bore inside the ball where the pressure recovers to a certain extent. The flow then moves to the second orifice, where another pressure drop occurs, followed by another pressure recovery. This two-step process is beneficial in that lower process velocities are created by the dual pressure drops, which is important with slurry applications. The flow rate of a full-port valve is determined by the decreasing flow area of the ball's hole as the valve moves through the quarter-turn motion, providing an inherent equal-percentage characteristic with a true circular opening. As the area of the flow passageway diminishes as the valve approaches closure, the sliding action of the ball against the seal creates a scissorslike shearing action. This action is ideal for slurries where long entrained fibers or particulates can be sheared off and separated at closing. On the other hand, globe-valve trim and butterfly disks do not have this shearing action and can only attempt to separate the fibers by pinching them between seating or sealing surfaces. In many cases, the fibers stay intact and do not allow for a complete seal, creating unplanned leakage.

At the full-closed position, the entire face of the ball is fully exposed to the flow, as the flow hole is now perpendicular to the flow, preventing it from continuing past the ball.

With the characterized segmented-ball design, only one pressure drop is taken through the valve—at the orifice where the seal and ball come in contact with each other. When the segmented ball is in the full-open position, the flow is restricted by the shape of the flow passageway. In essence, this creates a better throttling situation, since a pressure drop is taken through the reduction of flow area. As the segmented ball moves through the quarter-turn action, the shape of the V-notch or parabolic port changes with the stroke, providing the flow characteristic. Like the full-port design, the sliding seal of the characterizable ball provides a shearing action for separating slurries easily.

6.4.4 Ball-Control-Valve Installation

The ball control valve should be visually examined before installation to check for damage and correct features for that particular process. If possible, the signal and power source should be attached to determine correct function prior to installing the valve in-line. Be especially careful to keep hands and clothing away from moving parts—especially the regulating element, which can shear off fingers quite easily. If the valve is damaged, incorrectly built, or not functioning properly, it should be corrected before installation. The shipping box or packaging should contain installation and maintenance instructions, which should be kept for reference purposes. To avoid personal injury, large ball valves [2 in (DN 50) or larger] should not be lifted by hand—rather, lifting straps and a hoist or other mechanical lifting device should be used. Lifting straps should be secured around the valve-body subassembly as well as the actuator. The straps should avoid contact with the tubing or actuator accessories. With any large and heavy equipment, caution should be taken not to allow the valve to drop, which could damage the equipment or cause personal injury.

Before installation, the upstream pipe should be cleaned of any foreign material, such as dirt, welding chips, scale, or spent welding rods. If not removed from the line, this debris may become caught in the valve and possibly damage the ball and seal. With ball valves, side and/or top clearance must be available for the body to be removed from the piping. Side and/or top clearance must also be available to allow for disassembly and calibration of the actuator. Most certified dimensional drawings provide the necessary clearances for a particular valve. If limited space is available, in some situations the valve can be installed if enough room is available for the removal of the actuator. Once the actuator is removed, space is then available for the removal of the body subassembly. Although not ideal, this solution will allow for installation of the valve.

Ball valves can be installed in either a horizontal or vertical pipeline. Because of the shearing action of the ball, ball valves do not have problems with closing against particulates that have settled in a horizontal pipeline unlike butterfly valves. If an actuator is used to throttle the valve, it should be installed right-side up (unless space restrictions are present), allowing for easier maintenance and calibration. If space constrictions do not allow this orientation, the second best orientation is to install the valve with the positioner facing upward, which allows for easier access for calibration.

A flow plate or cast arrow on the body indicates flow direction for the ball valve. Unlike butterfly valves, the flow direction is not very critical to the correct function of the ball valve since the ball itself is outside of the influence of the direct flow path. Because the movement of the ball is influenced only slightly by the direction of the flow, the valve can fail using only the internal failure mode of the actuator. However, flow direction may be important in two cases: first, when the seal design is such that the upstream fluid pressure is needed to assist with the seal; and second, in erosive applications where the area most likely to be eroded is determined by the flow direction of the valve. The ball-valve manufacturer may prefer to have the shaft upstream to allow the erosion to occur in the flow area of the seal retainer, which is more easily replaced and less expensive than a new body.

With many throttling rotary-action valves, the actuator's failure mode is determined by how the actuator is mounted to the transfer case (Fig. 6.28). Mounting the actuator on one side of the transfer case will produce a downward motion toward closure, while mounting on the opposite side will produce a downward motion toward opening.

When installing the ball valve between piping-flange end connections, care should be taken to ensure that sufficient space exists between the pipe flanges to allow for the face-to-face of the body, the width of the gaskets, and some clearance to allow the valve to slide into place. If the piping is fixed and cannot be moved, too much clearance between flanges may make it hard to sufficiently tighten the flange bolting to prevent gasket leakage. A tight space means the piping flanges will need to be expanded to install the valve, making both installation and removal difficult. Before the gaskets and valve are positioned between the flanges, the user should ensure that the gasket surfaces of the body are free of foreign materials, which may cause leakage following installation.

If the valve requires through-bolting, the easiest method of installation is to loosely install two studs and nuts in the bottom two (or four in larger sizes) bolt holes, thus establishing a cradle for the valve body to sit in. The flange gaskets should be placed in position against the serrated valve body faces before slipping the body subassembly into place. With the valve faces concentric with the piping flanges, the remainder of the bolting can then be reinstalled.

To avoid leakage, the flange gaskets must be compressed equally, using a criss-cross pattern that involves tightening one bolt and then the one opposite. This method allows the flanges to stay square with the valve body and gaskets. Manufacturer or industry torque values should be met when tightening the bolting. Low-strength bolting is specified for flanged connections, while intermediate- or high-strength bolting is required for through-bolted joints. Long through-bolted joints associated with larger valve sizes should use high-strength bolting. Typical ANSI flange-bolting specifications are found in Table 3.2 in Chap. 3.

If a ball valve is equipped with integral flanges, the end flanges on the piping flanges should match up, avoiding any offset or misaligned hole patterns. Because the flanges on the body are fixed, the hole pattern lines up with the vertical axis of the valve. If the piping flanges are true to each other, yet not square with the vertical axis, the ball valve will lean slightly, which is only a minor concern with some users because most ball valves are designed to operate in nearly any position. Some minor play does exist between the size of the flange hole and the required stud, which can be used to straighten the misalignment. A major problem can occur if one of the two flanges is misaligned with the vertical axis. In this case, installation may be difficult or impossible unless the misaligned flange is corrected. Rather than cut out a flange and weld it correctly in place, some users widen the holes on either the piping flange or valve flange to allow for alignment. For the most part, this practice is discouraged because it weakens the structural integrity of the flange.

Some ball valves can be designed with separable flanges, which can be rotated to easily match up with misaligned piping flanges. However, the user should take care to ensure that the flanges are sufficiently tightened to prevent the valve from rotating, especially since the weight of the actuator creates a side load. To ensure that rotation does not occur, some users prefer to apply several tack welds between the separable flange and the body.

The actuator, actuation system, or positioner can now be connected to the instrumentation signal and power supply. The actuator or actuation system has a sticker or tag that indicates the maximum power supply (either pneumatic, hydraulic, or electric) that the unit is designed to handle. If a pneumatic actuator is used, a pressure regulator may be necessary to lower the plant air supply to the required levels—especially with diaphragm actuators. With spring cylinder actuators, high air-supply pressure up to 150 psi (10.3 bar) can be handled without a pressure regulator. Unless the plant air supply is exceptionally clean, an air filter should be installed before the actuator or positioner to prevent water, oil, dirt, or other debris from entering and fouling the operator. Most air filters are designed with a filter cartridge, which should always point downward to allow entrapment of any foreign materials.

When the valve is fully stroked, the stroke indicator should be observed on the transfer case or the actuator itself to see if the shaft turns in a smooth, rotary motion. If the shaft moves erratically, an internal problem with the ball or seal, shaft guides or bearings, packing tightness, linkage, or actuator could exist. At this point, valve operation should be discontinued until the cause is thoroughly investigated and corrected. During the initial stroking of an automated ball valve, the calibration of the range should be checked by making the entire instrument signal change—from full-open to full-close—such as from 3 to 15 psi (0.2 to 1.0 bar) for pneumatic signals or from 4 to 20 mA for electrical signals. All actuator connections (either electrical or pneumatic) should be visually inspected, looking for any air leakage or loose connections.

Sometimes packing consolidation occurs between the factory and the final installation. If leakage is indicated through the packing, the packing box should be retightened according to the manufacturer's procedures. Caution should be taken not to overtighten the packing, since this will cause increased stem friction, erratic shaft motion, and premature wear of the shaft and packing. If the process has high temperatures or wide temperature swings, packing tightness should be rechecked periodically during the first few days of installation. When high temperatures exist, the pipe-flange bolting and flange gaskets should be checked for possible leakage from thermal expansion. This is particularly important if straight-through bolting is used with the longer face-to-face associated with ball valves. If the actuated valve is designed to fail to or in a particular position, the power supply should be disconnected from the actuator and the valve should fail to the desired position (open, closed, or at last position).

6.4.5 Ball-Control-Valve Preventative Maintenance

The service life of a ball control valve is improved by periodically checking the valve for proper operation, as well as performing preventative maintenance. If problems exist with the internal parts of the body subassembly, the valve must be removed from the line for inspection and repair. In this case, the process will either need to be interrupted or blocked around the valve. Servicing should be done according to the specific methods and procedures outlined by the valve manufacturer, although a number of general guidelines are outlined in this section.

Some troubleshooting can take place before the valve is removed from the line. First, the piping-flange connections should be checked for signs of leakage. If leakage has occurred, the gaskets may have failed or thermal expansion may have affected the bolting tightness. The packing box, body end plate or plug, and any other pressure-retaining connections should also be checked for process leakage. With liquid processes, leakage is evident if stains or moisture are found, while gas leakage is discovered by using a "sniffer" sensing device.

If leakage is found between the piping flanges and the body, the piping-flange bolting should have been tightened in an even manner until the leak is eliminated. If that does not totally eliminate the leak, the gasket may have failed and will need replacement. If leakage is found at the packing box, the gland-flange bolting should be tightened to the manufacturer's recommended torque value or procedure until the leakage stops. The user should be careful not to overtighten the packing as this may crush it or cause consolidation and extrusion.

Following inspection, the valve should be thoroughly cleaned. If leakage has occurred, any process buildup should be cleaned from the valve. If metal surfaces show signs of severe oxidation, they should be scraped or brushed to remove the oxidation and repainted using a rust inhibitor. Some environments, such as locations near seawater, are more corrosive than others and special care should be taken to ensure that the life of the valve is not shortened by oxidation. Also, preventative measures should be taken to ensure that process drippings from nearby lines are not falling on the ball valve or the actuator. If so, those piping leaks should be corrected or the valve itself shielded from the process drips.

If possible—even if it is used for midrange throttling—the valve should be fully stroked so that the user can be assured that it is capable of a full and smooth stroke. If the shaft rotates erratically, an internal galling, packing tightness or actuator or positioner problem may be present. If graphite packing is used for a high-temperature service, such packings are known to grab the shaft even when properly compressed.

If the ball valve has been removed from the line, it should be inspected for any obvious porosity or fractures that may have occurred during service, which may lead to more serious leakage problems. The inner flow area (including the ball, retainer, and inside of the body) should also be inspected for signs of erosion or cavitation. Sometimes if the crack or porosity is minor, the affected area can be ground out and filled with a weld. If this is done, the user should make sure that it is done according to established industry methods so that the integrity of the pressure-retaining vessel will not be jeopardized.

If the process allows (or if the ball valve is blocked from the process), the power supply from the actuator or actuation system should be removed and the failure action verified. Most rotary valve actuators have transfer cases with a cover that can be removed, allowing the linkage and bolting to be checked. Excessive line vibration can occasionally loosen linkage connections, leading to increased mechanical hysteresis or malfunction of the valve itself. Some valve designs use the transfer case cover to help guide the end of the shaft. With that particular design, the valve should not be stroked when the cover is off, since the shaft is unsupported and this may cause undue stresses on the shaft.

The actuator should be checked for any leaks of the power supply. With pneumatic actuators, air leaks can be detected using a soap solution around the tubing connections. If the pneumatic actuator is equipped with an air filter, the cartridge should be checked and replaced, if necessary. If the valve uses a hydraulic or electrohydraulic actuator, it should be visually inspected for any hydraulic fluid leaks. Manual handwheels should be checked for adequate lubrication around the handwheel stem. If a lubrication fitting exists on the handwheel assembly, lubrication should be added as a routine matter.

6.4.6 Ball-Control-Valve Troubleshooting

Generally, because of their complexity and wider application, ball control valves are usually more prone to problems than their manual-ball-valve counterparts, especially when actuation is involved. This section deals with troubleshooting ball control valves, explaining the possibilities for a particular malfunction—although these guidelines are not intended to replace the manufacturer's specific instructions for troubleshooting or repairing the problem.

The most common malfunction of a ball valve involves leakage through the valve seal. In this case, a leakage malfunction is defined as any leakage beyond the desired leakage rate. A worn seal or ball will show leakage only after the valve has been operating satisfactorily for a reasonable period. Probable causes are process erosion, mechanical failure of the seal, frictional wear between the two mating seating surfaces of the seal and the ball, damage from a foreign object caught between the seal and the ball, or cavitation damage to the ball. The soft seal or seat gasket (metal seat) may also have failed, causing leakage to occur through the joint between the body and the seal retainer. If the shaft guides or bearings are worn, the ball may be slightly off center from the seal. This misalignment will also result in increased leakage and increased wear of the regulating element. Leakage can also be caused by a galled shaft, preventing full motion of the ball. In other cases, an auxiliary handwheel or limit-stop may be incorrectly positioned, which may inadvertently limit the travel of the ball.

Persistent leakage from the piping flanges, even after tightening, is indicative of a parallel misalignment between the upstream and downstream piping or can be caused by a flange gasket failure. In some cases, leakage occurs if the gaskets or gasket surfaces are dirty or if the valve was installed with particulates or other foreign objects on the gasket surfaces. If tightening the packing box fails to stop the leakage from that region, the packing may have completely consolidated or extruded. When this happens, the packing cannot provide a full-contact seal around the shaft or body bore despite additional compression. Therefore, the only option is to rebuild the packing box using new packing. Some packings, such as braided graphite rings, are known to become more resistant as more thrust is applied, creating a very erratic control situation. One problem with high-temperature applications is that the recommended packing material—graphite—forms a nearly perfect seal with the shaft, creating greater friction than the softer, more pliant packings (such as PTFE). Some packing materials are more abrasive than others, such as graphite. Because a ball valve has a quarter-turn motion, abrasive packing has a tendency to wear the shaft where it makes continual contact. If the packing is highly abrasive, these frictional losses can eventually cause leakage unless the packing is tightened. This wear will lead to the eventual replacement of the shaft.

As with butterfly control valves, another common problem associated with ball valves is erratic or impeded shaft movement. Generally the most common cause is overtightened packing, which causes the packing rings to bind to the shaft and significantly increase breakout torque. Other possible causes of erratic shaft travel are tight or misadjusted linkage between the shaft and actuator, actuator failure, or worn or damaged shaft bearings.

If the shaft moves with the signal, but the valve does not respond in kind, the shaft–ball connection has failed or the shaft has been sheared into two pieces. Poor throttling function or limited closure of the valve may be caused by the pneumatic actuator with a low air supply or a malfunctioning positioner. When an actuator fails, such as when a diaphragm bursts or stem seal fails, the actuator will not be able to hold a position and the valve will move to the failure position. This can also occur when the linkage between the actuator and shaft fails. If the valve is providing poor flow control, the most likely causes are a malfunctioning positioner or actuator, or an incorrect flow characteristic cam in the positioner. As discussed in Sec. 2.2, the valve's flow characteristic can be affected by piping and pump effects that will change the inherent flow characteristic to an installed flow characteristic.

6.4.7 Ball-Control-Valve Servicing

The general servicing guidelines in this section are not intended to supersede any ball-control-valve manufacturer's specific maintenance and servicing instructions. The manufacturer's instructions should be followed exactly as they are intended.

If an internal problem is found, the ball valve will most likely require disassembly to remedy the problem. The valve will most likely need to be removed from the line. If the process cannot be interrupted, bypass block valves should be used to divert the flow around the ball valve. That portion of the line (or the entire pipeline, if not bypassed) should be completely depressurized and decontaminated, if necessary, prior to removing the piping-flange bolting. If the ball valve is automated, the actuator power supply (air, hydraulic fluid, or electric power) should be removed and the valve allowed to move to its failure position.

Before the piping-flange bolting is removed, the valve should be completely supported using a hoist or another mechanical device, followed by removal of the piping-flange bolting. If the valve is fixed in place, an expanding device may be required to separate the piping flanges slightly to allow for removal of the valve body. The actuator should not be used as a lever to pry the flanges apart, because this could place strain on the shaft, linkage, etc. As the ball valve is removed from the pipeline, the gasket surfaces should be kept square with the pipeline to avoid scoring or damaging those surfaces.

The valve actuator may need to be removed at this point, because the body subassembly is generally easier to disassemble without it. To do this, the actuator should be completely depressurized if it is pneumatically or hydraulically driven. If the actuator has a failure spring, it should be decompressed for safety reasons. The linkage bolting that connects the shaft to the actuator will need to be loosened, after which the operator can be removed from the body. Sometimes the clamping device used to hold on to the splined or flattened end of the shaft continues to hold the shaft firmly even after the bolting is removed. In this case, the halves of the clamp can be spread by a wedge until the shaft is freed.

At this point, the seal should be removed from the body. If the body is a one-piece design, the seal retainer should be removed to gain access to the seat itself. Seal retainers are fastened to the body by threads or bolting. If threaded, a tool is sometimes necessary to turn the retainer out of the body. This is especially true with larger valves that have retainers with large outside diameters, which require some torque to overcome the large frictional forces. If the retainer is held in place by bolting, the bolting should be removed. The main problem with bolting is that process corrosion may create removal problems, resulting in the technician having to drill out the bolting and retap the threaded hole.

With split-body designs, the body bolting will need to be removed to divide the body subassembly into two halves, exposing the seal (or seals if the ball valve uses a floating-ball design) and the ball. If the full-bore ball valve has two seals, disassembling the body will expose one seal. At that point, the ball will need to be removed to reveal the other opposite seal. In some designs, the packing and the shaft will need to be removed first to allow for removal of the ball. In other designs, where the stem is installed through the inside of the body (to prevent blowout), the ball must be turned to the closed position to allow the ball to slide away from the key portion of the stem.

After the seal is exposed, careful attention should be paid to the order and orientation of each part (seal, inserts, gaskets, O-rings, etc.) as it is removed. Some seal retainers use O-rings as sealing devices, which should be examined for aging or damage and replaced if necessary. The packing should be completely decompressed by removing the gland flange and bolting, allowing the shaft to turn freely. The shaft should be turned so that the connection between the ball and shaft is exposed. The pins, key, or bolting used to hold the ball to the shaft should be removed, being careful not to damage or deform the shaft, pins, or guides or bearings. A safe practice with larger balls is to support them (other than with hands) to prevent them from falling loose during disassembly and damaging critical seating surfaces or injuring nearby personnel. With the packing decompressed and the ball loose, the shaft should be slid out the body, being careful of the shaft surfaces that interface with the packing and guides or bearings. If the ball is trunnion-mounted, the supporting shaft should also be removed through the blind end of the body. The user should not twist the shaft so that it binds on the guides or bearings. After the shaft is removed, the ball can now be removed from the body. The user should be cautious not to scratch any polished surface during removal. Scratches on the ball will lead to increased leakage and seal wear. The entire ball sealing surface should be checked for any signs of wear, erosion, or damage.

With the ball and shaft removed from the body, the bearings or guides can now be removed. Some bearings or guides are pressed into the body and thus require mechanical force to remove them. If mechanical force is used, the user should be careful not to damage the body or the guides unless replacement of the guides is necessary. A wooden dowel of similar size is a convenient tool to push the packing out of the packing box from inside the body. As each ring is removed, it should be inspected for signs of consolidation or extrusion. If extensive extrusion is discovered, the packing box should be rebuilt using antiextrusion rings.

Before reassembling the valve-body subassembly, all parts should be thoroughly examined and cleaned. Areas of severe oxidation should be removed and painted with an antioxidation paint. Any damaged or worn parts should be replaced. If the ball has minor scratches, light rubbing of a polishing compound or cloth may be possible to minimize the damage without compromising the seal. The shaft, especially around the packing and guiding surfaces, should be free of damage and not sufficiently worn to prevent a good packing seal while allowing free rotation of the stem.

The guides should be reinstalled, using a mechanical press if necessary. The ball is now repositioned inside the body so that it intersects the shaft and post (trunnion-mounted design). The shaft and post can now be reinserted through the body, intersecting the ball. Using the necessary pins, keys, or bolting, the ball can now be reattached to the shaft and post. The user should note that some posts rotate while others are stationary. Regardless of the design, the moving part of the post should be thoroughly lubricated before installation. If a post is used with a trunnion-mounted ball, the end plug or flange should be reinstalled to seal the blind end of the body. New gaskets or O-rings are recommended to ensure pressure retention.

If a floating ball is used, the seats must be replaced inside the two halves of the body, followed by the shaft. Some shafts are mounted through the inside of the valve, because the shaft is designed with a special step to prevent blow-out. At the end of the shaft is an integral key, which must be turned parallel with the body, allowing the ball to be slipped into place using the notch at the top of the ball. If the shaft does not have this blow-out feature, the ball is installed first, followed by the shaft through the outside of the body, making sure the slot between the ball and shaft lines up. Once the ball is in place, the two halves of the body can be reunited using the body bolting.

After the ball and shaft are in place, the packing box should be rebuilt using new packing rings. The user should take precautions not to damage the individual rings as they are placed over the end of the shaft, which may have sharp edges from the splines or flats. Each ring should be slid down the shaft into the body bore of the packing box. The user should ensure that the correct number and order of packing rings, spacers, antiextrusion rings, and packing are used. After the packing box has been packed, the gland flange and bolting should be reinstalled and tightened properly. If the bolting is corroded and cannot be cleaned up adequately, it should be replaced to avoid a false torque reading. The user should be careful not to overcompress the packing, which can lead to excessive stem friction and wear.

On characterizable segmented balls that use seal retainers, the seal should be replaced, using the correct order and orientation of the seal, inserts, O-rings, etc. The retainer should then be reinstalled. The effectiveness of the seal is sometimes dependent upon the retainer's compression of the seal. When this is the case, the manufacturer provides torque values for a thread retainer or retainer bolting. With a threaded retainer, a compatible grease should be used on the threads to ensure easy assembly and future disassembly.

Before remounting the actuator or actuator system onto the body subassembly, the user should verify that the ball is in the correct position (full-open or full-closed) according to the failure mode of the actuator. The actuator can now be remounted to the valve. The connection end of the shaft should fully engage the linkage. The actuator should be then securely bolted to the body subassembly and the linkage clamp tightened to the shaft. If the end of the shaft is supported by a guide in the transfer case, the user should ensure that the shaft is supported by the guide before operation occurs. All moving parts of the linkage should be lubricated. If the actuator is provided with limit-stops, the valve should be slowly stroked to ensure that ball travel is limited to the parameters of the application. After correct operation is verified, the valve is then reinstalled in the line and returned to service.

6.5 Eccentric Plug Control Valves

6.5.1 Introduction to Eccentric Plug Control Valves

One control valve design that is growing in demand is the *eccentric plug valve* (sometimes called *eccentric rotating plug valve*), which combines many of the positive aspects of the globe, butterfly, and ball valves. In simple terms, the eccentric plug valve is a rotary valve that uses an offset plug to swing into a seat to close the valve, much like an eccentric butterfly valve. However, the eccentric movement of the plug swings out of the flow path, similar to a segmented-ball valve. Overall this design provides minimal breakout torque, as well as tight shutoff without excessive actuator force. Figure 6.33 shows the internal construction of an eccentric plug valve.

Eccentric plug valves can typically handle pressure drops from 1450 psi (100 bar). The eccentric motion also avoids water-hammer effects and the poor rangeability inherent with butterfly valves. Unlike a ball valve where the ball is in constant contact with the seal, the plug lifts off the seat upon opening. Seat contact and partwear only occur when the valve is closed (Fig. 6.34)—a feature similar to globe-valve trim.

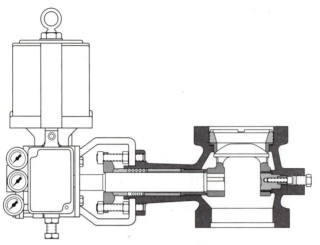

Figure 6.33 Eccentric plug valve. (*Courtesy of Sereg/Valtek International*)

Because the plug swings out of the flow area—as does a segmented-ball valve—it allows for greater flow capacity and avoids erosion from the process.

With the stability of the plug design, eccentric plug valves provide exceptional stability, providing rangeabilty of greater than 100:1, compared to 50:1 for globe valves and 20:1 for butterfly valves. Only the ball control valve has better rangeability (up to 400:1). Because the shaft and plug do not directly intersect the flow, the flow capacity is slightly

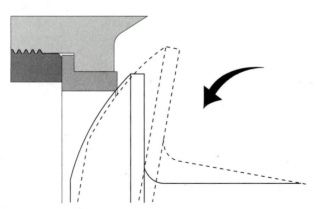

Figure 6.34 Seating path of eccentric plug design. (*Courtesy of Sereg/Valtek International*)

less than ball valves but is better than most high-performance globe and butterfly valves. Its design permits a reasonable pressure drop to be taken across the valve. Eccentric plug valves are best applied in applications with moderate pressure drops. In normal applications, the eccentric plug valve operates equally well in either flow-to-close or flow-to-open applications. The design of the plug permits the flow direction to assist with the closure or opening of the valve. As the eccentric plug valve opens, the flow characteristic is an inherent linear characteristic. With the regulating element outside on the outside boundaries of the flow, very little process turbulence is created.

Eccentric plug valves are typically available in sizes from 1 in (DN 25) to 12 in (DN 300), in ANSI Classes up through Class 600 (PN 100), and handle temperatures typically from $-150°F$ ($-100°C$) to $800°F$ ($430°C$).

6.5.2 Eccentric-Plug-Control-Valve Design

The body design of an eccentric plug valve is very similar to a characterizable segmented ball valve in many aspects. The valve body and packing box are similar in shape and function, although the shaft alignment with the seal is different. With a ball valve, the centerline of the shaft is aligned exactly with the seal so that the ball is always in direct contact with the seal, whereas the shaft of an eccentric plug valve is slightly offset from the seat. This offset keeps the rotating plug away from any seating surfaces until closure occurs. Overall, this is similar in concept to the offset of an eccentric and cammed disk in high-performance butterfly valves. With fail-closed situations, the offset design positions the plug correctly upon failure, reducing the actuator failure spring requirements.

Although a segmented ball and an eccentric plug look similar at first appearance, each is designed differently. Where the ball is spherical in design, the plug is designed more like the plug head of a globe valve that is attached at a right angle with the shaft. The contour of the face of the rotary plug is similar to a modified quick-open plug contour in a globe valve, although the major difference is that the contour of the eccentric plug is also the seating surface. The seat construction is similar to the seat retainer in a ball valve, which can be threaded in place. Newer designs use a two-piece construction featuring a floating, self-centering seat with a threaded seat retainer that, when tightened, fixes the seat in place. On the other hand, one-piece seats have difficulty achieving tight shutoff because of the possibility of misalignment between the plug and seat. Seats can be either metal (providing ANSI Class IV shutoff) or provided with a soft seat elastomer (providing ANSI Class VI shutoff).

One design attribute of the eccentric plug valve that is similar to globe valves is its ability to provide reduced trims by simply changing the seat to one with a smaller opening. Because the eccentric plug has one large seating surface, it can be used with a variety of smaller seats, providing a reduced trim option that is not normally available in other rotary valves.

Eccentric plug valves utilize straight-through bolting or flanged end connections.

6.5.3 Eccentric-Plug-Control-Valve Operation

The eccentric ball valve strokes through a quarter-turn motion, with 0° at full-closed and 60° to 80° at full-open. Maximum rotation (80°) is preferred because it provides increased controllability and resolution. When less than full rotation is required, some actuators have limit-stops that can prevent the full motion.

When the valve is in the full-open position, the plug is located nearly perpendicular to the seat (Fig. 6.35) and parallel to the flow. As the flow moves through the body, it is restricted by the diameter of the seat and geometric shape of the plug, taking a reasonable pressure drop.

In fail-open applications, the flow assists the opening of the plug since the shaft is downstream from the flow and the plug swings with

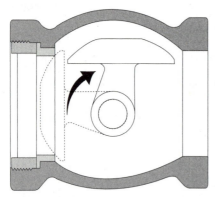

Figure 6.35 Eccentric plug in the open position. (*Courtesy of Sereg/Valtek International*)

the flow until it is perpendicular to the seat. The process flows through the seat, taking a small pressure drop, and then slightly recovers inside the body. The majority of the flow moves through the center of the valve body and the horseshoe-shaped opening of the plug, encountering minimal flow resistance. As the flow exits the valve body, the pressure recovery is completed. As the valve begins to close, the plug moves against the flow, restricting the flow by degrees until the plug is approaching the closed position. At that point, the offset shaft aligns the plug exactly with the seat, seating surfaces meet, and the valve closes.

In fail-closed applications, the shaft is upstream from the flow and the plug must open against the flow, moving perpendicular to the seat. Flow moves through the body and the plug opening to the seat, taking a small pressure drop at the plug opening and a larger pressure drop at the seat, with pressure recovering in the downstream piping. As the valve fails, the direction that the plug swings to close is the same as the flow, using the flow pressure to assist with the closure. A feature unique to automatic rotary valves in general is the ability to mount the valve on either side of the pipeline so that the shaft orientation (shaft upstream or shaft downstream) and the failure mode (fail-open and fail-closed) can operate in tandem with the air-failure action of the actuator. Figure 6.28 is a good reference illustration for showing the four common orientations (fail-closed, shaft upstream, and air-to-open; fail-open, shaft upstream, and air-to-close; fail-open, shaft downstream, and air-to-close; and fail-closed, shaft downstream, and air-to-close).

6.5.4 Eccentric-Plug-Control-Valve Installation

Before the eccentric plug valve is installed, it should be visually examined to check for damage and for proper application. The user should check for correct actuator and valve function by attaching the signal and power source and stroking the valve. Hands and clothing should be kept away from all moving parts, especially the plug and seat. Damage or improper operation is best repaired before the valve is installed. The shipping box or packaging usually contains installation and maintenance instructions, which should be kept for future reference purposes.

To avoid personal injury, large eccentric plug valves [3 in (DN 80) or larger] should be lifted using straps and a hoist or similar mechanical lifting device, with the straps secured around the valve-body subassembly as well as the actuator. The user should make sure the straps avoid contact with the tubing or actuator accessories to prevent damage. When lifting the valve, caution should be taken not to allow the valve to drop, which could damage the equipment or cause personal injury.

Special care should be taken to ensure that the upstream pipe is cleaned of any foreign material, such as dirt, welding rods or chips, and scale. If such materials are not removed before start-up, they may become caught in the valve and may damage the plug and seat surfaces. As with all rotary valves with actuators or actuation systems, side and/or top clearance must be available for the body to be removed from the piping. Side and/or top clearance must also be available to allow for disassembly and calibration of the actuator. To determine the disassembly clearances, the valve's certified dimensional drawing should be referenced. Eccentric plug valves can be installed in either a horizontal or vertical pipeline. If an actuator is used to throttle the valve, it should be installed right-side up (unless space restrictions are present), allowing for easier maintenance and calibration. If space constrictions do not.allow this orientation, the second best orientation is to install the valve with the positioner facing upward, which allows for easier access for calibration.

The eccentric plug valve body's flow orientation is indicated by a metal plate or cast arrow on the body. Maintaining the correct flow direction is critical to the failure mode for the eccentric plug valve, since the flow direction must coincide with the correct shaft orientation as well as the failure mode of the actuator. With many eccentric valves, the actuator's failure mode is determined by how the actuator is mounted to the transfer case. Mounting the actuator on one side of the transfer case may produce a downward motion toward closure, while mounting on the opposite side will produce a downward motion toward opening.

When the eccentric plug valve is installed between the piping flange end connections, sufficient space must be provided between the piping flanges to allow for the face-to-face of the body and the width of the gaskets, as well as some additional clearance to allow the valve to slide into place. If the piping is fixed and cannot be adjusted, insufficient space for the valve's face-to-face means the piping flanges will need to be expanded to install the valve, making both installation and removal difficult. Too much space between flanges may make it hard to sufficiently tighten the flange bolting to prevent gasket leakage. Before the gaskets and valve body are positioned between the flanges, the gasket surfaces of the body should be examined by the user to make sure they are free of foreign materials such as grit or loose welding scale. Any contamination or irregularity on the gasket surfaces may eventually cause leakage.

If the eccentric plug valve is flangeless and requires through-bolting, two studs and nuts should be installed in the bottom two bolt holes (or four in larger sizes), thus providing a cradle to support the valve body. The flange gaskets should be positioned against the serrated valve-body faces before slipping the body subassembly into place. With the valve faces concentric and true with the piping flanges, the remainder of the bolting can then be installed. To avoid leakage, the flange gaskets should be compressed equally, using a criss-cross pattern by tightening one bolt and then the opposite bolt. This method allows the flanges to stay square with the valve body and gaskets. The manufacturer's torque values should be achieved when tightening the bolting. Low-strength bolting is specified for flanged connections, while intermediate- or high-strength bolting is used with through-bolting. Long through-bolted joints associated with larger valve sizes (6 in or DN 150) should use high-strength bolting. Table 3.2 in Chap. 3 provides typical ANSI flange-bolting specifications for low- and intermediate-strength studs. If the eccentric plug valve is equipped with integral flanges, the end flanges on the piping flanges should match up evenly. If possible, the piping flanges should not have offset or misaligned hole patterns. Because the flanges on the body are fixed, the valve is manufactured so that the hole pattern lines up with the vertical axis of the valve. If the piping flanges are true to each other, yet not square with the vertical axis, the eccentric plug valve will lean, which is only a minor concern because most eccentric plug valves are designed to operate in nearly any position. Some minor play does exist between the inside diameter of the flange hole and the outside diameter of the stud, which allows some ability to correct the misalignment. If one of the flanges is misaligned with the vertical axis, installation may be difficult unless the misaligned flange is corrected. The user has the difficult option of cutting the flange from the pipe and then welding it correctly in place, or widening the holes on either the piping flange or valve flange, to allow for alignment. Generally, this latter practice is discouraged because it weakens the structural integrity of the flange.

With some rotary-valve designs, separable flanges may be an option. Because they can be rotated individually, separable flanges can easily correct the problem of misaligned piping flanges. However, the user should take precautions to ensure that the flanges are sufficiently tightened to prevent the valve from rotating. This is especially critical with rotary valves because the actuator is usually installed to the side of the body, creating an unequal side load. To ensure that rotation does not occur, several tack welds may be used between the separable flange and the body. However, this should only be done if the two materials—the flange and the body—are compatible for welding.

Finally, the actuator, actuation system, or positioner should be connected to the instrumentation signal and power supply. The actuation system indicates the maximum power supply (pneumatic, hydraulic, or electric) that it is designed to handle. If a pneumatic diaphragm actuator is used, a pressure regulator may be necessary to lower the plant air supply to the required level. With spring cylinder actuators, high air-pressure supply up to 150 psi (10.3 bar) can be handled without a pressure regulator. Unless the plant air supply is exceptionally clean, an air filter should be installed before the actuator (or positioner, if necessary) to prevent water, oil, dirt, or other debris from entering and fouling the actuator or positioner. Most air filters are designed with a filter cartridge, which should always point downward to entrap and collect any foreign materials.

Following the installation of the eccentric plug valve and the attachment of the power supply, the valve should be checked for correct operation before being placed in full service. The first step is to stroke the valve through its full range. If the actuator has an auxiliary handwheel, the handwheel should also be used to stroke the valve through the full stroke. In either case, the stroke indicator should be observed on the transfer case to see if the shaft turns in a smooth, rotary motion. If the shaft moves erratically, an internal problem with the plug and seat, shaft guides or bearings, packing tightness, linkage, or actuator may be occurring. If a problem exists, the user should discontinue stroking the valve until the problem can be thoroughly investigated and fixed.

During the initial stroking of an automated eccentric plug valve, the signal range and calibration can be checked by making the entire instrument signal change (full-open to full-close). The most common signal ranges are 3 to 15 psi (0.2 to 1.0 bar) for pneumatic signals or 4 to 20 mA for electrical signals. All actuator connections (either electrical or pneumatic) should be visually checked for any air leakage or loose connections and corrected if necessary. An easy method of checking pneumatic air-line connections is by applying a soap solution to the connection itself.

Over time, packing can consolidate while under compression, even if the valve has not been in use. If process leakage at the shaft is evident, the packing box should be retightened according to the manufacturer's procedures. Particular caution should be taken not to overtighten the packing, since this will cause increased stem friction, erratic shaft motion, and premature wear of the shaft and packing. When the service conditions include high operating temperatures or wide temperature swings, packing tightness should be rechecked periodically during the first few days of operation. Pipe flange bolting and flange gaskets should be checked for possible leakage due to thermal expansion. This is particularly important when straight-through bolting is used with the longer face-to-face associated with eccentric plug valves.

Some automatic control valves are designed to fail to a particular position upon loss of the actuator's power supply or signal. To check for correct failure function, the power supply should be disconnected from the actuator and the valve's failure action observed. During this test, hands and tools should be kept away from the closure element.

6.5.6 Eccentric-Plug-Control-Valve Preventative Maintenance

The life of an eccentric plug valve is improved dramatically by occasionally checking the valve for proper function and by performing preventative maintenance. With most rotary valves, if internal problems occur within the body subassembly, the valve will need to be removed from the line for detailed inspection and repair. If the service cannot be interrupted during this procedure, the system will need to be blocked around the valve—if block valves are provided for that purpose.

Servicing should be done according to the specific methods and procedures outlined by the eccentric plug valve manufacturer. A number of general guidelines and troubleshooting procedures are outlined in this section for consideration. The piping-flange connections can be checked for signs of leakage. If leakage has occurred, the gaskets may have failed because of a flaw in the gasket material, improper installation, or thermal expansion affecting the tightness of the bolting. All pressure-retaining connections—the packing box, body end plate or plug, etc.—should also be visually checked for process leakage, if possible. Leakage from some gas services may require the use of a "sniffer" sensing device.

When leakage is discovered between the piping flanges and the body, the piping-flange bolting should be tightened in an even manner until the leak is eliminated. If tightening does not eliminate the leak, the line gasket has probably failed, and the piping connection will need to be disassembled and rebuilt using a new gasket. If leakage is found at the packing box, the gland flange bolting should be tightened to the manufacturer's recommended torque value or procedure until the leakage stops. Special precautions should be taken not to over-tighten the packing. Overtightening may crush the packing, or cause consolidation and extrusion.

If leakage has occurred, any process buildup should be cleaned from the valve. If metal surfaces show signs of severe oxidation, they should be scraped or brushed to remove the oxidation and repainted using a rust inhibitor. Some environments, such as applications near seawater, are more corrosive than others and special care should be taken to ensure the life of the valve is not shortened by oxidation. Also, nearby piping and equipment should be examined to ensure that process drippings from nearby lines are not falling on the eccentric plug valve or the actuator. If occurring, those piping leaks should be corrected or a shield placed over the valve to protect it.

The rotary actuator should be checked for full and smooth stroke operation, if possible, even if the valve is used for midrange throttling. If the shaft moves erratically, the most likely causes are internal galling between the shaft and guides, packing overtightness, or an actuator or positioner malfunction. If graphite packing is used for a high-tempera-ture service, the user should remember that such packings are known to cause jerky shaft motion even when properly compressed.

When the eccentric plug valve body is removed for servicing, the entire body should be examined for porosity, stress fracture, cavita-tion, or erosion that may lead to more serious leakage problems. The plug, seat, and inside of the body should be closely examined for ero-sion or cavitation. Sometimes if the crack or hole is minor, that area can be ground out and filled with a weld. If this is done, it should be done according to established industry methods so that the integrity of the pressure-retaining vessel will not be weakened.

When the valve is built to fail to a particular position, the power supply should be removed from the actuator or actuation system to verify that the ball valve fails in the correct position. Because the valve is still in line at this point, the user should do this only if the process allows or if the eccentric plug valve is blocked from the process. With actuated eccentric plug valves, most have transfer cases with a cover that can be removed, allowing the linkage and associated bolting to be checked. Line vibration can loosen linkage connections, leading to increased mechanical hysteresis, dead band, or malfunction of the valve–actuator connection. Some valve designs use the transfer-case cover to help guide the end of the shaft; therefore the valve should not be stroked when the cover is off, since the shaft is unsupported and may cause undue stresses on the shaft.

If the pneumatic actuator is equipped with an air filter, the cartridge should be checked and replaced if necessary. If the valve uses a hydraulic or electrohydraulic actuator, it should be visually inspected for any hydraulic fluid leaks. If a lubrication fitting exists on the auxiliary handwheel, adequate lubrication should be added.

6.5.6 Eccentric-Plug-Control-Valve Troubleshooting

If the eccentric plug valve is leaking excessively through the seat, a number of potential problem areas should be investigated. A seat or plug that is worn or damaged will show leakage only after the valve has been operating satisfactorily for a reasonable period. The most probable causes of a worn regulating element are process erosion, frictional wear between the two mating seating surfaces of the seat and the plug, an internal mechanical failure, damage from a foreign object caught between the seating surfaces, or cavitation damage. The elastomer seat or seat gasket (metal seats) may also have failed, allowing leakage to occur through the joint between the body and the seat. If the shaft guides or bearings are worn, the ball may be slightly off center from the seal, also resulting in increased leakage and wear. Leakage can also be caused by a galled shaft, which may prevent the plug from reaching the seat. If an auxiliary handwheel or limit-stop is included with the actuator, it may be incorrectly positioned—inadvertently limiting the travel of the plug.

After retightening, a persistent leak between the piping flanges and the body may indicate a parallel misalignment between the upstream and downstream piping. A persistent leak can also be caused by a flange-gasket failure. If the flange gasket was installed on dirty gasket surfaces, the failure may be occurring where foreign objects on the surface are creating a leak path.

When the packing box is leaking, the usual cause is packing failure through consolidation or extrusion. When this happens, additional compression has no effect in providing a full-contact seal around the shaft or body bore. When consolidation or extrusion is evident, the packing box must be rebuilt using new packing. Packing leaks can also be caused by a worn shaft. Some packing materials (such as graphite) are more abrasive than others. Because of the quarter-turn motion of an eccentric plug valve, any packing with abrasive characteristics has a tendency to wear the shaft where it makes continual contact. These material losses to friction can eventually cause leakage unless the packing is tightened. Over time, this wear will lead to the eventual replacement of the shaft.

Erratic shaft movement can be caused by the type of packing specified for the application. Some high-temperature packings—such as braided graphite rings—are known to become more resistant as more thrust is applied, creating a very erratic shaft rotation. One problem with graphite packing is that it forms a strong seal with the shaft, creating higher friction than softer, more pliant packings. In general, erratic shaft rotation can also be caused by overtightened packing, which causes the packing to bind to the shaft and increase breakout torque. Other probable causes of erratic shaft travel can be tight or misadjusted linkage between the shaft and actuator, actuator failure (such as internal galling), or worn or damaged shaft bearings.

When the actuator or shaft moves and yet the valve's regulating element does not respond in kind, the shaft has most likely sheared or the connection between the plug and the shaft has failed.

Overall valve performance can also be hampered by a failure with the actuator or actuation system. With a pneumatic actuator, poor throttling function or limited closure of the body subassembly may be caused by a low air supply, a leaking air chamber, or a malfunctioning positioner. When an actuator diaphragm bursts or stem seal fails, the actuator will move to its failure position.

In a situation where the valve is providing only marginal or poor flow control, the most likely causes are a malfunctioning positioner or actuator or an incorrect flow characteristic cam in the positioner. As discussed in Sec. 2.2, the valve's flow characteristic can be affected by piping and pump effects that will change the inherent flow characteristic to an installed flow characteristic.

Another common problem is water hammer, occurring when the plug slams into the seat. Water hammer typically occurs when an undersized or underpowered actuator is used. (A more detailed discussion about water hammer can be found in Sec. 11.6.) With a low-thrust actuator, the insufficient stiffness cannot hold a low-flow position close to the seat ring when the flow is shaft upstream. As the actuator moves the plug toward the seal, the process pressure may overcome the actuator and the plug is sucked into the seat, which is commonly called the "bathtub stopper effect."

If the valve fails in the wrong direction, the most common cause is that the valve is installed incorrectly, with either the wrong flow direction or rotary actuator orientation.

6.5.7 Eccentric-Plug-Control-Valve Servicing

The general servicing guidelines in this section are not intended to supersede any eccentric-plug-control-valve manufacturer's specific maintenance and servicing instructions. The manufacturer's instructions should be followed exactly as they are intended.

Following troubleshooting, the user may have a better idea of how to correct the problem. In most cases, when a problem exists, the eccentric plug valve-body subassembly will need to be removed from the line and disassembled. If the service cannot be interrupted and the system is equipped with bypass block valves, they should be used to divert the flow around the valve. That portion of the line (or the entire pipeline, if not bypassed) should be completely depressurized and decontaminated prior to removing the piping-flange bolting. The actuator power supply (air, electric, or hydraulic) should be removed, which will cause the valve to move to its failure position. Special caution should be taken to keep hands and tools away from the regulating element during this procedure.

Prior to removal, the valve should be supported using straps and a hoist or another mechanical device before the piping-flange bolting is removed. If the valve does not break free, an expanding device may be needed to separate the piping flanges slightly to allow for removal of the valve body—the actuator itself should not be used as a lever to pry the flanges apart, because this could place undue strain on the shaft, linkage, etc. As the eccentric plug valve is removed from the pipeline, the gasket surfaces should be kept square with the pipeline to avoid scoring or damaging those surfaces.

In most cases, the body subassembly is easier to disassemble after the actuator or actuation system is removed. To do this, the actuator should be depressurized (if the actuator is pneumatically or hydraulically driven). If the actuator has a failure spring, it should be decompressed for safety reasons. The linkage that connects the shaft to the actuator should be disconnected, allowing the actuator to be removed from the body. Sometimes the clamping device used to hold on to the splined or flatted end of the shaft continues to hold the shaft firmly even after the bolting is removed. In this case, the halves of the clamp should be wedged apart carefully until the shaft is freed.

The seat is removed from the body by first removing the seat retainer. The seat retainer is often threaded to the body and may require a tool from the manufacturer to turn the retainer out of the body. If the retainer is held in place by bolting, the bolting should be removed. This sometimes requires hex-head wrenches, if socket-head bolts are used. The main problem with bolting is that process corrosion may create removal problems, resulting in the technician having to drill out the bolting and retap the threaded hole. After the seat is exposed, the seat, spacers (or shims), or gaskets should be removed from the body, paying careful attention to the order and orientation of each part as it is removed. After all parts are removed, they should be examined for aging or damage and set aside for replacement if necessary.

The packing should be decompressed by removing the gland-flange bolting and the gland flange itself. If a connection exists between the plug and the shaft (pin, key, etc.), it should be removed. Before the shaft is removed, care should be taken to fully support the plug to prevent it from falling loose during disassembly and damaging critical sealing surfaces or injuring nearby personnel. With the packing decompressed and the plug and shaft disconnected, the shaft should be carefully slid out the body. The post opposite the shaft can now be removed. This may require some disassembly of the plug or shaft installed at the blind end of the body. With both the post and shaft removed, the plug can be lifted carefully from the body. During this procedure, the user should be careful not to scratch any portion of the seating surface. Scratches on the seat surfaces will lead to increased leakage rates. The entire parameter of the seating surface should be examined for any signs of wear, erosion, or damage. With the ball and shaft removed from the body, the bearings or guides that support the shaft should be removed. Some bearings or guides are pressed into the body and require mechanical force to remove them. If mechanical force is used, care should be taken not to damage the body or the guides unless the bearing or guides will be replaced.

A wooden dowel of similar size can be used to push the packing out of the packing box from inside the body. As each ring is removed, it should be inspected for signs of consolidation or extrusion. If extensive extrusion is discovered, the packing box should be rebuilt using new packing and antiextrusion rings. Even if the packing appears to be in good condition, it should always be replaced. Before reassembling the body subassembly, all parts should be thoroughly cleaned. Areas of severe oxidation should be removed and painted with an antioxidation paint. Any damaged or worn parts should be replaced.

If applicable, the guides or bearings should be reinstalled in the body. As found in the disassembly procedure, some guides or bearings can be placed by hand, while others will need to be pressed into place in the body's shaft bore. Taking care not to damage the seating surfaces of the eccentric plug, the plug should be positioned inside the body's flow area, aligning the plug correctly in relation to the body and shaft. Large plugs may need to be supported by a mechanical device, rather than held by hand. The support post can now be reinstalled in the body. The shaft should then be inserted through the packing end of the body without scratching or scoring the sealing or guiding surfaces of the shaft. Using the pins, keys, or other fasteners, the plug can be secured to the shaft. The fasteners should be securely placed so that they will not work free during normal service. The shaft should be turned through the entire quarter-turn motion to ensure that no binding of guiding surfaces or misalignment of the plug with the body exists.

The packing box should be rebuilt using new packing rings without damaging the individual rings as they are placed over the end of the shaft, which may have sharp edges from the splines or flats. Each ring should be slid down the shaft into the body bore of the packing box. The user should ensure that the correct number and order of packing rings, spacers, antiextrusion rings, and packing are used for that particular packing box design. The gland flange and bolting can now be reinstalled. If the bolting is corroded and cannot be adequately cleaned up, it should be replaced at this time. Corroded bolting may provide a false torque reading if torque values are required for correct packing compression. The packing should be compressed according to the value provided by the manufacturer. Overcompressed packing can also lead to excessive stem friction and wear.

The seat can now be reinstalled. The user should make certain that the seat is correctly positioned, since some seat designs can only be installed one way, while others are made to be universal. With the seat in place, the retainer or metal seal can be reinstalled, using the fasteners needed to secure the seat assembly. The eccentric plug should then be turned into the seat to ensure full contact of the seating surfaces.

Before remounting the actuator, actuator system, or handwheel onto the body subassembly, the eccentric plug must be in the correct position (full-open or full-closed) according to the failure mode of the actuator. The actuator can now be remounted onto the shaft with the connection end of the shaft fully engaging the linkage. The actuator should be securely rebolted to the body subassembly and the linkage clamp tightened to the shaft. If the end of the shaft is supported by a guide in the transfer case, the user should ensure that the shaft is supported by the transfer case before operation. Joints of the linkage should be lubricated before the cover plate is replaced. The valve should now be slowly stroked and checked for proper operation. If the actuator is provided with limit-stops, they should be checked to ensure that plug travel is limited to within the parameters of the application. The valve may now be reinstalled in the line and returned to service.

7

Manual Operators and Actuators

7.1 Introduction to Manual Operators and Actuators

7.1.1 Purpose of Manual Operators and Actuators

With most valves, some mechanical device or external system must be devised to open or close the valve, or to change the position of the valve if it is to be used in throttling service. Manual operators, actuators, and actuation systems are those mechanisms that are installed on valves to allow this action to take place.

7.1.2 Definition of Manual Operators

A *manual operator* is any device that requires the presence of a human being to provide the energy to operate the valve, as well as to determine the proper action (open, closed, or a throttling position). Manual operators require some type of a mechanical device that allows the human being to easily transfer muscle strength to mechanical force inside the valve, usually through a handwheel or lever that provides mechanical leverage. Since the beginning of process industry, manual operators have been in use and are very commonplace, although over the past three decades, their use has declined somewhat in favor of automatic control actuators. The reason is simply the cost as well as imperfections of the human operator. A human being must be dispatched to the valve with a manual operator and complete the action on the valve. With simple on–off control, this action may be adequate. However, with the accuracy required in today's process systems, the human operator may not be fast enough to reach a valve—or stroke it—when an action is required. With throttling situations, a human operator can only guess at an approximate position of the valve's closure element, which may not be exact enough for a critical service. Even an extra half turn of a handwheel may create too much or too little flow, pressure, or temperature for some applications, especially with some inherent or installed flow characteristics. In addition to the slowness and inaccuracies of human beings, some applications have high internal forces that manual operators cannot overcome because of the physical limitations of the human being, even with extraordinarily long levers or wide handwheels. Also, in business terms, human beings are expensive. The days are over when runners on bicycles were dispatched from the control room to turn handwheels. Nearly all plants today are looking for technology to replace human beings, not only because of the human resource cost, but also for the greater accuracy, efficiency, and productivity associated with higher technology.

7.1.3 Definition of Actuators and Actuation Systems

Automatic control of valves requires an *actuator,* which is defined as any device mounted on a valve that, in response to a signal, automatically moves the valve to the required position using an outside power source. The addition of an actuator to a throttling valve, which has the ability to adjust to a signal, is called a *control valve.* Some say that by the pure definition of actuator, a manual operator is an actuator. However, when most people associated with valves discuss the term actuator, they are referring to a power-actuated operator using an outside signal and power source rather than a human being. Typical classifications of actuators include pneumatic actuators (diaphragm, piston cylinder, vane, etc.), electronic motor actuators, and electrohydraulic actuators. *Actuation systems* are special actuators that are commonly mounted on manually operated valves and can be used in either on–off or throttling applications.

Actuators are critical elements in the *control loop,* which consists of a sensing device, controller, and an actuator mounted on a valve. With a control loop, a sensing device in the process system—such as a temperature sensor or a flow meter—is installed downstream from the control valve and is set to measure a particular variable in the process. The sensor reports its finding to a controller, which compares the actual data against the predetermined value required by the process. If the measured value is different from the predetermined value, the controller sends a correction signal to control valve's actuator. This signal can be sent using one of three methods: increasing or decreasing air pressure, varying electric voltage, or increasing or decreasing hydraulic pressure. The actuator receives this signal and moves accordingly to vary the position of the closure element until the controller determines that the measured value is equal to the predetermined value. At that point, the signal increase or decrease stops, and the actuator—and subsequently, the closure element—holds its position.

Not only must the actuator have the ability to adjust to a changing signal, but it must also have enough power to overcome the internal forces of the process, the effects of gravity, and friction in the valve itself. The majority of applications requiring actuators today require the use of compressed air, with nine out of ten actuators pneumatically driven. Air is by far the preferred power medium, since it is relatively cheap and is available in nearly all plants. In addition, it does not contaminate the environment and can be regulated easily. Typical plant compressed air supply is generally between 60 and 150 psi (between 4 and 10 bar), which is sufficient to run a large portion of the pneumatic actuators available today. When a valve must overcome exceptionally high pressures or when the valve must stroke quickly, bottled nitrogen is often used, allowing pressures up to 2200 psi (150 bar). Not only does a bottle allow for high pressures of nitrogen, it also relatively moisture-free and extremely free of particulates and other foreign material. In general, the disadvantage of air-driven actuators is that, because of the compressibility of gases, some exactness is lost through that medium.

Other power sources can include electrical (both ac and dc power) as well as hydraulics (and to a far lesser extent, steam). Although electromechanical and electrohydraulic actuators are more expensive than pneumatic actuators, they do have the advantages of extremely good accuracy and the ability to operate in environments experiencing low temperatures (where typical air lines can freeze from condensed water) or when high thrusts are required.

If a signal is sent separately from the power supply, pneumatic or electric signals are the industry preference. Prior to 1980, the majority of actuators received pneumatic signals. These signals were typically 3 to 15 psi (0.2 to 1 bar), although 3 to 9 psi (0.2 to 0.6 bar) and 9 to 15 psi (0.6 to 1 bar) were also commonplace. However, with the arrival of the precise control associated with electropneumatic and digital control systems, the pendulum has swung in favor of the electric signals (4 to 20 mA or 10 to 50 mA).

Actuators are described as either single or double acting. A *single-acting actuator* uses a design in which the power source is applied to only one side of an *actuator barrier* (piston, diaphragm, vane, etc.) and the opposite side is not opposed by the power sources. A spring may be added to the opposite side to counteract the single action. A related term is the *direct-acting actuator,* which refers to a design in which the power source is applied to extend the stem. On the other hand, a *reverse-acting actuator* refers to an actuator where the power source causes the actuator stem to retract. *Double-acting actuator* is a term used for actuators that have power supplied to both sides of an actuator barrier. By varying the pressure on either side of the actuator barrier, the barrier moves up or down. Pneumatic double-acting actuators nearly always require the use of a positioner to provide the varying power to the chambers above and below the barrier.

An actuator is normally a separate subassembly from the body, meaning it can be removed from the body for servicing without disassembly of the body subassembly. On the other hand, the body can be serviced without disassembly of the actuator.

7.2 Manual Operators

7.2.1 Introduction to Manual Operators

As discussed in Sec. 7.1, manual operators require the strength and positioning ability of a human being in order to operate the valve. Generally, manual operators are associated with the operation of on–off applications, as well as simple throttling applications not requiring undue accuracy or immediate feedback. The majority of the valves described as manual valves in Chap. 3 uses manual operators.

The advantage of a manual operator lies in its mechanical simplicity—minimal moving parts and no sealed chambers to leak or fail. A human being moves one part (such as a handwheel or a lever) and the valve is opened, closed, or placed in a midstroke position. Design simplicity also means that troubleshooting, maintenance, and disassembly are easier. The disadvantage of manual operators is slow response, since response depends upon a human being operating the manual operator—which in some cases may take some time. For example, a linear handwheel may require 30 or more revolutions to close a valve with a 4-in stroke. And, because a human being must be dispatched to a manually operated valve, the travel time to the valve makes for even slower response. Also, if the valve requires throttling (a midstroke) position, the position of the valve depends upon the judgment of the operator, which may vary widely. In some applications this may not be a problem, but as systems have become more exact over the years, finding the right throttling point has become much more difficult with a manual operator.

7.2.2 Manual Operator Design

Generally, manual operators are divided into two categories: linear motion and rotary motion. Linear-motion manual handwheels use a threaded connection between a fixed-position part of the handwheel assembly, such as a yoke or housing, and a dynamic part (usually a handwheel stem). Multiple turns of a hand-held part mechanism—in most cases, a handwheel—cause linear movement of the dynamic stem, which is connected to a linear-motion closure element.

One of the more common designs is shown in Fig. 7.1, which shows an independent linear handwheel operator that is mounted directly to a body subassembly and is not an integral part of the valve. The actuator uses a yoke to support the handwheel mechanism and to attach the operator to the valve. The connection to the body is made with an inside diameter of the lower portion of the yoke, called the *spud*. The yoke's spud fits over the bonnet and is secured with a yoke nut or other clamping device. The closure device's stem—such as a plug stem, compressor stem, or gate stem—is threaded to the bottom of the handwheel stem. The upper portion of the yoke houses the handwheel nut, which turns with the handwheel. Some designs allow the handwheel and nut to be one integral part, while others make them separate because of material considerations. When the handwheel is separate, a key or locking bolt is used to secure the handwheel to the handwheel nut. The handwheel nut is retained in position, allowing rotational movement, and is internally drilled and tapped to receive the handwheel stem. The matching external threads of the handwheel stem are threaded into the handwheel nut, allowing for several threads to be engaged at any given position. Generally, ACME threads are used for manual operators. To avoid problems with constant contact between similar metals, which can lead to galling, the handwheel stem and handwheel nut are made from dissimilar materials. The most common combination is brass or bronze for the nut and stainless steel for the stem. As the handwheel is turned, the retained handwheel nut turns the engaged threads of the handwheel stem, extending or retracting the stem, depending on which direction the handwheel was turned. The extension or retraction of the stem then operates the linear

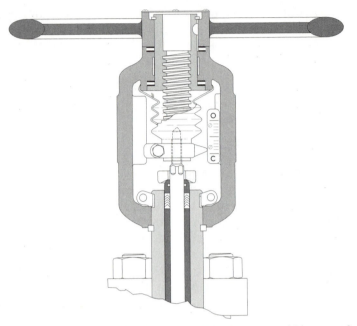

Figure 7.1. Independent linear handwheel operator. (*Courtesy of Valtek International*)

motion of the closure element. In some larger designs or high-pressure applications, rollers or races are placed between the handwheel and the upper portion of the yoke to minimize friction between mating parts, providing easier turning of the handwheel.

The chief advantage of the independent operator is that the valve does not need to be disassembled to service the operator. The disadvantage is that the overall valve has a greater height than other designs.

The other common linear manual-operator design is the dependent linear handwheel operator, which has the handwheel mechanism built directly into the bonnet cap of the valve, as shown in Fig. 7.2. In this case, instead of a yoke, the bonnet cap retains the handwheel nut. The one-piece stem has dual duty of operating both the closure element and the handwheel. The obvious advantage of this design over the independent operator is that the height of the valve is far lower. The disadvantage is that operator problems require some valve disassembly.

Linear operators are also divided into two design categories: the rising-stem and nonrising-stem designs. The *rising-stem* design uses a

Figure 7.2. Integral linear handwheel operator.
(*Courtesy of Orbit Valve Company*)

handwheel nut to retract the handwheel stem. As the handwheel nut is turned, the handwheel stem rises above the handwheel. A majority of manual linear-motion valves use rising-stem operators. On the other hand, the *nonrising-stem* design is typically used with dependent operators. The handwheel turns the retained and threaded stem, which engages the closure element (such as a wedge gate). As the handwheel is turned, the stem turns with it. The closure element is designed to be fixed by guiding so that it cannot rotate; therefore the closure element has a tendency to rise or lower with the stem rotation.

As noted earlier, the most common way of handling a linear manual operator is through a handwheel. Handwheels come in all different surface finishes, from smooth to rough, depending on the work conditions and the type of construction. Many are spoked to save weight, although some petroleum and refining applications require solid handwheels to ensure that they stay intact during a fire. Spoked handwheels have the added advantage of greater security, by allowing a locking mechanism to be placed on the operator to prevent accidental or intentional tampering with the valve's position. Another common handwheel design is the chain wheel. A *chain wheel* is a handwheel with teeth or grooves to accommodate a circular length of chain, allowing for the user to operate an out-of-reach valve.

Rotary-motion manual operators are used with quarter-turn valves, such as plug, ball, and butterfly valves. The most efficient method to turn a quarter-turn closure element is through a right-angle extension of the stem, which allows for better leverage. The two most common types of rotary-motion manual operators are the *handle* and the *wrench*. Many technicians refer to the two terms interchangeably, but a difference does exist. Handles are bolted to the stem of the closure element (Fig. 7.3) and are commonplace with smaller sizes in the lower-pressure classes. Handles are specified with soft-seated ball valves in sizes up to 6 in (DN 150) and butterfly valves in sizes up to 8 in (DN 200). On the other hand, wrenches are not permanently secured to the stem and can be moved from valve to valve (Fig. 7.4), allowing for the operator to place the valve in a particular position and leave it alone without fear of accidental or intentional tampering. Wrenches are normally equipped with plug valves up through 4 in (DN 100) with sleeved plugs and 6 in (DN 150) with lubricated plugs. In some ball and butterfly manual-valve designs, the handle is integral to the stem, but the most common and inexpensive design is a separable handle in which the handle (or wrench) has an opening that is cut to the shape of the plug stem. A square stem allows for the positioning of the handle or

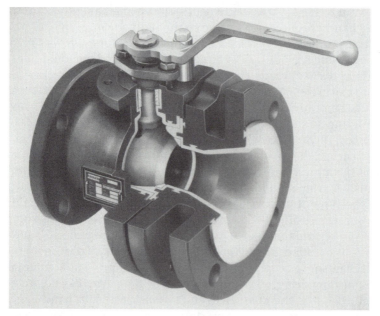

Figure 7.3. Quarter-turn handle mounted on lined ball valve. (*Courtesy of Atomac/The Duriron Company, Valve Division*)

Figure 7.4. Quarter-turn wrench mounted on plug valve. (*Courtesy of The Duriron Company, Valve Division*)

wrench in any one of the four quadrants, while a two-sided flatted stem allows for positioning in one of two positions, front and back. Handles are secured to the stem using a bolt and locking washer.

Handles and wrenches are usually made from ductile iron, although stamped stainless-steel plate is used also. A plastic or rubber grip is placed on the end for comfortable turning. Most manufacturers supply a standard length that handles most applications within the pressure or temperature range of the valve, although longer lengths are sometimes offered to allow for easier operation. Longer lengths, however, may cause problems where space is restricted, not allowing the full quarter-turn motion.

Below the wrench is a *collar-stop* that is used to limit the motion of the closure element to a 90° (or quarter-turn) range. Turning the wrench moves the stem, which in turn moves the plug, ball, or disk, until the collar stops the travel. When the travel is stopped, the closure element should either be in its full-open or full-closed position.

Because of the large forces that can act upon a disk in some applications, butterfly valves may require handlevers for manual operation. A *handlever* is a two-piece, spring-loaded operator that can be positioned in a number of preset slots (Fig. 7.5). The handlever has a fixed upper lever and a movable lower lever. In the static position, the spring load-

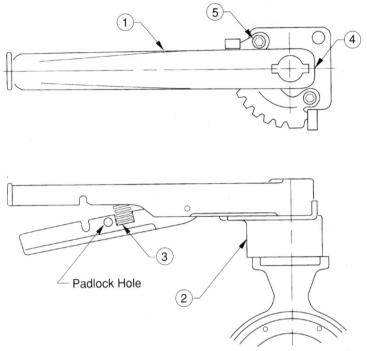

Figure 7.5. Quarter-turn lever operator. Numbered parts are as follows: (1) lever, (2) rachet plate, (3) spring, (4) set screw, (5) socket
head cap screw. (*Courtesy of Flowseal, a unit of the Crane Valve
Group*)

ing of the lower lever allows it to seat in one of multiple slots in the
collar. By squeezing the upper and lower levers, the lower lever disen-
gages the slot, allowing rotational movement to another desired slot.
When the handlever it released, the lower lever seats into the slot,
locking the valve in that particular position. The range of slots can
vary according to the number of positions required. A typical handlever has a minimum of three positions, full-open, full-closed, and
midstroke position, although any number of positions can be planned
for as long as room exists for the desired number of slots in the collar.

In larger linear and rotary valves, or in higher-pressure classes, the
use of conventional handwheels, handles, and wrenches is not desirable. The circumference of the handwheel or length of the wrench or
handle would be so long to handle the leverage that the arc and the
weight of the operator would be impractical. In this case, gear operators are used. As shown in Figs. 7.6 and 7.7, *gear operators* (sometimes
called *gearboxes*) use gearing to translate handwheel torque into high-

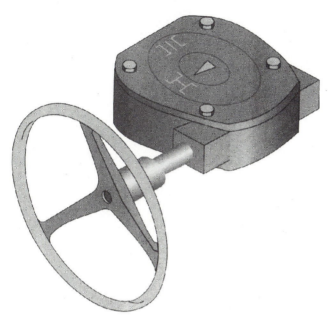

Figure 7.6. Quarter-turn worm-gear operator. (*Courtesy of Flowseal, a unit of the Crane Valve Group*)

output thrust, which is necessary to overcome the greater thrust requirements of larger flows or higher pressures. Linear-motion gearboxes use spur or beveled gearing, while rotary-motion gearboxes use rack-and-pinion or worm gearing. Gear operators use gears with ratios anywhere between 7:1 and 3:1. Both handwheels and cranks are used to turn the gears. The gearing is protected by the gearbox, which not only protects nearby personnel from the turning gears but also minimizes contact with atmospheric or outside conditions. Gear operators are normally bolted onto the bonnet or bonnet cap of linear-motion and some quarter-turn valves and bolted onto the body of butterfly and some ball valves. With linear-motion valves, the stem is threaded directly to the operator stem. With rotary-motion valves, the shaft end may be splined or squared and may intersect with the internal opening of a gear inside the gearbox. When a valve is installed in the line, its position may be difficult to determine without some type of positioner indicator. Most operators have a position indicator consisting of an arrow and a matching position plate, which shows the position of the valve.

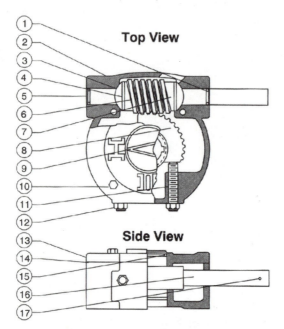

Figure 7.7. Internal view of quarter-turn worm-gear operator. Numbered parts are as follows: (1) seal-input shaft, (2) housing, (3) bearing, (4) washer, (5) plug, (6) worm gear, (7) worm pin, (8) gear segment, (9) indicator cap, (10) cover bolt, (11) stop adjustable screw, (12) hex nut, (13) cover, (14) gasket cover, (15) O-ring, (16) worm shaft, (17) roll pin. (*Courtesy of Flowseal, a unit of the Crane Valve Group*)

7.2.3 Manual-Operator Troubleshooting

In linear valves the most common problems involving manual operators deal with sticking stem travel or limited stroke. The first step in troubleshooting manual operators is to stroke the valve from full-open to full-closed several times (if the process will allow), observing the motion and stroke of the valve. In some cases, sticking may not be evident by observation; however, a problem may be felt through the handwheel, wrench, or handle. Sticking stem travel may indicate a problem with the valve itself rather than the manual operator. Applicable body problems affecting the actuator may be packing tightness, galling of mating parts, or misalignment of the closure element with the valve. In these cases, the valve will need to be disassembled and serviced, or the packing will need to be tightened to the correct compression.

A lack of lubrication between threads or gears can cause binding in linear manual operators or gear operators. Most threaded or gear-operated manual operators provide a method of injecting grease through a grease fitting or by removing the housing to gain access to the threads or gears. If the actuator has similar materials between threaded parts or gears, galling is a possibility, especially if the lubrication has diminished through use, time, or exposure to the atmospheric conditions. Minor galling can be corrected through lightly grinding away or buffing the affected area. Major galling between parts requires new parts or a new manual-operator assembly. Also, if rollers or races are used with the handwheel, sticking may occur if they are worn or damaged

Problems with a limited stroke can also be tied to internal valve problems, such as foreign-material buildup, an object caught in the closure element, or binding of mating parts—all of which require disassembly. With the manual operator, the problem could be a failed gear or damaged threads of the stem or handwheel nut. In some gear operators and manual operators, a limit-stop is included in the design, which limits the travel of the stem or shaft. A maladjustment of the limit-stop could also cause this problem. If limited stroke occurs following reassembly of the valve for servicing, oftentimes the handwheel stem is not fully engaged with the handwheel nut according to the manufacturer's instructions. Improper engagement between these two parts will also limit the full extension or retraction of the stem.

Because of the simplicity of the design, rotary quarter-turn operators are often problem-free, since the wrench or handle is directly connected to the stem of the closure element. If the valve sticks or has less than desired travel, the likely cause could be associated with the closure element or packing box. Sometimes the stroke stops on the collar can be damaged, which may provide an incorrect stroke limitation. Another common problem is a loose handle, which is easily corrected by tightening the bolt connecting the handle and the stem.

7.2.4 Manual-Operator Servicing

Because of the wide variance of manual operators, both independent and dependent designs, individual step-by-step instructions for all possible linear and rotary designs would be impractical to outline in this format. For this reason, the user should always refer to the specific servicing instructions provided for that particular operation. However, a few general maintenance guidelines that apply to servicing and repair manual operators are covered in this section.

The entire operator should be examined for damage caused by process drippings or corrosive fumes and the problem corrected through sealing the leaks or by placing shields over the valve. Process leakage can affect the actuator by depleting the lubrication or by corroding the threads.

If the manual operator is an independent design, meaning that it can be removed from the valve without disassembling the body, the process line will need to be depressurized to avoid process upset or personal injury when the handwheel is removed. If the operator is a dependent design, the entire line will need to be depressurized and decontaminated, since the valve's top-works will need to be removed from the valve and disassembled. If the process line must continue to operate, block valves should be used to channel the flow around the valve to be serviced.

During disassembly, the user should note how much engagement exists between the stem and handwheel nut in linear handwheels, since this will be necessary to remember for reassembly. The threads of both the handwheel nut and stem should be carefully inspected for damage or galling. Rollers or races should be examined for wear, especially if the operator was used frequently.

Before reassembly takes place, all parts should be cleaned and repainted if any areas were damaged by severe oxidation. All threaded parts or gears should be cleaned of old grease and thoroughly lubricated. The user should make sure that any threaded parts are fully engaged as noted from disassembly.

If a dependent design is rebuilt, new gaskets should be used with the bonnet cap to ensure a tight seal and the packing readjusted according to the manufacturer's specifications. If an independent operator is rebuilt, the user should ensure that the connection (nut, clamps, etc.) between the valve and operator is secure. Also, the connection with the closure element's stem or shaft should be checked to make sure that it is securely fastened. The plug, gate, or compressor stem should be correctly engaged with the handwheel stem, since this affects the adjustment of the closure element. For example, with globe manual valves, threading a plug stem too far into the handwheel stem may not allow the plug to reach the seat, causing seat leakage despite full extension of the operator.

After the operator is rebuilt, it should be turned to stroke the valve several times to ensure smooth, full stroke operation. If necessary, lubrication should be added to the threads or gears through the grease fitting.

7.3 Pneumatic Actuators

7.3.1 Introduction to Pneumatic Actuators

The most commonly applied actuator is the pneumatically driven actuator, because the power source—compressed air—is relatively inexpensive when compared to a human resource or electrical or hydraulic power sources. For that reason, approximately 90 percent of all actuators in service today are driven by compressed air. When compared to the cost of electromechanical and electrohydraulic actuators, pneumatic actuators are relatively inexpensive as well as easy to understand and maintain. Most are available as standard off-the-shelf products in a number of predetermined sizes corresponding to maximum thrust. Only in special services are special-engineered actuators produced, such as those applications requiring exceptionally long strokes, high stroking speeds, or severe temperatures. From a maintenance standpoint, pneumatic actuators are more easily serviced and calibrated than other types of actuators. Some pneumatic actuators are designed to be *field-reversible,* meaning that they can be converted from air-to-extend to air-to-retract (or visa versa) in the field without special tools or maintenance procedures. Although not as powerful as hydraulic actuators, pneumatic actuators can generate substantial thrust to handle a majority of applications, including high-pressure and high-pressure-drop situations. While air lines are not easy to install, the cost is less than installing electrical conduit and electrical lines as well as hydraulic hoses. Pneumatic actuators also bleed compressed air to atmosphere, which is environmentally safe, when compared to hydraulics. When pneumatic positioners are used with a pneumatic actuator, they are ideal for use in explosive and flammable environments since they do not depend upon electrical signals or power, which could potentially spark a fire if not explosion-proof or intrinsically safe.

The chief disadvantage of pneumatic actuators is that some response and stiffness are lost because of the compressibility of gases—especially with pneumatic actuators that use elastomers with large areas, such as diaphragms. This is not a factor, however, in the majority of applications that do not require a high degree of stiffness or response. With

Table 7.1. Actuator Stroking Times*

Actuator Size (piston area in inches²/cm²)	Seconds to Maximum Stroke (0.25-inch/6 mm tubing)	Stroke Length (inches/cm)	Seconds Per Inch (2.5 cm) of Stroke
25/161	1.2	1.5/3.8	0.8
50/323	3.5	3/7.6	1.2
100/645	9.6	4/10.2	2.4
200/1290	20.8	4/10.2	5.2
300/1936	31.3	4/10.2	7.8

*Data courtesy of Valtek International. Data based upon cylinder actuator design.

larger actuators, speed is an issue since the volume of the actuator must be filled with compressed air and/or bled to atmosphere to move. For this reason, larger actuators take longer to stroke from full retraction to full extension than smaller actuators, as shown in Table 7.1. Also, pneumatic actuators must be close to an air supply and are dependent upon the continued operation of a compressor unless a separate backup system or volume tank arrangement is installed. Although some designs are better than others, pneumatic actuators do have limits on the amount of thrust available, making some designs unlikely choices for high-pressure applications in large line sizes. Low thrust is commonly associated with diaphragm actuators since the diaphragms can only handle so much air pressure without failing, thus limiting their thrust capabilities.

7.3.2 Pneumatic Actuator Design

The most commonly applied pneumatic actuator over the past 40 years has been the *diaphragm actuator* (Fig. 7.8). Most diaphragm actuators are designed for linear motion, although some rotary-motion designs exist. By definition, a typical diaphragm actuator is a single-acting actuator that provides air pressure to one side of an elastomeric barrier (called the *diaphragm*) to extend or retract the actuator stem, which is

Figure 7.8. Single-acting diaphragm actuator.
(*Courtesy of Fisher Controls International, Inc.*)

connected to the closure element. The diaphragm is sandwiched between upper and lower casings, either of which can be used to hold air pressure, depending on the style of the actuator.

In the single-acting design, the air chamber on one side of the diaphragm is opposed on the other side of the diaphragm by an internal spring, called the *range spring,* that allows the actuator to move in the opposite direction when the air pressure in the chamber is lessened. The range spring also acts as a fail-safe mechanism, allowing the actuator to return to either an open or closed position when the air supply to the actuator is interrupted. Depending on the configuration, the spring is installed next to the diaphragm or the diaphragm plate. The actuator stem is connected to the diaphragm plate and is supported through the top of the yoke with the assistance of a guide. As the diaphragm moves with increasing air pressure, the plate moves in a corresponding manner. That linear motion is directly transferred to the actuator stem, which moves the closure element in the valve. A *yoke* attaches the actuator to the valve body to show the position of the actuator and valve, to support the actuator stem, and to make the actuator-stem to valve-stem connection. It also provides a convenient place to attach accessories. With diaphragm actuators, the most common connection between the body and the actuator is a threaded yoke nut. A clamp is used to prevent the accidental rotation of the actuator stem with the valve stem. The clamp can also be equipped with a pointer that can indicate actuator or valve position.

With conventional single-acting diaphragm actuators, the air signal from the controller to the actuator has a dual role. First, it provides a positioning signal. Second, it provides the power to generate the thrust necessary to overcome the process forces, friction, gravity, the weight of the closure element, and the opposing force generated by the range spring.

Diaphragm actuators have both direct-acting and reverse-acting designs. With the direct-acting design (Fig. 7.9), air pressure is sent to the actuator, which extends the actuator stem and allows the valve to close. This also means that the actuator will retract its stem upon loss of air, allowing the valve to open and remain open. With the reverse-acting design (Fig. 7.10), as the air pressure is sent to the actuator, the stem retracts and the valve opens. If the supply or signal air pressure is interrupted, the actuator moves to the extended position, allowing the valve to close.

With the direct-acting design, air is introduced to the upper casing located above the diaphragm. Beneath the diaphragm are the diaphragm plate and the range spring. The range spring bottoms out in

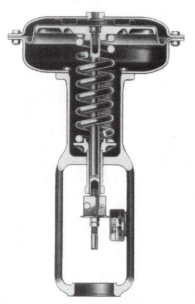

Figure 7.9. Direct-acting diaphragm actuator.
(*Courtesy of Fisher Controls International, Inc.*)

Figure 7.10. Reverse-acting diaphragm actuator. (*Courtesy of Fisher Controls International, Inc.*)

the bottom of the yoke, allowing the upper end of the spring to push against the diaphragm plate and subsequently the diaphragm. In this relaxed (or failure) position, the diaphragm is pushed into the area of the upper casing. As air is introduced into the upper casing and pressure builds, the diaphragm and plate push against the spring. As the signal pressure increases, the air pressure overcomes the opposing forces and the diaphragm and plate move downward. This movement allows the actuator stem to extend and the valve to move toward the closed position. Eventually as the full signal air pressure is reached and the resulting air pressure is introduced into the chamber, the diaphragm and plate reach their full travel. On the other side of the plate, the range spring is nearly fully compressed. At this point, the stem is at its full extension and the valve is closed at the full pressure end of the signal. As the signal is lessened, resulting in lower air pressure in the chamber, the counterforce of the range spring begins to take effect, and the actuator moves to its relaxed state and the valve is opened.

With the reverse-acting design, the lower chamber is used to provide the air pressure to retract the actuator stem, while a reverse-acting spring is used to provide the counterforce, as well as the failure mode.

The upper casing is static and only needs to retain the diaphragm and to vent displaced air volume to atmosphere. With this configuration, the lower casing is pressure retaining and requires an air connection to inject air into that chamber. The diaphragm plate is installed above the diaphragm. The range spring, which is still located below the diaphragm, is seated below the lower casing and is not in direct contact with the diaphragm and plate assembly. Instead, the range spring is seated on a retainer on the lower portion of the actuator stem. Because the range spring bottoms out (or in this case, tops out) at the bottom of the lower casing, as the actuator stem retracts with air to the lower chamber, the spring's resistance increases proportionately. As the actuator stem retracts, the valve begins to open. When the air signal is at the high end of the range, the actuator stem is fully retracted, and the range spring is almost completely compressed. When the signal changes and moves to the lower end of the range, the air pressure to the lower chamber is lessened. At that point, the range spring's counterforce begins to push the actuator stem to the relaxed (extended) state until the full extension is reached and the valve closes.

When positioners are used to improve the overall response of the actuator, three-way positioners can be installed that supply or exhaust air pressure to only one side of the diaphragm. Three-way positioners can be mounted on the actuator's yoke leg or can be integrally mounted inside the actuator, as shown in Fig. 7.11.

Diaphragm actuators are produced in several sizes, with a different diaphragm area for each size as well as several range-spring options. Each size has a given range of thrust that is available to overcome process forces, frictional forces, gravitational forces, and the range spring. Therefore, the actuator size has less to do with the process' line size than the service conditions. Whether the valve is used primarily for on–off service or throttling service has some bearing on the actuator size. With diaphragm actuators, the instrument signal can vary widely to accommodate power considerations. Although 3 to 15 psi (0.2 to 1.0 bar) is considered standard, diaphragm actuators can have signal ranges as high as 3 to 27 psi (0.2 to 1.9 bar) or 6 to 30 psi (0.4 to 2.1 bar). Diaphragm actuators are sized according to the square inches of the diaphragm. For example, a size 125 diaphragm actuator has a diaphragm of 125 square inches (in^2).

The chief advantage of diaphragm actuators is that they are relatively inexpensive to produce and are commonly seen through the entire process industry. Although limited in high-thrust requirements, they are well suited to a good portion of applications in lower-pressure ranges, where thrust requirements are not so demanding. The basic

Figure 7.11. Diaphragm actuator with integral three-way positioner. (*Courtesy of Kammer Valves*)

single-acting design and method of operation are simple to understand. Because the positioning signal is also conveniently used to power the actuator, the expense of a positioner and tubing is not necessary. Without a positioner, an involved calibration process and the potential for mechanical difficulties associated with that device are not necessary. The lack of positioner also means that less moving parts, such as a positioner-to-actuator linkage, are involved that may cause potential maintenance problems. When used with linear-motion valves, the entire movement of the actuator stem is transferred directly to the valve's closure element. Because no tight dynamic seals, such as O-rings, are involved with the diaphragm, no breakout force is necessary during positioning, providing immediate and accurate response. Generally, diaphragm actuators are ideal for those applications in which precise positioning and immediate response are important and in which medium to low thrust is acceptable to overcome the process and valve forces.

Several disadvantages of the design should be noted. Because the diaphragm is relatively large, the subsequently large casing may present weight and height problems, especially when mounted on smaller valve sizes. This can cause problems with stress at the connection point between a small valve and an oversized actuator. Because of the restrictions in the elasticity of the diaphragm, its stem travel is limited. Strokes are somewhat short, when compared to other types of actuators. This poses a problem with special severe service trims in which a long stroke is necessary to provide a particular flow characteristic or provide a greater flow capacity through a stack or other trim device. Most diaphragm actuators have strokes of 2 in (5.1 cm) or less, although 4-in (10.2-cm) strokes are possible in some special designs. The largest drawbacks are the thrust and air-pressure restrictions of the diaphragm itself. Because the amount of force produced by the diaphragm actuator is proportional to the size of the actuator, the physical size required for high thrusts is limited by the size of the diaphragm. Most diaphragms are rated for operation in the 20- to 30-psi (1.4- to 2.1-bar) range, therefore limiting the amount of air pressure acting on the diaphragm. For example, a size 125 diaphragm actuator operating with 30 psi (2.1 bar) air pressure can produce a maximum of 3750 lb of thrust (1700 kgf). For that reason, the only way to increase the thrust is to increase the size of the diaphragm, which results in a larger actuator and air chamber. In turn, this larger volume produces slower actuator speed and decreases overall response. The air-pressure limitations of the diaphragm also require the use of air regulators because the air pressure supplied by most plant compressors is between 80 and 125 psi (between 5.5 and 8.6 bar). If diaphragms could handle such high air pressures, the thrust capabilities of the example above would increase dramatically to 10,000 lb (4400 kgf) of thrust. Unfortunately, no diaphragm material has been developed that can provide such strength yet provide the required resilience to move through the full stroke. The thrust limitations of a diaphragm actuator can be overcome by using it with valve designs that can balance the process flow conditions, such as double-seated valves or pressure-balanced trim. Although the cost of such valve bodies may be higher than unbalanced designs, the cost may be negated by the smaller actuator.

Generally, diaphragm actuators—because of the limitations of the diaphragm—do not provide exceptional stiffness and therefore have problems with fluctuations in the process flow. They also experience problems when throttling close to the seat, not having enough power to prevent the closure element from being pulled into the seat. The stiffness value of a diaphragm actuator is usually constant throughout the entire stroke. When the closure element is close to the seat, a sudden change or fluctuation in the process flow can cause the valve to slam shut, causing water-hammer effects.

From a maintenance standpoint, the life of diaphragm actuators is somewhat limited by the life of the diaphragm. If the diaphragm develops even a minor failure, the actuator is inoperable. Since the two casings are bolted together with numerous bolts, disassembly can be somewhat laborious and time consuming. Diaphragm actuators are not field-reversible, because different parts are required for the direct- and reserve-acting designs. Diaphragm actuators have about one-third more parts than other types of pneumatic actuators, which increases their cost somewhat.

Although the diaphragm actuator is the most common pneumatic actuator, the piston cylinder actuator (Fig. 7.12) is gaining widespread acceptance, especially as processes become more advanced and demanding. As shown in Fig. 7.13, the *piston cylinder actuator* uses a sliding sealed plate (called the *piston*) inside a pressure-retaining cylinder to provide double-acting operation. With the double-acting design, air is supplied to both sides of the piston by a positioner. As with all double-acting actuators, a positioner must be used to take the pneumatic or electric signal from the controller and send air to one side of the piston while bleeding the opposite side until the correction position is reached. An opposing range spring is not necessary with the piston cylinder actuator, although a spring may be included inside the

Figure 7.12. Piston cylinder actuator. (*Courtesy of Valtek International*)

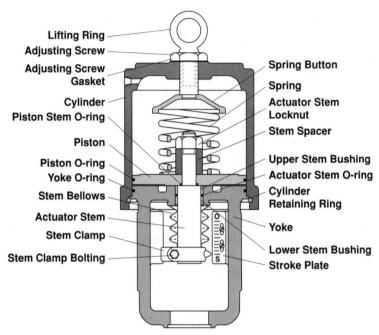

Lifting Ring
Adjusting Screw
Adjusting Screw Gasket
Cylinder
Piston Stem O-ring
Piston
Piston O-ring
Yoke O-ring
Stem Bellows
Actuator Stem
Stem Clamp
Stem Clamp Bolting

Spring Button
Spring
Actuator Stem Locknut
Stem Spacer
Upper Stem Bushing
Actuator Stem O-ring
Cylinder Retaining Ring
Yoke
Lower Stem Bushing
Stroke Plate

Figure 7.13. Internal view of piston cylinder actuator. (*Courtesy of Valtek International*)

cylinder to act as a fail-safe mechanism. More information about the use and operation of positioners is found in Sec. 7.6.

Like diaphragm actuator designs, piston cylinder actuators can be used with either linear or rotary valves. Linear designs are the most efficient since the entire movement of the actuator stem is transferred directly to the valve stem. On the other hand, the rotary design must use some type of linear- to rotary-motion linkage. This can create some hysteresis and dead band because of the lost motion caused by the use of linkages or slotted levers.

The design of the linear cylinder actuator involves a cast yoke, which is used to make the connection to the valve body. It also provides room for the connection between the valve's stem and the actuator stem, attaches the cylinder mechanism to the valve, supports the actuator stem, and allows the installation of the positioner and other accessories. The cylinder can be made from either aluminum (for weight and machining considerations) or steel, based on the application. Fire-sensitive applications prefer the higher melting point of steel over aluminum. The inside of the cylinder is machined to a polished finish to allow for a good seal. The piston itself is a flat disk that is machined nearly to the inside diameter of the cylinder. An O-ring (or similar elastomer seal) fits inside a groove along the sealing edge of the piston. When the O-ring and piston are installed inside the cylinder, the cylinder wall is lubricated to allow a strong, sliding seal. If a fail-safe spring is required, it can be installed either above or below the piston. Unlike the diaphragm actuator that requires a different range spring for different opposing forces, the piston cylinder actuator spring is only needed for fail-safe operation. Therefore, only one heavy-duty spring is needed to cover most applications with the thrust requirements of that actuator size. For extremely high-pressure-drop-applications, a nested spring configuration (one spring inside another) can be used, as shown in Fig. 7.14. Spring compression is applied by the introduction of an adjusting bolt, which compresses the spring to the required return force. Adjusting bolts of different lengths can be used to vary the spring compression. The cylinder is installed above the yoke with either a snap-ring arrangement or bolting. The actuator stem is attached to the piston and is supported by the top of the yoke with guides. It is sealed from the lower chamber with an O-ring. With piston cylinder actuators, the most common connection between the body and the actuator is a two-piece yoke clamp (Fig. 7.15). This permits a tight connection without larger threads to contend with, which can be a problem with atmospheric corrosion. A clamp is used to pre-

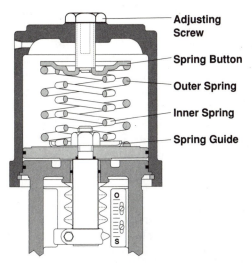

Adjusting Screw

Spring Button

Outer Spring

Inner Spring

Spring Guide

Figure 7.14. Piston cylinder actuator with dual springs. (*Courtesy of Valtek International*)

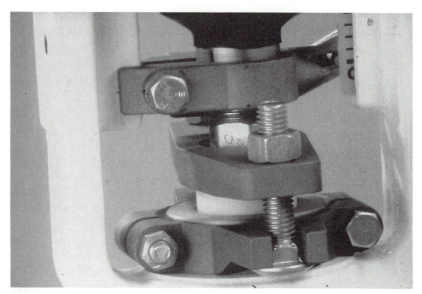

Figure 7.15. Two-piece yoke clamp connection between yoke and bonnet. (*Courtesy of Valtek International*)

vent the accidental rotation of the actuator stem with the valve stem. The clamp can also be equipped with a pointer to indicate actuator or valve position.

Most rotary designs use some type of linkage to transfer linear motion to rotary action. Figure 7.16 shows one common design in which a splined lever is attached to the valve's shaft and has a pivot point on the actuator stem to minimize hysteresis. Such a design requires a sliding seal to allow for the rocking motion of the piston, which will rock slightly as the actuator stem rotates with the travel of the lever. As shown in Fig. 7.17, another common rotary piston cylinder design uses a slotted lever that intersects a pinned actuator stem. This design avoids the rocking piston and its requirement for a sliding seal, although it does have potential for some slight hysteresis and dead band because of the slotted-lever design. With this design, the heavy-duty return spring is placed in a separate housing, opposite the cylinder.

Piston cylinder actuators are reversible, meaning that the same actuator can be modified for either air-to-close (actuator stem extends) or air-to-open (actuator stem retracts), as shown in Fig. 7.18. With air-to-close designs, the spring is placed below the piston and is held in place by a ringed groove in the top of the yoke.

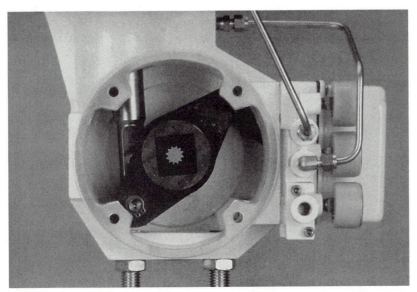

Figure 7.16. Splined clamp connection between rotary actuator and shaft. (*Courtesy of Valtek International*)

The operation of piston cylinder actuators is quite simple. As an air-to-close signal is sent from the controller to the positioner, the positioner sends air to the cylinder's upper chamber above the piston, while the positioner bleeds a comparable amount of air from the lower chamber below the piston. The changing pressures in these two chambers cause the piston to move downward. Subsequently the actuator stem moves downward, as does the valve stem. As the signal changes to "open," the air pressure in the lower chamber builds, while the air

Figure 7.17. Slotted-lever and pinned actuator-stem connection between rotary actuator and shaft. (*Courtesy of Automax, Inc.*)

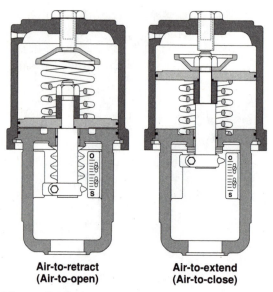

Air-to-retract   Air-to-extend
(Air-to-open)   (Air-to-close)

Figure 7.18. Air-to-retract and air-to-extend configurations for piston cylinder actuators. (*Courtesy
of Valtek International*)

pressure in the upper chamber is bled off, allowing the piston to move
upward. Therefore the valve's closure element opens. If the signal or
power supply is lost, the piston is assisted by the fail-safe spring and
moves to its relaxed position. In air-to-close configurations, the relaxed
state is with the stem retracted. In air-to-open configurations, the
relaxed state is with the stem extended.

The primary advantage of cylinder actuators is the higher thrust
capability, size for size, over comparable diaphragm actuators. Because
the cylinder actuator with a positioner does not need to use air supply
as a signal, the plant's full air-supply pressure can be used to power
the actuator. The piston with its sliding O-ring seal is much more capable of handling greater air pressure than the diaphragm. To demon-
strate the significance of this difference, a piston cylinder actuator with
a piston of 25 in² (161 cm²) used with an 80-psi (5.5-bar) air supply is
capable of producing 2000 lb of thrust (910 kgf). Assuming a 6- to 30-
psi (0.4- to 2.1-bar) range, a comparable diaphragm actuator would
only generate 750 lb (340 kgf) of thrust using the 30-psi (2.1-bar) air
supply. A far larger diaphragm actuator would be needed to provide
the same thrust requirement as the piston cylinder actuator.

Piston actuators, which have smaller chambers to fill with higher pressures of air, have faster stroking speeds than diaphragm actuators, which must fill larger chambers with lower pressures of air. For example, a size 25 piston cylinder actuator can stroke 1.5 in (3.8 cm) in less than 1 s, while a diaphragm actuator takes over 2 s to stroke the same distance.

Generally, cylinder actuators can be operated with air supplies as high as 150 psi (10.3 bar) or as low as 30 psi (2.1 bar). A side benefit to a piston cylinder actuator handling up to 150-psi (10.3-bar) plant air is that air regulators are not required. For diaphragm actuators such regulators are necessary since they cannot handle plant air normally beyond 40 psi (2.8 bar).

Placing air pressure on both sides of the piston also permits greater actuator stiffness, meaning that the actuator can hold a position without being influenced by fluctuation of the process flow. This is especially important with globe or butterfly valves when the plug or disk is being throttled close to the seat and the "bathtub stopper effect" (Sec. 11.6) can take place. Single-acting actuators have difficulty with the bathtub stopper effect because the range spring (which provides the counterforce) may not be strong enough to prevent it from happening. Stiffness of piston cylinder actuators can be calculated by using the following equation:

$$K = \frac{kPA^2}{v}$$

where K = stiffness
k = ratio of specific heat
P = supply pressure
A^2 = piston area
v = cylinder volume under the piston

To illustrate how drastic the stiffness rates vary between piston cylinder actuators and diaphragm actuators, a comparison can be made using a piston cylinder actuator with a 25-in^2 (161-cm^2) piston, which is typical for a 2-in (DN 50) globe valve. With a supply pressure of 100 psi (6.9 bar) and a 0.75-in (1.9-cm) stroke, the stiffness value at midstroke would be 9333 lb/in (1667 kg/cm). In comparison, a diaphragm actuator with a 46-in^2 diaphragm (296 cm^2), which is required for a 2-in valve, only has a stiffness value of 920 lb/in (164 kg/cm). In addition, as the closure element approaches the closed position with a very close throttling position, the reduced volume in the bottom of the cylinder provides for increased and exceptional stiffness. With the 25-in actuator example used earlier, if the plug in a globe valve is 0.125 in (0.3 cm) away from the seat, the piston is only 0.375 in (1 cm) away from the top of the yoke. That would yield over 18,000 lb/in (3214 kg/cm) of stiffness. For that reason, piston cylinder actuators are preferred when process fluctuations occur or if throttling close to the seat is required by the application.

As a general rule, piston cylinder actuators are much more compact, being smaller in height and weight, than diaphragm actuators—an important consideration with installation, maintenance, and seismic requirements. Of course, the size difference is highly accentuated when larger-diaphragm actuators are needed to generate higher thrusts. A height comparison of comparable actuators is shown in Fig. 7.19.

Another consideration is the length of the stroke. With spring cylinder actuators, the stroke is only limited by the height of the cylinder, permitting longer strokes that diaphragm actuators, which are restricted by the resilience limitations of a diaphragm.

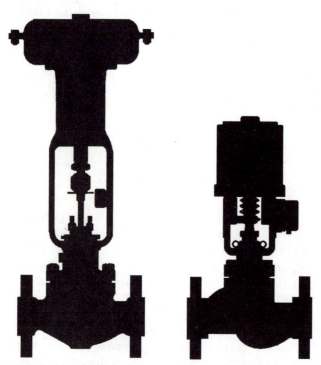

Figure 7.19. Height comparison between comparable diaphragm (left) and piston (right) cylinder actuators. (*Courtesy of Valtek International*)

Due to the accuracy associated with the positioner, piston cylinder actuators generally perform better than diaphragm actuators, with virtually no hysteresis, highly accurate signal response, and excellent linearity.

Piston cylinder actuators have some drawbacks. First, if the actuator remains in a static position for some time, some breakout force may be necessary to move the piston when a signal is eventually sent. When considering the added thrust and response associated with piston cylinder actuators, this breakout torque may not be noticeable. The requirement of a positioner does add expense to the actuator— although with less parts, the actuator itself is less expensive than a diaphragm actuator. A positioner also requires calibration. As discussed in Sec. 7.6, positioners can present problems with exposed linkage and fouled air passages.

A recent modification of the piston cylinder actuator, a similar design that features a canister assembly and an integral positioner, is shown in Fig. 7.20. Instead of using a dynamic piston, the piston is static and the chambers are dynamic. As shown in Fig. 7.21, the entire

Figure 7.20. Piston cylinder actuator with canister assembly and integral positioner. (*Courtesy of Fisher Controls International, Inc.*)

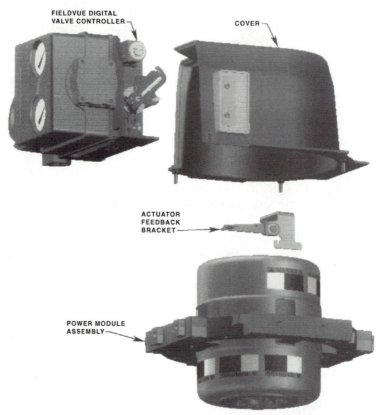

FIELDVUE DIGITAL
VALVE CONTROLLER

COVER

ACTUATOR
FEEDBACK
BRACKET

POWER MODULE
ASSEMBLY

Figure 7.21. Internal view of piston cylinder actuator with canister assembly and integral positioner. (*Courtesy of Fisher Controls International, Inc.*)

canister assembly is held in place by the upper and lower casings. As the upper chamber moves, the integral positioner (which is encased in the upper casing) has a follower arm that can receive position feedback by the top of the chamber. Instead of tubing, special air chambers channel air to either the lower or upper chamber.

This design provides a low-profile, compact actuator without the problems associated with external linkage between the actuator and the positioner. With internal air passages, tubing is eliminated—reducing the possibility of damaged tubing or leaking connections. The only disadvantage of this design is that the canister assembly is not designed to be disassembled. The need for a spare part involves the entire assembly, which is far more costly than replacing typical soft goods.

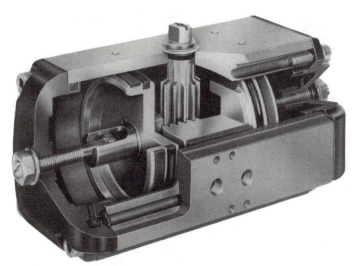

Figure 7.22. Double-acting rack-and-pinion rotary actuator. (*Courtesy of Automax, Inc.*)

Another commonly applied pneumatic actuator is the *rack-and-pinion actuator*, which is used to effectively transfer the linear motion of piston cylinder actuators to rotary action. Rack-and-pinion actuators are used extensively for actuating quarter-turn valves (ball, plug, and butterfly valves). As shown in Fig. 7.22, two pistons are placed on each end of a one-piece housing, typically extruded aluminum or stainless steel. Each piston is connected to a *rack,* a series linear teeth, that move in a linear motion with the piston. In most cases, the rack is an integral part of the piston itself. Sandwiched between the two racks is the *pinion,* which is a shaft equipped with linear teeth. The shaft is connected directly to the valve stem. With direct-acting rack-and-pinion actuators, as air is applied to the two outer pressure chambers, the pistons move toward the inner chamber, exhausted to atmosphere. As shown in Fig. 7.23, when the two pistons move toward each other, the attached racks move in opposite directions, allowing the rack teeth to drive the teeth of the pinion in a counterclockwise rotational manner. As shown in Fig. 7.24, when increasing air pressure is directed to the inner chamber and the outer chambers are exhausted, the pistons move away from each other and the pinion is driven in a clockwise direction.

Rack-and-pinion actuators can be equipped with internal springs to allow the actuator to achieve a failure mode (fail-clockwise, fail-coun-

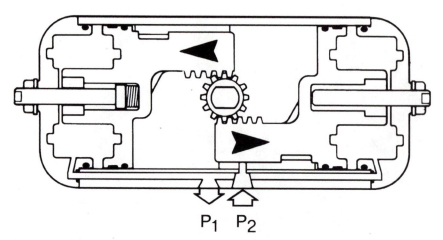

Figure 7.23. Counterclockwise action of rack-and-pinion actuator. P_1 = upstream pressure; P_2 = downstream pressure. (*Courtesy of Automax, Inc.*)

terclockwise) when the air supply or signal is lost. They are also field-reversible by removing the end caps and rotating the pistons 180°. Rack-and-pinion actuators can also be provided with travel stops to allow for precise adjustment of the open and closed positions of the valve.

Overall, rack-and-pinion actuators are ideal for automating manually operated rotary valves: They are compact, allow for field reversibili-

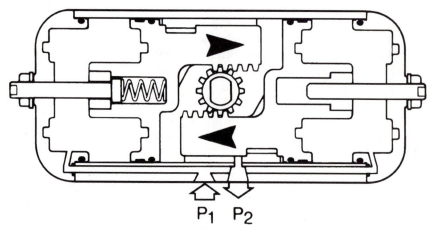

Figure 7.24. Clockwise action of rack-and-pinion actuator. P_1 = upstream pressure; P_2 = downstream pressure. (*Courtesy of Automax, Inc.*)

ty, provide adequate torque for most standard operations, and are easy to maintain and to understand.

Another common, inexpensive double-acting actuator is the *vane actuator*, which uses a pie-shaped pressure-retaining housing and a rectangular piston, called the *vane*, to seal between the two pressure chambers (Fig. 7.25). As with rack-and-pinion actuators, vane actuators are commonly used with quarter-turn valve applications.

The housing is divided into two halves and is pie-shaped to allow the vane to move the 90° required for quarter-turn operation. The vane is pinned to the actuator shaft, avoiding excessive hysteresis and dead band. The vane seals the two pressure chambers with an O-ring. Generally the design does not permit the inclusion of a spring. Instead, a pneumatic fail-safe system is often used in place of the spring. The double-acting design requires the use of a positioner for throttling applications; each pressure chamber has an air connection for increasing or exhausting air pressure.

The operation of the vane actuator can be reversed by simply removing the actuator from the valve and installing it upside down

Figure 7.25. Vane rotary actuator. (*Courtesy of Xomox/Fisher Controls International, Inc.*)

(since both ends of the actuator have universal mounting). Limit-stops can be included on both ends of the housing to limit the motion of the vane.

The advantages of the vane actuator are its simple design with few moving parts, no hysteresis, low cost, minimal weight, and compact size. The chief disadvantage of the vane actuator is that it only generates relatively low torque values when compared to other designs; therefore, vane actuators are commonly applied to low-pressure applications. In addition, the two-piece housing with a joint down the middle provides a possible leak path between air chambers.

7.3.3 Pneumatic Actuator Installation

Prior to installation, valves equipped with pneumatic actuators should be connected to the power source and signal and fully stroked to ensure proper operation. With single-acting actuators, the signal air pressure should be connected to the air connection on the top casing, for air-to-extend (or air-to-close) the valve. With air-to-retract (air-to-open) the valve applications, the air connection should be directed to the bottom casing. If a three-way positioner is used with a single-acting actuator, the air supply and signal should be connected directly to the positioner. The positioner is already tubed to the correct air chamber depending on the air action. In double-acting actuators with positioners, the air or electric signal should be attached to positioner's pneumatic or electropneumatic connection, and the plant air supply should also be connected.

Most actuators will indicate the maximum air pressure that it is designed to handle. This is especially important with single-acting (diaphragm) actuators, since overpressurization can cause the diaphragm to fail. The opposite is true with double-acting piston cylinder actuators: the higher the plant air pressure (up to 150 psi or 10.3 bar), the greater the thrust and stiffer the performance. In some cases, service technicians who are well acquainted with diaphragm actuators require the use of pressure regulators to limit the air pressure to below 40 psi (2.8 bar) and routinely limit the plant air pressure to those same limits for piston cylinder actuators and other high-thrust designs. Not only does this limit the thrust capabilities of the piston cylinder design but also degrades the higher performance associated with those designs.

When connecting air lines to the pneumatic actuator, the installer should leave some slack in the line so that line vibration or thermal expansion will not place undue stress on the pneumatic tubing (or electrical conduit if an electric signal is use). One common method is to bend a loop into the tubing, which can then flex with any fluctuating energy caused by the valve or process line. After all air lines are installed to the actuator, a soap solution can be used to check for leaks in any connections.

After the signal and/or air supply is connected, the actuator should be operated to observe the valve's stroke. Using the position indicator, the stroke should be checked for any jerky or erratic motion or limited motion (less than required), which may indicate an internal valve problem. The valve should be checked for correct calibration so that it begins opening at the opening signal [such as 4 mA or 3 psi (0.2 bar)], as well as closing at the end of the signal range [such as 20 mA or 15 psi (1.0 bar)]. If the calibration is incorrect, the positioner will need recalibration or a new range spring. If a single-acting actuator is used, a different range spring may be needed.

Most pneumatic actuators have an internal spring, which is needed to assist with the failure position of the valve (fail-open or fail-closed upon loss of pneumatic air power). After stroking the valve to test for full stroke, the air pressure should be disconnected and the stroke indicator observed to ensure that the valve moves to the correct failure position. During installation of the valve, sufficient room must be provided around the actuator for disassembly of the actuator or valve, as well as access to the positioner or other actuator accessories for calibration and maintenance.

Limit switches should be checked to ensure full contact with the actuator's stroke mechanism, making sure that contact is made at the appropriate portion of the stroke. As mentioned earlier, air regulators may or may not be required, depending upon the type of pneumatic actuator used. However, a good recommendation is to install an air filter before the single-acting actuator or positioner. Not only will the filter trap any particulates that could foul a positioner or damage a diaphragm, but it will also trap moisture from condensation in the air line. Without an air filter, the user may disassemble a malfunctioning actuator and find it filled with water.

7.3.4 Pneumatic Actuator Troubleshooting

One of the most common problems associated with a pneumatic actuator is its lack of response to a signal with a corresponding movement to the failure position. In this case, the most likely problem is a failed diaphragm in diaphragm actuators or a failed piston O-ring in piston cylinder actuators or rack-and-pinion actuators. Because of the failure of the chief sealing device in the actuator, the chamber cannot hold pressure in a single-acting actuator, or the pressure is balanced in both chambers of a double-acting actuator. In this case, the actuator must be disassembled to replace the failed elastomer.

If the actuator is performing in an erratic manner, the problem is most likely overtightened packing or binding or galling of the regulating element. In high-temperature applications graphite packing has a tendency to grab the valve stem and cause erratic stem behavior. If the problem occurs with the regulating element, the body assembly must be disassembled and repaired. If the problem is with the actuator, probable causes could be galling or binding in the linear-to-rotary mechanism of a rotary pneumatic actuator or between the actuator stem and guides of a linear-motion actuator. If the actuator has a positioner, erratic behavior can be caused by a problem with the actuator-to-positioner linkage, such as loose connections, worn guiding surfaces, or dirt in the positioner itself. If dirt is found in the positioner, pushing the clean-out plunger, if available, should purge the positioner of the foreign material. If an air filter is not installed, one should be installed to ensure against future contamination.

If the valve does not move on a command signal or move to its failure position on loss of air supply, the most likely cause is the valve itself—a galling problem may have caused the regulating element to seize up. However, if the problem is in the actuator, the cause could be a broken spring or binding between the actuator stem and the guides.

If the actuator has exceptionally high air consumption, the user should check for air leaks in the supply or instrument signal system. If a positioner is supplied, it may have failed and will need repair. In addition, the seals (O-rings or diaphragm) to the pressure chamber may not be sealed properly or might have a small leak. In some cases, if an actuator stem is open to the atmosphere, it can become damaged or corroded, which can eventually wear away the actuator-stem O-ring that seals the lower chamber.

7.3.5 Pneumatic Actuator Servicing

To service an actuator, the process must be depressurized before disassembling or removing the actuator; otherwise, the valve may inadvertently move to an unwanted position when released from the actuator. This could upset the process, damage equipment, or injure personnel.

Prior to disassembly, the signal and air-supply lines should be shut off or depressurized and disconnected from the actuator, unless the signal and air pressure are needed to remove the actuator from the valve. If the actuator is under compression from a range spring or failure spring, the actuator should be decompressed when possible. Some designs use a spring-compression bolt, which releases the spring compression when removed. If the actuator must be disassembled while under compression, a warning tag is normally found on the actuator—warning the user not to disassemble the actuator without special tooling or equipment. In some cases, a special press is needed to hold the entire assembly together while the casing is unbolted. After the bolting is removed, the press is slowly retracted, allowing the compression to be slowly released. Sometimes the tag warns the user not to disassemble the actuator because of this internal compression. In this case, the actuator will need to be serviced by a certified valve repair operation or by the manufacturer.

If the actuator must be removed from a valve for servicing, the entire valve can be removed from the line and the actuator later removed, or the valve can remain in-line and the actuator disconnected from the valve. In either case, a signal and air pressure may be needed to avoid damage to the valve body during disassembly. With air-to-retract (air-to-close) linear actuators, the best procedure is to send a signal to the actuator to slightly lift the valve off the seat. In this way, when the valve stem is unthreaded from the actuator stem, it will not fall the entire stroke. With air-to-extend (air-to-open) actuators, the regulating element is in the closed position. In some designs, such as a globe valve with a plug and a seat, rotation of the plug while unthreading it from the actuator stem could cause galling between similar materials. In this case, a signal can be used to slightly lift the plug up from the seat to allow removal without galling. In rotary actuators, the linear-to-rotary mechanism is usually clamped or connected to the shaft. This connection must be released before the rotary actuator can be unbolted from the valve.

Most linear actuators use a threaded yoke nut or clamps to secure the actuator to the valve's bonnet. These must be removed to slide the yoke off of the bonnet. The problem with the threaded yoke nut is that removal may be difficult when the actuator is severely corroded or oxidized. In this case, force may be required to break the yoke nut free. In a worst-case scenario, the yoke nut must be cut off. With clamps, corrosion may cause the bolting to seize up. Cutting and removing bolting is much easier than struggling with unthreading rusted parts. After the bolting is cut away, the clamps are easily removed.

Some pneumatic actuators can be partially disassembled while attached to the valve, allowing access to the internal soft goods of the actuator. To replace a diaphragm, the actuator casing bolting must be removed, allowing the upper casing to be removed (refer again to Figs. 7.9 and 7.10). With direct-acting diaphragm actuators, the diaphragm is exposed at this point and can be removed. If the spring needs to be replaced, the diaphragm plate and spring guide will need to be removed from the actuator stem, allowing for removal of the spring. With reverse-acting diaphragm actuators, after the upper casing is exposed, the diaphragm plate should be unbolted from the actuator stem to remove the plate from the actuator. At that point the diaphragm can be replaced. If the range spring must be replaced, the bottom casing must be unbolted from the yoke, allowing for access to the spring. After total disassembly takes place, the diaphragm should be carefully examined to determine the cause of failure. An old or worn diaphragm is characterized by stiffness and cracks throughout the entire diaphragm exposed to the pressurized chamber. On the other hand, a flawed diaphragm or overpressurized actuator fails in only one portion of the diaphragm.

Before reassembly takes place, all parts should be thoroughly cleaned. Any areas that are severely corroded or oxidized should be replaced, repaired, and repainted. In general, dynamic parts inside the diaphragm actuator do not require a lubricant. When a diaphragm actuator is reassembled, the parts are reinstalled in the reverse order in which they were removed. All bolted connections should be securely tightened with a wrench, especially if vibration is a constant problem with the application. When the casing bolting is reinstalled, the bolting should be evenly tightened in a criss-cross manner until an adequate seal is achieved. The actuator can now be reconnected to the valve (if removed), and the signal air reattached to the actuator. The diaphragm actuator should be stroked several times to ensure smooth operation, proper calibration, and correct stroking speed.

Before disassembly takes place, the piston cylinder actuator should be fully decompressed as described in the beginning of this section. Referring to Fig. 7.26, the bolting or retaining ring that holds the cylinder to the yoke should be removed and the cylinder slid off the yoke by lifting straight up. This exposes the piston, which allows access to the piston and yoke O-rings, which may need to be replaced. If the actuator stem O-ring must be replaced, the actuator stem must be removed from the valve stem. After unbolting the piston and removing it along with the spring, the actuator stem can be lifted out of the yoke, exposing the O-ring. All O-rings should be checked for damage or wear. If an O-ring appears to be cut or ragged, a burr may exist on the cylinder wall or actuator stem. The damaged part must be removed and repaired, or it

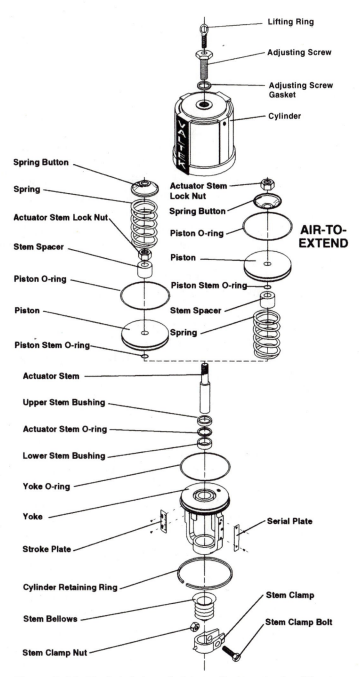

Figure 7.26. Exploded view of piston cylinder actuator. (*Courtesy of Valtek International.*)

should be replaced. Following disassembly, all parts should be thoroughly cleaned. Corroded or severely oxidized parts should be repaired or replaced. The inside diameter of the cylinder must be lubricated (following the manufacturer's recommendation) to ensure smooth stroking. All parts of the piston cylinder actuator should be reassembled in the reverse order in which they were removed. All bolted connections should be securely tightened, especially when the valve is installed in a process that experiences vibration. When the cylinder is reinstalled on the yoke, the bolting should be evenly tightened in a criss-cross manner and the retaining ring should be fully engaged within the lip of the cylinder. If previously removed, the actuator can now be reinstalled on the valve, and the air signal and air lines reattached to the actuator. The user should then stroke the actuator several times to ensure smooth stroking, proper calibration, and correct stroking speed.

With rack-and-pinion actuators, the actuator will need to be removed from the valve, since the pinion must freely rotate to remove the piston or rack parts. This will necessitate depressurizing the process line prior to disconnecting the power and signal lines and removing the actuator. If failure springs are used with the rack-and-pinion actuator, spring pressure must be released according to the manufacturer's written instructions. Referring to Fig. 7.27, the end caps should be removed, followed by fail-safe springs, if applicable. The pinion should be rotated either clockwise or counterclockwise to drive the racks off the pinion, pushing the pistons themselves toward the ends of the actuator housing. Once the piston or racks are free of the piston, they can be pulled out of the actuator housing. The pinion is held in place by a snap ring, which should be released to allow for the removal of the pinion from the actuator housing. All soft goods should be carefully inspected, including any elastomeric guides or O-rings. Any worn parts should be replaced. The teeth of the racks and pinion should be checked for unusual wear. As with piston cylinder actuators, the inside of the housing, pinion, racks, and the piston's O-ring seals should be lightly lubricated using the grease recommended by the manufacturer. At this point, the pinion should be reinstalled, followed by the piston or racks. The piston or racks must be installed according to the correct rotation of the pinion (clockwise or counterclockwise). The fail-safe springs (if applicable) and end caps should then be reinstalled, providing the correct spring compression according to the manufacturer's instructions. The actuator should now be installed on the valve, and the air signal and air lines reattached to the actuator. The actuator should be stroked several times to ensure smooth stroking, proper operation, and correct stroking speed.

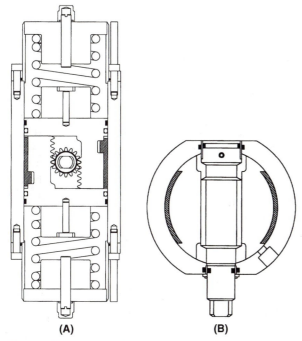

(A)   (B)

Figure 7.27. Internal view of rack-and-pinion actuator.
(A) Top view. (B) Bottom view. (*Courtesy of Automax, Inc.*)

Servicing of vane actuators is fairly simple. After the process line is depressurized, the signal and power lines are removed from the actuator, and the actuator is removed from the valve. Once the housing bolting is removed, the housing can be removed to expose the vane. At this point the vane and actuator-stem assembly can be removed from the lower housing and the O-ring seal removed from the vane.

After cleaning and lubricating the inside walls of the housing and the O-ring seal, the actuator is reassembled in the reverse order of the parts previously removed for disassembly. The vane actuator can then be reinstalled on the valve, according to the correct orientation, and stroked several times to ensure smooth operation.

7.4 Nonpneumatic Actuators

7.4.1 Electric Actuators

Electric motors installed on process valves were one of the first types of actuators used in the process industry. Such electric actuators have been used since the 1920s, although the designs have improved dramatically since those early days, especially in terms of performance, reliability, and size. In basic terms, the electric actuator consists of a reversible electric motor, control box, gearbox, limit switches, and other controls (such as a potentiometer to show valve position).

The chief applications for electric actuators are in the power and nuclear power industries, where high-pressure water systems require smooth, stable, and slow valve stroking.

The main advantages of electric actuators are the high degree of stability and constant thrust available to the user. In general, the thrust capability of the electric actuator is dependent on the size of the electric motor and the gearing involved. The largest electric actuators are capable of producing torque values as high as 500,000 lb (225,000 kgf) of linear thrust. The only other comparable actuator with such thrust capabilities is the electrohydraulic actuator, although the electric actuator is much less costly.

Stiffness is far better with electric actuators, because no compressibility of air is involved with the electric actuator. One additional benefit of an electric actuator is that it always fails in place upon loss of electrical power, whereas a pneumatic actuator requires a complex fail-in-place system. Since fluids (such as air or hydraulics) are not required to power the actuator, leaks and tubing costs are not factors.

The disadvantage of electric actuators is their relative expensive cost when compared to the more commonly applied pneumatic actuators. Also, they are much more complex—involving an electric motor, electrical controls, and a gearbox—therefore much more can go wrong. An electric motor is not conducive to flammable atmospheres unless stringent explosion-proof requirements are met. When high amounts of torque or thrust are required for a particular valve application, an electric actuator can be quite large and heavy, making it more difficult to remove from the valve. Depending upon the gear ratios involved and the pressures involved with the process, an electric actuator can be quite slow, when compared to electrohydraulic actuators or even pneumatic actuators. It can also generate heat, which may be an issue in enclosed spaces. If the torque or limit switches are not set correctly, the force of the actuator can easily destroy the regulating element of the valve.

Based on the thrust requirements, electric actuators are available in compact, self-contained packages (Fig. 7.28), as well as larger units with direct-drive handwheels (Fig. 7.29). As shown in Figs. 7.30 and 7.31, the basic design of the electric actuator consists of the electric motor, the gearbox or gearing, the electrical controls, limit or torque

Figure 7.28. Compact electric actuator.
(*Courtesy of Kammer Valves*)

Figure 7.29. Electric actuator with direct-drive handwheel.
(*Courtesy of Rotork Controls Inc.*)

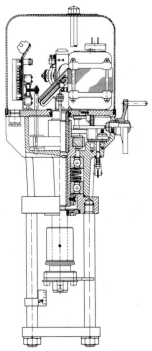

Figure 7.30. Internal view of compact electric actuator. (*Courtesy of Kammer Valves*)

switches, and the positioning device. By design, electric motors are more efficient at their maximum speed; therefore, most electric actuators use some type of mechanical device, such as a hammer blow yoke nut, to engage the load after the motor has achieved its full speed. This is especially important since the largest amount of thrust or torque is required at the opening or closing of the valve. For the actuator to operate in both directions, the motor must be reversible to open and close the valve. For efficiency reasons, electric motors operate best at high revolutions per minute (1000 to 3600 r/min). Therefore, gearing is used to reduce the stroking speed for use with valves. The gearbox uses worm gearing to make the reduction and is totally encased in an oil bath for maximum life of the gears.

Because of the exceptional stiffness and torque associated with electric actuators, the valve can overstroke if the actuator is not adjusted correctly—and possibly damage or destroy the regulating element or limit the stroke of the valve. To avoid overtravel, limit switches are used to shut off the motor when the open or closed position is reached.

Figure 7.31. Internal view of electric actuator with direct-drive handwheel. (*Courtesy of Rotork Controls Inc.*)

Torque switches can also be used to shut off the motor when the torque resistance increases as the closed or open positions are reached. The added benefit of the torque switch is that if an object is caught in the regulating element or if the valve is binding, the actuator will shut off rather than apply thrust to reach the closed position and further damage the valve. Ideally, torque switches are best used with valves that have floating seats (such as ball or wedge gate valves), while limit switches are best used with valves with fixed seats (such as globe or butterfly valves).

The electrical controls can be accessed on the valve itself or controlled at a remote location using extended electrical lines. Either handlevers or buttons are provided to operate the electric motor. With the handlever, turning the lever clockwise extends the actuator stem, while counterclockwise retracts the stem. Placing the handlever in the middle position shuts off the motor and maintains that particular valve position. With button controls, three buttons are used in the normal configuration: one to extend the actuator stem, one to retract, and another to stop the motor. Red and green lights are used to show the user if the valve is in the open position (usually green) or closed position (usually red). When the motor is in operation, both lights are on.

Electric actuators, in smaller sizes, operate using 110 to 120 V ac, 60-Hz, single-phase power, drawing anywhere between 3 and 30 A. Larger electric actuators use 220 to 240 V, three-phase, 50- or 60-Hz power supply—or 125 or 250 V dc. This may require drawing up to 300 A. Exceptionally large actuators may require even greater voltage (up to 480 V ac).

When manual operation or manual override is needed, most electric actuators allow for the electric motor to be disengaged. A declutchable handwheel can then be used to position the valve manually. Because of the complex electrical and mechanical nature of electric actuators, most calibration adjustments and recommended servicing are made at the manufacturer's factory or an authorized service center.

7.4.2 Hydraulic and Electrohydraulic Actuators

When exceptional stiffness and high thrust are required—as well as fast stroking speeds—hydraulic and electrohydraulic actuators are specified. *Hydraulic actuators* use hydraulic fluid above and below a piston to position the valve. Hydraulic pressure can be supplied by an external plant hydraulic system (Fig. 7.32). Its design is similar to a cylinder actuator, with a cylinder and a piston acting as a divider between the two chambers. Hydraulic actuators do not have a failure spring, so providing a failure action requires a series of tripping systems, which are very complex and require special engineering. On the other hand, an *electrohydraulic actuator* uses a hydraulic actuator—rather than use an external hydraulic system, it has a self-contained hydraulic source that is a physical part of the actuator. An electrical signal feeds to an internal pump, which uses hydraulic fluid from a reservoir to feed hydraulic fluid above or below the piston.

The advantage of using hydraulic and electrohydraulic actuators is that they are exceptionally stiff because of the incompressibility of liquids. This is important with those throttling applications that can be unstable when the regulating element is close to the seat. In some cases, these actuators are used in valves with traditionally poor rangeability, such as butterfly valves. When specially engineered, they can be designed to have exceptionally fast stroking speeds, sometimes closing long strokes in under a second—which makes them ideal for safety management systems. The chief disadvantages of hydraulic and electrohydraulic actuators are that they are expensive, large and bulky, highly complex, and require special engineering.

Figure 7.32. Hydraulic actuator mounted on a severe service valve. (*Courtesy of Valtek International*)

7.5 Actuator Performance

7.5.1 Performance Nomenclature

A number of technical terms are used to describe the performance capabilities of an actuator.

Hysteresis is a common term used to describe the amount of position error that occurs when the same position is approached from opposite directions. *Repeatability* is similar to hysteresis, although it records the maximum variation of position when the same position is approached from the same direction. Typically hysteresis and repeatability readings can be anywhere between 0.25 and 2.00 percent of the full stroke of the actuator. *Response level* is the maximum amount of input change required to create a change in valve-stem position (in one direction only). Typically response levels can be anywhere between 0.1 to 1.0 percent of full stroke. *Dead band* is a term used to describe the maximum amount of input that is required to create a reversal in the movement of the actuator stem. Typical dead-band measurements can fall between 0.1 and 1.0 percent of the full stroke. *Resolution* describes the smallest change possible in a valve-stem position. Typical resolution is between 0.1 and 1.0 percent full stroke.

Steady-state air consumption applies to actuators with positioners in which the positioner consumes a certain amount of air pressure to maintain a required position. Depending on the positioner design, typical steady-state air consumption can vary anywhere between 0.2 and 0.4 SCFM (standard cubic feet per minute) (between 1.6 and 3.2 cm^3/min) at 60 psi (4.1 bar). *Supply-pressure effect* describes the change of the actuator stem's position for a 10-psi (0.7-bar) pressure change in the supply [for example, if a 50-psi (3.5-bar) supply is increased suddenly to a 60-psi (4.1-bar) supply]. Typical supply-pressure effects can vary anywhere between 0.05 and 0.1 percent of the full stroke of the actuator. *Open-loop gain* is the ratio of the imbalance that occurs when an instrument signal change is made and the actuator stem is locked up. Typical open-loop gains can be anywhere between 550:1 to 300:1 at 60-psi (4.1-bar) supply. *Stroking speed* is defined as the amount of time, in seconds, that an actuator requires to move from the fully retracted to the fully extended position. Stroking speed depends on the length of the stroke, the volume of the pressure chambers, the air supply, and internal resistance of the actuator itself.

Frequency response is a response to a system or device to a constant-amplitude sinusoidal input signal. In other words, it is a measurement of how fast a system can keep up with a changing input signal. When frequency response is calculated, the output amplitude and phase shifts are recorded at a number of frequencies. They are then recorded as a function of input signal frequency. *Independent linearity* is the maximum amount that an actuator stem will deviate from a true straight linear line. Typical linearity can vary anywhere between ±1.0 and ±2.0 percent.

Maximum flow capacity is the volume of air pressure that can flow into an actuator during a particular time period. This is recorded in standard cubic feet per minute (SCFM) or in cubic centimeters per minute.

7.6 Positioners

7.6.1 Introduction to Positioners

By definition, a *positioner* is a device attached to an actuator that receives an electronic or pneumatic signal from a controller and compares that signal to the actuator's position. If the signal and the actuator position differ, the positioner sends the necessary power—usually through compressed air—to move the actuator until the correct position is reached. Positioners are found in one of two designs. *Three-way positioners* (Fig. 7.33) send and exhaust air to only one side of a single-acting actuator that is opposed by a range spring. *Four-way positioners*

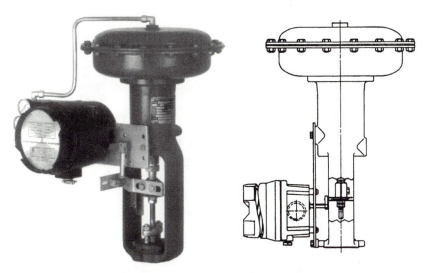

Figure 7.33. Three-way electropneumatic positioner mounted on a diaphragm actuator. (*Courtesy of Fisher Controls International, Inc.*)

(Fig. 7.34) send and exhaust air to both sides of the an actuator, which is required for double-acting actuators. A four-way positioner can be used as a three-way positioner by plugging one of the positioner-to-actuator air-supply lines on the positioner itself.

When a position signal is sent from a controller, positioners can receive either electronic signals with ranges of 4 to 20 mA and 10 to 50 mA or pneumatic signals with ranges of 3 to 15 psi (0.2 to 1.0 bar) or 6 to 30 psi (0.4 to 2.1 bar). The term *range* is used to show the region between the lower and upper signal limits. A *span* is defined as the difference between the lower and upper limits of the signal. For example, for a range of 3 to 15 psi (0.2 to 1.0 bar), the span is 12 psi (0.8 bar). Internal feedback springs (sometimes called *range springs*) are used inside the positioner to help determine the correct span. *Split range* is the term used to indicate a partial use of a range, such as a 3- to 9-psi (0.2- to 0.6-bar) signal or a 12- to 20-mA signal. In some designs, a split range can be achieved by adjusting a zero or range adjustment on the positioner, while in others a new range spring is required.

As the use of distributive control systems has increased in the past decade, so has the need for electropneumatic (I/P) positioners to handle the milliampere-current control signals. I/P positioners are capable of converting the milliampere signal to an equivalent pneumatic signal, which can then operate the pilot valve of the positioner.

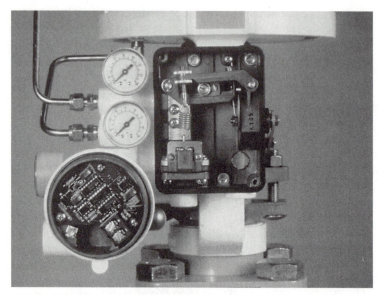

Figure 7.34. Four-way electropneumatic positioner mounted on a piston cylinder actuator (without covers). (*Courtesy of Valtek International*)

7.6.2 Positioner Operation

Positioning is based on balancing the force between the incoming signal from the controller and the actuator positioner. In other words, the positioner works to balance two forces: first, the force proportional to the incoming instrument signal, and second, the force proportional to the actuator's stem position. As shown in Fig. 7.35, an incoming instrument signal is received by the positioner. If this signal is a milliampere signal, a conversion to a pneumatic signal must take place through the use of a *transducer*. The transducer consists of a feedback loop of a pressure sensor, electromagnetic pressure modulator, and necessary electronics. The pressure modulator consists of a flapper that can open or close an air nozzle. The flapper itself moves when attracted by an electromagnet. As the signal moves the electromagnet, the flapper moves accordingly, creating a proportional air signal to the positioner. The transducer can also include a small air regulator to assist in providing the proper air pressure for the pneumatic signal. If the positioner accepts a pneumatic signal, that signal is sent directly to the positioner.

As the pneumatic signal changes, the air pressure inside the instrument signal capsule also changes, causing a repositioning of the pilot

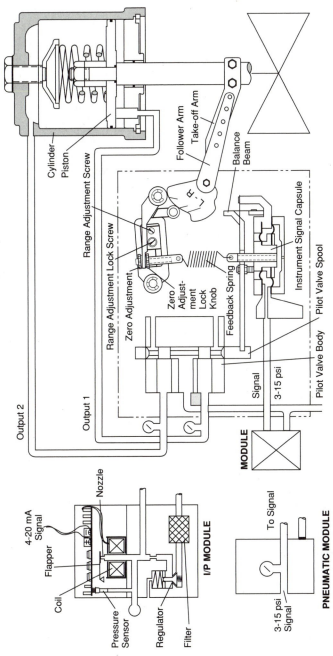

Figure 7.35. Positioner schematic for linear actuator (air-to-retract). (*Courtesy of Valtek International*)

373

valve. As the pilot valve opens, air is supplied or exhausted to one side of the actuator (three- and four-way positioners). In four-way positioners working with double-acting actuators, the opposite action occurs on the opposing side. If air is increased on one side, the other side must exhaust.

The change in air pressures to the upper and lower chambers of the actuator causes the actuator stem to move either upward or downward. The motion of the actuator stem is transmitted to the positioner through some type of internal or external linkage or lever. As this feedback motion is received by the positioner, the stretch and force of the feedback spring are increased or decreased, which changes the counterforce to the instrument signal capsule. At this point, when the correct actuator position is achieved, the instrument signal capsule and pilot valve return to their state of equilibrium and the air flow to the actuator discontinues.

With valves that only have inherent flow characteristics, such as a butterfly valve, a characterizable cam (Fig. 7.36) can be used with the positioner to provide a modified flow characteristic.

7.6.3 Positioner Calibration

Positioners normally come from the factory calibrated to the requirements of the actuator and valve application; however, shipping and handling may cause the calibration to shift. Prior to service, the positioner should be connected to the signal and supply lines and should then be operated. If significant inaccuracy occurs, the positioner calibration should be examined. The two most common adjustments with positioners are the zero and the span. The zero adjustment is used to vary the point where the actuator begins its stroke, normally 3 psi (0.2 bar) or 4 mA for most common applications. After the zero has been calibrated, the span adjustment is used to increase or decrease the span from the zero point, normally 12 psi (0.8 bar) for a 3- to 15-psi (0.2- to 1.0-bar) pneumatic signal or 16 mA for a 4- to 20-mA electronic signal. Some span adjustments allow for certain split ranges without changing the feedback spring. For example, a 3- to 15-psi (0.2- to 1.0-bar) feedback spring may allow the span to be adjusted to a 3- to 9-psi (0.2- to 0.6-bar) or a 9- to 15-psi (0.6- to 1.0-bar) split range. After the span adjustment has been made, the user should return to the zero point to ensure that it stayed true during the span adjustment. Locking nuts or other locking devices are installed to prevent the calibration from shifting during service.

The zero and span adjustments, as well as a number of split ranges available, depend on the type of the feedback spring being used.

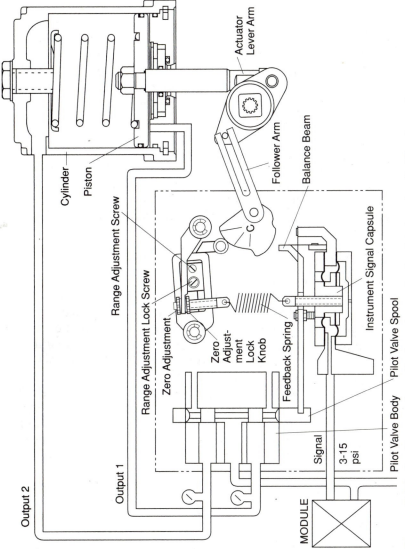

Figure 7.36. Positioner schematic for rotary actuator (air-to-retract). (*Courtesy of Valtek International*)

Actuator Lever Arm

Follower Arm

Balance Beam

Instrument Signal Capsule

Pilot Valve Spool

Pilot Valve Body

3-15 psi

Signal

MODULE

Feedback Spring

Zero Adjustment Lock Knob

Zero Adjustment

Range Adjustment Lock Screw

Range Adjustment Screw

Output 1

Output 2

Cylinder

Piston

Significant range changes, such as changing from a 3- to 15-psi (0.2- to 1.0-bar) range to a 6- to 30-psi (0.4- to 2.1-bar) range would require a new feedback spring.

7.6.4 Positioner Troubleshooting

As explained earlier, one disadvantage of using a positioner with an actuator is that another device is added to the actuation system that can malfunction or fail. One of the best investments in keeping a positioner operating is the installation of an air filter before the positioner. An air filter will help to keep impurities from fouling the small passageways inherent to a positioner.

If the actuator does not stroke and if no excessive air is exhausting from the positioner, the most likely causes of problems are tubing piped to the incorrect pressure chambers of the actuator, stuck or failed linkage between the actuator and the positioner, a plugged air filter, or a stuck pilot spool valve. If the actuator moves to its full signal position despite the signal input, the problem could be a failed feedback spring, disconnected or failed linkage, or a stuck pilot spool valve. If the calibration shifts constantly, the positioner itself may be loose from its actuator mounting. Other causes are loose or worn linkage and loose zero or span adjustments. The zero and span adjustments should always be checked to see that they are locked in place, especially when vibration is present. Excessive air consumption can be caused by worn or failed soft goods (O-rings or diaphragms) inside the positioner, loose or leaking air connections, or internal actuator leakage, such as a worn piston O-ring.

If the actuator strokes more slowly in one direction than the other during normal operation, the most likely cause is a restriction in the tubing or a plugged passageway inside the positioner spool. Some positioners have a self-cleaning plunger device that, when pushed, uses air pressure to purge dirt or water from the internal air passages of the positioner. Erratic positioner operation can be caused by a plugged passageway in the positioner, a bent spool in the pilot spool valve, or worn or failed linkage.

7.7 Auxiliary Handwheels
7.7.1 Introduction to Auxiliary Handwheels

Occasionally manual operation of an actuated valve is preferred or required; therefore, an *auxiliary handwheel* is attached to the actuator to allow for manual operation of the actuated valve in case of an emergency or when a major power interruption or failure occurs. Not only do auxiliary handwheels allow for manual operations, but some designs can be set in a position so that the handwheel acts as a stop to limit the stroke of the valve.

If an auxiliary handwheel is used while the actuator is still under signal, a three-way bypass valve is installed before the actuator or positioner to shut off the air supply and bleed or neutralize the pressure chamber(s). To prevent accidental or intentional manual operation, some manufacturers provide a locking bar that can be placed around a leg of the handwheel and locked. If this feature is not provided, a simple chain and lock can prevent movement of the handwheel.

7.7.2 Auxiliary-Handwheel Designs

Designs of auxiliary handwheels vary widely. Designs are sometimes based upon the linear or rotary motion of the actuator and/or valve. Some are an integral part of the actuator, while others are an addition to the existing actuator design, following minor modification for attachment. Auxiliary handwheels can be mounted above the actuator (called *top-mounted handwheels*) or on the side of the actuator (called *side-mounted handwheels*).

The most common auxiliary-handwheel design for linear actuators is the *continuously connected handwheel*, which is an assembly attached to the actuator stem with a neutral range that accommodates the full stroke of the actuator without interference from the handwheel. When the handwheel is turned, the handwheel nut (or similar device) moves out of a neutral range and engages either an upper or lower stop. As the handwheel continues to turn, the handwheel nut pushes against the stop, causing the actuator stem to move in that direction. The advantage of the continuously connected design is that it does not require a declutching mechanism to engage or disengage the handwheel in order to operate the actuator. In addition, when the handwheel is left in a non-neutral position, it can act as a limit-stop for that direction. Continuously connected handwheels that are integral to the actuator can be either top- or side-mounted designs (Figs. 7.37 and 7.38).

Side-mounted continuously connected handwheels can also be designed as a separate unit, which is then added to an existing actuator with slight modifications (Figs. 7.39 and 7.40), such as a special yoke. The attachment of the handwheel to the actuator stem is made external to the cylinder or diaphragm case. Therefore, the chief advan-

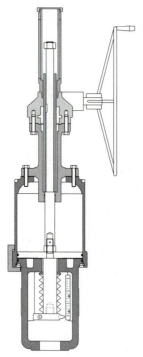

Figure 7.37. Top-mounted continuously connected handwheel mounted on a linear actuator.
(*Courtesy of Valtek International*)

tage of this design is that the handwheel can be used to lock the stem
position, allowing for disassembly of the cylinder or diaphragm casing
for maintenance while the valve remains in operation.

Another common auxiliary-handwheel design for linear actuators is
the *push-only handwheel,* which is commonly seen with both diaphragm
and piston cylinder actuators (Figs. 7.41 and 7.42). This design is top-
mounted and very simple in concept. When the handwheel is turned,
the handwheel stem—which is threaded to the top of the actuator—
lowers until the handwheel stem makes contact with the piston or
diaphragm plate and pushes it until the valve is closed or reaches a
midstroke point. The push-only design requires a spring on the opposite side of the piston or diaphragm to ensure a counterforce. Not only
can the handwheel be used to close or throttle the valve, but it can also
be used as an upper limit-stop. A modified design is available for
reverse-acting diaphragm actuators (Fig. 7.43).

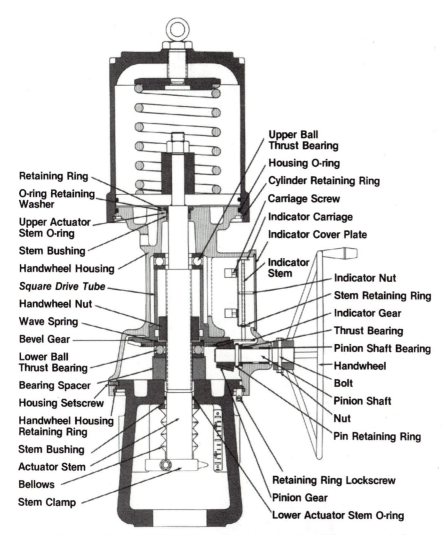

Figure 7.38. Side-mounted continuously connected handwheel mounted on a linear actuator. (*Courtesy of Valtek International*)

Rotary-motion valves can also be equipped with auxiliary handwheels (Fig. 7.44), although the rotation of the shaft does not normally permit the continuously connected design. Instead, a declutchable handwheel is used that allows the user to engage or disengage the handwheel from making a positive connection with the shaft. The main problem with the declutchable handwheel is that forces on the

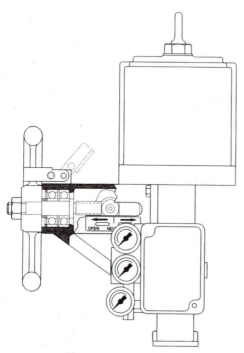

Figure 7.39. Internal view of auxiliary side-mounted handwheel mounted on a piston cylinder actuator. (*Courtesy of Valtek International*)

Figure 7.40. Auxiliary side-mounted handwheel mounted on a diaphragm actuator. (*Courtesy of Fisher Controls International, Inc.*)

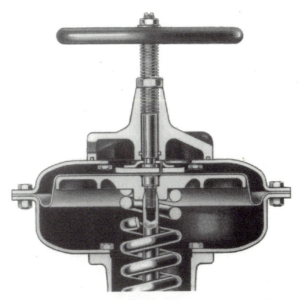

Figure 7.41. Top-mounted handwheel mounted on a direct-acting diaphragm actuator. (*Courtesy of Fisher Controls International, Inc.*)

handwheel during operation make it difficult to disengage. Also, the user must be careful to remember to disengage the auxiliary handwheel after use, since automatic operation of the actuator and valve will turn the handwheel, creating potential safety and eventual maintenance problems.

7.7.3 Auxiliary-Handwheel Troubleshooting

The chief maintenance issue with auxiliary handwheels is keeping the threads fully lubricated. If the threads are dry, the torque needed to turn the handwheel is increased, which can cause premature wear. The location of the threads determines the life of the lubrication on the threads. Some handwheels are completely self-contained, meaning that the handwheel stem and nut threads are totally encased inside the actuator. During assembly in the factory, this region is packed with sufficient grease to meet the needs of the life of the actuator or to the next maintenance cycle. Other designs provide grease fittings for injection of grease to threads, bearings, or other moving parts.

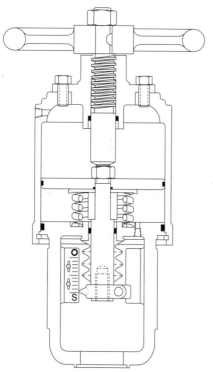

Figure 7.42. Top-mounted push-only hand-wheel mounted on a piston cylinder actuator. (*Courtesy of Valtek International*)

Lubrication is a maintenance issue when the lubricated threads are exposed to the atmosphere, such as with top-mounted push-only handwheels, or external side-mounted handwheels. In these two cases, the threads should be thoroughly cleaned before the handwheel threads are lubricated. During maintenance checks, this lubrication should be checked periodically and reapplied, if necessary. Some newer designs have overcome this problem by using a solid film lubricant, which does not require continual relubrication and does not attract dirt. Eventually, in high-usage situations, the moving parts of the handwheel (bearings, threads, guides, pins, shafts or stems, etc.) will wear. This can be detected by pulling on the handwheel and checking for any looseness, which may require replacement of the worn part.

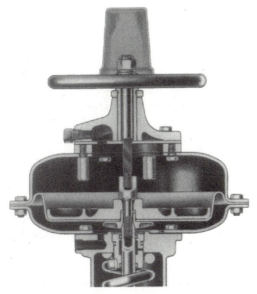

Figure 7.43. Top-mounted handwheel mounted on a reverse-acting diaphragm actuator. (*Courtesy of Fisher Controls International, Inc.*)

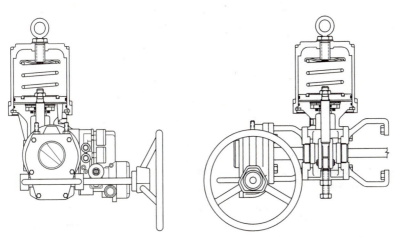

Figure 7.44. Auxiliary rotary declutchable handwheel mounted on a rotary actuator (two views). (*Courtesy of Valtek International*)

7.8 External Failure Systems

7.8.1 Introduction to External Failure Systems

In some situations, the conditions of a service are greater than the capability of an actuator's fail-safe spring. In other applications, an actuator with a heavy-duty spring may not be practical, either mechanically or economically. In these cases, an external failure system (called an *air spring*) may be added to a pneumatic actuator. An air spring is a self-contained, pressurized system that has enough pneumatic power to force the closure element to move to a particular position when the actuator's power supply is interrupted. In most cases, this failure action is to close the valve, although some applications exist that require a fail-open action. The volume of air required for this action can sometimes be provided by the actuator, or in other cases, by an external volume tank.

Occasionally the design of the valve will permit a smaller air spring. For example, with globe valves, a flow-over-the-plug design allows the plug to remain in the seated position because of the process forces; therefore the air spring needs to generate only enough force to move the plug to the seated position. If the valve is a flow-under-the-plug design, the air spring must not only have the capability of seating the valve, but also maintaining that position, which may require a larger volume of air and larger external volume tanks. Obviously, if the air spring is designed to open the valve upon failure, a flow-under-the-plug design would help that situation. The point to remember is that sometimes modifying the design of the valve itself can sometimes overcome the need for a huge external failure system.

Occasionally, the application will require that the valve remain in its last position upon loss of power, which requires a different failure system configuration. In that case, the design of the valve has no bearing on the size of the failure system, because the system must be able to handle any throttling position between full-open and full-closed.

7.8.2 Air Springs Using Cylinder Volume

For applications where the service conditions are moderate in nature, yet the failure spring cannot overcome the process, an air spring can be applied, using the air volume from the actuator. This system (air spring using cylinder volume) requires the use of a positioner. As shown in the two schematics for fail-closed and fail-open in Figs. 7.45 and 7.46, the air

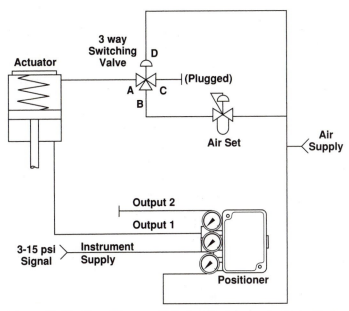

Figure 7.45. Signal-to-open (fail-closed) air spring using cylinder volume schematic. (*Courtesy of Valtek International*)

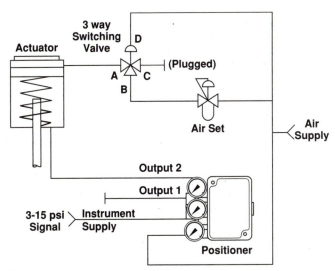

Figure 7.46. Signal-to-close (fail-open) air spring using cylinder volume schematic. (*Courtesy of Valtek International*)

spring uses a three-way switching valve and an airset. The positioner acts as a three-way positioner, providing air to only one side of the actuator. The airset is used to supply a constant air pressure on the opposite side of the actuator. It is preset at the factory to provide the necessary pressure to overcome the unbalanced forces for that particular application while still allowing the actuator to stroke normally. The three-way switching valve is used to monitor the air supply and is preset at a level close to the expected air supply—yet low enough to avoid problems with normal swings of the supply pressure. When the air supply fails or decreases below a certain preset point, the constant-pressure side of the actuator drives the actuator to its failure position. When the air supply is restored to normal levels, the three-way switching valve opens to allow normal operation of the actuator.

When air springs using cylinder volume are required, the set pressure must be calculated, using the following equation:

$$P_{A1}V_{C1} = P_{A2}V_{C2}$$

where P_{A1} = initial air pressure (absolute)
P_{A2} = final air pressure (absolute)
V_{C1} = initial volume of the actuator's pressure chamber
V_{C2} = final volume of the actuator's pressure chamber

The user must then evaluate the worst-case scenario for the required actuator force (F_A) and the actuator's piston or diaphragm area (A), which can be obtained from the manufacturer. After the force and the area are known, the following equation is used to determine the final air pressure required in the actuator (P_{A2}) for the proper failure operation:

$$P_{A2} = \frac{F_A}{A} + 14.7$$

where F_A = required actuator force
A = area of the piston or diaphragm (square inches)

To determine the switching valve setpoint (also known as the initial actuator pressure), the following equation should be used:

$$P_{SVS} = \frac{P_{A2}V_M}{V_M - AS} - 14.7$$

where P_{SVS} = switching valve setpoint (psig)
V_M = maximum volume of the actuator side that requires air to move actuator to failed position (in^3)
S = length of valve stroke (inches)

If the switching valve setpoint (P_{svs}) exceeds 80 percent of the air supply pressure, then the air volume of the actuator is not capable of handling the failure mode and an external volume tank must be used.

7.8.3 Air Springs Using a Volume Tank

When the air volume inside an actuator is not large enough to drive the actuator to its failure position, an external volume tank is provided with the valve to supply the necessary air volume. The typical air spring using a volume tank system involves an external volume tank, a three-way switching valve, two pilot-operated three-way lock-up valves, and a check valve. A four-way positioner is necessary for this arrangement, which acts to supply air to both sides of the actuator. The purpose of the check valve is to maintain the air pressure inside the volume tank if the air supply should fail.

As shown in the two schematics for fail-closed and fail-open cases in Figs. 7.47 and 7.48, the three-way switching valve monitors the air supply and is preset to a level close to the expected air supply yet low enough to avoid problems with normal swings of the supply pressure. During normal operation, the lock-up valves allow air to flow normally between the positioner and the actuator. When the air supply decreases or falls below the preset value, the pressure from the pilot of

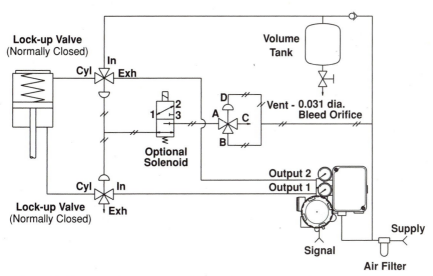

Figure 7.47. Signal-to-open (fail-closed) external volume tank schematic. (*Courtesy of Valtek International*)

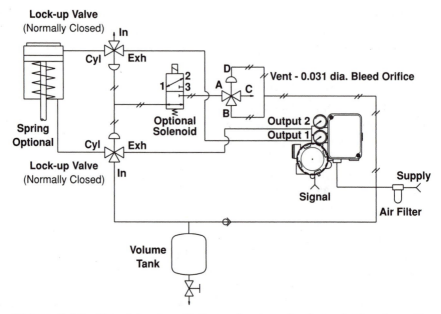

Figure 7.48. Signal-to-close (fail-open) external volume tank schematic. (*Courtesy of Valtek International*)

the three-way switching valve causes the two lock-up valves to be released. One lock-up valve channels air from the volume tank to one side of the actuator, while the other lock-up valve exhausts the other side of the actuator to atmosphere. Air from the volume tank drives the actuator to its failure position. Unless air leakage is occurring through the tubing, connections, lock-up valve, or check valve between the volume tank and the actuator, the actuator should maintain its position indefinitely. The seal between the two sides of the actuator must also be leak-free.

If the tank volume must be calculated, the following equation should be used:

$$P_{A1}V_{T1} = P_{A2}V_{T2}$$

where V_{T1} = initial volume of the external volume tank
V_{T2} = final volume of the external volume tank

The user must then evaluate the worst-case scenario for the required actuator force (F_A), and the actuator's piston or diaphragm area (A), which can be obtained from the manufacturer. After the force and the area are known, the following equation should be used to determine the final air pressure required in the actuator (P_{A2}) for the proper failure operation:

$$P_{A2} = \frac{F_A}{A} + 14.7$$

To determine the switching valve setpoint (also known as the initial actuator pressure), the following equation should be used:

$$P_{SVS} = \frac{P_{A2}V_M}{V_M - AS} - 14.7$$

If the initial pressure exceeds 80 percent of the air supply pressure, an external volume tank must be used. The following calculations help determine the correct size of volume tank.

Fail-closed actuators:

$$V_T = \frac{P_{A2}V_M}{P_{SVS} + 14.7 - P_{A2}}$$

Fail-open actuators:

$$V_T = \frac{P_{A2}V_M A}{P_{SVS} + 14.7 - P_{A2}}$$

where V_T = volume of the external volume tank (cubic inches)

7.8.4 Lock-Up Systems

Some applications require that the valve remain in place on loss of power supply. In these situations, a *lock-up system* is used. As shown in Fig. 7.49, the typical lock-up system requires a three-way switching valve, two pilot-operated three-way lock-up valves, and a four-way positioner. The three-way switching valve monitors the air supply and is preset to a level close to the expected air supply yet low enough to avoid problems with normal swings of the supply pressure. During normal operation, the lock-up valves allow air to flow normally between the positioner and the actuator.

When the air supply decreases or falls below the preset value, the pilot pressure from the three-way switching valve to the lock-up valves is released, causing both lock-up valves to close. This traps the existing air pressure on both sides of the actuator. The exhaust ports of the two lock-up valves must be plugged; otherwise, the existing air to the actuator bleeds out, creating an unstable condition.

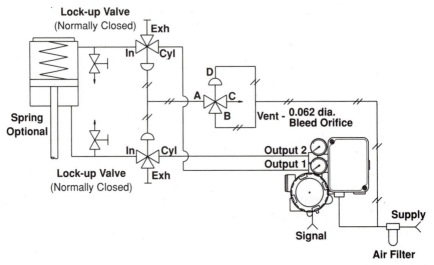

Figure 7.49. Signal-to-open, fail-in-place lock-up system schematic. (*Courtesy of Valtek International*)

7.9   Common Accessories

7.9.1   Introduction to Accessories

Some special actuation systems or actuators require fast stroking speeds, signal conversions from one medium to another, position transmissions, etc. In these applications, *accessories* are included with the actuator to help perform these special functions. Ideally, accessories are mounted directly onto the valve to ensure that the user is aware of the location of the device—although sometimes the accessory is not mounted directly onto the valve and the user must determine the location of the device.

Accessories may be produced directly by the valve manufacturer; however, in most cases they are produced by a separate manufacturer and purchased by the valve manufacturer. Rather than recreate the original vendor instructions, valve manufacturers normally include them with the valve shipment. These instructions are either attached to the accessory or included with the valve's or actuator's instructions. Keeping this literature for both the valve or actuator and accessories is important, since it details installation and servicing instructions. If instructions about the accessory are not included in the shipment, the valve manufacturer should be contacted.

7.9.2 Filters

One of the most basic accessories for actuators, whether pneumatic or hydraulic, is the filter. The *filter* is designed to screen the power supply medium of impurities or other foreign fluids or objects that may contaminate an actuation system, positioner, or other accessory. As shown in Fig. 7.50, filters are installed between the source of the power supply and the actuator or positioner. Generally, the filter is mounted immediately upstream from the accessory to ensure that the fluid is screened just prior to entering the actuator or positioner. Most are nipple- or bracket-mounted to the actuator or positioner. Filters have either a filter cartridge that has minute openings or a series of screens (screen openings are typically 5 μm in diameter). These filters or screens trap any particles of a larger diameter that can clog the inside small passages of a positioner, foul a metal moving part (such as a piston), or damage an elastomer (such as an actuator stem O-ring).

Because compressed air, especially in humid environments, has a tendency to produce water condensation, air filters have a drip well and a drain valve to allow for draining of any water. Water can foul the passageways in a positioner or cause bacterial growth that can lead to erratic performance. In single-acting valves without an air filter, the

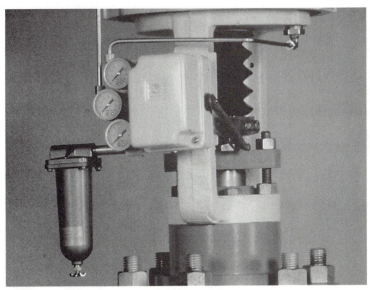

Figure 7.50. Air filter installed before a pneumatic positioner. (*Courtesy of Valtek International*)

pressure chamber can fill with water, causing slow operation or eventually no actuation at all. Through the air pressure of the system itself, the drain can also be used to remove oil and large particulates, which may be present in the air line.

7.9.3 Pressure Regulators

A *pressure regulator* (also known as an *airset*) is used to regulate or limit the air supply to the actuator. A typical pressure regulator is shown in Fig. 7.51. While many plants provide air pressures between 60 and 80 psi, some actuators cannot operate at such pressures without an internal failure. As discussed in Sec. 7.3, single-acting actuators are limited to the lower range of air pressure (usually limited to 40 psi or 2.8 bar) and require the installation of pressure regulators.

A common problem found in plants that use both single- and double-acting actuators is that some technicians, as a routine procedure, install pressure regulators on all valves regardless of the style—thereby limiting the pressure to all actuators. However, some actuators, such as piston cylinder actuators, actually operate better at higher

Figure 7.51. Pressure regulator, including air filter and moisture trap. (*Courtesy of Fisher Controls International, Inc.*)

pressures, providing greater thrust, faster stroking speeds, better stiffness, etc. In addition, placing a pressure regulator on an actuator unnecessarily can lead to possible misadjustments or add one more device that can possibly fail. Manufacturers commonly provide a sticker or tag on the actuator, notifying the user as to the pressure limits of the actuator. The general rule is to install pressure regulators only on those actuators that can only perform with lower air pressures.

7.9.4 Limit Switches

When an electrical signal must be sent indicating an open, closed, or midstroke position of an actuator or valve, an electrical switching device—called a *limit switch*—is used. Limit switches are normally used to sound alarms or operate signal lights, electric relays, or small solenoid valves. A typical signal-at-open and signal-at-closed limit-switch design is shown in Fig. 7.52, while a cammed limit switch is shown in Fig. 7.53. Limit switches are mounted directly to the actuator or rotary-transfer case and use energized arms to make a connection with the moving stem or shaft through a stop plate or similar device. Limit switches come in two basic styles: a single-pole–double-throw

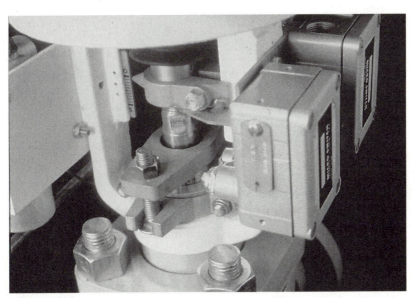

Figure 7.52. Signal-at-open and signal-at-closed limit switch schematic. (*Courtesy of Valtek International*)

Figure 7.53. Cammed limit switch. (*Courtesy of Fisher Controls International, Inc.*)

style that allows one signal to be sent to one receiver, and a double-pole–double-throw style that allows for two signals to be sent to two receivers. Cammed limit switches are capable of operating anywhere between two and six switches with one unit. Both ac and dc voltage models are available.

7.9.5 Proximity Switches

When a mechanical connection between the limit switch and the stem or shaft is not desirable, a proximity switch is used. A *proximity switch* is a limit switch that use a magnetic sensor instead of a mechanical arm. The switch's sensor is placed close to the stem or shaft, and a metal protrusion is used to trigger the switch when it approaches the sensor.

7.9.6 Position Transmitters

A *position transmitter* is a device that provides a continuous signal indicating the position of the valve or actuator, allowing for signal indication, monitoring actuator performance, logging data, or controlling associated instrumentation or equipment. A potentiometer inside the position transmitter is directly linked to the actuator stem or rotary linkage through an energized arm or linkage (Fig. 7.54). Separate zero and span adjustments are provided, allowing for special modifica-

Figure 7.54. Position transmitter (without cover).
(*Courtesy of Valtek International*)

tions, such as monitoring only a critical portion of an actuator stroke. Position transmitters can also be designed with up to four limit switches. Most position transmitters operate off of a two-wire loop, using a 4- to 20-mA dc power supply, and can be made explosion-proof with a special housing.

From a performance standpoint, position transmitters typically provide linearity and hysteresis of between ±1 and ±2 percent of full scale and repeatability between ±0.25 and ±1 percent of full scale.

7.9.7 Flow Boosters

Flow boosters are used to increase the stroking speed of larger pneumatic actuators. Because of their increased volumes, large actuators have difficulty making fast and immediate stokes. Overall, flow boosters respond quickly to sizable changes in the input signal while allowing for smooth response when the actuator receives small signal changes. A common flow-booster arrangement is shown in Fig. 7.55.

Flow boosters are typically used with positioners with the flow booster being mounted between the positioner and actuator. The flow booster is tubed to the air supply, allowing for the full air pressure to be used to stroke the actuator in the event a larger signal increase or

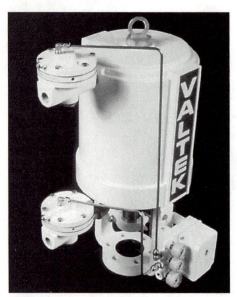

Figure 7.55. Flow boosters mounted to double-acting actuator. (*Courtesy of Valtek International*)

decrease is given. The flow booster utilizes the full air supply only if a large signal is received; otherwise, the normal air flow from the positioner moves through the booster unaided. The air flow is preset using a bypass valve inside the booster. However, when a larger signal is received, the booster inlet or exhaust port opens. If the booster inlet opens, full air supply is sent unregulated to the desired air chamber. At the same time, another booster's exhaust port opens, allowing the opposite air chamber to vent. Both boosters remain in these positions until the pressure differential reaches the dead-band limits of the bypass valve in the booster. When the bypass valve opens, the supply inlet or exhaust ports close and the flow boosters return to normal operation.

To illustrate the advantage of using flow boosters, the following example is provided. A standard 50-in^2 (322-cm^2) actuator requires nearly 4 s to stroke 3 in (7.6 cm), using 0.25-in (0.6-cm) tubing and a 80-psi (5.5-bar) air supply. With flow boosters, this same actuator can stroke in under 1 s. In larger actuators, the example is even more dramatic. A 300-in^2 (1935-cm^2) actuator with a 4-in (10.2-cm) stroke, using 0.375-in (1-cm) tubing and a 80-psi (5.5-bar) air supply, requires over 30 s to stroke. However, with the aid of flow boosters, the stroking time is decreased to under 3 s.

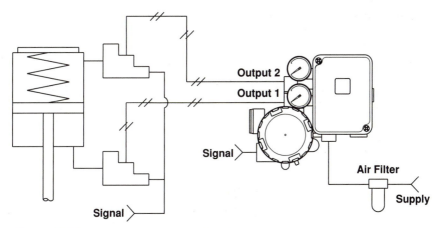

Figure 7.56. Signal-to-open, fail-closed flow-booster schematic. (*Courtesy of Valtek International*)

Figures 7.56 and 7.57 show flow-booster schematics for both signal-to-open and signal-to-close arrangements. For exceptional situations, two flow boosters can be installed on each side of the actuator—as long as both flow boosters are connected parallel to the cylinder port, positioner output tubing, and the air supply.

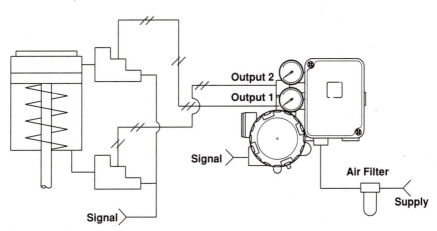

Figure 7.57. Signal-to-close, fail-open flow-booster schematic. (*Courtesy of Valtek International*)

7.9.8 Solenoids

A *solenoid* is an electrical control device that receives an electrical signal (usually a 4- to 20-mA signal) and, in response, channels air supply directly to the actuator. Two types of solenoids, three-way and four-way, are commonly used with actuators and positioners. *Three-way solenoids* are sometimes used to operate single-acting actuators, such as diaphragm actuators, since they are designed to only send air to one air chamber in the actuator. With double-acting actuators, three-way solenoids are used to interrupt or override an instrument signal to a pneumatic positioner.

Four-way solenoids are used in lieu of positioners to provide on–off operation of double-acting actuators, providing a positive two-direction action. As shown in Figs. 7.58 and 7.59 (showing both closed and open actions), upon deenergization the four-way solenoids send full air supply to one side of the actuator, while exhausting the other side to atmosphere.

7.9.9 Quick Exhaust Valves

Quick exhaust valves are pressure-sensitive venting devices that are used with double-acting actuators in on–off applications where positioners are not required. When triggered, quick exhaust valves almost instantaneously vent one side of the double-acting actuator to atmosphere, allowing the valve to move to the full-closed or full-open position. Quick exhaust valves are installed between the air supply and the actuator. As long as a normal air supply is provided to the actuator,

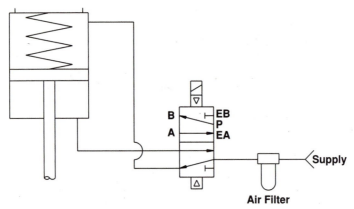

Figure 7.58. Deenergized-to-close, fail-closed four-way solenoid schematic. (*Courtesy of Valtek International*)

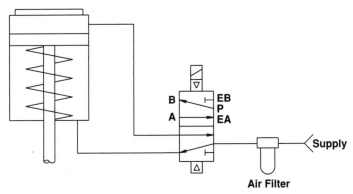

Figure 7.59. Deenergized-to-open, fail-open four-way solenoid schematic. (*Courtesy of Valtek International*)

normal operation continues. However, when the air supply fails or is interrupted, the quick exhaust valve reacts to the significant differential pressure. An internal diaphragm diverts the exhaust flow coming from the actuator through an enlarged orifice, allowing the internal pressure of the actuator to vent much more quickly. A needle valve must be installed parallel to the quick exhaust valve so that the trip point of the quick exhaust valve can be adjusted, allowing it to react only to large signal demands.

Quick exhaust valves are especially helpful with on–off applications, where exceptional stroking speeds are required in both directions (see Figs. 7.60 and 7.61). Another common application for quick exhaust

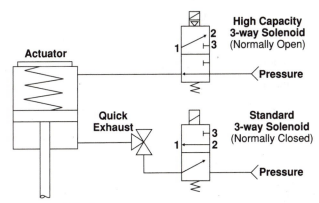

Figure 7.60. Fast-closing, fail-closed on–off system with quick exhaust schematic. (*Courtesy of Valtek International*)

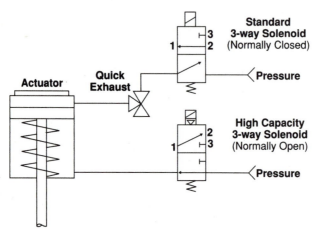

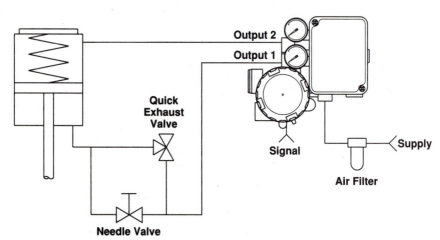

Figure 7.61. Fast-opening, fail-open on–off system with quick exhaust schematic. (*Courtesy of Valtek International*)

valves is when a double-acting actuator with a positioner must provide a fast stroke in one direction (as shown in Figs. 7.62 and 7.63).

7.9.10   Speed Control Valves

Speed control valves are used to limit the stroking speed of an actuator by restricting the amount of air flow to or from the actuator. These small valves can be mounted between the tubing and the actuator and

Figure 7.62. Signal-to-open, fail-closed positioner with quick exhaust schematic. (*Courtesy of Valtek International*)

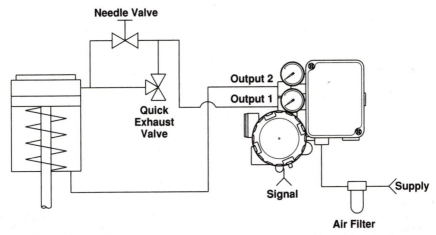

Figure 7.63. Signal-to-close, fail-open positioner with quick exhaust schematic. (*Courtesy of Valtek International*)

are available in sizes that match common tubing sizes. They can only be used in one direction; therefore, if stroking speeds must be controlled in both directions, two speed control valves must be used (one in each direction). A typical application using speed control valves is found in Fig. 7.64.

7.9.11 Safety Relief Valves

When volume tanks are used or if high-pressure actuators must be used to handle the service conditions, some local codes require the

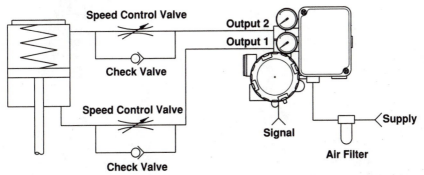

Figure 7.64. Signal-to-open, fail-closed speed control system schematic. (*Courtesy of Valtek International*)

installation of safety relief valves on these high-pressure vessels as protection against overpressurization. By definition, *safety relief valves* are designed to open to atmosphere when a particular pressure is exceeded. Because of the differing codes for local governing bodies, valve manufacturers normally defer to the user to install safety relief valves.

7.9.12 Transducers

Transducers are devices that convert an electrical signal to a pneumatic signal, which may be required to operate a positioner with a pneumatic actuator. Transducers have become more commonplace as the popularity of I/P signals has increased with newer control systems, and existing positioners must be converted from pneumatic to electrical signals. The most common transducer is one that converts a 4- to 20-mA signal to a 3- to 15-psi (0.2- to 1.0-bar) pneumatic signal. The pneumatic output signal coming from the transducer normally follows a linear characteristic. Transducers can be mounted directly on the actuator or installed separately, if vibration is a problem (Fig. 7.65).

Figure 7.65. Transducer separately mounted from actuator. (*Courtesy of Valtek International*)

7.10 Electrical Equipment Certifications

7.10.1 Introduction to Electrical Certification

The installation of electrical equipment (such as electric actuators, limit switches, transducers, etc.) on valves poses a problem for plants that have a potentially explosive or highly flammable process. This is particularly a problem in the petroleum production and chemical industries, where both conditions exist and must be protected against. Electrical equipment has the potential to spark and cause a flammable reaction and must be certified to be safe for those types of environments. Classifications may be listed according to the following common terms: intrinsically safe, explosion-proof, flame-proof, ingress protection (IP), and nonincendive.

As defined by The National Electric Code (National Fire Protection Association), a *hazardous location* is defined as one in which an explosion or fire may result due to the existence of flammable vapors in the atmosphere or due to the presence of highly flammable liquids, fibers, or dust. The intent of designating an area as hazardous is to ensure that all electrical devices use certain safeguards to ensure that an electrical spark will not cause a fire. To create a flame or an explosion, three elements must be present. First, a fuel must be present, consisting of a flammable gas, vapor, powder, dust, or fiber. Second, an oxygen source must be present, which includes air and oxygen services. The third element is the most controllable, the ignition source itself. Therefore, in most hazardous locations, where two of the three elements are present (fuel and oxygen), a barrier or relocation of the ignition source must be used to avoid a fire or explosion.

7.10.2 Hazardous Classifications

In most industrial areas of the world, classification systems have been established to differentiate between different types of hazardous locations. For example, in North America, hazardous locations are grouped according to class, division, and group. The *hazardous class* deals with the type of material present surrounding the device. Class I is used for all flammable vapors or gases. Class II is used for combustible dust, while Class III is used to describe all fibers or flyings. *Flyings* are defined as airborne particles. The *hazardous division* is divided into two areas that deal with the likelihood that an atmosphere conducive to explosions is present. Division I is defined as normal operating conditions that include the probability of an explosive mixture. Division II is defined as an abnormal operating condition that would be conducive to an explosion.

European codes [such as the IEC (International Electrotechnical Commission)] use the designation *hazardous zone* instead of *hazardous division* with Zone 0 defined as a continuous hazard, Zone 1 as an intermittent hazard, and Zone 3 as a hazard under irregular conditions. *Hazardous groups* are more descriptive in nature than classes. Under Class I (flammable gases and vapors), those materials placed in Group A are defined as the most flammable, while those materials placed in Group D are defined as the least flammable (although still flammable in nature). Under Class II (combustible dust), Group E is the most combustible, while Group G is the least. Under Class III (ignitable fibers), materials are not grouped. Under the European standard (IEC), the properties of the gas or vapor determine which group it is placed in. These properties include the auto-ignition temperature (AIT), minimum igniting current (MIC), etc. Figure 7.66 provides an easy reference table for understanding the hazardous location class, division, and group, as well as listing common gases, dusts, or flyings for each group.

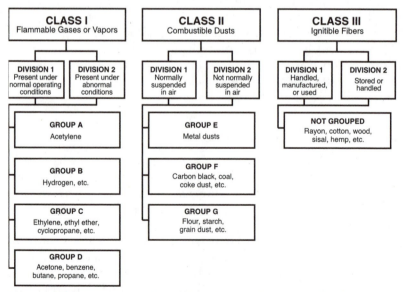

Figure 7.66. International hazardous location by materials. (*Courtesy of Valtek International*)

7.10.3 Temperature Classifications and Codes

Electrical devices that exist in hazardous conditions are required by code to list the surface temperature the device can produce when placed in a malfunction or fault condition. This is common throughout the industrial areas of the world. Each type of dust, vapor, or gas has an *auto-ignition temperature* (AIT), which is the surface temperature that will cause the flammable material to spontaneously ignite. For example, a gas that has an auto-ignition temperature of 270°F (132°C), would require a T4 rating on any electrical device placed in that environment. Table 7.2 provides a listing of North American and European classifications and the maximum surface temperature of each T classification. Note that the higher the classification, the lower the allowable surface temperature.

A number of protection methods have been devised to ensure that electrical devices operate safely in a hazardous location. The two most common methods are explosion-proof and intrinsically safe. A device is considered *explosion-proof* if it has the capability to withstand an internal explosion of a defined vapor or gas. In addition, the device must be designed such that it will not permit the ignition of a vapor in the surrounding atmosphere due to a flash, spark, or internal explosion inside the device. The device also must not produce a surface temperature that will ignite the flammable atmosphere. An electrical device is considered to be *intrinsically safe* when the device and its wiring do not have the capability of creating electrical heat that could cause an ignition under normal or abnormal conditions. This requirement means that an intrinsically safe device must have a safety barrier. A *safety barrier* is defined as a special assembly located outside of the hazardous location, which is designed to restrict the voltage and current being sent to the device during both normal and abnormal operation. Generally, explosion-proof devices are more commonplace in North America, while intrinsically safe devices are common to Europe.

7.10.4 Testing Agencies and Codes

Electrical equipment can be certified by a number of governing bodies, which have jurisdiction over certain types of equipment or industries. Some of the most common are National Electrical Manufacturers Association (NEMA), the National Fire Protection Agency, and Underwriters Laboratories (UL) in the United States, the Canadian Standards Association (CSA), European Committee for Electrotechnical Standardization (CENELEC), Physikalisch-Technische

Table 7.2. Temperature Classifications (According to
Maximum Temperature)*

Europe Classification	North America Classification	Maximum Surface Temperature
T1	T1	842° F 450° C
T2	T2	572° F 300° C
	T2A	536° F 280° C
	T2B	500° F 260° C
	T2C	446 ° F 230° C
	T2D	420° F 215° C
T3	T3	392° F 200° C
	T3A	356° F 180 ° C
	T3B	329° F 165° C
	T3C	320° F 160° C
T4	T4	275° F 135° C
	T4A	248° F 120° F
T5	T5	212° F 100° C
T6	T6	185° F 85° C

Data Courtesy of Valtek International.

Bundesanstalt (PTB) in Germany, the International Electrotechnical Commission (IEC), the Health and Safety Executive (BASEEFA) in the United Kingdom, and the Sira Certification Service (SCS) in the United Kingdom. Each agency performs a number of rigorous tests on a device to understand its capabilities and protect the device against adverse conditions, as well as protecting a volatile atmosphere from any sparking caused by the electrical device.

Testing agencies designate approved equipment as explosion-proof or protected for particular applications by the *Ex* symbol. On the other hand, the symbol *EEx* is used to show that the equipment was tested and approved by one of the European standards of CENELEC. Equipment with the EEx designation is accepted by members of the European Union (EU). CENELEC has developed a classification system to show the level of explosion protection for a particular device. For example, an electrical device with the classification of *EEx d IIC T6* would indicate the following: *E*, certified to European standards by CENELEC; *Ex*, explosion protection; *d*, flame-proof protection provided; *IIC*, acetylene or hydrogen gas service; and *T6*, a maximum surface temperature of 185°F (85°C). Tables 7.3 and 7.4 provide common types of protection and gas group designations.

The IEC classifies the degree of protection a certain enclosed device provides against the entrance of solids and liquids. The IEC system (Standard 529) uses the designation *IP*, followed by a two-digit number to show this degree of protection against solids (first digit) and liquids (second digit). For example, a device with the designation *IP 65* would indicate the following: *IP* indicates that the device is protected against the entrance of solids and liquids, *6* indicates that the device is dust-tight, and *5* indicates that the device is protected against water jets. Table 7.5 lists the IEC codes for these two-digit codes.

NEMA has established a common classification system for designating the type of protection a housing will provide against specific environmental or atmospheric conditions. A listing of NEMA codes and the atmospheric or environmental conditions for each is provided in Tables 7.6 and 7.7.

Table 7.3. CENELEC Protection Designations*

Designation	Type of Protection
o	Oil immersion
p	Pressurization
q	Powder filling
d	Flame-proof
e	Increased safety
i ia ib	Intrinsically safe Zones 0, 1 and 2 Zones 1 and 2
m	Encapsulation
n	Non-incendive (Zone 2 only)

Data courtesy of Valtek International.

Table 7.4. CENELEC Gas Group Designations*

Designation	Gas Group
IIA	Propane
IIB	Ethylene
11C	Acetylene, hydrogen

Data courtesy of Valtek International.

Table 7.5. IEC IP Codes*

First Digit	Dust Protection	Second Digit	Water Protection
0	None	0	None
1	Objects 50 mm or larger	1	Vertically falling water drops
2	Objects 12.5 mm or larger	2	Vertically falling water drops with enclosure tilted up 15°
3	Objects 2.5 mm or larger	3	Spraying water
4	Objects 1.0 mm or larger	4	Splashing water
5	Dust protected	5	Water jets
6	Dust tight	6	Powerful water jets
		7	Temporary immersion in water
		8	Continuous immersion in water

Data courtesy of Valtek International.

Table 7.6. NEMA Hazardous Location Designations*

NEMA Rating Type	Environment	Hazard Locations
7	Inside	Locations classified as Class I; Groups A, B, C, or D (National Electrical Code)
8	Inside/Outside	Locations classified as Class I; Groups A, B, C, or D (National Electrical Code)
9	Inside	Locations classified as Class II; Groups E, F or G (National Electrical Code)
10	Not Applicable	Enclosures constructed to meet the applicable requirements of the Mine Safety and Health Administration (USA)

Data courtesy of Valtek International.

Table 7.7. NEMA Enclosure Designations*

Nema Rating Type	Environment	Protection Against
1	Inside	Contact with the enclosed equipment
2	Inside	Limited amounts of falling water and dirt
3	Outside	Windblown dust, rain and sleet, and to remain undamaged by the formation of ice on the enclosure
3R	Outside	Falling rain and sleet, and to remain undamaged by the formation of ice on the enclosure
4	Outside	Windblown dust, rain, splashing water, and hose-directed water, also to remain undamaged by the formation of ice on the enclosure
4X	Outside	Same as NEMA rating 4 except enclosure is corrosion-resistant
5	Inside	Settling airborne dust, falling dirt, and dripping non-corrosive liquids
6	Inside/Outside	Entry of water during occasional temporary submersion at a limited depth
6P	Inside/Outside	Entry of water during prolonged submersion at a limited depth
11	Inside	Oil immersion, enclosed equipment against the corrosive effects of liquids and gases
12	Inside	Dust, falling dirt, and dripping non-corrosive liquids
12K	Inside	Enclosures with knockouts used to provide protection against dust, falling dirt and dripping non-corrosive liquids (other than at knockouts)
13	Inside	Dust, spraying of water, oil and non-corrosive coolant

Data courtesy of Valtek International.

8

Smart Valves and Positioners

8.1 Process Control

8.1.1 Introduction to Process Control

Although the majority of valves and actuators are used with analog systems, the face of process control is changing such that digital technology may quickly overtake those analog systems that are so prevalent today. A *smart* valve is defined as any valve with a microprocessor integrated into the valve design. Smart final control elements—such as intelligent systems mounted on valves or digital positioners used with actuators—have less or no moving parts to fail, and the performance associated with digital communications is far and away better than the 4- to 20-mA signal found with I/P analog systems.

To understand the terminology and abilities of smart products, a number of common instrumentation and control principles and terms must be generally understood.

8.1.2 Controllers and Distributive Control Systems

A wide majority of control systems that link process sensors and final control elements, such as control valves and actuators, use controllers or distributive control systems to provide intelligence in the control loop. A *controller* is a microprocessor that receives input from a process sensor—such as a pressure or temperature sensor or flow meter—and compares that signal against a predetermined value. After the comparison is made, it sends a correcting signal to a final control element until the predetermined value is reached. A common controller seen in today's systems has a three-way mode that allows for loop tuning—in other words, the adjustment by the user of the proportional, integral, and derivative settings, which is commonly called *PID control*. With PID control, these three settings can be adjusted to optimize the control loop or to provide certain control loop characteristics. For example, with PID control, variations between the set point and process variable can be automatically corrected, or the system speed can be increased to improve system response.

Related to a controller, but on a much larger scale, is the *distributive control system* (or DCS). The DCS is a central microprocessor designed to receive data from a number of devices and control the feedback to several final control elements. With a DCS, all wiring for the input devices and final control elements lead to one central area, usually in a control room where the DCS is located.

8.1.3 Analog Process Control Systems

With a conventional analog system, the process sensing device transmits a 4- to 20-mA signal to a controller or DCS. The signal is sent through a dedicated line, which is typically a shielded two-wire line. Because the controller or DCS is simply a process computer that utilizes digital signals, the analog information coming from the field must be converted to a digital signal for the controller or DCS to use. This is accomplished through an analog input–output interface card, which converts the analog signal to a digital signal for the microprocessor to use, as shown in Fig. 8.1. If the information received from the transmitted signal is different from the value needed by the process, the controller or DCS sends a correcting signal to the final control element, which can be a control valve. Once again, because of the analog communication lines involved, the controller or DCS will send a digital signal, which is then converted to an analog signal and transmitted across a dedicated analog line to the control valve. The control valve responds by moving its position until the correct process value is achieved.

Analog devices—such as a flow meter, a limit switch, or a positioner—are used to generate process information or react to feedback from the controller and create an analog signal through mechanical means. For example, a limit switch depends on the mechanical movement of the shaft to make contact with the lever arm of the limit switch, which causes the contacts of the switch to meet and send the analog signal.

The main advantage of an analog process control system is that,

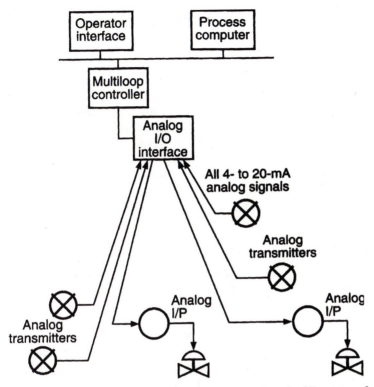

Figure 8.1. Analog process communication network. (*Courtesy of Fisher Controls International, Inc.*)

because of the analog input–output interface, any analog device—whether it be a flow meter or control valve—can communicate with the controller or DCS, making equipment interchangeability easy. A secondary advantage is that the analog system has general acceptance around the world. Instrumentation people are familiar with it and the majority of process devices use it.

Analog systems have a number of disadvantages. First, they must have dedicated lines—or in other words, one line per device. If two devices are placed on one 4- to 20-mA line, the signals are apt to interact adversely with one another and confuse the controller or DCS. Of course, electrical lines can be influenced adversely by magnetic fields and radio frequencies. In addition, wires can be damaged or break. Analog devices must have moving parts to create the analog signal, which can wear, fail, or hang up. Also, because analog devices have mechanical adjustments, calibration can wander or drift from the necessary settings, especially where vibration occurs.

8.1.4 Digital Process Control Systems

Because of the disadvantages of the analog process control system, coupled with the advent of microprocessor-based controllers and distributive control systems, the interest in using digital communication has grown significantly throughout the 1990s. Ideally, a digital process control system would not only utilize the digital communications associated with the controller or DCS, but would also use the same digital communications with the process sensors and final control elements. This would do away with the analog-to-digital interface conversion as well as some of the mechanical parts and motion associated with analog devices. It would greatly improve product reliability, with a minimal amount of moving parts to fail or wires to break. It would also ensure that exact information is received by the controller and that the final control element follows the feedback perfectly. With digital systems, hysteresis, repeatability, and other control problems are minimal as compared to analog systems. Although lines would still be needed between the controller or DCS, as well as the process sensor and final control element, digital communications would allow a number of devices to use a single line. This is because each device can have an electrical signature that would allow it to identify itself to the DCS or controller without signal interference.

Digital communications are dependent on a standardized communication all-digital language, called *fieldbus*. With a standardized fieldbus, field devices could not only communicate with the controller or DCS but also with other field devices. Up until the mid-1990s, the problem with a standardized fieldbus has been the lack of a general agreement as to which communication language would best serve the needs of the process industry. Early fieldbus developers each produced a different communication language. However, for a user to have full digital communication, all of the smart devices had to operate off of the same fieldbus, which limited the options of the user since several fieldbus standards were proposed. This debate is finally being resolved among those promoting various fieldbus languages. In the near future, a standardized fieldbus will be designated, allowing for all smart process equipment to use the same language and a true digital relationship to exist between all products. The fieldbus must also provide a reasonable power supply to run the complex functions of smart equipment.

   With a digital system, analog input–output interfaces will be replaced by a fieldbus digital interface, which can receive a number of signals from multiple devices connected to one digital line. The main advantage of digital communications is that the signals sent by any device are easily identified through an instrument signature and can be separated from competing signals. This allows the DCS to sort the information according to one device and send feedback input to another device, all on one line, as shown in Fig. 8.2.

   The most obvious advantage of a digital system is the improved accuracy and response of the system. With digital communications, no portion of a signal is lost. The lack of moving parts or linkages means better performance, less maintenance and recalibration, and lower spare part inventories. Once a full digital communication link is in place, interchangeability among all devices is possible, which was one of the benefits of the analog system. If PID control is included with the

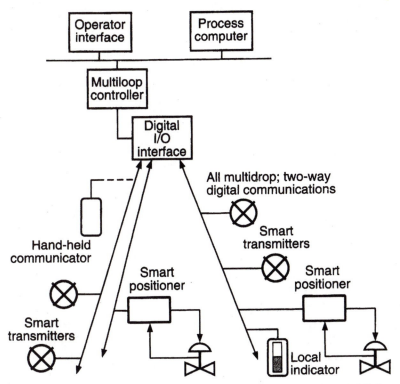

Figure 8.2. Digital fieldbus communications network. (*Courtesy of Fisher Controls International, Inc.*)

system, the digital system will allow for automatic loop-tuning, improving the performance of the control loop. Information about the performance of equipment can lead to equipment and process diagnostics, which assists with planned maintenance and eliminates maintenance surprises. With fieldbus, power is available within the communication lines to run the digital equipment, eliminating outside power sources.

8.1.5 Role of Smart Valves

The term *smart valves* has been applied to those control valves that use on-board microprocessors or digital positioners to communicate with either analog or digital systems. As final control elements, control valves must have the ability to communicate digitally with the controller or DCS, as well as to interact with other digital field instruments and to take advantage of the positive aspects of digital communications. As a minimum this requires a digital positioner.

This development of smart products, however, has been slowed by the lack of a standardized fieldbus—although some smart valves and positioners available today have been developed so that they can handle a number of proposed fieldbus versions. Today's smart valves vary widely according to the capabilities of microprocessor and design. For example, intelligent systems provide complete single-loop control when placed on a valve—which requires process sensors, a controller, and a digital positioner. This allows for a wide range of functions, from process control to data acquisition to self-diagnostics. In addition, with some smart-valve designs, PID control can be added to automatically loop-tune the process so that it is more efficient. On the other hand, a digital positioner has the microprocessor included with the positioner and is used only to assist the valve with its ability to act as the final control element. Overall, both smart valves and positioners can provide various levels of valve self-diagnostics and management of safety systems, such as a controlled shutdown.

Nearly all existing plants today are wired with analog lines, each attached to an individual input device or final control element. To replace these analog lines with digital lines would be time consuming and expensive. That is why an open protocol has been developed. The open protocol allows smart products to utilize existing analog lines for both communication and power needs. That means that those smart devices that use the existing 4- to 20-mA lines must use the worst-case scenario—4 mA—as the main power source. The problem with such low power is that the device can only have a limited amount of electronics and, therefore, the smart capabilities of such devices are limited. As mentioned earlier, one advantage of a fieldbus is that the power could be increased to expand the capabilities of smart devices in general.

Smart valves are primarily linear-motion valves, with globe valves being the primary focus, although some rotary-motion valves have been modified to smart service. An advantage of using smart products with a rotary valve (which has an inherent flow characteristic) is that a modified flow characteristic can be custom programmed, providing better flow control for the user. Also, a smart valve can correct the problems associated with a positioner's linear-to-rotary motion, which does not produce a true linear signal because of the swing arc of the positioner take-off arm.

8.2 Intelligent Systems for Control Valves

8.2.1 Introduction to Intelligent Systems

As discussed earlier, the most sophisticated smart valve is a control valve that is equipped with an intelligent system with process sensors. The *intelligent system* is a microprocessor-based controller that is capable of providing local process control, diagnostics, and safety management. Process input to the intelligent system comes through process sensors mounted on the body, as shown in Fig. 8.3. The system also has internal sensors to monitor valve-stem position and pressures on both sides of the pneumatic actuator.

Placing a controller and process sensors on a control valve allows for *single-loop control*—defined simply as an input sensor sending information to the controller, which sends a correcting signal to a final control element until the correct value is achieved. By monitoring the upstream pressure, downstream pressure, temperature, and the stem position, the intelligent system can calculate the flow rate for the valve and compare that against the predetermined set point—and make any necessary position adjustments to provide the correct flow rate. The intelligent system can be configured to handle single-loop control for the pressure differential, upstream pressure, downstream pressure, temperature, flow rate, stem position, or another auxiliary process loop. Because the intelligent system can be programmed to handle local control and measurement of the process, the DCS can be used to handle more demanding control situations elsewhere in the plant or to

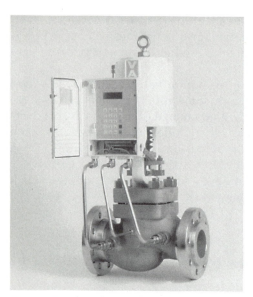

Figure 8.3. Intelligent control system with an integral digital positioner mounted on a globe control valve. (*Courtesy of Valtek International*)

provide an overall process supervising function. With its local controller, the intelligent system is then capable of monitoring and creating a record of the upstream and downstream pressure, pressure differential, process temperature, and the flow rate. The controller of intelligent systems can be equipped with PID control that uses a value from an external transmitter or internal process parameters as the control variable. This allows the process to be tuned for more efficient process control in a number of wide-ranging applications.

Intelligent systems can be used in either analog or digital systems with digital or conventional analog positioners (Fig. 8.4). They can respond to PID operation with a 4- to 20-mA analog signal, with a digital signal, or through a preprogrammed set point. Intelligent systems sometimes require the use of a personal computer or the DCS to set the tuning and operating parameters of the smart valve—although some of the newer versions come equipped with an on-board keypad, which allows for direct operation.

The user communicates with the intelligent system through a number of operator interfaces: DCS input–output interface card, personal computer, hand station and recorder, or personal computer. When a personal computer is used to communicate with the intelligent system,

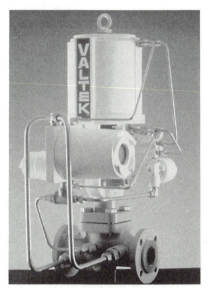

Figure 8.4. Intelligent control system combined with an analog positioner mounted on a globe control valve. (*Courtesy of Valtek International*)

interface software (provided by the manufacturer) must be installed.

The close proximity of the process sensors and control valve to the controller greatly reduces the dead time or lag time, significantly increasing the response to process changes. When a digital positioner is included in the intelligent system, the problems associated with hysteresis, linearity, and repeatability are greatly reduced. The intelligent system has the capability of collecting and issuing flow and process data to the DCS, which provides the user with current engineering analysis of the process. Remote sensors can also be tied to the intelligent system for improved control of the other parameters of the process without having to channel the data through the DCS.

An important side benefit of an intelligent system is that line penetrations are reduced significantly—an important consideration in this age where fugitive emissions are a critical concern. Because the process sensors are installed on the valve itself, the single-point installation of the valve eliminates separate line penetrations for the flow meters as well as the temperature and pressure sensors. Therefore, instead of having four or five line penetrations as part of the control loop, only one (the smart valve) exists, which eliminates a number of potential leak paths as well as decreasing functions that need to be reported to the Environmental Protection Agency (EPA) or other governing body.

Intelligent systems allow for valve and process self-diagnostics through their ability to record a signature of the valve or process. When the valve is first installed, a signature can be taken of the valve's initial start-up performance or of the process itself by plotting the flow against certain travel characteristics. As the valve continues in operation, periodic monitoring of the valve's and system's performance can be compared against the initial start-up signature. When this performance begins to falter through normal wear or through an unexpected failure, the intelligent system can warn the user of pending or existing problems, allowing for preventative maintenance or corrective action to take place before a major system or valve failure. For example, the system can take a signature of the leakage through the seat in a closed position (by monitoring the downstream pressure). Over time the intelligent system can compare the initial signature against the current body leakage signature. If the current reading exceeds the ANSI leakage class (a preset condition) due to a damaged or worn closure or regulating element, the system can warn the user that servicing of the closure element is needed. By monitoring the upper and lower pressure chambers of the actuator, intelligent systems can also evaluate a loss of packing compression and actuator seals or recognize jerky stem travel, which may point to a problem with the closure or regulating element. If an analog positioner is used with the system, hysteresis, repeatability, and linearity can be monitored.

Since a process signature is possible, the system's overall performance, which can be affected by associated upstream or downstream equipment, can be monitored and evaluated as well. For example, if an upstream pump begins to slow, the upstream pressure will decrease and fall below acceptable limits at a certain point. When the intelligent system finds the pressure dropping below the preset value, it can alert the user, who can then schedule the necessary valve or actuator maintenance.

Safety management is another use for intelligent systems, since they are capable of programmable settings that can notify the user when process limits are violated by a system upset. In addition, the systems can be used to monitor and analyze the process during start-up and shutdown, warning of any sudden departures from the normal service conditions. Multiple failure modes can be programmed into the intelligent system, which will provide a different mode for a variety of failures: loss of air supply or power, process failure, loss of command signal, etc.

Data logging is another advantage of intelligent systems, as they have the ability to record process conditions through user-specified intervals. For example, some intelligent systems are capable of recording up to 300 lines of process conditions at intervals anywhere between 1 s and 3 h apart. This data log is normally provided so that the user can evaluate the process, looking for any abnormalities or upsets.

The wide range of benefits of an intelligent system is often reflected in the price of the intelligent system, which may produce some "sticker shock" to those only accustomed to the cost associated with other actuator accessories. However, the user should look at the larger picture: The intelligent system takes the place of a controller, individual pressure and temperature sensors, a flow meter, limit switches, tubing and wiring, etc. Taken together, the cost of an intelligent system mounted directly on a control valve is less than the sum of the individual pieces of equipment. The only evident problem with an intelligent system is that it requires a separate 24-V dc power supply to run the electronics, which may require some additional wiring and a conversion box if only standard ac power is available.

A simplification of the intelligent system is to install the system to the actuator without including the process sensors in the body (using existing sensors already installed in the system)—in essence, creating a very powerful digital positioner. This allows the intelligent system to function with many of the advantages discussed earlier, but without the on-board single-loop control. The advantage is that the cost is less, yet the system offers many of the smart technology benefits associated with the full intelligent system.

8.2.2 Intelligent System Design

Shown in Fig. 8.5 is a schematic of a typical intelligent system. Power is supplied by a separate 24-V dc source as well as a compressed air source. Pressure sensors are mounted directly to the body on the upstream and downstream sides of the closure element. The location of the pressure sensors on the body is critical to ensuring proper pressure readings without being affected by an increase of velocity as the flow moves through the closure or regulating element or any other narrowed section of the body. The temperature sensor is placed between the pressure sensors and as close to the closure or regulating element as necessary to determine the best process temperature reading. The wiring for the sensors is tubed directly to the intelligent system. Pneumatic lines feed air from the digital positioner (in this case, the digital positioner is part of the intelligent system) to the upper and lower chambers of the actuator.

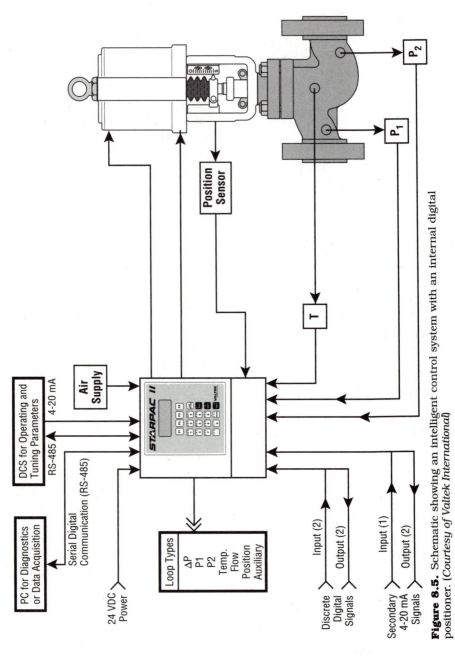

Figure 8.5. Schematic showing an intelligent control system with an internal digital positioner. (*Courtesy of Valtek International*)

Operating or tuning input, as well as data acquisition, takes place through either the supervisory DCS or through a personal computer via a serial digital communication line, which is a designated electrical signal, such as RS-485. A separate 4- to 20-mA line is linked from the DCS to the intelligent system for any stand-alone command signals. With the single-loop control associated with an intelligent system, this line is often not necessary, but is available if needed.

Input and output lines are provided for discrete digital signals that act as switches, allowing the user to toggle between manual and automatic operation of the intelligent system or for other custom configurations. The secondary 4- to 20-mA signal inputs are used for any auxiliary input, such as from a remote flow meter to control downstream pressure. The secondary 4- to 20-mA outputs are used to communicate with another supervisory device, such as another controller.

As noted earlier, intelligent systems can be used with stand-alone positioners. A schematic of an intelligent system with an analog positioner is shown in Fig. 8.6.

8.3 Digital Positioners

8.3.1 Introduction to Digital Positioners

Following the introduction of the intelligent system for control valves, a logical step was to move toward *digital positioners*, which are devices that use a microprocessor to position the pneumatic actuator and to monitor and record certain data (Figs. 8.7 and 8.8).

Digital positioners do not provide single-loop control as intelligent systems do; therefore, they must be installed in a more conventional process loop, with a controller and process sensors. Although they are not equal to intelligent systems, digital positioners can perform some of the same functions. For example, a digital positioner can measure and transmit actuator-stem position, providing alarm signals (similar to limit switches) when a certain position is reached or exceeded and eliminating any requirement for an independent position transmitter. PID control and tuning are also possible.

Because the pressures to the actuator are monitored, changes in actuator operation pressures can allow self-diagnostics of the actuator and certain aspects of the valve, such as changes in packing compression or a binding closure element. As with all smart devices, digital positioners have an electronic signature that allows for remote identification. The positioner can be characterized and calibrated remotely

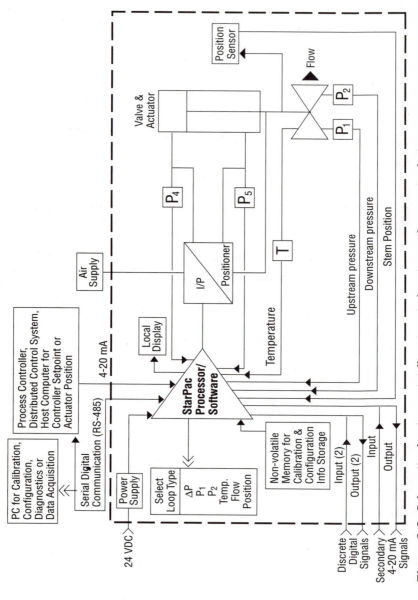

Figure 8.6. Schematic showing an intelligent control system with separate electro-pneumatic positioner. (*Courtesy of Valtek International*)

Figure 8.7. Computer interface with a digital positioner. (*Courtesy of Valtek International*)

Figure 8.8. Two-way communication link with a digital positioner. (*Courtesy of Fisher Controls International, Inc.*)

through input from the DCS or a personal computer. No characterizable cam is required to modify an inherent valve characteristic; instead, the electronics can be used to provide a modified or customized flow characteristic.

As stated earlier, digital positioners have far lower hysteresis and better repeatability and linearity than analog positioners. However, because digital positioners still have some moving parts—such as a spool valve and a linear-to-rotary linkage at the actuator stem—some hysteresis, repeatability, and linearity problems can exist. The advantage to using smart electronics is that such errors can be zeroed out, allowing the positioner to take such problems into account. Both an advantage and disadvantage of the digital positioner is its reliance upon two-wire 4- to 20-mA signal and power sources. The obvious advantage is that an analog positioner receiving an electrical signal could be replaced with a digital positioner. The disadvantage is that only 4 mA is available to run the positioner, which limits the amount of electronics that can be run through the power source.

8.3.2   Digital-Positioner Design and Operation

A typical digital-positioner schematic is shown in Fig. 8.9. The command 4- to 20-mA signal provides the power source to the electronics. Compressed air is also required to provide the power to the pneumatic actuator. The actuator's feedback position is provided by a special take-off arm that provides a mechanical-to-electronic function: The linear motion of the actuator stem turns a rotating potentiometer, which provides position feedback to the positioner's electronics and compares that feedback to the signal. If a discrepancy occurs either through a changing signal or through an incorrect actuator position, a correcting electronic signal is sent to a pressure modulator. The pressure modulator then positions an inner spool, which sends air to one side of the actuator and exhausts the other side. This action moves the position of the actuator and continues until the correct position is reached. At this point, the feedback is equal to the signal and the pressure modulator places the inner spool in a holding position.

A key element in the correct operation of digital positioners is the placement of pressure sensors in the electronics that can monitor the air pressure sent to the actuator. This information is important in recording an initial signature for the actuator's function, as well as providing future signatures that can be used for self-diagnostics.

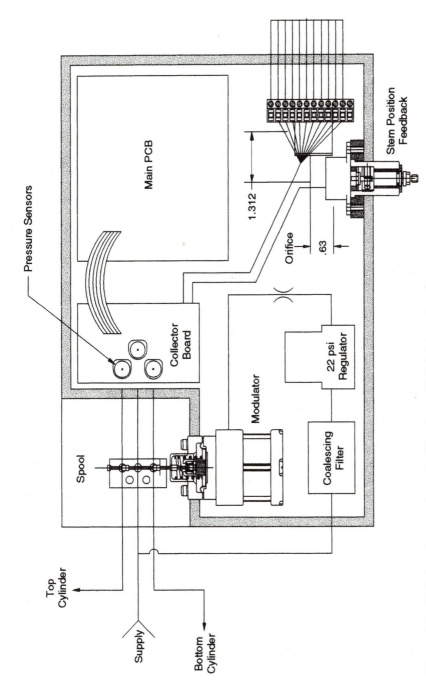

Figure 8.9. Digital positioner schematic. (*Courtesy of Valtek International*)

Pressure Sensors

Main PCB

Stem Position Feedback

1.312

Orifice

.63

Collector Board

Modulator

22 psi Regulator

Coalescing Filter

Spool

Top Cylinder

Supply

Bottom Cylinder

9
Valve Sizing

9.1 Introduction to Valve Sizing

9.1.1 The Importance of Correct Valve Sizing

Generally, valve sizing is based on the standard thermodynamic laws of fluid flow. The application of these laws is affected by the particular function of the valve plus the type and severity of the service. Simple on–off block valves are expected to pass nearly 100 percent of the flow without a significant pressure drop, since they are not expected to control the flow other than to shut it off. On the other hand, throttling services are expected to produce a certain amount of flow at certain positions of opening and take a particular pressure drop. Therefore, the science of valve sizing is almost always directed toward sizing throttling valves.

With manually operated on–off block valves, the valve is often expected to pass full flow. If the valve's internal flow passage or closure element is sized smaller than the upstream piping, flow will be restricted from that point forward. This will cause the valve to take a pressure drop and pass less flow, defeating the major purpose of the on–off valve. If the on–off block valve is sized larger than the upstream piping, installation costs are more expensive (since increasers are required). The larger valve is also more expensive. On the other hand, throttling valves, which are intended to take a pressure drop and to reduce the flow, may have a seat that is significantly less in diameter than the upstream port. Determining the flow through this diameter is the science behind valve sizing. If a throttling valve is sized too small, the maximum amount of flow through the valve will be limited and will inhibit the function of the system. If a throttling valve is sized too large, the user must bear the added cost of installing a larger valve. Another major disadvantage is that the entire flow control may be accomplished in the first half of the stroke, meaning that a minor change in position may cause a large change in flow. In addition, because regulation occurs in the first half of the stroke, flow control is extremely difficult when the regulating element is operating close to the seat. The ideal situation is for the throttling valve to utilize the full range of the stroke while producing the desired flow characteristic and maximum flow output.

Throttling valves are rarely undersized because of the number of safety factors built into the user's service conditions and the manufacturer's sizing criteria. Because of these safety factors, a large number of throttling valves actually end up being oversized. This happens because the user provides a set of service conditions that are usually the maximum conditions of the service (temperature, pressure, flow rate, etc.). The manufacturer then adds its own safety factors into the sizing equations. The valve manufacturer does this to avoid the error of undersizing, which is less forgiving than oversizing. Although not ideal, an oversized valve is still workable.

9.1.2 Valve-Sizing Criteria for Manual Valves

The basic function of manual on–off block valves is quite simple: to pass full flow while the valve is open or to shut off or divert the full flow when closed. Therefore, the valve size can sometimes be determined simply by the size of the piping, which has already been sized by the system engineers. Manual-valve manufacturers often provide sizing charts that indicate the relationship between the flow-rate requirement (Q) are the minimum and maximum valve size that can pass the given flow rate.

An important choice in manual-valve sizing is whether the valve should be full bore or reduced bore. In many cases this is more a function of the valve's purpose to pass full flow or to take a slight pressure drop. If the valve is installed in an application that must allow the passage of a pig to clean or scour the pipeline, the valve chosen must be full bore, since the pig is the same size as the inside diameter of the pipe. Another application calling for full-bore manual valves is one installed in slurries or services with entrained materials or particulates. If the valve has a reduced bore, these particulates or slurries have a tendency to settle and become trapped at the narrowed constriction. A full-bore valve has no such restriction, allowing for free passage of the foreign material without collection. Full-bore manual valves are also chosen for services with high velocities, for which a restriction would increase the chance of erosion as well as increase the velocity further.

The service conditions generally required for correct manual-valve sizing are maximum and minimum temperatures, pressures, flow rates, and specific volume (steam applications). Not only are the extremes important, but also the average operating conditions are important. The specific volume is normally provided to the user by commonly published steam tables, which show the specific volume in cubic feet per pound. Most steam tables provide the data in *pounds per square inch absolute* (psia), which does not take atmospheric pressure (14.7 psi or 1.03 bar) into account. On the other hand, *pounds per square inch gage* (psig) accounts for this adjustment for the atmospheric pressure. The metric equivalent for psig is *barg*.

9.1.3 Valve-Sizing Criteria for Check Valves

The most critical element of check-valve sizing is that a sufficient pressure drop and minimum flow exist for the check valve to open. Without a pressure drop, the closure element will not open and the valve will remain closed, which is what happens when a pump fails to maintain a proper flow or flow reverses.

The minimum pressure drop required for check valves to open is typically 1 psi (0.07 bar). This minimum pressure drop is needed to maintain the open position of the closure element without failing. If the pressure drop falls to less than 1 psi, the closure element will float back and forth, which is commonly called "flutter." As the disk moves toward the seat, the opening narrows and pressure rebuilds, which causes the disk to open higher. This low-pressure drop situation will cause this cycle to repeat until the pressure drop is increased, causing wear of the moving parts and shortening the life of the check valve. The maximum pressure drop is approximately 10 psi (0.7 bar), depending on the size of the check valve. Higher pressure drops lead to severe erosion of the check valve's closure element.

Check-valve manufacturers provide the cracking pressure of their check valves. The cracking pressure is the minimum pressure required to open the check valve and is a fixed number associated with the style and size of the check valve. It can vary anywhere from 0.1 to 0.5 psi (0.01 to 0.03 bar). Generally, cracking pressures are of little concern unless the pressures in the process are extremely low or the pressure drop is small (less than 1 psi). However, the cracking pressure can be important if the valve is installed in a vertical line, where the check valve must open against gravitational forces in addition to the process pressure. Smaller lines have higher cracking pressures than larger lines. This is because the larger the line, the larger the process force must be against the component's mass in the check valve.

Unless the flow experiences a wide range of flow during the service, check valves are sized for minimum flow, which in turn determines the valve size. This is done using manufacturer's sizing charts. If the size provided for the minimum flow is equivalent to or greater than the pipeline size, the pipeline size should be used for the valve size. For example, if the manufacturer's literature calls for a 4-in check valve, yet the pipe size is 3-in line, a 3-in check valve should be satisfactory. The larger, oversized valve will not benefit the flow rate yet is more expensive and would require the installation of increasers. If the suggested valve size for the minimum flow is smaller than the pipeline, reducers must be installed and the smaller-sized check valve installed.

The user should ensure that the flow rates are within the parameters of the check-valve design. High flow rates can increase the frequency of vortices and currents, which will increase the pressure drop across the valve as well as cause valve wear. Insufficient flow will cause the valve to flutter. The flow must be sufficient to overcome the closed position of the check valve—whether it be gravity, weight of the closure element, line orientation, or spring force.

As a general rule, the maximum liquid flow velocity for check valves is 11 ft/s (3.4 m/s). The minimum liquid flow velocity is normally 6 to 7 ft/s (1.8 to 2.1 m/s), although some designs (such as a double-disk check valve) can operate at 3 ft/s (0.9 m/s).

9.1.4 Valve-Sizing Criteria for Throttling Valves

Throttling valves require a systematic method of determining the required flow through the valve, as well as the size of the valve body, the body style, and materials that can accommodate (or tolerate) the process conditions, the correct pressure rating, and the proper installed flow characteristic. The industry standard for determining the flow capacity of a throttling valve is ANSI/ISA Standard S75.01, which contains the equations required to predict the flow of incompressible (liquid) and compressible (gas) process fluids. Because of the compressibility issues between liquids and gases, equations have been formulated for each and are included in this chapter.

Proper selection of the valve is based on the service conditions of the process. For correct sizing, the following conditions are needed: the upstream pressure; the maximum and minimum temperatures; the type of process fluid; the flow rate that is based upon the maximum flow rate, the average flow rate, and the minimum flow rate; vapor pressure; pipeline size, schedule, and material; the maximum, average, and minimum pressure drop; specific gravity of the fluid; and the critical pressure.

9.2 Valve-Sizing Nomenclature

9.2.1 Upstream and Downstream Pressures

In process systems, most valves are designed to either pass or restrict the flow to some extent. In order for the process to flow in a particular direction through a valve, the upstream and downstream pressures must be different; otherwise, the pressure would be equal and no flow would occur. By definition, the *upstream pressure* is the pressure reading taken before the valve, while the *downstream pressure* is the pressure reading taken after the valve.

9.2.2 Pressure Drop

The resulting difference between the upstream and downstream pressures is called the *pressure drop* (or the *pressure differential*). The pressure drop allows for the flow of fluid through the process system from the upstream side of the valve to the downstream side. In theory, the greater the pressure drop, the greater the flow through the valve.

9.2.3 Flow Capacity

The most commonly applied sizing coefficient is known as the *valve coefficient* (C_v), which is defined as one U.S. gallon (3.8 liters) of 60°F (16°C) water that flows through a valve with 1.0 psi (0.07 bar) of pressure drop. This general equation is written several ways, but two of the most common methods are

$$C_v = Q\sqrt{\frac{S_g}{\Delta P}}$$

or

$$C_v = \frac{Q}{\sqrt{\dfrac{\Delta P}{S_g}}}$$

where C_v = required flow coefficient for the valve
       Q = flow rate (in gal/min)
       S_g = specific gravity of the fluid
       ΔP = pressure drop (psi)

When calculated properly, C_v determines the correct trim size (or area of the valve's restriction) that will allow the valve to pass the required flow while allowing stable control of the process throughout the stroke of the valve.

9.2.4 Actual Pressure Drop

Another term for pressure drop, *actual pressure drop* (ΔP), is defined as the difference between the upstream (inlet) and downstream (outlet) pressures. When the choked and actual pressure drops are compared and the actual pressure drop is smaller, it is used in the C_v sizing equation.

9.2.5 Choked Pressure Drop

As the C_v equation is examined, the assumption is made that if the pressure drop is increased, the flow should increase proportionately. A point exists, however, where further increases in the pressure drop will not change the valve's flow rate. This is what is commonly called *choked flow*.

As illustrated in Fig. 9.1, with liquid applications having a constant upstream pressure, the flow rate Q is related to the square root of the pressure drop with a proportional and constant C_v. When the valve begins to choke, the flow-rate curve falls away from the linear relationship. Because of the choked condition, the flow rate will reach a maximum condition due to the existence of cavitation in liquids or sonic velocity with gases.

Depending on the valve style, this departure from the linear relationship will occur at different regions of the line, with some being more gradual and others more abrupt. For example, globe-style valves tend to handle higher pressure drops without choking, as opposed to rotary valves, which tend to choke and cavitate at smaller pressure drops.

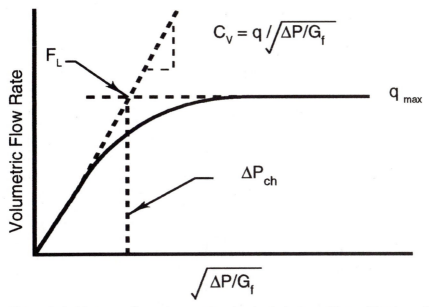

Figure 9.1. Maximum flow rate occurring due to choked conditions. (*Courtesy of Valtek International*)

For simplicity, the term *choked pressure drop* ΔP_{choked} is used to show the theoretical point where choked flow occurs, intersecting the linear lines of the constant C_v and the maximum flow rate Q_{max}. This point is known as the liquid pressure-recovery factor F_L, which is discussed in more detail in this section. The ANSI/ISA sizing equations for liquids use F_L to calculate the theoretical point where choked flow occurs (ΔP_{max}) so that the valve can be sized without the difficulty of the process being choked.

For gas applications, the terminal pressure-drop ratio x_T is used to describe the choked pressure drop for a particular valve.

9.2.6 Allowable Pressure Drop

The *allowable pressure drop* ΔP_a is chosen from the smaller of the actual pressure drop or the choked pressure drop and is used in the determination of the correct C_v. When determining the C_v of a liquid application, the following must be considered to see if the allowable pressure drop should be used: first, if the inlet pressure P_1 is fairly close to the vapor pressure; second, if the outlet pressure P_2 is fairly close to the vapor pressure; and third, if the actual pressure drop is fairly large when compared to the inlet pressure P_1. If any one of the above three conditions exists, the user should calculate the allowable pressure drop and compare it against the actual pressure drop, using the smaller value.

9.2.7 Incipient and Advanced Cavitation

With liquid applications, when the fluid passes through the narrowest point of the valve (vena contracta), the pressure decreases inversely as the velocity increases. If the pressure drops below the vapor pressure for that particular fluid, vapor bubbles begin to form. As the fluid moves into a larger area of the vessel or downstream piping, the pressure recovers to a certain extent. This increases the pressure above the vapor pressure, causing the vapor bubbles to collapse or implode. This two step-process—creation of the vapor bubbles and their subsequent implosion—is called *cavitation* and is a leading cause of valve damage in the form of erosion of metal surfaces.

As the pressure drops, the point where vapor bubbles begin to form is called *incipient cavitation*. The pressure level where cavitation is occurring at its maximum level is called *advanced cavitation*. During advanced cavitation, the flow is choked and cannot increase, which affects the flow capacity of the valve as well as its function. The point where advanced cavitation occurs can be predicted. To do this, the pressure drop must be determined, using the liquid cavitation factor F_i. A detailed discussion about the causes and effects of cavitation is found in Sec. 11.2.

9.2.8 Flashing Issues

When the downstream pressure does not recover above the vapor pressure, the vapor bubbles remain in the fluid and travel downstream from the valve, creating a mixture of liquid and gas. This is called *flashing*. Problems typically associated with flashing are higher velocities and erosion of valve components. Section 11.3 provides a more detailed discussion about flashing and its effects.

9.2.9 Liquid Pressure-Recovery Factor

A critical element in liquid sizing is the *liquid pressure-recovery factor F_L*, which predicts the effect the geometry of a valve's body will have on the maximum capacity of that valve. F_L is used to predict the amount of pressure recovery occurring between the vena contracta and the outlet of the body.

The liquid pressure-recovery factor is determined by the manufacturer through flow testing that particular valve style. F_L factors can vary significantly depending on the internal design of the valve. Valves from the same basic design (for example, butterfly valves) may have varying F_L factors depending on the unique internal designs of the manufacturer. Generally, rotary valves, especially ball and butterfly valves, allow for a high recovery of the fluid following the vena contracta. Therefore, they tend to cavitate and choke at smaller pressure drops than globe valves. For the most part, globe valves have better F_L factors and are able to handle severe services when compared to rotary valves.

9.2.10 Liquid Critical-Pressure Ratio Factor

The *liquid critical-pressure ratio factor* F_F is important to liquid sizing because it predicts the theoretical pressure at the vena contracta, when the maximum effective pressure drop (or in other words, the choked pressure drop) occurs across the valve.

9.2.11 Choked Flow

With liquid services, the presence of cavitation or flashing expands the specific volume of the fluid. The volume increases at a faster rate than if the flow increased due to the pressure differential. At this point, the valve cannot pass any additional flow, even if the downstream pressure is lowered.

With gas and vapor services, choked flow occurs when the velocity of the fluid achieves sonic levels (Mach 1 or greater) at any point in the valve body or downstream piping. Following the basic laws of mass and energy, as the pressure decreases in the valve to pass through restrictions, velocity increases inversely. As the pressure lowers, the specific volume of the fluid increases to the point where a sonic velocity is achieved.

Because of the velocity limitation [Mach 1 for gases and 50 ft/s (12.7 m/s) for liquids], the flow rate is limited to that which is permitted by the sonic velocity through the vena contracta or the downstream piping.

9.2.12 Velocity

As a general rule, smaller valve sizes are better equipped to handle higher velocities than larger-sized valves, although the actual sizes vary according to the valve style. For liquid services, the general guideline for maximum velocity at the valve outlet is 50 ft/s (12.7 m/s), while gas services are generally restricted to Mach 1.0. When cavitation or flashing is present, creating a higher velocity associated with the liquid–gas mixture, the maximum velocity is usually restricted to 500 ft/s (127 m/s). Some exceptions exist, however, for liquids. In services where temperatures are close to saturation point, the velocity must be less—approximately 30 ft/s (7.6 m/s). This lower velocity prevents the fluid from dropping below the vapor pressure, which will lead to the formation of vapor bubbles. The rule of 30 ft/s (7.6 m/s) is also applicable to those valve applications that must have a full flow rate with minimal pressure drop. A valve in which the pressure drop falls below the vapor pressure and advanced cavitation is occurring should be restricted to 30 ft/s (7.6 m/s) to minimize the cavitation damage that would spread from the valve into the downstream piping. Ideally, the user would try to restrict the pressure recovery and allow the subsequent cavitation damage to be contained in the body and not downstream into the piping. In essence, the valve body is sacrificed and the piping is saved.

9.2.13 Reynolds-Number Factor

Some processes are characterized by nonturbulent flow conditions in which laminar flow exists (such as oils). Laminar fluids have high viscosity, operate in lower velocities, or require a flow capacity requirement that is extremely small. The *Reynolds-number factor F_R* is used to correct the C_v equation for these flow factors. In most cases, if the viscosity is fairly low (for example, less than SAE 10 motor oil), the Reynolds-number factor is insignificant.

9.2.14 Piping-Geometry Factor

The flow capacity of a valve may be affected by nonstandard piping configurations, such as the use of increasers or reducers, which must be corrected in the C_v equation using the *piping-geometry factor F_P*.

Standardized C_v testing is conducted by the valve manufacturer with straight piping that is the same line size as the valve. The use of piping that is larger or smaller than the valve, or the close proximity of piping elbows, can decrease these values and must be considered during sizing.

9.2.15   Expansion Factor

With gas services, the specific weight of the fluid varies as the gas moves from the upstream piping and through the valve to the vena contracta. The *expansion factor Y* is used to compensate for the effects of this change in the specific weight of the gas. The expansion factor is important in that it takes into account the changes in the cross-sectional area of the vena contracta as the pressure drop changes in that region.

9.2.16   Ratio of Specific Heats Factor

Because the C_v equation for gases is based upon air, some adjustment must be made for other gases. The *ratio of specific heats factor F_K* is used to adjust the C_v equation to the individual characteristics of these gases.

9.2.17   Terminal Pressure-Drop Ratio

With gases, the point where the valve is choked (which means that increasing the pressure drop though lowering the downstream pressure cannot increase the flow of the valve) is predicted by the *terminal pressure-drop ratio x_T*. Similar in many respects to the liquid pressure-recovery factor F_L, the terminal pressure-drop ratio is affected by the geometry of the valve's body and varies according to valve style and individual size.

9.2.18   Compressibility Factor

Because the density of gases varies according to the temperature and pressure of the fluid, the fluid's compressibility must be included in the C_v equation. Therefore the *compressibility factor Z* is included in the equation and is a function of the temperature and pressure.

9.3   Body Sizing of Liquid-Service Control Valves

9.3.1   Basic Liquid Sizing Equation

The liquid C_v sizing equation is a general-purpose equation for most liquid applications, using the actual pressure drop (upstream pressure minus downstream pressure) to calculate the flow capacity. For non-laminar liquids,

$$C_v = \frac{Q}{F_P} \sqrt{\frac{S_g}{\Delta P_a}}$$

where C_v = valve-sizing coefficient
F_P = piping geometry function
Q = flow rate (gal/min)
S_g = specific gravity (at flowing temperature)
ΔP_a = allowable pressure drop across the valve (psi)

For sizing purposes, the liquid C_v equation can be determined step by step by following Secs. 9.3.2 through 9.3.14.

9.3.2 Actual-Pressure-Drop Calculation

Before the allowable pressure drop is determined, the actual pressure drop should be determined by using the following equation:

$$\Delta P = P_1 - P_2$$

where ΔP = actual pressure drop (psi)
P_1 = upstream pressure (at valve inlet, psia)
P_2 = downstream pressure (at valve outlet, psia)

9.3.3 Choked Flow, Cavitation, and Flashing Determination

The choked flow point must be predicted using the following equation:

$$\Delta P_{choked} = F_L^2(P_1 - F_F P_V)$$

where ΔP_{choked} = choked pressure drop
F_L = liquid pressure-recovery factor
F_F = liquid critical-pressure-ratio factor
P_V = vapor pressure of the liquid (at inlet temperature, psia)

The liquid pressure-recovery factor F_L is usually provided by the manufacturer. Table 9.1 provides typical F_L values for throttling linear globe and rotary valves.

Table 9.1 Typical F_L Factors*,†

Valve Style	Flow Direction	Trim Area	F_L
Linear globe	Over seat	Full area	0.85
	Over seat	Reduced area	0.80
	Under seat	Full area	0.90
	Under seat	Reduced area	0.90
Butterfly	60° open	Full area	0.76
	90° open	Full area	0.56
Ball	60° open	Full area	0.78
	90° open	Full area	0.66

*Data courtesy of Valtek International.
†Note: All values provided are full-open.

Continuing with the ΔP_{choked} equation, the liquid critical-pressure ratio factor F_F is determined by using the following equation:

$$F_F = 0.96 - 0.28 \sqrt{\frac{P_V}{P_C}}$$

where P_C = critical pressure of the liquid (psia)

Critical pressures for common gases and liquids are found in Table 9.2. A more complete list of critical pressures is found in Appendix B.

If the calculation for the choked pressure drop ΔP_{choked} is a smaller value than the actual pressure drop ΔP, the ΔP_{choked} value should be used for the actual pressure drop ΔP_a in the C_v equation. To determine at what pressure drop advanced cavitation begins, the following equation should be used:

$$\Delta P_{cavitation} = F_i^2 (P_1 - P_V)$$

Table 9.2 Critical Pressures for Common
Process Fluids

Liquid	Critical Pressure *psia/bar*
Ammonia	1636.1/112.8
Argon	707.0/48.8
Benzene	710.0/49.0
Butane	551.2/38.0
Carbon Dioxide	1070.2/73.8
Carbon Monoxide	507.1/35.0
Chlorine	1117.2/77.0
Dowtherm A	547.0/37.7
Ethane	708.5/48.8
Ethylene	730.5/50.3
Fuel Oil	330.0/22.8
Gasoline	410.0/28.3
Helium	32.9/2.3
Hydrogen	188.1/13.0
Hydrogen Chloride	1205.4/83.1
Isobutane	529.2/36.5
Isobutylene	529.2/36.5
Kerosene	350.0/24.1
Methane	667.3/46.0
Nitrogen	492.4/33.9
Nitrous Oxide	1051.1/72.5
Oxygen	732.0/50.5
Phosgene	823.2/56.8
Propane	615.9/42.5
Propylene	670.3/46.2
Refrigerant 11	639.4/44.1
Refrigerant 12	598.2/41.2
Refrigerant 22	749.7/51.7
Seawater	3200.0/220.7
Water	3208.2/221.2

Table 9.3 Typical F_i Factors*,†

Valve Style	Flow Direction	Trim Area	F_i
Linear globe	Over seat	Full area	0.75
	Over seat	Reduced area	0.72
	Under seat	Full area	0.81
	Under seat	Reduced area	0.81
Butterfly	60° open	Full area	0.65
	90° open	Full area	0.49
Ball	60° open	Full area	0.65
	90° open	Full area	0.44

Data courtesy of Valtek International.
†*Note: All values provided at full open.*

where $\Delta P_{cavitation}$ = pressure drop with advanced cavitation
F_i = liquid cavitation factor

Typical liquid cavitation factors for common valve styles are found in Table 9.3.

9.3.4 Specific-Gravity Determination

The value for the fluid's specific gravity S_g should be determined using the operating temperature. Appendix B provides specific-gravity data for over 400 process fluids.

9.3.5 Approximate-Flow-Coefficient Calculation

Using the values calculated to this point, the approximate flow capacity should be calculated, using the C_v sizing equation for liquids from

Sec. 9.3.1. For this calculation, the assumption should be made that the piping-geometry factor F_p is 1.0. When the valve is not operating in a laminar flow—due to high viscosity, low velocity, or low flow—the effects of nonturbulent flow can be ignored.

9.3.6 Approximate Body Size Selection

Using the manufacturer's C_v tables, the smallest-sized body that can accommodate the calculated C_v should be selected. Typical C_v data are found in Fig. 9.2.

9.3.7 Reynolds-Number-Factor Calculation

The following equation can be used to determine the Reynolds-number factor:

$$\text{Re}_V = \frac{N_4 F_d Q}{\nu \sqrt{F_L C_v}} \left(\frac{F_L^2 C_v^2}{N_2 d^4} + 1 \right)^{0.25}$$

where Re_V = valve Reynolds number
N_4 = 17,300 (when Q is in gal/min and d in inches)
F_d = valve style modifier (see Table 9.4)
ν = kinematic viscosity (centistokes, μ/S_g)
C_v = valve flow coefficient (from Sec. 3.9.1)
N_2 = 890 (when d is in inches)
d = valve inlet diameter (inches)

Valve Type: Mark One, Unbalanced
Body Rating: Class 900-1500
Trim Characteristics: Quick Open
Flow Direction: Flow Over

SIZE	TRIM NO	STROKE	F_L	PERCENT OPEN										SEAT AREA
				100	90	80	70	60	50	40	30	20	10	
1.00	.81	.75	.87	9.0	8.9	8.9	8.7	8.6	8.5	7.5	5.7	3.5	1.9	.52
1.50	1.25	1.00	.85	24	24	24	24	24	21	18	13	8.7	4.9	1.23
2.00	1.62	1.50	.87	41	41	40	40	39	39	34	26	15	8.1	2.06
3.00	2.00	2.00	.86	106	105	105	104	104	94	81	62	39	21	5.41
4.00	3.50	2.50	.87	185	185	183	181	178	162	139	105	68	37	9.62
6.00	5.00	3.00	.85	382	382	381	380	355	317	270	210	140	75	19.63

Note: All Cv values are shown at 100% open. For each vavle size below, the full area values are shaded. Reduced trim values follow, in descending order.

Figure 9.2. Typical manufacturer's C_v data. (*Courtesy of Valtek International*)

Table 9.4 Valve Recovery Coefficient and Incipient Cavitation Factors*

Valve Type	Flow Direction	Trim Size	F_L	F_i	x_T	F_d
Globe	Over Seat	Full Area	0.85	0.75	.70	1.0
	Over Seat	Reduced Area	0.80	0.72	.70	1.0
	Under Seat	Full Area	0.90	0.81	.75	1.0
	Under Seat	Reduced Area	0.90	0.81	.75	1.0
Valdisk	60° Open	Full	0.76	0.65	.36	.71
Rotary Disc	90° Open	Full	0.56	0.49	.26	.71
ShearStream	60° Open	Full	0.78	0.65	.51	1.0
Rotary Ball	90° Open	Full	0.66	0.44	.30	1.0
CavControl	Over Seat	All	0.92	0.90	N/A	$.2\sqrt{d}$
MegaStream	Under Seat	All	~1.0	N/A	~1.0	$(n_s/25d)^{2/3}$**
ChannelStream	Over Seat	All	~1.0	0.87 to 0.999	N/A	.040*
Tiger-Tooth	Under Seat	All	~1.0	0.84 to 0.999	~1.0	.035*

* Typical   ** n_s = number of stages
NOTE: Values are given for full-open valves. See charts below for part-stroke values

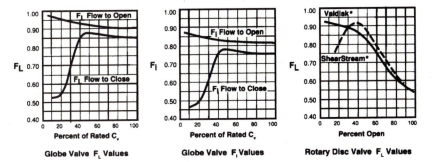

Globe Valve F_L Values   Globe Valve F_i Values   Rotary Disc Valve F_L Values

Courtesy of Valtek International.

If the valve Reynolds number (Re_V) is equal to or greater than 40,000 ($Re_V \geq 40{,}000$), 1.0 should be used for the Reynolds-number factor F_R. The following equation is used to find F_R, if the valve Reynolds number is less than 40,000 ($Re_V \leq 40{,}000$):

$$F_R = 1.044 - 0.358\left(\frac{C_{vS}}{C_{vT}}\right)^{0.655}$$

through use of these equations:

$$C_{vS} = \frac{1}{F_S}\left(\frac{Q\mu}{N_S\Delta P}\right)^{0.667}$$

$$F_S = \frac{F_d^{0.667}}{F_L^{0.333}}\left(\frac{F_L^2 C_v^2}{N_2 d^4} + 1\right)^{0.167}$$

where C_{vS} = laminar flow C_v
  C_{vT} = turbulent flow C_v (the C_v is used from the liquid C_v equa-
     tion in Sec. 9.3.1)
  F_S = laminar or streamline flow factor
  N_2 = 890 (when d is in inches)
  N_4 = 17,300 (when Q is in gal/min and d is in inches)
  N_S = 47 (when Q is in gal/min and ΔP is in psi)
  μ = absolute viscosity (centipoise)

9.3.8 Flow-Coefficient Recalculation

Flow is considered to be laminar when the Reynolds-number factor F_R is less than 0.48 ($F_R < 0.48$). That means that the C_v is the same as the C_{vS}, which is determined from the equation in Sec. 9.3.7.

If F_R is larger than 0.98 ($F_R > 0.98$), the flow is determined to be turbulent and assumed to be equal to 1.0 ($F_R = 1.0$). At this point, the C_v is determined from the standard C_v liquid sizing equation found in Sec. 9.3.1. The piping-geometry factor F_p is not required in this situation and should not be figured into the C_v equation.

If F_R falls between 0.48 and 0.98, the flow is determined to be in a transitional stage, which is calculated using the following equation:

$$C_v = \frac{Q}{F_R}\sqrt{\frac{S_g}{\Delta P}}$$

where F_R = Reynolds-number factor
  S_g = specific gravity (at flowing temperature)

9.3.9 Piping-Geometry-Factor Calculation

The inside diameter of the piping is required to determine the piping-geometry factor F_p. In the event that the pipe size is not provided or known, the body size determined from Sec. 9.3.6 should be used to determine the pipe size. Tables 9.5 and 9.6 can be used to find the piping-geometry factors. Table 9.7 provides F_p for valves with reducers (or increasers) on both the inlet and outlet of the valve. Table 9.8 provides F_p for a valve with the reducer (or increaser) on the valve outlet only. The maximum effective pressure drop (defined as ΔP_{choked}) can be affected by the use of increasers and reducers.

Table 9.5 Piping-Geometry Factors for Valves with Reducers and Increasers on Both Ends*,†

C_v/d^2	d / D				
	0.50	0.60	0.70	0.80	0.90
4	0.99	0.99	1.00	1.00	1.00
6	0.98	0.99	0.99	1.00	1.00
8	0.97	0.98	0.99	0.99	1.00
10	0.96	0.97	0.98	0.99	1.00
12	0.94	0.95	0.97	0.98	1.00
14	0.92	0.94	0.96	0.98	0.99
16	0.90	0.92	0.95	0.97	0.99
18	0.87	0.90	0.94	0.97	0.99
20	0.85	0.89	0.92	0.96	0.99
25	0.79	0.84	0.89	0.94	0.98
30	0.73	0.79	0.85	0.91	0.97
35	0.68	0.74	0.81	0.89	0.96
40	0.63	0.69	0.77	0.86	0.95

Courtesy of Valtek International.

†*Note:* The maximum effective pressure drop (ΔP choked) may be affected by the use of reducers and increasers. This is especially true of butterfly valves.

9.3.10 Final-Flow-Coefficient Calculation

After the piping-geometry factor F_p is determined, it should be applied to the liquid C_v equation (Sec. 9.3.1) and the final C_v calculated.

9.3.11 Valve Exit-Velocity Calculation

As discussed in Sec. 9.2.12, the general rule for velocities in liquids is that the velocity should be limited to 50 ft/s (15.2 m/s), although this may vary according to the size of the valve—smaller valves can handle

Table 9.6 Piping-Geometry Factors for Valves with Reducers and Increasers on Outlet Only*,†

C_v / d^2	d / D				
	0.50	0.60	0.70	0.80	0.90
4	1.00	1.00	1.00	1.00	1.00
6	1.01	1.01	1.01	1.01	1.01
8	1.01	1.02	1.02	1.02	1.01
10	1.02	1.03	1.03	1.03	1.02
12	1.03	1.04	1.04	1.04	1.03
14	1.04	1.05	1.06	1.05	1.04
16	1.06	1.07	1.08	1.07	1.05
18	1.08	1.10	1.11	1.10	1.06
20	1.10	1.12	1.12	1.12	1.08
25	1.17	1.22	1.24	1.22	1.13
30	1.27	1.37	1.42	1.37	1.20
35	1.44	1.65	1.79	1.65	1.32
40	1.75	2.41	3.14	2.41	1.50

*Courtesy of Valtek International.

†Note: d = valve port inside diameter in inches; D = internal diameter of the piping in inches.

higher velocities, while larger valves handle lower velocities. To calculate the exit velocities from the valve, the following equation is used:

$$V = \frac{0.321Q}{A_V}$$

where V = velocity (ft/s)

A_V = flow area of valve body port (square inches) from Table 9.9

If the exit velocity exceeds the acceptable velocity for that given application, a larger valve size may be chosen to prevent damage from erosion. If a larger body size is chosen, the piping-geometry factor F_p will have to change, requiring a new C_v calculation.

Table 9.7 Piping-Geometry Factors, with Reducers or Increasers on Both Inlet and Outlet of Valve*,†

C_v/d^2	0.50 d/D	0.60 d/D	0.70 d/D	0.80 d/D	0.90 d/D
4	0.99	0.99	1.00	1.00	1.00
6	0.98	0.99	0.99	1.00	1.00
8	0.97	0.98	0.99	0.99	1.00
10	0.96	0.97	0.98	0.99	1.00
12	0.94	0.95	0.97	0.98	1.00
14	0.92	0.94	0.96	0.98	0.99
16	0.90	0.92	0.95	0.97	0.99
18	0.87	0.90	0.94	0.97	0.99
20	0.85	0.89	0.92	0.96	0.99
25	0.79	0.84	0.89	0.94	0.98
30	0.73	0.79	0.85	0.91	0.97
35	0.68	0.74	0.81	0.89	0.96
40	0.63	0.69	0.77	0.86	0.95

*Data courtesy of Valtek International.
†Note: d = inside diameter of valve port (inches); D = inside diameter of piping (inches).

9.3.12 Trim-Size Selection

Control-valve manufacturers provide tables that outline the C_vs for a certain valve style, flow direction, body pressure rating, flow characteristic, size of the valve seat or the seal, and length of stroke. Some charts may be broken down to percentages of opening, since some throttling services may not utilize the entire stroke.

Using the manufacturer's C_v table based upon the correct criteria (body size, flow characteristic, flow direction, etc.), the correct size of the valve opening (of the seat or the seal) should be chosen. This open-

Table 9.8 Piping-Geometry Factors with
Reducer or Increaser on Outlet of Valve*,†

C_v/d^2	0.50 d/D	0.60 d/D	0.70 d/D	0.80 d/D	0.90 d/D
4	1.00	1.00	1.00	1.00	1.00
6	1.01	1.01	1.01	1.01	1.01
8	1.01	1.02	1.02	1.02	1.01
10	1.02	1.03	1.03	1.03	1.02
12	1.03	1.04	1.04	1.04	1.03
14	1.04	1.05	1.06	1.05	1.04
16	1.06	1.07	1.08	1.07	1.05
18	1.08	1.10	1.11	1.10	1.06
20	1.10	1.12	1.12	1.12	1.08
25	1.17	1.22	1.24	1.22	1.13
30	1.27	1.37	1.42	1.37	1.20
35	1.44	1.65	1.79	1.65	1.32
40	1.75	2.41	3.14	2.41	1.50

Data courtesy of Valtek International.
†*Note: d* = inside diameter of valve port (inches); *D* = inside diameter of piping (inches).

ing and its dimension are often called the *trim number*. A globe-style valve will have a number of trim-number options, including one that is a *full-area trim number*, the largest sized diameter opening for that particular size. The valve may also have several *reduced-area trim numbers*, which are progressively smaller in diameter and allow smaller C_vs in the same body size.

9.3.13 Flashing-Velocity Calculation

As described in Sec. 9.2.8, if the valve outlet pressure is lower than the vapor pressure, the vapor bubbles that are formed remain in a gaseous state, providing a downstream flow that has a combined liquid–gas mixture. This results in increased velocity and difficult control situations. Since the application is found to be flashing, certain measures

Table 9.9 Valve Port Areas*,†

Valve Size (inches)	Valve Outlet Area, A_v (Square Inches)						
	Class 150	Class 300	Class 600	Class 900	Class 1500	Class 2500	Class 4500
1/2	0.20	0.20	0.20	0.20	0.20	0.15	0.11
3/4	0.44	0.44	0.44	0.37	0.37	0.25	0.20
1	0.79	0.79	0.79	0.61	0.61	0.44	0.37
1 1/2	1.77	1.77	1.77	1.50	1.50	0.99	0.79
2	3.14	3.14	3.14	2.78	2.78	1.77	1.23
3	7.07	7.07	7.07	6.51	5.94	3.98	2.78
4	12.57	12.57	12.57	11.82	10.29	6.51	3.98
6	28.27	28.27	28.27	25.97	22.73	15.07	10.29
8	50.27	50.27	48.77	44.18	38.48	25.97	19.63
10	78.54	78.54	74.66	69.10	60.13	41.28	28.27
12	113.10	113.10	108.43	97.12	84.62	58.36	41.28
14	137.89	137.89	130.29	117.86	101.71	70.88	50.27
16	182.65	182.65	170.87	153.94	132.73	92.80	63.62
18	233.70	226.98	213.82	194.83	167.87	117.86	84.46
20	291.04	283.53	261.59	240.53	210.73	143.14	101.53
24	424.56	415.48	380.13	346.36	302.33	207.39	143.14
30	671.96	660.52	588.35	541.19	476.06	325.89	
36	962.11	907.92	855.30				
42	1320.25	1194.59					

*Data courtesy of Valtek International.

†Note: To find approximate fluid velocity in the pipe, use the equation $V_P = V_V A_V/A_P$, where V_P = velocity in pipe, A_V = valve outlet area. V_V = velocity in valve outlet, and A_P = pipe area.

To find equivalent diameters of the valve or pipe inside diameter use $d = \sqrt{4A_V/\pi}$, $D = \sqrt{4A_P/\pi}$.

must be taken to prevent undue damage and premature wear to the valve, such as using special trims or hardened materials. Flashing applications must be limited to a velocity of 500 ft/s (152 m/s), unless special modifications are made to the valve-body design to accommodate the increased volume and velocity. Either of the following equations can be used to calculate flashing velocity, depending on the flow-rate measurement (lb/h or gal/min):

$$V = \frac{0.040}{A_V}\, w\left[\left(1 - \frac{x}{100\%}\right)V_{f2} + \frac{x}{100\%}V_{g2}\right]$$

$$V = \frac{20}{A_V}\, Q\left[\left(1 - \frac{x}{100\%}\right)V_{f2} + \frac{x}{100\%}V_{g2}\right]$$

where V = velocity (ft/s)

$\quad w$ = liquid flow rate (lb/h)

$\quad V_{f2}$ = saturated liquid specific volume (ft^3/lb at outlet pressure P_2)

$\quad V_{g2}$ = saturated vapor specific volume (ft^3/lb at outlet pressure P_2)

$\quad x$ = percentage of liquid mass flashed to vapor (Sec. 9.3.14)

9.3.14 Percentage of Flashing Calculation

To calculate the percentage of the liquid flashing into gas, the user should have access to steam tables, which provides a listing of enthalpies and specific volumes. To make this calculation, the following equation should be used:

$$x = \left(\frac{h_{f1} - h_{f2}}{h_{fg2}} \right) \times 100\%$$

where h_{f1} = enthalpy of saturated liquid at inlet temperature

$\quad h_{f2}$ = enthalpy of saturated liquid at outlet pressure

$\quad h_{fg2}$ = enthalpy of evaporation at outlet pressure

9.3.15 Liquid Sizing Example A

For this example, the following service conditions are given in Imperial units:

Liquid	Water
Critical pressure P_C	3206.2 psia
Temperature	250°F
Upstream pressure P_1	314.7 psia
Downstream pressure P_2	104.7 psia
Flow rate Q	500 gal/min
Vapor pressure P_V	30 psia
Specific gravity S_g	0.94
Kinematic viscosity v	0.014 cS
Pipeline size	4 in (ANSI Class 600)
Valve	Globe, flow-to-open
Flow characteristic	Equal percentage

The actual pressure drop ΔP is calculated using the C_v equation for liquids (Sec. 9.3.1):

$$\Delta P = P_1 - P_2 = 314.7 \text{ psia} - 104.7 \text{ psia} = 210 \text{ psi}$$

Choked flow can be checked by finding the liquid pressure-recovery factor F_L from Table 9.4, which is 0.90. Then, the liquid critical-pressure ratio factor (F_F) is calculated by using the equation found in Sec. 9.3.3.

$$F_F = 0.96 - 0.28 \sqrt{\frac{P_V}{P_C}} = 0.96 - 0.28 \sqrt{\frac{30}{3206.2}} = 0.93$$

After determining F_L and F_F, these numbers are used in the choked pressure drop $(\Delta P_{\text{choked}})$ equation from Sec. 9.3.3:

$$\Delta P_{\text{choked}} = F_L^2 (P_1 - F_F P_V) = (0.90)^2[314.7 - (0.93)(30)] = 232 \text{ psi}$$

A comparison should be made between the actual pressure drop ΔP of 210 psi and the choked pressure drop ΔP_{choked} of 232 psi. Since the actual pressure drop is smaller than the choked pressure drop, the actual pressure drop will be used to size the valve.

By using the equation in Sec. 9.3.3, the advent of incipient cavitation should be checked:

$$\Delta P_{\text{cavitation}} = F_L^2(P_1 - P_V) = (0.81)^2(314.7-30) = 187 \text{ psi}$$

In this example, the actual pressure drop (ΔP) of 210 psi is greater than the pressure drop associated with incipient cavitation $(\Delta P_{\text{cavitation}})$ of 187 psi. This can be interpreted to mean that, although cavitation is occurring in the service, the cavitation is not causing the flow to choke. In this case, the user should begin considering methods to deter the cavitation damage, such as special trims or hardened materials. With a specific gravity of 0.94 and assuming the piping-geometry factor F_P is 1.0 (Sec. 9.3.9), the C_v should be calculated using the original liquid sizing equation (Sec. 9.3.1):

$$C_v = \frac{Q}{F_P} \sqrt{\frac{S_g}{\Delta P_a}} = \frac{500}{1} \sqrt{\frac{0.94}{210}} = 33.4$$

The required valve is a globe valve with flow-under-the-plug trim design, equal-percentage flow characteristic, and ANSI Class 600 pressure class. The manufacturer's C_v tables should be examined to determine the smallest valve available that would allow the flow of $33.4C_v$ through the flow area of the seat or seal. In this case, the assumption is made that, according to the charts, a 2-in valve body would be the smallest size with a trim number available to pass the required C_v.

At this point, the Reynolds-number factor F_R is calculated by using the equation from Sec. 9.3.7:

$$
\mathrm{Re}_V = \frac{N_4 F_d Q}{v\sqrt{F_L C_v}} \left(\frac{F_L^2 C_v^2}{N_2 d^4} + 1 \right)^{0.25}
$$

$$
= \frac{(17{,}300)(1)(500)}{(0.014)\sqrt{(0.90)(33.4)}} \left(\frac{(0.\,90)^2(33.4)^2}{(890)(2)^2} + 1 \right)^{0.25}
$$

$$
= 114 \times 10^6
$$

Because the Reynolds-number factor F_R is significantly larger than 40,000 (114×10^6 versus 40,000), the calculated C_v remains 33.4 and is used in further calculations. With a 2-in body tentatively chosen for this application and a 4-in pipeline, the calculation of the piping-geometry factor F_P is made using Table 9.5 with the following numbers:

$$
\frac{d}{D} = \frac{2}{4} = 0.5
$$

and

$$
\frac{C_v}{d^2} = \frac{33.4}{2^2} = 8.35
$$

According to the table, the piping-geometry factor (F_P) should be 0.97. Now, the F_P of 0.97 can be inserted into the C_v equation to determine the final C_v:

$$
C_v = \frac{Q}{F_P}\sqrt{\frac{S_g}{\Delta P_a}} = \frac{500}{0.97}\sqrt{\frac{0.94}{210}} = 34.5
$$

Using Table 9.9, for a 2-in valve in ANSI Class 600 service, the valve outlet area A_V is 3.14 in². Using this number and a flow rate of 500 gal/min (1892 liters/m), the velocity through the valve can be calculated as

$$V = \frac{0.321Q}{A_V} = \frac{0.321(500)}{3.14} = 51 \text{ ft/s (130 m/s)}$$

The velocity of 51 ft/s exceeds the limit of 50 ft/s for liquids. Since the service is cavitating, damage will most likely occur to the valve body. At this point, the only option to lower the velocity is to chose the next larger valve size, a 3-in body with reduced trim. Using a 3-in body and an A_V of 7.07, the velocity is significantly lowered to acceptable levels:

$$V = \frac{0.321Q}{A_V} = \frac{0.321(500)}{7.07} = 23 \text{ ft/s (5.8 m/s)}$$

Despite the lower velocity with the 3-in body, cavitation remains a concern and some material or design action should be taken to prevent damage. Another option that may reduce the cost of a larger valve would be to use an expanded outlet body—for example, a 2 × 4-in expanded outlet valve (since the piping is 4 in). Because of the velocity issue, which required the changing of the valve size to 3 in, the C_v equation will need to be recalculated using a new piping-geometry factor F_p:

$$\frac{d}{D} = \frac{3}{4} = 0.75$$

and

$$\frac{C_v}{d^2} = \frac{33.4}{3^2} = 3.71$$

With a piping-geometry factor F_p of 1.00 (interpolated from Table 9.5), the revised C_v for a 3-in body is

$$C_v = \frac{Q}{F_p} \sqrt{\frac{S_g}{\Delta P_a}} = \frac{500}{1} \sqrt{\frac{0.94}{210}} = 33.4$$

9.3.16 Liquid Sizing Example A (with Flashing)

For this example, the same service conditions as the previous example are provided, except that the temperature is increased by 100°F from 250 to 350°F. Using the saturated steam temperatures in the steam tables, the saturation pressure for water at 350°F is 134.5 psia. Because the saturation pressure (134.5 psia) is significantly higher than the downstream pressure of the valve (104.7 psia), the service is flashing. Because of the flashing, the percent flash x must be calculated:

$$x = \left(\frac{h_{f1} - h_{f2}}{h_{fg2}}\right) \times 100\% = \left(\frac{321.8 - 302.3}{886.4}\right) \times 100\% = 2.2\%$$

where $h_{f1} = 321.8$ Btu/lb at 350°F (from the saturation temperature table)

$h_{f2} = 302.3$ Btu/lb at 105 psia (from the saturation pressure table)

$h_{fg2} = 886.4$ Btu/lb at 105 psia (from the saturation pressure table)

The equation from Sec. 9.3.13 must then be used to determine the velocity from a 3-in valve:

$$V = \frac{20}{A_V} Q\left[\left(1 - \frac{x}{100\%}\right)V_{f2} + \frac{x}{100\%} V_{g2}\right]$$

$$V = \frac{(20)(500)}{7.07}\left[\left(1 - \frac{2.2\%}{100\%}\right)0.0178 + \frac{2.2\%}{100\%} 4.234\right] = 156 \text{ ft/s}$$

where $V_{f2} = 0.0178$ ft³/lb at 105 psia (from the saturation pressure table)

$V_{g2} = 4.324$ ft³/lb at 105 psia (from the saturation pressure table)

From Sec. 9.2.12, the maximum velocity for flashing services is 500 ft/s. The calculated velocity of this service is 156 ft/s, which is far below the maximum level. Once again, however, the presence of flashing should be considered by selecting hardened materials or special trim features.

9.3.17 Liquid Sizing Example B

In this second liquid example, the following service conditions are provided in Imperial units:

Liquid   Ammonia
Critical pressure P_C 1638.2 psia

Temperature	20°F
Upstream pressure P_1	149.7 psia
Downstream pressure P_2	64 psia
Flow rate Q	850 gal/min
Vapor pressure P_V	465.6 psia
Specific gravity S_g	0.65
Kinematic viscosity v	0.02 cS
Pipeline size	3 in (ANSI Class 600)
Valve	Globe, flow-to-close
Flow characteristic	Linear

The actual pressure drop ΔP is calculated as follows:

$$\Delta P_a = P_1 - P_2 = 149.7 \text{ psia} - 64.7 \text{ psia} = 85 \text{ psi}$$

Choked flow is checked by determining the liquid pressure-recovery factor F_L from Table 9.4, which is 0.85. The liquid critical-pressure ratio factor F_F can then be calculated by using the following equation found in Sec. 9.3.3.

$$F_F = 0.96 - 0.28 \sqrt{\frac{P_V}{P_C}} = 0.96 - 0.28 \sqrt{\frac{45.6}{1638.2}} = 0.91$$

After determining that F_L (= 0.85) and F_F (= 0.91), these numbers are inserted in the choked-pressured-drop ΔP_{choked} equation from Sec. 9.3.3:

$$\Delta P_{\text{choked}} = F_L^2(P_1 - F_F P_V) = (0.85)^2[149.7 - (0.91)(45.6)] = 78.2 \text{ psi}$$

In comparing the actual pressure drop ΔP of 85.0 psi and the choked pressure drop ΔP_{choked} of 78.2 psi, the choked pressure drop is smaller than the actual pressure drop. Therefore, the smaller of the two numbers—the choked pressure drop—is used to size the valve. Because the valve is choked, the service is also cavitating. Therefore, checking for incipient cavitation $\Delta P_{\text{cavitation}}$ is not necessary. In this case, the user should plan to use special anticavitation trim inside the valve as well as hardened materials to avoid the erosion of metal parts associated with cavitation.

With a specific gravity of 0.65 and assuming a piping-geometry factor F_P of 1.0 (Sec. 9.3.9), a preliminary C_v can be calculated using the original liquid sizing equation (Sec. 9.3.1):

$$C_v = \frac{Q}{F_P}\sqrt{\frac{S_g}{\Delta P_a}} = \frac{850}{1}\sqrt{\frac{0.65}{78.2}} = 77.5$$

From the conditions of this example, the preferred valve is a globe valve with flow-over-the-plug trim design, a linear flow characteristic, and ANSI Class 600 pressure classification. The manufacturer's C_v tables can then be examined to estimate the smallest valve available that would allow the flow of $77.5C_v$ through the flow area of the seat. In this case, the assumption is made that the manufacturer's C_v tables show that a 3-in valve body would be the smallest size with a trim number that would pass the required C_v.

Because the flow is cavitating, it is turbulent when exiting the valve. Because of the turbulent flow, the Reynolds-number factor F_R is assumed to be $F_R = 1.0$ and no further calculations are required.

Since a 3-in body was chosen initially for this application and the pipeline is determined to be a 3-in line, the piping-geometry factor F_p will be 1.0 (no reducers or increasers are required). Because $F_p = 1.0$, the C_v calculation made earlier does not change because of the piping geometry and remains at 77.5.

Using Table 9.9, for a 3-in valve in ANSI Class 600 service, the valve outlet area A_V is 7.07 in^2. Using this number and a flow rate of 850 gal/min, the velocity through the valve can be calculated as

$$V = \frac{0.321Q}{A_V} = \frac{0.321(850)}{7.07} = 39 \text{ ft/s}$$

The velocity of 39 ft/s is below the limit of 50 ft/s for liquids. Therefore, a 3-in body is acceptable for this application, although the cavitating service will need to be dealt with through modifications to the valve, such as special trim or hardened materials.

9.4 Body Sizing of Gas-Service Control Valves

9.4.1 Basic Gas Sizing Equations

The basic difference between liquid sizing and gas sizing deals with the compressibility of gases. Because of their compressibility, gases have a tendency to expand as the pressure drop occurs through the vena contracta. In turn, this lowers the specific weight of the gas. This changing specific weight must be taken into account during the sizing process using a special factor called the *expansion factor Y*.

Depending on the given service conditions or variables, one of four gas sizing equations is used. The numerical constants included in each equations deal with unit conversion factors.

$$w = 63.3 F_p C_v Y \sqrt{x P_1 \gamma_1}$$

$$Q = 1360 F_p C_v P_1 Y \sqrt{\frac{x}{G_g T_1 Z}}$$

$$w = 19.3 F_p C_v P_1 Y \sqrt{\frac{x M_W}{T_1 Z}}$$

$$Q = 7320 F_p C_v P_1 Y \sqrt{\frac{x}{M_W T Z}}$$

where w = gas flow rate (lb/h)
F_P = piping-geometry factor
C_v = valve sizing coefficient
Y = expansion factor
x = pressure-drop ratio
γ_1 = specific weight at inlet service conditions (lb/ft^3)
Q = gas flow (scfh)
G_g = specific gravity or gas relative to air at standard conditions
T_1 = absolute upstream pressure (°R = °F + 460)
Z = compressibility factor
M_W = molecular weight
P_1 = upstream absolute pressure (psia)

One of the four gas sizing equations should be selected based on the available data for the given service conditions.

9.4.2 Choked-Flow Determination

The terminal pressure-drop ratio x_T is determined by taking the appropriate value from Table 9.10. The ratio of specific heats factor F_K can be calculated by using the following equation:

$$F_K = \frac{k}{140}$$

Table 9.10 Typical x_T Factors*,†

Valve Style	Flow Direction	Trim Area	X_T
Linear globe	Over seat	Full area	0.70
	Over seat	Reduced area	0.70
	Under seat	Full area	0.75
	Under seat	Reduced area	0.75
Butterfly	60° open	Full area	0.36
	90° open	Full area	0.26
Ball	60° open	Full area	0.51
	90° open	Full area	0.30

*Data courtesy of Valtek International.
†Note: All values provided at full-open.

where F_K = ratio of specific heats factor
    k = ratio of specific heats

The ratio k of specific heats can be found for common gases in Table 9.11, which is provided for quick reference. A more complete listing is found in App. B.

The ratio x of actual pressure drop to absolute inlet pressure is determined by using the following equation:

$$x = \frac{\Delta P_a}{P_1}$$

where x = ratio of actual pressure drop to absolute inlet pressure
    ΔP = actual pressure drop (psi)
    P_1 = upstream pressure (at inlet, psia)
    P_2 = downstream pressure (at outlet, psia)

If the value for x is less than the value for $F_K x_T$, choked flow is not occurring. Inversely, when x reaches or exceeds the value of $F_K x_T$, the

Table 9.11 Physical Data for Common Gas Services

Gas	Molecular Weight (M$_w$)	Critical Temperature (°R*)	Critical Pressure (psia/bar)	Ratio of Specific Heats (k)
Air	28.97	227°	492/33.9	1.40
Ammonia	17.00	730°	1636/112.8	1.31
Argon	39.95	271°	707/48.8	1.67
Carbon Dioxide	44.01	547°	1070/73.8	1.29
Carbon Monoxide	28.01	239°	507/35.0	1.40
Ethane	30.07	549°	709/48.9	1.19
Ethylene	28.10	508°	731/50.4	1.24
Helium	4.00	9°	33/2.3	1.66
Hydrogen	2.02	59°	188/13.0	1.40
Methane	16.04	343°	667/46.0	1.31
Natural Gas	16.04	343°	667/46.0	1.31
Nitrogen	28.00	227°	492/33.9	1.40
Oxygen	32.00	278°	732/50.5	1.40
Propane	44.10	665°	616/42.5	1.31
Steam	18.02	1165°	3208/221.2	1.33

*°R = °F + 460.

flow is choked. If the flow is choked, the value $F_K x_T$ should be used instead of x, if x is used in the chosen gas sizing equation.

9.4.3 Expansion-Factor Calculation

Because of the compressibility of gases, the expansion factor Y must be determined by using the following equation. If choked flow is occurring, the value $F_K x_T$ should be used instead of x.

$$Y = 1 - \frac{x}{3F_K x_T}$$

where Y = expansion factor

x_T = terminal pressure-drop ratio

9.4.4 Compressibility-Factor Determination

The compressibility factor Z is determined by calculating the reduced-pressure value P_r and the reduced-temperature value T_r:

$$P_r = \frac{P_1}{P_c}$$

where P_r = reduced pressure

$$T_r = \frac{T_1}{T_c}$$

where T_r = reduced temperature

T_1 = absolute upstream temperature

T_C = absolute critical temperature

Once the reduced pressure P_r and reduced temperature T_r are known, the compressibility factor Z can be determined with either Fig. 9.3 or 9.4.

9.4.5 Flow-Coefficient Calculation

Using the factors determined to this point, a preliminary C_v is calculated by using the applicable gas sizing C_v equation. For this equation, the piping-geometry factor F_P should be assumed to be 1.0.

9.4.6 Approximate Body-Size Selection

Using the manufacturer's C_v tables, the smallest sized body is selected that can accommodate the calculated preliminary C_v.

9.4.7 Piping-Geometry-Factor Calculation

When the pipeline size has not been determined or is unknown, for calculation purposes the body size that was determined from Sec. 9.4.6 is used as pipeline size. The inside diameter of the piping is required

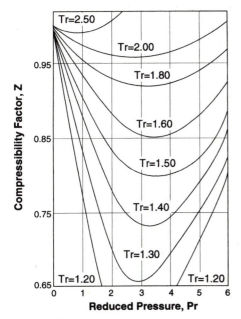

Figure 9.3. Compressibility factors, reduced pressures 0 to 6. (*Courtesy of Valtek International*)

to determine the piping-geometry factor F_p. Tables 9.5 and 9.6 are used to find the piping-geometry factors. Table 9.5 provides F_p for valves with reducers (or increasers) on both the inlet and outlet of the valve. Table 9.6 provides F_p for a valve with the reducer (or increaser) on the valve outlet only.

9.4.8 Final-Flow-Coefficient Calculation

Using the piping-geometry factor F_p, the final C_V is calculated, using one of the four equations provided. Usually, the C_V will be close to the preliminary C_V chosen earlier. Therefore, the body size will most like stay the same, unless high velocities are present.

9.4.9 Valve Exit Mach-Number Calculation

With the flow coefficient known, as well as the body size, the exit velocity of the gas from the valve is determined in Mach numbers. The

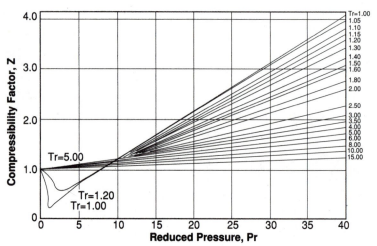

Figure 9.4. Compressibility factors, reduced pressures 0 to 40. *(Courtesy of Valtek International)*

following two equations are used for calculating velocities in gas services:

$$M_{gas} = \frac{Q_a}{5574 A_V \sqrt{\dfrac{kT}{M_W}}}$$

$$M_{gas} = \frac{Q_a}{1036 A_V \sqrt{\dfrac{kT}{G_g}}}$$

where M_{gas} = Mach number for gas service
$\quad Q_a$ = actual flow rate (cfh instead of scfh)
$\quad A_V$ = applicable flow area of body port (square inches) from Table 9.9
$\quad k$ = ratio of specific heats
$\quad T$ = absolute temperature (°R or °F + 460)
$\quad M_W$ = molecular weight
$\quad G_g$ = specific gravity at standard conditions relative to air

The following velocity equation is used for air service:

$$M_{air} = \frac{Q_a}{1225 A_V \sqrt{T}}$$

where M_{air} = Mach number for air service

To convert scfh to cfh, the following equation is used:

$$\frac{P_a Q_a}{T_a} =$$

where P_a = actual operating pressure
Q_a = actual volume flow rate (cfh)
T_a = actual temperature (°R or °F + 460)
P_S = standard pressure (14.7 psi)
Q = standard volume flow rate (scfh)
T_s = standard temperature (520°R)

The following velocity equation is used for steam service:

$$M_{steam} = \frac{wv}{1514 A_V \sqrt{T}}$$

where M_{steam} = Mach number for air service
w = mass flow rate (lb/h)
v = specific volume at flow conditions (ft³/lb)

Once the exit velocity has been calculated and is found to exceed Mach 0.5, the possibility of excessive vibration and noise will become evident because of the turbulence caused in the valve. The velocity limit for valves is near Mach 1. If noise is occurring in the valve and a special antinoise trim is used in the valve, the velocity is normally limited to Mach 0.33. If the high velocity exceeds the Mach-0.5 limit for noise generation, a larger valve body will need to be chosen. If the velocity approaches Mach 1.0 in this situation, a larger body size should also be chosen.

9.4.10 Trim-Size Selection

Valve manufacturers provide tables that outline the flow coefficients for a certain valve style, flow direction, body-pressure rating, flow characteristic, size of the valve seat (either full or reduced area) or the seal, and stroke. Some charts may be broken down to percentages of opening, since some throttling services may not utilize the entire stroke. Depending on the style of the valve, a trim number is offered with a predetermined flow area that allows the passage of flow equal to the C_v maximum.

9.4.11 Gas Sizing Example A

For this example, the following service conditions and equipment requirements are given in Imperial units:

Gas	Steam
Temperature	450°F
Upstream pressure P_1	140.0 psia
Downstream pressure P_2	50.0 psia
Flow rate Q	10,000 lb/h
Critical pressure P_C	3206.2 psia
Critical temperature T_C	705.5°F
Molecular weight M_W	18.03
Specific volume	10.41
Ratio k of specific heats	1.33
Pipeline size	2 in (ANSI Class 600)
Valve	Globe, flow-to-open
Flow characteristic	Equal percentage

Of the four C_V equations given for gas sizing (Sec. 9.4.1), the following equation is appropriate for the provided service conditions:

$$w = 19.3 F_p C_v P_1 Y \sqrt{\frac{x M_W}{T_1 Z}}$$

From Table 9.4, the pressure-drop ratio x_T for a globe valve with flow-to-open action is 0.75. The user should check for choked flow by calculating the ratio of specific heats factor F_K:

$$F_K = \frac{k}{1.40} = \frac{1.33}{1.40} = 0.95$$

The ratio of actual pressure drop to absolute inlet pressure x is now calculated with the following equation:

$$x = \frac{\Delta P}{P_1} = \frac{140 - 50}{140} = 0.64$$

The value $F_K x_T$ can then be calculated as

$$(0.95)(0.75) = 0.71$$

Because the ratio of actual pressure drop to absolute inlet pressure x is less than the combined value $F_K x_T$, choked flow is not occurring and the value x is used with the remaining calculations.

The expansion factor Y is now calculated using the following equation:

$$Y = 1 - \frac{x}{3(F_K x_T)} = 1 - \frac{0.64}{3(0.71)} = 0.70$$

The compressibility factor Z can be determined by using the equations for the reduced-pressure factor P_r and the reduced-temperature factor T_r.

$$P_r = \frac{P_1}{P_C} = \frac{140}{3208.2} = 0.04$$

$$T_r = \frac{T_1}{T_C} = \frac{450 + 460}{705.5 + 460} = 0.78$$

With the aid of these two numbers and Fig. 9.3, the compressibility factor Z is found to be 1.0. Assuming that the piping-geometry factor F_p is 1.0, the appropriate C_v equation should be used to calculate a preliminary C_v:

$$w = 19.3 F_p C_v P_1 Y \sqrt{\frac{x M_W}{T_1 Z}} \quad \text{or} \quad C_v = \frac{w}{19.3 F_p P_1 Y} \sqrt{\frac{T_1 Z}{x M_W}}$$

$$C_v = \frac{10,000}{(19.3)(140)(0.70)} \sqrt{\frac{(910)(1.0)}{(0.64)(18.02)}} = 47.0$$

From the manufacturer's C_v tables, the smallest valve body should be chosen that will pass the required C_v of 47. For assumption purposes, a 2-in valve is the smallest size that will accommodate a C_v of 47. Because the 2-in body is the same size as the pipeline size, the piping-geometry factor F_p is 1.0 and C_v remains the same. In this case, the preliminary C_v becomes the final C_v.

At this point, the exit velocity should be calculated to ensure that it is within the velocity limits of Mach 0.5 for noise or Mach 1.0 for maximum velocity. The valve outlet of a 2-in valve is 3.14 (from Table 9.9). From the steam tables, v is found to be 10.41 ft^3/lb and T is 414°F.

Therefore, the following velocity equation should be used for steam service:

$$M_{\text{steam}} = \frac{wv}{1514\, A_V \sqrt{T}} = \frac{(10{,}000)(10.41)}{(1515)(3.14)\sqrt{414 + 460}} = 0.74$$

Because Mach 0.74 is greater than the noise limit of Mach 0.5, the turbulence will most likely create noise in the valve, and preventative measures may be necessary, such as special trim, insulation, or isolation of the valve. Because the velocity did not exceed the limit of Mach 1.0, a larger valve size is not necessary and the final C_v remains the same.

9.4.12 Gas Sizing Example B

For the second gas example, the following service conditions and equipment requirements are provided in Imperial units:

Gas	Natural gas
Temperature	65°F
Upstream pressure P_1	1314.7 psia
Downstream pressure P_2	99.7 psia
Flow rate Q	2,000,000 scfh
Critical pressure P_C	672.9 psia
Critical temperature T_C	342.8°F
Molecular weight M_W	16.04
Ratio k of specific heats	1.31
Pipeline size	Unspecified (ANSI Class 600)
Valve	Globe, flow-to-open
Flow characteristic	Linear

Of the four C_v equations given for gas sizing from Sec. 9.4.1, the following equation is best for the provided service conditions:

$$Q = 7320\, F_p C_v P_1 Y \sqrt{\frac{x}{M_W T}}$$

Referring to Table 9.4, the pressure-drop ratio x_T for a globe valve with flow-to-open action is 0.75. A choked-flow condition should be checked first by calculating the ratio of specific heats factor F_K:

$$F_K = \frac{k}{1.40} = \frac{1.31}{140} = 0.94$$

The ratio x of actual pressure drop to absolute inlet pressure is determined by using the following equation:

$$x = \frac{\Delta P}{P_1} = \frac{1314.7 - 99.7}{1314.7} = 0.92$$

The value $F_K x_T$ can then be calculated as follows:

$$(0.94)(0.75) = 0.70$$

Because the combined value $F_K x_T$ is less than the ratio of actual pressure drop to absolute inlet pressure x, choked flow is occurring and $F_K x_T$ is used with the remaining calculations. The expansion factor Y is now calculated using the following equation:

$$Y = 1 - \frac{x}{3(F_K x_T)} = 1 - \frac{0.70}{3(0.70)} = 0.67$$

Before the compressibility factor Z can be determined, the reduced-pressure factor P_r and the reduced-temperature factor T_r must be calculated with the following equations:

$$P_r = \frac{P_1}{P_C} = \frac{1314.7}{667.4} = 1.97$$

$$T_r = T\frac{1}{T_C} = \frac{65 + 460}{342.8} = 1.53$$

Using the P_r and the T_r factors with Fig. 9.3, the compressibility factor Z is found to be approximately 0.86. With the assumption that the piping-geometry factor F_P is 1.0 and that x is now replaced by the combined value $F_K x_T$, the chosen C_v equation is used to calculate a preliminary C_v:

$$Q = 7320 F_P C_v P_1 Y \sqrt{\frac{F_K x_T}{M_W T Z}} \quad \text{or} \quad C_v = \frac{Q}{7320 F_P P_1 Y} \sqrt{\frac{M_W T Z}{F_K x_T}}$$

$$C_v = \frac{2,000,000}{(7320)(1.0)(1314.7)(0.667)} \sqrt{\frac{(16.04)(525)(0.86)}{0.70}} = 32$$

From the manufacturer's C_v tables, the user should find the smallest valve body that will pass the required C_v of 32. For this example, a 1.5-in valve is assumed to be the smallest size that will accommodate the preliminary C_v of 32. Because the pipeline size is unspecified, the user must assume that the piping-geometry factor F_p is 1.0 and the final C_v remains the same as the preliminary C_v.

The exit velocity is now calculated to ensure that the 1.5-in body will handle the velocity limit of Mach 1.0. If the velocity exceeds Mach 0.5, noise will most likely be generated. From Table 9.9, the valve outlet area A_V of a 1.5-in body is 1.77. Since the fluid is natural gas, the following velocity equation for gas service is used after converting scfh to cfh (Sec. 9.4.9):

$$M_{gas} = \frac{Q_a}{5574 \, A_V \sqrt{\frac{kT}{M_W}}} = \frac{297,720}{(5574)(1.77)\sqrt{\frac{(1.31)(65+460)}{16.04}}} = \text{Mach } 4.61$$

Because a Mach number exceeding sonic velocity (Mach 1.0) at the outlet of the valve is not possible, a larger valve size must be chosen to lower the velocity to below Mach 1.0.

The chosen valve would ideally handle a velocity of Mach 0.5 or less. To find the correct valve size to handle the process at Mach 0.5, the velocity equation should be used—except the user should solve for the unknown factor, which is the valve outlet area A_V:

$$A_V = \frac{Q_a}{5574 M_{gas}\sqrt{\frac{kT}{M_W}}} = \frac{297,720}{(5574)(0.5)\sqrt{\frac{(1.31)(65+460)}{16.04}}} = 16.3 \text{ in}^2$$

The valve outlet area A_V can then be used to solve the size of the valve:

$$A_V = pd^2 \quad \text{or} \quad d = \sqrt{\frac{4A_V}{\pi}} = \sqrt{\frac{(4)(16.3)}{3.14}} = 4.6 \text{ in}$$

Because a 4-in valve would be too small and a 5-in valve does not exist, a 6-in valve is necessary. This valve will need a reduced trim to accommodate $32C_v$.

9.5 Pressure-Relief-Valve Sizing

9.5.1 Introduction to Pressure-Relief-Valve Sizing

The proper sizing of pressure-relief valves involves determining the size of the valve itself as well as calculating the pressure loss in the inlet pipeline and the back pressure in the downstream piping. Sizing should also determine the discharge reactive force. This section will deal with the sizing of both relief and safety valves and their respective liquid and gas applications.

In simple terms, the basic formula for sizing pressure-relief valves is

$$A = \frac{W}{BKG}$$

where A = flow area of the pressure-relief valve
W = required mass flow rate
B = sizing constant
K = coefficient of discharge
G = mass flux

To accurately find the mass flow rate and appropriate size of the pressure-relief valve, each of these variables must be determined using appropriate equations. As with throttling valve sizing, some key differences exist between gas and liquid sizing equations.

The coefficient of discharge K is a ratio of experimental flow to theoretical flow, provided and certified by the manufacturer. The sizing constant is a fixed number that is used depending on the numerical system used (metric or Imperial units).

9.5.2 Gas Sizing of Nonchoked Flow

The first step in sizing safety valves in gas service is to determine if the flow is choked or not. The flow is not choked if the ratio of the outlet and inlet pressures (absolute) is greater than the critical pressure, as shown by the following equation:

$$\frac{P_2}{P_1} > \left(\frac{2}{k+1} \right)^{k/(k-1)}$$

where P_1 = upstream pressure at inlet
P_2 = downstream pressure at outlet
k = isentropic coefficient

The isentropic coefficient k is a physical constant that is determined at atmospheric pressure and 449°R, although it can vary significantly when the pressures or temperatures change. The isentropic coefficient can equal the ratio for specific heats if the assumption is made that the fluid is pure isentropic flow. If the isentropic coefficient is not known, the ratio for specific heats can be used, which is included in App. B.

For nonchoked flow, the equation for the mass flux G is written as

$$G = P_1 \left\{ \frac{1}{RT_1Z} \frac{2k}{k-1} \left[\left(\frac{P_2}{P_1}\right)^{2/k} - \left(\frac{P_2}{P_1}\right)^{(k+1)/k} \right] \right\}^{-0.5}$$

where G = mass flux
R = gas constant*
T_1 = temperature at inlet
Z = compressibility factor

Substituting this equation into the simplified sizing equation, the following equation can be used for gas sizing in nonchoked flow applications:

$$A = \frac{W}{BKG} = \frac{W}{BKP_1} \left\{ \frac{1}{RT_1Z} \frac{2k}{k-1} \left[\left(\frac{P_2}{P_1}\right)^{2/k} - \left(\frac{P_2}{P_1}\right)^{(k+1)/k} \right] \right\}^{0.5}$$

where A = flow area of the safety valve (square inches for Imperial units or square meters for metric)
W = mass flow rate
B = 20,420 (Imperial units) or 1.0 (metric)
K = coefficient of discharge
Z = compressibility factor (Figs. 9.3 and 9.4)

The compressibility factor Z provides an allowance for any deviation between the properties of the real gas from the ideal gas. If a compressibility factor is not known, the user may substitute a value of 1.0 instead. (Section 9.4.4 provides a more detailed explanation.) To calculate Z, the

*For metric units,

$$R = \frac{8314.3}{M} \frac{J}{kg \cdot K}$$

For imperial units,

$$R = \frac{1545}{M} \frac{ft \cdot lbf}{lb \cdot °R}$$

user must first determine the reduced pressure P_r and the reduced temperature T_r, which are determined by the following equations:

$$P_r = \frac{P_1}{P_C}$$

where P_r = reduced pressure
 P_C = critical pressure

$$T_r = \frac{T_1}{T_C}$$

where T_r = reduced temperature
 T_1 = upstream temperature at inlet (absolute)
 T_C = critical temperature (absolute)

9.5.3 Gas Sizing of Choked Flow

The flow is considered to be choked when the ratio of the outlet and inlet pressures (absolute) is less than the critical pressure, which is determined by the following equation:

$$\frac{P_2}{P_1} < \left(\frac{2}{k+1}\right)^{k/(k-1)}$$

For nonchoked flow, the equation for the mass flux G is written as

$$G = P_1\left[\frac{2}{ZRT_1}\left(\frac{2}{K+1}\right)^{(k+1)/(k-1)}\right]^{0.5}$$

Substituting the above equation for mass flux into the simplified sizing equation, the following equation can be used for gas sizing for choked-flow applications:

$$A = \frac{W}{BKG} = \frac{W}{BKK_bP_1}\left[\frac{1}{ZRT_1}k\left(\frac{2}{k+1}\right)^{(k+1)/(k-1)}\right]^{-0.5}$$

where B = 1702 (Imperial units) or 1.0 (metric)
 K_b = capacity-correction factor

The capacity-correction factor K_b is a compensation factor for the loss of capacity because of back-pressures. This factor varies according to the particular manufacturer's design.

9.5.4 Steam Sizing

For lower-pressure ranges of less than 1600 lb/in^2 (11 MPa), the following equations are used for sizing safety valves used in steam services. For metric units:

$$A = \frac{686W}{KP_1}$$

For Imperial units:

$$A = \frac{W}{51.5KP_1}$$

When pressures range between 1600 lb/in^2 (11 MPa) and 3200 lb/in^2 (22 MPa), the equations above provide a gradual shift toward oversizing the valve. Therefore, to compensate for this shift, a capacity-correction factor is included in the ASME code for boilers and pressure vessels. For the upstream pressure P_1 the following factors are provided, depending on the units used.

For values of MPa (absolute):

$$K_n = \frac{27.644P_1 - 1000}{33.242P_1 - 1.061}$$

For values of lb/in^2:

$$K_n = \frac{0.1906P_1 - 1000}{0.2292P_1 - 1.061}$$

With superheated steam applications, the equation above must be corrected by a correction factor for superheat K_{SH}, which is defined as the ratio of the maximum flow of the nozzle (based on the inlet service conditions) to the flow as calculated by the steam flow equation.

9.5.5 Liquid Sizing

For sizing of liquid applications, once again, the basic formula for sizing relief valves is used:

$$A = \frac{W}{BKG}$$

With liquid conditions, the mass flux G is determined by the following equation:

$$G = \left(\frac{1}{2(P_1 - P_2)\rho}\right)^{-0.5}$$

where ρ = density of the liquid

Substituting the above equation in the sizing equation, the flow area can be calculated:

$$A = \frac{W}{BKK_WK_V}\left(\frac{1}{2(P_1 - P_2)\rho}\right)^{0.5}$$

where B = 1702 (Imperial units) or 1.0 (metric)
K_W = back-pressure capacity-correction factor
K_V = viscosity-correction factor

The liquid-sizing equation introduces two new factors. The back-pressure capacity-correction factor K_W adjusts the equation for the flow loss associated with decreased valve lift due to the back-pressure of the liquid, as shown in Table 9.12.

The viscosity-correction factor K_V takes into account the flow loss associated with fluids with a Reynolds number less than 60,000. (If the Reynolds number is higher than 60,000, the factor is 1.0.) Typical K_V values can be found by using the American Petroleum Institute Standard API RP520 Part 1 (1976).

9.5.6 Sizing Example

For this example, the following service conditions are given in metric units:

Fluid	Oxygen
Molecular weight M	32.0
Isentropic coefficient k	1.4
Critical pressure P_C	1.08
Critical temperature T_C	154.8 K

Table 9.12 Back-Pressure
Capacity-Correction Factor

Percent of Back Pressure	K_w
10%	1.0
15%	0.98
20%	0.95
25%	0.90
30%	0.86
35%	0.82
40%	0.77
45%	0.72
50%	0.68

Mass rate of flow W	5.5 kg/s
Upstream pressure P_1	30×10^6 Pa
Upstream temperature T_1	200°C (or 473 K)
Atmospheric discharge	100×10^3 Pa
Valve type	Safety valve
Coefficient of discharge K	0.9

In order to determine the compressibility factor, the reduced pressure P_r and reduced temperature T_r are calculated:

$$P_r = \frac{P_1}{P_C} = \frac{30}{1.08} = 27.78$$

$$T_r = \frac{T_1}{T_C} = \frac{473}{154.8} = 3.06$$

Using Fig. 9.3, the compressibility factor Z is found to be 1.76. Assuming choked flow and using the expanded sizing equation, the flow area of the pressure-relief valve is

$$A = \frac{W}{BKG} = \frac{W}{BKK_b P_1} \left[\frac{1}{ZRT_1} k \left(\frac{2}{k+1} \right)^{(k+1)/k-1)} \right]^{-0.5}$$

$$= \frac{5.5}{(1.0)(0.9)(0.975)(30 \times 10^6)} \left[\frac{32.0}{(10^3)(8314.3)(47\ 3)(1.76)} 1.4 \left(\frac{2}{2.} 4 \right)^{2.4/0.4} \right]^{-0.5}$$

$$= 142 \times 10^{-6}\,\text{m}^2$$

According to API Standard 526, the correct nozzle size for a flow area of 142×10^{-6} m^2 (0.2201 in^2) is an E orifice, which offers a flow area of 0.2279 in^2.

10

Actuator Sizing

10.1 Actuator-Sizing Criteria

10.1.1 Introduction to Actuator Sizing

With the automation of process systems, the use of actuators on throttling valves and actuation systems on manual on–off valves has increased dramatically. Generally, actuator sizing is a complex science, involving a number of factors that must be considered to match the correct actuator with the valve. For the valve to open, close, and/or throttle against process forces, proper actuator selection and sizing are critical.

Some users equate valve-body size with the actuator size; for example, a false assumption can be made that a 3-in valve always uses a certain size actuator, whose standard actuator yoke connection matches the valve connection. If all process service conditions and valve designs were equal, this might be possible. However, processes vary widely in terms of pressures, pressure drops, temperatures, shutoff requirements, etc. Valves vary according to motion (linear and rotary), packing friction, balancing (nonbalanced versus pressure-balanced), etc. Because of all the variables between the process and the valve, one valve size may have a number of actuator size options. For this reason, the user cannot simply place any spare actuator on a valve and expect it to work correctly—the actuator will most likely be undersized or oversized for that valve and the process. If the actuator is undersized, the major problem is that it will not be able to overcome the process and valve frictional forces. If the actuator is slightly undersized, it will struggle to overcome the forces working against it, providing sluggish and erratic stroking, as well as possibly not meeting the shutoff requirement. In addition, if the actuator is not stiff enough to hold its position close to the seat or seal, the "bathtub stopper" effect will take place and the closure element will slam into the seat or seal, causing a water-hammer effect. If the actuator is extensively undersized, it will not be able to open or close or throttle correctly.

If the actuator is oversized, the main disadvantage is that the actuator cost is higher. In addition, the oversized actuator is heavier and taller, which may create seismic, space, or maintenance concerns. From a performance standpoint, the larger actuator may be more sluggish in terms of speed and response. Larger actuators also produce greater thrust, which may damage the internal parts of the valve if the process forces are not present to counter that thrust. For this reason, oversized actuators require the use of a pressure regulator, which may create additional problems of incorrect settings and even slower response.

Generally, actuators have a tendency to be oversized because of the buildup of safety factors that the user and manufacturer add to the design process to ensure adequate "worst-case scenario" protection. If the calculations show a certain actuator size to be marginally or slightly undersized for a given process and valve, most users tend to move to the next larger size. However, because of the safety factors already built into the sizing process, the smaller size may function just as well, if not better, with that process.

10.1.2 Basic Actuator-Sizing Criteria

Actuator-sizing methods vary from manufacturer to manufacturer, depending on the basic design; however, several basic concepts are central to any actuator sizing. First and most importantly, the actuator must have the thrust to overcome the process forces that are operating inside the valve—in particular, the upstream and downstream pressures. In some services, a valve is working in an unbalanced situation where the upstream pressure is working against one side of the closure element, and the downstream pressure is working against the opposite side. These forces can be significant and will require a larger actuator as the force increases. Other valves permit a pressure-balanced design in which the upstream pressure is allowed to act on both sides of the closure element. This allows a minimal amount of process force to act against the element, permitting a smaller actuator.

The actuator must also provide enough force to overcome the process pressures in order to close the closure element, as well as to maintain the shutoff requirements indefinitely, according to the seat leakage classification (Sec. 2.3). The tighter the shutoff requirement, the greater the force must be provided by the actuator. If tight shutoff is not a main consideration, or if the valve is expected to throttle and close rarely, a lower shutoff classification may suffice that will allow the use of a smaller actuator.

The actuator must also overcome any frictional forces between the valve's stem and packing box. This friction can vary from a number of factors: number of rings, packing material, linear versus rotary motion, and packing compression requirements.

The final factor that may create a need for additional force is the design criteria of the valve itself. For example, a linear globe valve may be designed with pressure-balanced trim. Although the process forces are minimized, the seals of the pressure-balanced plug will increase the frictional forces, as well as add to the weight of the plug. In extremely large valves, the weight of the closure or regulating element (especially with globe-style plugs) must be taken into consideration.

Therefore, the forces that must be considered to determine the size and subsequent thrust of the actuator are written as

$$F_{total} = F_{process} + F_{seat} + F_{packing} + F_{miscellaneous}$$

where F_{total} = total force (or actuator thrust) required to open, close, or throttle valve

$F_{process}$ = force to overcome unbalanced process pressure
$F_{packing}$ = force required to overcome packing friction
F_{seat} = force to provide correct seat load
$F_{miscellaneous}$ = force to overcome special design factors, weight, etc.

Another design criteria is the speed requirement of the actuator. In some cases, such as applications in which the process or personnel safety is a concern, the user may want the valve to close in a short time, such as less than a second, as opposed to several seconds. However, excessively fast actuator speed can present multiple problems, including water-hammer effects and position overshoot. Pneumatic actuators are subject to a number of factors that affect air capacity, such as pressure fluctuations, piping and tubing bends, filters, etc. For these reasons, high-speed actuation systems are normally hydraulic or electrohydraulic designs.

10.1.3 Free Air

Because the majority of actuators or actuation systems are pneumatically driven, certain principles concerning air compressibility and volume changes occurring with pressure changes must be understood. The specifications for pneumatically driven equipment, including actuators, are provided using the term *free air*. By definition, free air is the flow or volume rate at standard atmospheric temperature [70°F (21°C)] and pressure [14.7 psia (1 bar)]. Using free air avoids any misunderstanding regarding changes in volume. Typically, absolute pressure is designated as *psia*, gauge pressure as *psig*, and differential pressure as simply *psi*. For most equipment, the free-air flow rate is expressed in standard cubic feet per minute (scfm).

Because air volume can vary according to changes in pressure, the amount of free air contained in a vessel can be written as

$$V_1 = V_2 \frac{P_2}{P_1}$$

where V_1 = free-air volume (standard cubic feet)
V_2 = vessel volume
P_1 = atmospheric pressure (14.7 psia)
P_2 = absolute vessel pressure (psia)

10.1.4 Supply Flow Rates

For pneumatically driven actuators, determining the correct air supply rate to the actuator is critical to ensure that enough air will be available to operate the actuator and provide the thrust necessary for the application. The relationship between flow rate and pressure drop is demonstrated by the following equations:

$$\Delta P = \frac{LQ^2}{kC_R d^{5.31}} \quad \text{or} \quad Q = \sqrt{\frac{\Delta P k C_R d^{5.31}}{L}}$$

where ΔP = pressure drop (psi)
L = length of tubing or piping (ft)
Q = standard air flow rate (scfm)
k = constant of 35,120
C_R = ratio of line pressure (psia) to atmospheric pressure (14.7 psia)
d = inside diameter of piping or tubing (inches) from Tables 10.1 and 10.2

For example, if a given actuator operates best at 80 psi and must have 4.3 scfm to operate at the required speed, the following parameters apply:

Line pressure	85 psia
Length L of tubing	100 ft
Tubing size	0.25 in

Table 10.1 Piping Values for d and $d^{5.31*,\dagger}$

Factor	0.25 in.	0.375 in.	0.5 in.	0.75 in.	1.0 in
d	0.364	0.493	0.622	0.824	1.049
$d^{5.31}$	0.0047	0.0234	0.0804	0.3577	1.2892

*All dimensions are outside diameter.
†*Data courtesy of Automax, Inc.*

The pressure drop ΔP is 5 psi (85−80 psi) and the ratio C_R of line pressure to atmospheric pressure is 6.78, which is shown as

$$C_R = \frac{P_1 + 14.7}{14.7} = \frac{99.7}{14.7} = 6.78$$

Using the calculations above, the flow rate for 0.25-in tubing (from Table 10.2, $d = 0.204$) can be calculated as follows:

$$Q = \sqrt{\frac{\Delta P k C_R d^{5.31}}{L}} = \sqrt{\frac{(5)(35,120)(6.78)(0.204)^{5.31}}{100}} = 1.6 \text{ scfm}$$

Because 1.6 scfm is less than the 4.3 scfm required for the speed requirement, a larger tube size must be chosen. A 0.375-in tube would produce 5.7 scfm, which is more than adequate:

$$Q = \sqrt{\frac{\Delta P k C_R d^{5.31}}{L}} = \sqrt{\frac{(5)(35,120)(6.78)(0.329)^{5.31}}{100}} = 5.7 \text{ scfm}$$

Table 10.2 Tubing Values for d and $d^{5.31*,\dagger}$

Factor	0.25 in.	0.375 in.	0.5 in.	0.75 in.	1.0 in
d	0.204	0.329	0.430	0.555	0.680
$d^{5.31}$	0.0002	0.0027	0.0113	0.0439	0.129

*All dimensions are outside diameter.
†*Data courtesy of Automax, Inc.*

10.1.5 Air Usage and Consumption

The user must ensure that the air-supply capacity can meet the needs of all the pneumatic operators involved with a typical process system. This means that the compressor must be sized according to the air requirements of the actuators, which requires knowledge of the air usage and consumption. Correct calculations of the air usage and consumption allow for a more accurate prediction of air requirements and proper sizing of the compressor. In those cases where the air requirements exceed the capacity of the compressor or if the compressor is undersized, the pressure will not be adequate. Overall, this results in sluggish response or not enough thrust to operate the valve.

The term *air usage* refers to the amount of air used by a pneumatic actuator to stroke the valve. After the valve is stroked, the air usage stops until the valve is stroked again. The term *air consumption* refers to those pneumatic instruments that bleed air constantly, such as is the case with positioners. For spring-return (single-acting) diaphragm actuators, no air is used on the spring side of the diaphragm. Therefore, when the actuator is fully stroked, the air usage is the amount of the actuator's free-air volume at the pressure given. For example, using the free-air equation from Sec. 10.1.3, the assumption is made that a single-acting actuator has a volume of 2.1 ft^3 at 60 psi of air supply and will stroke six times per hour. The usage per cycle in standard cubic feet is

$$V_1 = V_2 \frac{P_2}{P_1} = 2\left(\frac{60 + 14.7}{14.7}\right) = 10.2 \text{ scf}$$

The usage per hour (standard cubic feet per hour) involving six strokes per hour can then be calculated:

$$10.2 \text{ scf} \times 6 = 61.2 \text{ scfh}$$

Double-acting actuators use air on both sides of the diaphragm or piston, depending on the design. Ideally, the air volume on both sides would be equal, but this is not the case because one side has less volume due to the actuator stem, travel stop, or fail-safe spring. In this case, the total volumes of the two sides are calculated separately and added together to present air volume per cycle. For example, a rack-and-pinion actuator has 500 in^3 on one side and 300 in^3 on the opposite side. It will be stroked 12 times an hour with 80 psi of air supply.

Using the conversion factor, cubic inches are converted to cubic feet for the first side:

$$\text{scf} = \frac{\text{in}^3}{1728} = \frac{500}{1728} = 0.29 \, \text{scf}$$

The opposite side is converted likewise:

$$\text{scf} = \frac{\text{in}^3}{1728} = \frac{300}{1728} = 0.17 \, \text{scf}$$

The combined air volume for the actuator is then 0.46 scf (0.29 + 0.17) and the air usage per cycle is calculated as

$$V_1 = V_2 \frac{P_2}{P_1} = 0.46 \left(\frac{80 + 14.7}{14.7} \right) = 2.96 \, \text{scf}$$

The usage per hour (standard cubic feet per hour) involving 12 strokes per hour can then be calculated:

$$2.96 \, \text{scf} \times 12 = 35.52 \, \text{scfh}$$

If a positioner is used with a single-acting actuator, the air usage can vary considerably since the actuator is throttling between the open and closed positions. Depending on the position movement, which can be large or small, the air usage is directly proportional to the movement. The air usage for an actuator with a positioner can be determined by the following equation:

$$\text{scfh} = \frac{V}{A} [P_S(M_2 - M_1) + 0.4PM_1)]N$$

where V = actuator volume (ft^3)
P_S = supply pressure (psia)
M_1 = starting position (fraction of stroke)
M_2 = finished position (fraction of stroke)
A = atmospheric pressure (14.7 psia)
N = number of strokes per hour

For example, a single-acting actuator with a positioner has an air volume of 500 in^3 and is stroked between 10 and 50 percent open. It is required to stroke eight times an hour using 60 psi of air supply. This would be calculated as

$$\text{scfh} = \frac{V}{A}[P_S(M_2 - M_1) + 0.4\,PM_1]N$$

$$= \frac{\dfrac{500}{1728}}{14.7}\{[(60 + 14.7)(0.5 - 0.1)] + [0.4(60 + 14.7)(0.1)]\}8$$

$$= 5.26 \text{ scfh}$$

This calculation provides only the air usage. Because some positioners bleed continually, they provide air consumption, which must be figured into the total air requirement when sizing the compressor. If a positioner is used with a double-acting actuator, as most are, the above equation is modified slightly:

$$\text{scfh} = \frac{V}{A}\{[2P_S(M_2 - M_1)] + [0.4P(1 - M_2 - M_1)]\}N$$

For example, a double-acting actuator with a positioner has an air volume of 300 in^3 and is stroked between 20 and 70 percent open. It is required to stroke 12 times an hour using 80 psi of air supply. This would be calculated as

$$\text{scfh} = \frac{V}{A}\{[2P(M_2 - M_1)] + [0.4\,P(1 - M_2 - M_1)]\}N$$

$$= \frac{\dfrac{300}{1728}}{14.7}\{[2(80 + 14.7)(0.7 - 0.2)] + [0.4(80 + 14.7)(1 - 0.7 - 0.2)]\}12$$

$$= 11.82 \text{ scfh}$$

Once again, the user should remember that any bleeding of air from the positioner (air consumption) must also be added to the air usage calculation.

10.2   Sizing Pneumatic Actuators

10.2.1   Actuator Force Calculation for Linear Valves

To determine what size of actuator is required for a linear-motion valve, such as a globe control valve, the user must examine the force that the process is applying inside the valve. This force value is known as $F_{process}$. A major factor in determining the process force is calculating the unbalanced area. The *unbalanced area* is defined as the area of the cage (or sleeve) minus the stem area. The unbalanced area must be greater than the area of the seat. In equation form, it is written as

$$A_{unbalanced} = A_{cage\ or\ sleeve} - A_{stem} > A_{seat}$$

where $A_{unbalanced}$ = unbalanced area
$A_{cage\ or\ sleeve}$ = area of the cage or sleeve*
A_{stem} = area of the plug stem
A_{seat} = area of the seat

Formulas for calculating the process force are based upon the service conditions as well as three design criteria: The first determination is whether the flow assists with the opening or the closing of the valve. The second determination is whether the valve is unbalanced or pressure-balanced (globe or double-ported valves only). And the third determination is whether the flow is under or over the closure or regulating element (assumed to be a globe valve plug). The following formulas apply for the following valve configurations:

Pressure assists opening, unbalanced trim, flow under the plug:

$$F_{process} = (P_1 - P_2)\, A_V + P_2 A_{stem}$$

Pressure assists opening, unbalanced trim, flow over the plug:

$$F_{process} = (P_1 - P_2)\, A_V - P_1 A_{stem}$$

Pressure assists opening, balanced trim, flow under the plug:

$$F_{process} = (P_1 - P_2)\, A_{unbalanced} - P_2 A_{stem}$$

*If the valve does not have a cage or sleeve, the area of the top of the plug is used.

Pressure assists opening, balanced trim, flow over the plug:

$$F_{process} = (P_1 - P_2)A_{unbalanced} + P_2A_{stem}$$

Pressure assists closing, unbalanced trim, flow under the plug:

$$F_{process} = -[(P_1 - P_2)A_V + P_1A_{stem}]$$

Pressure assists closing, unbalanced trim, flow over the plug:

$$F_{process} = -[(P_1 - P_2)A_V - P_1A_{stem}]$$

Pressure assists closing, balanced trim, flow under the plug:

$$F_{process} = -[(P_1 - P_2)A_{unbalanced} - P_1A_{stem}]$$

Pressure assists closing, balanced trim, flow over the plug:

$$F_{process} = -[(P_1 - P_2)A_{unbalanced} + P_1A_{stem}]$$

where $F_{process}$ = force to overcome the process pressure unbalance
P_1 = upstream pressure at inlet (psia)
P_2 = downstream pressure at outlet (psia)
A_V = area of the valve port (in^2)
A_{stem} = area of the plug stem (in^2)
$A_{unbalanced}$ = unbalanced area (in^2)

When the actuator force is used to open the valve, three of the four forces oppose the actuator: the process force, the packing friction force, and any miscellaneous design forces. Because no actuator force is needed for seat loading, that value is not necessary. This can be written as

$$F_{open} = F_{process} + F_{packing} + F_{miscellaneous}$$

where F_{open} = total force (or actuator thrust) required to open valve
$F_{process}$ = force to overcome the process pressure unbalance
$F_{packing}$ = force required to overcome packing friction
$F_{miscellaneous}$ = force to overcome special design factors, weight, etc.

If the total force must close the valve, the process force must be a negative number, as demonstrated in the latter four equations. In other words, because the process pressure is assisting the valve to close, the process force actually decreases, rather than increases, the force required by the actuator. The actuator has to produce only enough force to overcome the combined force produced by the packing friction, seat load, and miscellaneous design forces, minus the process force. In this case, the actuator force requirement may be minimal. This can be written as

$$F_{close} = F_{seat} + F_{packing} + F_{miscellaneous} - F_{process}$$

where F_{close} = total force (or actuator thrust) required to close valve
F_{seat} = force required to provide correct seat load

In applications where the actuator must open and close the valve, both forces for opening and closing, F_{open} and F_{close}, must be calculated. The largest force of the two is then used to determine the size of the actuator.

After the process force has been determined, the next force to be calculated is the load required by the shutoff classification, which uses the following equation:

$$F_{seat} = F_{class} C_{port}$$

where F_{class} = required seat force of shutoff classification (see Table 2.7)
C_{port} = circumference of the valve port

The force required to overcome the packing friction, $F_{packing}$, is provided by the manufacturer. Packing friction is determined by the diameter of the stem and the packing material, assuming correct compression. Overcompressing the packing will add to the packing friction and the force required to overcome it.

After the cumulative forces are calculated, an actuator can be chosen from the manufacturer based on the thrust capabilities of the actuator. The final requirement is that the correct actuator can be mounted on the valve that has been sized for the service. In some applications involving large oversized actuators required for severe services, the yoke-to-bonnet connection may not be a standard and will require modifications.

With pneumatic actuators, the appropriate size and spring will need to be chosen for the application. Most manufacturers provide actuator tables that include the thrust that the actuator can generate. In addition, the user must chose the desired failure action (fail-open or fail-closed) and yoke-to-bonnet connection. The correct actuator is the smallest actuator that meets the thrust and mounting requirements.

10.2.2 Actuator Force Calculation for Butterfly Valves

A different actuator-sizing criteria must be considered with rotary valves. Critical to rotary actuator sizing is the butterfly valve's torque requirement, in other words, the amount of thrust that the actuator must apply to the shaft to produce a rotational force to operate the valve. In particular, the user must calculate the *seating torque*, which is the torque needed to close the valve against or with the process; the *breakout torque*, which is the torque needed to begin to open the valve; and the *dynamic torque*, which is the torque needed to throttle the valve. When these torque values are known, the correct rotary actuator can be chosen.

The first step in sizing an actuator for a butterfly valve is to determine the orientation of the shaft and the actuator stiffness requirements. Shaft orientation is critical with eccentric butterfly valves. When the shaft is placed on the upstream side of the flow, the process fluid forces the disk into the seal. On the other hand, when the shaft is placed on the downstream side of the flow, the process fluid forces the disk to open. In gas applications, when the butterfly valve is designated to fail-closed, the shaft is generally upstream. If the valve is designated to fail-open, the shaft is downstream. With liquid applications, the disk has a tendency to slam into the seal in fail-closed applications if the actuator is not stiff enough to withstand the process flow. A rotary actuator with insufficient stiffness is likely to cause water-hammer effects; therefore a stiffness calculation must be made by finding the ratio of the maximum pressure drop to the supply pressure:

$$A_S = \frac{P_1 - P_2}{P_S}$$

where A_S = required actuator stiffness
P_1 = upstream pressure at inlet (psia)
P_2 = downstream pressure at outlet (psia)
P_S = Supply pressure

Table 10.3 shows the maximum actuator stiffness values for three sizes of actuators. If the calculated value is larger than the table value, a larger actuator size must be chosen for that size of valve.

For example, a 4-in butterfly valve has an upstream pressure P_1 of 240 psia, and a downstream pressure P_2 of 60 psia, and a supply pressure of 80 psi. The required actuator stiffness ratio is

$$A_S = \frac{P_1 - P_2}{P_S} = \frac{240 - 60}{80} = 2.25$$

Looking at Table 10.3 for 4-in valves, the actuator stiffness is slightly larger than the maximum value for the smallest actuator, size A. Therefore, the next larger size, size B, would be required.

The chosen actuator must also have the necessary force to generate torque for the butterfly valve to close, to open (breakout torque), and

Table 10.3 Actuator Stiffness Factors

Valve Size	Actuator Size A	Actuator Size B	Actuator Size C
2-inch DN 50	4.2		
3-inch DN 80	3.1		
4-inch DN 100	2.0	6.7	
6-inch DN 150		4.3	7.0
8-inch DN 200		2.5	5.9
10-inch DN 250			4.1

to throttle between the open and closed positions. The following equations are used to determine seating and breakout torques:

Shaft downstream, torque required to close the valve:

$$T_{seat} = -T_P - T_S - T_H - \Delta P_{max}(C_B + C_O)$$

Shaft downstream, torque required to open the valve:

$$T_{breakout} = T_P + T_S + T_H + \Delta P_{max}(C_B - C_O)$$

Shaft upstream, torque required to close the valve:

$$T_{seat} = -T_P - T_S - T_H - \Delta P_{max}(C_B - C_O)$$

Shaft upstream, torque required to open the valve:

$$T_{breakout} = T_P + T_S + T_H + \Delta P_{max}(C_B + C_O)$$

where T_{seat} = seating torque required
$T_{breakout}$ = breakout torque required
T_P = packing torque
T_S = seat torque
T_H = handwheel torque*
ΔP_{max} = maximum pressure drop at shutoff
C_B = bearing (or guide) torque factor
C_O = off-balance torque factor

The packing torque T_P is the torque required to overcome the rotational friction of the packing on the shaft. The seat torque T_S is the torque required to overcome the friction of the seat on the disk. The bearing torque factor C_B indicates the relationship that as the pressure across the valve increases, the force on the bearing increases proportionally. The handwheel torque T_H is the torque required to overcome the friction of an attached handwheel. If a declutchable handwheel is used, this factor is considered only when the handwheel is in gear. The off-balance torque factor C_O shows the relationship of the off-balance forces between the disk and the mechanical connection in the actuator (which converts the actuator force to torque). Because these torques

*Handwheel torque is 0 if no handwheel exists.

and factors vary according to individual valve designs, they are provided by the manufacturer. If the final torque value is a negative value, this indicates that the butterfly disk will have a tendency to resist closing. Conversely, if the value is positive, the disk will have a tendency to resist opening.

When a high pressure drop is expected at any part of the quarter-turn stroke, the net torque output can vary dramatically throughout the shaft rotation. For this reason, the dynamic torque is calculated at various degrees of opening. When the shaft is downstream, a reversal of torque takes place at approximately 75 percent open, which can lead to control problems with the valve. If this happens, the user has the choice of changing the orientation of the shaft to shaft upstream (if possible), or placing the limit stops on the actuator to prevent rotation beyond 70 percent.

The following equations are used when calculating the dynamic torque for butterfly valves in gas services:

To close the valve:

$$T_D = -T_P - \Delta P_{eff}(C_{BT})$$

To open the valve:

$$T_D = T_P + (\Delta P_{eff}(C_{BT})$$

where T_D = dynamic torque
T_P = packing torque value (from manufacturer)
$\Delta P_{eff} = \Delta P_{actual}$ at the flowing condition at the degree of opening (limited to the ΔP_{choked})
C_{BT} = bearing or guide torque value (from manufacturer)

For liquid applications, the following equations are used:

To close the valve:

$$T_D = -T_P - \Delta P_{eff}(C_{BT} - C_D)$$

To open the valve:

$$T_D = T_P - \Delta P_{eff}(C_{BT} - C_D)$$

where C_D = dynamic torque factor (from Table 10.4)

Table 10.4 Butterfly-Valve Torque Factors*,†,‡

Valve Size (in.)	10° A	10° B	20° A	20° B	30° A	30° B	40° A	40° B	50° A	50° B	60° A	60° B	70° A	70° B	80° A	80° B	90° A	90° B
Dynamic Torque Factor vs. Disc Position (Degrees Open)																		
2	0	0	0	0	0	0	0	0	0	0	0	0	0	0	0	0	0	0
3	0	0	0	0	0	0	0	1	0	2	0	2						
4	0	0	0	0	0	0	0	0	0	1	1	1	1	4	0	6	-4	4
6	0	0	0	1	1	1	1	3	2	4	4	9	6	17	0	23	-19	19
8	0	0	0	1	1	2	2	4	3	7	4	12	1	19	-8	28	-23	23
10	0	1	1	2	2	6	5	9	7	17	8	28	4	46	-18	65	-55	55
12	1	2	2	7	5	13	10	20	16	38	18	61	8	101	-40	144	-121	121
14	1	3	3	10	6	18	15	28	22	54	25	85	12	142	-56	201	-168	168
16	2	4	4	12	8	23	19	36	27	67	32	106	15	177	-69	250	-211	211
18	5	10	10	29	19	54	44	83	64	156	73	247	34	411	-161	582	-489	489
20	5	10	10	31	21	58	47	89	68	168	78	264	6	440	-173	623	-524	524
24	7	14	14	42	28	78	64	121	92	227	106	358	50	596	-234	845	-710	710
30	24	49	49	146	97	268	219	414	316	779	365	1230	170	2046	-804	2899	-2436	2436

*A-shaft downstream; B = shaft upstream.

†Courtesy of Valtek International.

‡Note: When degrees of opening are not known, use highest value of C_d for valve size.

If the final number for the dynamic torque value is a negative number, the disk will resist closing with the flow moving the disk toward the open position. If the dynamic torque number is positive, the disk will resist opening—with the flow moving the disk toward the closed position. From the manufacturer's data and given the necessary available air supply, an actuator with sufficient torque can then be selected. This torque must overcome the seating and breakout torques—as well as the dynamic torque, which is required through the entire stroke of the valve. If the actuator's available torque is less than the dynamic torque, a larger actuator size with more torque force should be selected.

Following selection of the actuator, stiffness should again be checked to prevent the disk from slamming into the seat for those applications with the shaft downstream.

Consideration should be given to whether a spring is necessary to move the disk to a particular failure position (fail-open or fail-closed). For fail-closed applications that do not require a high degree of shutoff, the spring must have adequate torque to overcome the dynamic torque. If the valve requires tight shutoff, the spring must generate enough torque to overcome the required seating torque at the closed position. For fail-open applications, the spring must have enough torque to overcome the required breakout torque at the closed position, as well as to overcome any dynamic torque as it moves though the full stroke to the full-open position. If the spring is incapable of producing enough force to overcome the seating or breakout or dynamic torque, a volume tank could be specified to ensure adequate force to move the valve to the correct position upon loss of air supply.

10.2.3 Actuator Force Calculation for Ball Valves

Because of the ball valve's design with the ball moving into the flow stream, as opposed to a disk that is already in the flow stream, the forces acting on the ball valve (and the torques required) are somewhat different. This requires a different set of torque calculations for seating or breakout:

Shaft downstream, torque required to open the valve:

$$T_{\text{breakout}} = T_P + T_S + \Delta P_{\text{max}}(C_B + C_S) + T_H$$

Shaft upstream, torque required to open the valve:

$$T_{\text{breakout}} = T_P + T_S + \Delta P_{\max} C_B + (T_S - \Delta P\, C_S) + T_H$$

where ΔP = actual pressure drop $(P_1 - P_2)$
   C_S = seat torque factor

The packing torque T_P is the torque required to overcome the rotational friction of the packing on the shaft. The seat torque T_S is the torque required to overcome the friction of the seat on the disk. The bearing torque factor C_B shows the relationship that as the pressure across the valve increases, the force on the bearing increases proportionately. The handwheel torque T_H is the torque required to overcome the friction of an attached handwheel. If a declutchable handwheel is specified with the actuator, this factor is only considered when the handwheel is in gear. Because these torques and factors vary widely due to design differences, they are usually determined by the manufacturer.

With liquid services, the dynamic torque must also be calculated. As noted in the previous section, dynamic torque is the torque required to overcome the torque on the closure element caused by the fluid dynamic forces on the ball. To calculate dynamic torque, the following equation is used:

$$T_D = T_P + \Delta P_{\text{eff}} (C_D + C_B)$$

where ΔP_{eff} = actual pressure drop across the valve at the flowing condition that occurs when the valve is in the open position (ΔP_{eff} is less than or equal to ΔP_{choked})
   C_D = dynamic torque factor (from Table 10.5)
   C_B = bearing torque factor (from manufacturer)

Once the seating or breakout and dynamic torques have been calculated, the correct actuator with sufficient torque is then chosen from the manufacturer's tables.

If a spring is required to move the ball to a particular failure position (fail-open or fail-closed), special consideration should be given to sizing the correct spring that can overcome the process forces. For fail-closed applications that do not require a high degree of shutoff, the spring must have adequate torque to overcome the dynamic torque. If the ball valve requires tight shutoff, the spring must generate enough torque to overcome the required seating torque at the closed position. For fail-open applications, the spring must have enough torque to overcome the required breakout torque at the closed position as well as to overcome any dynamic torque as it moves through the full stroke

Table 10.5 Ball-Valve Torque Factors*,†

Valve Size	T_p=Packing Torque (in-lb)					T_s=Seat Torque	C_B=Bearing Torque Factor	C_s=Seat Torque Factor	C_D = Dynamic Torque Factor	
(in.)	(1)	(2)	(3)	(4)	(5)	(in-lb)	(6)	(7)	60°	90°
1	43	228	421	301	57	20	0.06	0.09	0.1	0.25	0.6
1½	50	280	477	350	63	40	0.06	0.09	0.1	0.5	1.0
2	50	280	477	350	63	60	0.06	0.09	0.15	1.0	2.1
3	57	333	533	399	71	150	0.19	0.28	0.42	4.5	8.0
4	57	333	533	399	71	360	0.38	0.58	0.82	10.0	17.0
6	71	438	646	496	92	540	0.97	1.45	1.64	19.5	30.5
8	71	438	646	496	92	670	1.58	2.37	2.62	52.0	75.5
10	104	648	870	691	151	1100	4.38	6.57	4.55	108.0	165.5
12	104	648	870	691	151	1300	5.61	8.41	6.05	191.0	218.5

Courtesy of Valtek International.
†(1)PTFE or filled PTFE V-ring packing, (2) grafoil, (3) twin grafoil, (4) asbestos-free packing (AFP), (5) braided PTFE, (6) PTFE lined bearings, (7) metal bearings.

to the full-open position. If the available springs are not capable of producing enough force to overcome the seating, breakout, or dynamic torque, a volume tank should be specified to ensure adequate force to move the valve to the correct position upon loss of air supply.

10.3 Sizing Electromechanical and Electrohydraulic Actuators

10.3.1 Introduction to Actuator Sizing for Electromechanical and Electrohydraulic Actuators

For the most part, electromechanical and electrohydraulic actuators are sized according to the thrust needed to overcome the forces inside the body as shown in the following equation from Sec. 10.1.2:

$$F_{total} = F_{process} + F_{seat} + F_{packing} + F_{miscellaneous}$$

where F_{total} = total force (or actuator thrust) required to open, close, or throttle valve
$F_{process}$ = force to overcome process pressure unbalance
$F_{packing}$ = force required to overcome packing friction
F_{seat} = force to provide correct seat load
$F_{miscellaneous}$ = force to overcome special design factors, weight, etc.

Individual sizing equations to determine actuator size vary widely depending on the design and application of the actuator and are not specifically included in this section.

10.3.2 Special Considerations

Typically, the application engineers for the electromechanical or electrohydraulic manufacturer will size the actuator based upon the process and frictional forces associated with the valve as well as include some additional thrust for safety considerations. With the high level of engineering required for these actuators, the prevailing thought is better too much actuator than not enough. However, in some cases, the accumulation of safety factors over the sizing process can add anywhere from 25 to 50 percent to the total thrust of the actuator. High costs are associated normally with electromechanical and electrohydraulic actuators. Therefore, if sizing formulas show the thrust requirement to be slightly more than a given size, all safety factors should be reconsidered to check for an impractical accumulation. If that is the case, the smaller actuator size can be considered.

Electromechanical and electropneumatic actuators are normally specified for those applications requiring faster stroking speeds or higher performance than provided normally by pneumatic actuators.

From a sizing standpoint, application engineers use specialized sizing equations to determine the stroking speed, frequency response, and level of precision positioning. Because most applications requiring electromechanical and electrohydraulic actuators are special or severe services, manufacturers have a tendency to size actuators based on flow rate and pressure drop.

11

Common Valve Problems

11.1 High Pressure Drops

11.1.1 Introduction to High Pressure Drops

Flow moves through a valve due to a difference between the upstream and downstream pressures, which is called the *pressure drop* (ΔP) or the *pressure differential*. If the piping size is identical both upstream and downstream from the valve and the velocity is consistent, the valve must reduce the fluid pressure to create flow by way of frictional losses. A portion of the valve's frictional losses can be attributed to friction between the fluid and the valve wall. However, this friction is minimal and is not sufficient to create enough pressure drop for an adequate flow. A more effective way to create a significant frictional loss in the valve is through a restriction within the body. Because many valves are designed to allow a portion of the valve to be more narrow than the piping, they can easily provide this restriction in the fluid stream. Because of the laws of conservation, as the fluid approaches the valve, its velocity increases in order for the full flow to pass through the valve, inversely producing a corresponding decrease in pressure (Fig. 11.1). The inverse relationship between pressure and velocity is shown by Bernoulli's equation, which is

$$\frac{\rho V_1^{\,2}}{2g_C} + P_1 = \frac{\rho V_{VC}^{\,2}}{2g_C} + P_{VC}$$

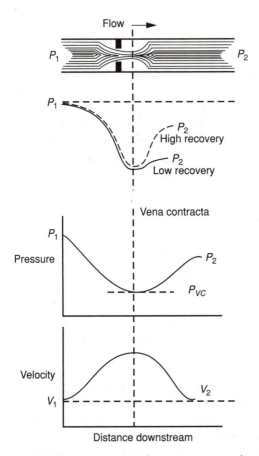

Figure 11.1 Location of vena contracta from point of orifice restriction and pressure and velocity curves. (*Courtesy of Fisher Controls International, Inc.*)

where ρ = density units

V_1 = upstream velocity

g_C = gravitational units conversion

V_{VC} = velocity at vena contracta

P_{VC} = pressure at vena contracta

P_1 = upstream pressure

The highest velocity and lowest pressure occur immediately downstream from the narrowest constriction, which is called the *vena contracta*. Figure 11.2 shows that the vena contracta does not occur at the

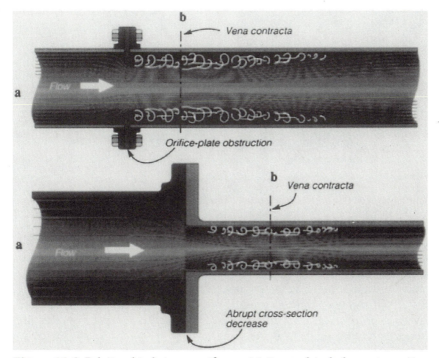

Figure 11.2 Relationship between orifice restriction and turbulence generation. (*Courtesy of Fisher Controls International, Inc.*)

restriction itself but rather downstream some distance from the restriction. This distance may vary according to the pressures involved. At the vena contracta the flow velocity is at a maximum speed, while the flow area of the fluid stream is at its minimum value.

Following the vena contracta, the fluid slows and pressure builds once again, although not to the original upstream pressure. This difference between the upstream and downstream pressures is caused by frictional losses as the fluid passes through the valve, and is called the *permanent pressure drop*. The difference in pressure from the pressure at the vena contracta and the downstream pressure is called the *pressure recovery*. A simplified profile of the permanent pressure drop and pressure recovery is shown in Fig. 11.3.

The flow rate for a valve can be increased by decreasing the downstream pressure. However, in liquid applications the flow can be limited by the pressure drop falling below the vapor pressure of the fluid, which will create imploding bubbles or pockets of gas (called cavitation or flashing, respectively). *Choked flow* occurs when the liquid flow

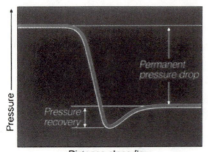

Distance along flow

Figure 11.3 Flow curve showing pressure recovery and permanent pressure drop. (*Courtesy of Fisher Controls International, Inc.*)

is saturated by the fluid itself mixed with the gas bubbles or pockets and can no longer be increased by lowering the downstream pressure. In other words, the formation of gas in a liquid crowds the vena contracta, which limits the amount of flow that can pass through the valve. With gases, as the velocity approaches sonic speeds, choked flow also occurs and the valve will not be able to increase flow despite a reduction in downstream pressure.

11.1.2 Effects of High Pressure Drops

As discussed in Sec. 11.1.1, the flow function of the valve is dependent on the existence of a pressure drop, which allows flow movement from the upstream vessel to the downstream vessel or to atmosphere. Because a pressure drop generated by the valve absorbs energy through frictional losses, the ideal pressure drop allows the full flow to pass through the body without excessive velocity, absorbing less energy. However, some process systems, by virtue of their system requirements, may need to take a larger pressure drop through the valve.

A high pressure drop through a valve creates a number of problems, such as cavitation, flashing, choked flow, high noise levels, and vibration. Such problems present a number of immediate consequences: erosion or cavitation damage to the body and trim, malfunction or poor performance of the valve itself, wandering calibration of attached instrumentation, piping fatigue, or hearing damage to nearby workers. In these instances, valves in high-pressure-drop applications require expensive trims, more frequent maintenance, large spare-part inventories, and piping supports. Such measures drive up maintenance and engineering costs.

Although users typically concentrate on the immediate consequences of high pressure drops, the greatest threat that a high pressure drop presents is lost efficiency to the process system. Because high pressure drops absorb a great deal of energy, that energy is lost from the system. In most process systems, energy is added to the system through heat generated by a boiler or through pressure created by a pump. Both methods generate energy in the system, and as more energy is absorbed by the system—including that energy lost by valves with high pressure drops—larger boilers or pumps must be used. Consequently, if the system is designed with few valves with high pressure drops, the system is more efficient and smaller boilers or pumps can be used.

11.2   Cavitation

11.2.1   Introduction to Cavitation

Cavitation is a phenomenon that occurs only in liquid services. It was first discovered as a problem in the early 1900s, when naval engineers noticed that high-speed boat propellers generated vapor bubbles. These bubbles seemed to lessen the speed of the ship, as well as cause physical deterioration to the propeller.

Whenever the atmospheric pressure is equal to the vapor pressure of a liquid, vapor bubbles are created. This is evident when a liquid is heated, and the vapor pressure rises to where it equals the pressure of the atmosphere. At this point, bubbling occurs. This same phenomenon can also occur by decreasing the atmospheric pressure to equal the vapor pressure of the liquid. In liquid process applications, when the fluid accelerates to pass through the narrow restriction at the vena contracta, the pressure may drop below the vapor pressure of the fluid. This causes vapor bubbles to form. As the flow continues past the vena contracta, the velocity decreases as the flow area expands and pressure builds again. The resulting pressure recovery increases the pressure of the fluid above the vapor pressure. This phenomenon is described in Fig. 11.4.

As a vapor bubble is formed in the vena contracta, it travels downstream until the pressure recovery causes the bubble to implode. This two-step process—the formation of the bubble in the vena contracta and its subsequent implosion downstream—is called *cavitation*. In simple terms, cavitation is a phase that is characterized by a liquid–vapor–liquid process, all contained within a small area of the valve and within microseconds. Minor cavitation damage may be con-

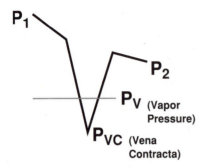

Figure 11.4 Flow curve showing pressure drop falling below the vapor pressure, which results in cavitation. (*Courtesy of Valtek International*)

sidered normal for some applications, which can be dealt with during routine maintenance. If unnoticed or unattended, severe cavitation can limit the life expectancy of the valve. It can also create excessive seat leakage, distort flow characteristics, or cause the eventual failure of the pressure vessels (valve body, piping, etc.). In some severe high-pressure-drop applications, valve parts can be destroyed within minutes by cavitation.

In general, five conditions must be present to produce cavitation. First, the fluid must remain a liquid both upstream and downstream from the valve. Second, the liquid must not be at a saturated state when it enters the valve or the pressure drop will create a residual vapor downstream from the valve. Third, the pressure drop at the vena contracta must drop below the vapor pressure of the process fluid. Fourth, the outlet pressure must recover at a level above the vapor pressure of the liquid. Fifth, the liquid must contain some entrained gases or impurities, which act as a "host" for the formation of the vapor bubble. This host is sometimes called the *nuclei*. The nuclei are contained in the process fluid as either microscopic particulates or dissolved gases. Since most process fluids contain either particulates or dissolved gases, the chances of forming vapor bubbles are very likely. In theory, if the liquid was completely nuclei-free, some experts believe that cavitation would not occur; however, this would be nearly impossible, especially considering the effects of thermodynamics.

The creation and implosion of the cavitation bubble involve five stages: First, the liquid's pressure drops below the vapor pressure as velocity increases through the valve's restriction. Second, the liquid expands into vapor around a nuclei host, which is either a particulate or an entrained gas. Third, the bubble grows until the flow moves away from the vena contracta and the increasing pressure recovery inhibits the growth of the bubble. Fourth, as the flow moves away from the vena contracta, the area expands—slowing velocity and increasing pressure. This increased pressure collapses or implodes the bubble vapor back to a liquid. Fifth, if the bubble is near a valve surface, the force of the implosion is directed toward the surface wall, causing material fatigue.

The bubbles created by cavitation are much smaller and more powerful than bubbles caused by normal boiling. This release of energy by the imploding bubbles can easily be heard as noise in the valve or in the downstream piping. The noise generated in the early stages of cavitation is described as a popping or cracking noise, while extensive cavitation produces a steady hiss or sizzling noise. Some describe the noise as gravel rolling down the piping. Noise is normally complemented by excessive vibration, which can cause metal or piping fatigue or miscalibration or malfunctioning of sensitive instrumentation. In some cases, the vibration can be minimized by anchoring the valve or piping securely to floors, walls, etc.

The most permanent damage caused by cavitation is the deterioration of the interior of the valve created by the imploding bubbles. As the bubbles expand in the vena contracta, they move into the downstream portion of the valve and then implode as the pressure recovery occurs. If the bubbles are near a metal surface, such as a body wall, they have a tendency to release the implosion energy toward the wall. This phenomenon occurs when unequal pressures are exerted upon the bubble. Since the fluid pressure is less on the side of the cavitation bubble closest to a nearby object, the energy of the implosion is channeled toward that surface (Fig. 11.5). This principle is identical to the implosion of a depth charge in antisubmarine warfare.

With cavitation, the real damage occurs in the second half of the process, when the bubbles implode. This energy burst toward the metal surface can tear away minute pieces of metal, especially if the pressure intensity reaches or surpasses the tensile strength of the valve material. These shock waves have been reported to be as high as 100,000 psi (6900 bar). This initial destruction is magnified since the drag in torn metal surface attracts and holds other imploding bubbles, causing even more cavitation damage. Valve parts damaged by cavitation have a pitted appearance or feel like a sandblasted surface (Fig. 11.6). The appearance of cavitation damage is far different from flashing or erosion damage, which appears smooth. Another possible long-term effect of cavitation is that it may attack a material's coating, film,

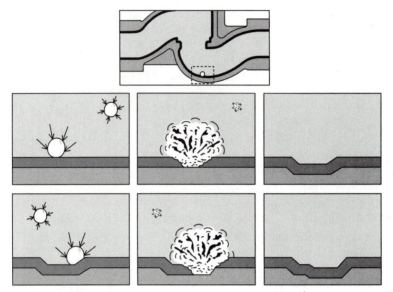

Figure 11.5 Implosion of cavitation bubbles by a valve-body wall. (*Courtesy of Valtek International*)

or oxide, which will open up the base material to chemical or corrosion attack.

The hardness of the metal plays a large role in how easily the metal can be torn by the cavitation bubbles. Soft materials, such as aluminum, yield easily to the forces generated by cavitation bubbles and

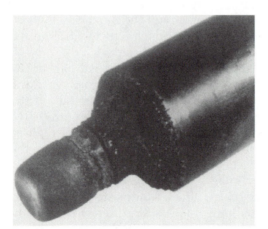

Figure 11.6 Plug damaged by cavitation. (*Courtesy of Fisher Controls International, Inc.*)

tear away quickly. Hardened materials are better able to withstand the effects of cavitation, and only after a period of time will they fatigue and begin to wear. No material can resist cavitation indefinitely. Even the hardest materials will eventually wear away.

Another serious side effect of cavitation is decreased performance in the valve and reduced efficiency in the process system. When cavitation occurs, the valve's ability to convert the entire pressure drop to mass flow rate is diminished. In other words, cavitation can cause less flow through the valve, generating a smaller C_v in actual service than what was originally calculated.

Cavitation can be controlled or eliminated by one of three basic methods: first, by modifying the system; second, by making certain internal body parts out of hard or hardened materials; or third, by installing special devices in the valve that are designed to keep cavitation away from valve surfaces or prevent the formation of the cavitation itself.

11.2.2 Incipient and Choked Cavitation

As the downstream pressure is lowered, creating a larger pressure drop, the advent of cavitation is called *incipient cavitation*. When damage occurs to the vessel, that stage is known as *incipient cavitation damage*. As the flow increases, it will eventually become choked, which is called *choked cavitation*. This linear relationship is shown in Fig. 11.7, which is based on the linear relationship between the flow rate Q and

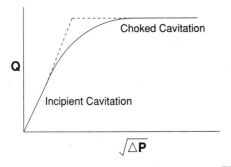

Figure 11.7 Fluid plot of flow vs. $\sqrt{\Delta P}$ and points of incipient and choked cavitation. (*Courtesy of Valtek International*)

the square root of the pressure drop $\sqrt{\Delta P}$. The constant of proportionality of this relationship is based upon the equation

$$Q = C_v \sqrt{\frac{\Delta P}{GS}}$$

where Q = flow rate
C_v = flow coefficient
ΔP = pressure drop
G_S = specific gravity

11.2.3 Cavitation Indices

Over the years, cavitation experts have developed a number of cavitation indices to predict the possibility of cavitation in process equipment, including valves. The ability to predict cavitation is critical to the design and application of the valve. For example, if cavitation exists, the valve can be fitted with special trim to minimize the effects or eliminate cavitation altogether. Certain parts, such as the plug or seat, can be made from hard or hardened materials, or the process system can be changed to minimize the pressure drop through the valve so that cavitation does not form.

For many years, the valve industry used the *flow curve cavitation index* K_C, which shows the effect of cavitation on the linear relationship between the flow rate and the square root of pressure drop. The index K_C is still in use today with some manufacturers and is occasionally used in calculations as

$$K_C = \frac{P_1 - P_2}{P_1 - P_V} = \frac{\Delta P}{P_1 - P_V}$$

where K_C = cavitation index
P_1 = valve inlet pressure
P_2 = valve outlet pressure
P_V = vapor pressure of liquid (at valve inlet and vena contracta)

The cavitation index assumes that a valve may function without cavitation at any pressure drop less than the pressure drop calculated with the index K_C. The basic problem associated with the cavitation index K_C is that it does not take into consideration any prechoked cavitation conditions, which may be just as damaging to the valve. Table 11.1 provides several common K_C values for a number of valve styles.

Table 11.1 Typical K_c Values†

Valve Style	K_c
Butterfly	0.50 K_M*
Ball	0.67 K_M
Rotary Plug	K_M
Globe with Hardened Trim (Cage Characterized)	K_M
Globe (Plug Characterized)	0.85
Globe with special trim	1.0

†*Data courtesy of Fisher Controls International, Inc.*
*K_M is equal to F_L^2 (valve recovery coefficient).

A more useful cavitation index for valves is σ, which was approved in 1995 by the Instrument Society of America. In general terms, σ is a ratio of forces that resist cavitation to forces that promote cavitation and is written as

$$\sigma = \frac{P_2 - P_V}{P_1 - P_2}$$

where σ = cavitation index
  P_1 = upstream pressure (measured one pipe diameter upstream from the valve)
  P_2 = downstream pressure (measured five pipe diameters downstream from the valve)
  P_V = liquid vapor pressure (at flowing temperature)

As σ becomes larger, less cavitation damage is occurring inside the valve. Inversely, as σ becomes smaller, cavitation damage is increasing. If σ is at zero or is a negative number, flashing is occurring. σ is expressed in two forms: *Incipient* σ is the value that indicates when cavitation is beginning. *Choked* σ is the value that indicates when

Table 11.2 Typical σ Values[†,‡]

Valve Style	Flow Direction	Trim Size	Incipient σ	Choked σ
Globe	over the plug	full area	0.73	0.38
	over the plug	reduced	0.93	0.56
	under the plug	full/reduced	0.52	0.52
Butterfly	60° open	full	1.40	0.73
	90° open	full	3.16	2.19
Ball	60° open	full	1.40	0.64
	90° open	full	5.20	2.19
Globe with special trim	under the plug	full/reduced	0.30 to 0.001	*

†*Data courtesy of Valtek International.*
‡*Note:* For estimation only; sigmas may vary according to particular valve design.
*Choking will not occur when properly applied.

choked flow or full cavitation is occurring. If the calculated σ falls between the incipient σ and choked σ values, some measures should be taken (using special trim, hard materials, or process changes) to avoid cavitation damage in the valve. Both incipient σ and choked σ values are determined through laboratory and field testing by the valve manufacturer. Examples of typical σ values for a given valve style are shown in Table 11.2.

11.2.4 σ **Example A**

To show an application of incipient σ and choked σ, the following example is used:

Fluid	Water
Temperature	80°F
Vapor pressure P_V	0.5 psia
Upstream pressure P_1	200 psia
Downstream pressure P_2	55 psia
Valve type	Single-seated globe valve, 100 percent open, flow-over-the-plug

The value for σ is

$$\sigma = \frac{P_2 - P_V}{P_1 - P_2} = \frac{55 - 0.5}{200 - 55} = 0.38$$

Referring to Table 11.2, incipient σ begins at σ = 0.73 (for a single-seated globe valve that is at 100 percent open with flow under the plug) and the choked σ occurs at σ = 0.38. In this example severe cavitation damage is occurring and the valve is choked and cannot increase flow any further.

11.2.5 σ **Example B**

Using the same valve in example A, new service conditions are applied to illustrate a cavitating, but nonchoking, situation:

Fluid	Water
Temperature	80°F
Vapor pressure P_V	0.5 psia
Upstream pressure P_1	500 psia
Downstream pressure P_2	200 psia
Valve type	Single-seated globe valve, 100 percent open, flow-over-the-plug

Using the σ index equation for these operating conditions, we find that σ is significantly higher:

$$\sigma = \frac{200 - 0.5}{500 - 200} = 0.67$$

This σ value is above the choked σ value for this valve (which is σ = 0.38) and indicates that the valve is not experiencing choked flow. However, this value is below the incipient σ value, which indicates that the valve is experiencing cavitation and damage may be occurring in the valve.

11.2.6 **System Modifications to Prevent Cavitation**

To eliminate the formation of cavitation, the answer lies in reducing the pressure from the upstream side to the downstream side, preventing the pressure at the vena contracta from falling below the vapor pressure. When this reduction is made, vapor bubbles are not formed and cavitation is avoided. This normally requires special trim or modifications of the system to provide a series of smaller pressure drops that result in the required downstream pressure. By taking a series of pressure drops, rather than one large drop, the service can be modified so that the pressure will not fall below the vapor pressure (Fig. 11.8).

In some cases, the process system and related service conditions, or the process equipment used in the system, can be modified to minimize the effects of cavitation. Even the type of valve or number of valves used in one system can modify cavitation effects. One system solution is the injection of air into the system. At first this may appear to worsen a bad situation as the addition of air will provide additional nuclei that can play host to vapor bubbles and increase the damage. However, cavitation studies have shown that at a certain point, additional air content to the process stream disrupts the explosive force of the imploding bubbles and can reduce the overall damage. This solution works best with large valves dumping into tanks or when large particulates in the flow stream prevent the use of cavitation-control trim, anticavitation trims, or downstream devices.

The intensity of cavitation can be modified by varying the downstream pressure, if possible. Increasing the downstream pressure may decrease the pressure drop sufficiently to avoid the pressure falling below the vapor pressure, although this will decrease the process flow capacity. Lowering the downstream pressure may not seem to be an option, since a greater pressure drop would create even more vapor

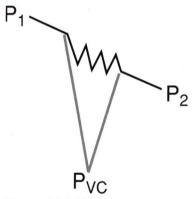

Figure 11.8 Flow curve showing gradual pressure reduction without dropping below vapor pressure. (*Courtesy of Valtek International*)

Figure 11.9 Back-pressure device used with globe valves. (*Courtesy of Fisher Controls International, Inc.*)

bubbles. However, the increased pressure differential provides less cavitation intensity.

A *downstream back-pressure device* is a device that is installed between the valve and the downstream piping to lower the pressure drop taken by the valve while increasing the resistance of the downstream and the downstream pressure. Although a wide variety of back-pressure devices are available today, a typical device is shown in Fig. 11.9. Because back-pressure devices may limit the flow capacity of the valve, a larger valve or different trim reduction may be required. The device must be examined periodically to ensure that it is not wearing out through erosion or minimal cavitation. A worn back-pressure device will ultimately decrease the downstream pressure, increase the pressure drop, and create cavitation. In addition, the user must be careful to use the back-pressure device within the limits of the flow range; otherwise cavitation can occur after the device in the downstream piping. A back-pressure device is commonly used with a rotary valve (Fig. 11.10), which cannot be designed to include an internal anticavitation device because of design limitations. Not only does the device perform the function of raising the downstream pressure, but it also controls existing cavitation by allowing it to occur in the small tubes where cavitation intensity is lower and can be absorbed by the tubes themselves. One caution applies when using a downstream cavitation-control device with a rotary valve: As the rotary valve begins to open (less than 30° open), the most severe cavitation intensity may occur in the outlet half of the body before it reaches the downstream device, causing serious damage between the valve and the device.

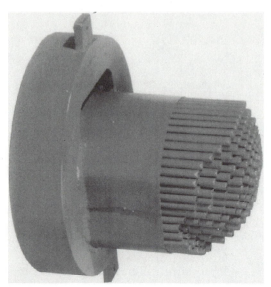

Figure 11.10 Back-pressure device used with rotary valves (*Courtesy of Fisher Controls International, Inc.*)

Some valve designs can be used to minimize cavitation damage. For example, while a globe-style linear valve exposes the bottom of the valve body to the cavitation, an angle-style linear valve may experience less damage since the flow continues straight down from the vena contracta and is directed into the center of the piping—no valve or piping surface is directly bombarded with vapor bubbles.

As a general rule, the face-to-face dimension of rotary valves—such as butterfly, eccentric plug, and ball valves—is far less than comparably sized globe valves. Therefore, the vena contracta generated by a rotary valve is most likely to occur not in the valve itself, but in the downstream piping. In this case, cavitation might be allowed and a segment of the downstream piping routinely replaced as part of periodic maintenance. Another option is to install two or three valves in lieu of one valve, allowing the pressure drop to be taken over more than one restriction and preventing a large pressure drop from falling below the vapor pressure. This option is more expensive in terms of additional valves, but may still be less expensive than obtaining a specially engineered valve. This solution has one disadvantage, however, that may occur when the first valve opens against a high upstream pressure. For a very short time, the first valve will take the entire pressure drop until the flow reaches the second valve. This may result in cavitation damage to the first valve in some unusual cases. In such an application, installing anticavitation trim in the valve may be a better option.

11.2.7 Materials of Construction

Cavitation easily attacks softer materials, which have a lower tensile strength than harder materials. One of the most common methods of dealing with cavitation is to make the valve out of hard or hardened materials (those materials exceeding a Rockwell hardness of 40). Generally, materials such as chrome–molybdenum and steel alloys (ASTM SA-217 Grade WC9 and C5) are used for the body, while solid alloy hard-facing, a solid alloy overlay with 316 stainless steel or 416 stainless steel, is used on trim parts.

One advantage to using angle-style valves in cavitating service is that one of three options—a hardened seat ring, an extended *Venturi seat ring*, or body liner—can be installed in the downstream portion of the valve. This part can then be replaced periodically after cavitation damage compromises the part. These liners can be made from Alloy 6 or 17-4ph stainless steels.

Because nonmetallic materials, such as PTFE liners or bodies made from plastic, have lower yield values than metal, they are more prone to cavitation damage and are not recommended for cavitating services.

11.2.8 Cavitation-Control Devices

Some valves can be equipped with special trims that will direct the cavitating flow, along with vapor bubbles, away from critical metal surfaces. Since cavitation-control trims are not as highly engineered as trims designed to prevent cavitation, they generally cost less and are simpler in concept.

The design shown in Fig. 11.11 illustrates how this principle works. In flow-over-the-plug applications, a special retainer with specially designed holes is placed inside the valve. As the close-fitting plug lifts out of the seat, the holes in the special retainer are exposed and allow the flow to pass through the seat. In this case, the holes in the retainer are the restrictions and cavitation occurs at that region. Because the holes are directly opposite each other, the cavitating flow from one hole impinges on the opposite hole's flow, thus keeping the cavitation in the center of the retainer. At this point, the only metal surface affected by the cavitation is the bottom of the plug, which can be made from hardened material. Since the middle of the bottom of the plug is flat

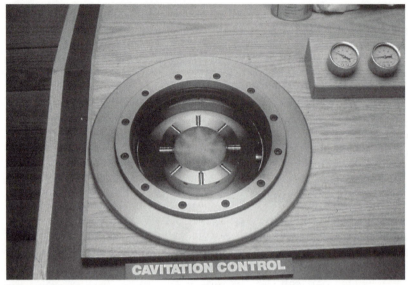

Figure 11.11 Laboratory experiment showing diversion of cavitation away from boundary surfaces using cavitation control trim. (*Courtesy of Valtek International*)

and not necessary for shutoff, it can be sacrificed over a period of time. Only when the deterioration reaches the plug's seating surface will the plug need replacement.

Such cavitation-control designs can be engineered with a wide range of C_vs and in either linear or equal-percentage flow characteristics. Because flow must always be over the plug, pressure-balanced trim is necessary in fail-open applications to prevent instability near the seat.

11.2.9 Cavitation-Elimination Devices

Some valves are designed to prevent the formation of cavitation altogether. Although it is a more expensive option, in some applications anticavitation design features are the only choice. Globe-style valves can be designed with special retainers or cages, which use either (or a combination of) a tortuous path, pressure-drop staging, and/or expanded flow areas to decrease the pressure drop through the valve and to prevent cavitation.

A *tortuous-path device* uses a series of holes and/or channels to increase the flow resistance through the trim (Fig. 11.12). This decreases the overall velocity through the valve, thereby reducing the pressure recovery. In addition, a tortuous path creates pockets of high and

Figure 11.12 Tortuous-path trim for velocity reduction. (*Courtesy of Control Components Inc., an IMI company*)

low pressures as the flow moves through the trim, creating substantial frictional losses. To illustrate the effect of frictional loses in this trim, the losses associated with a single 90° piping elbow are equal to 60 ft of straight pipe. The typical tortuous path uses a series of right-angle turns—similar in principle to a 90° elbow—to create frictional losses and lower velocities. Each turn reduces the velocity by one velocity head ($V_H = \rho V^2/2$). This velocity reduction can be calculated by changing the velocity equation as follows:

$$V = \sqrt{2SGV_H} \quad \text{to} \quad V = \sqrt{\frac{2SG\,V_H}{N}}$$

where V = required velocity (below sonic or generally below 300 ft/s)
$\quad SG$ = specific gravity
$\quad V_H$ = velocity head
$\quad N$ = number of turns (in series) in each passageway

Determining the number of turns is critical in the design of tortuous-path designs, since they determine the overall velocity-head loss, as well as the diameter of the stack.

Another method of decreasing the pressure drop is by *staged pressure reduction*, in which several smaller restrictions are taken through a trim rather than one large restriction. In effect, this creates a number of small pressure drops in lieu of one large pressure drop (refer again to Fig. 11.8). As the flow moves through the trim, it reaches the first restriction or stage, absorbing a certain amount of energy and taking a small pressure drop. As the flow continues, it provides a lower inlet pressure to the next stage where another pressure drop is taken, and so forth. The net result is that the entire pressure drop is taken over a series of small pressure drops without falling below the vapor pressure at the vena contracta, yet the overall pressure drop remains unchanged. In some cases, for whatever reason, systems pressures may change. This change may exceed the operating parameters of the valve and create cavitation in the valve, even if a staged-pressure-reduction trim is used. In this case, although cavitation is occurring, the anticavitation trim will continue to modify the pressure differential and the cavitation will not be as severe.

Related to the staged-pressure-drop concept is the *expanded flow-area* concept, in which the flow continues through several restrictions in the trim, the flow area is increased at each stage (Fig. 11.13). With compressible fluids, as dictated by the law of conservation of mass flow, the flow area must increase as the fluid pressure and density are reduced. In this concept, the largest portion of the pressure drop is taken at the first restriction, and then succeeding smaller portions of the pressure drop are taken over the following restrictions. When the flow reaches the last restriction, a minimal pressure drop is taken and

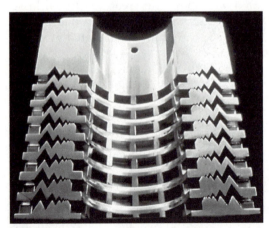

Figure 11.13 Expanding tooth trim for staged pressure reduction. (*Courtesy of Valtek International*)

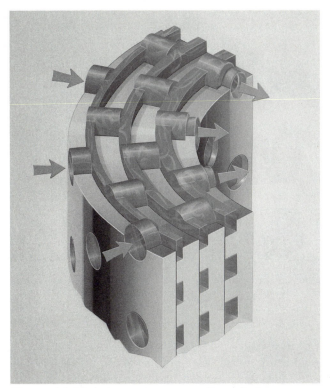

Figure 11.14 Anticavitation trim with multiple pressure-reduction mechanisms. (*Courtesy of Valtek International*)

the pressure recovery at that point is significantly decreased, preventing cavitation from occurring.

Valve manufacturers have developed a variety of sophisticated trims that use one or a combination of these concepts (tortuous path, staged pressure reduction, and expanded flow areas). For example, Fig. 11.14 shows a flow-over-the-plug trim that directs the flow through a series of close-fitting cylinders with each cylinder acting as a stage. The flow must follow a tortuous path as it travels through a series of 90° angles via the narrow channels and drilled holes, increasing the frictional losses. Pressure reduction is staged through the number of cylinders, allowing the pressure to remain above the vapor pressure. In addition, the channels become progressively deeper and the number of holes increase with each stage, providing expanded flow areas.

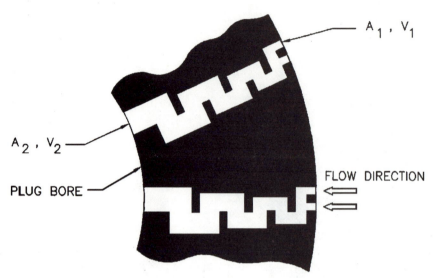

Figure 11.15 Expanding flow area of tortuous-path trim. (*Courtesy of Control Components Inc., an IMI company*)

Another common design that uses these principles is the *expanding tortuous-path trim*. In addition to the velocity control through the right-hand turns, the tortuous pathways can be enlarged, allowing for expanded flow areas (Fig. 11.15). The tortuous path can follow either a horizontal direction with etched disks (Fig. 11.16) or disks made from a punched plate (Fig. 11.17).

Most anticavitation trims follow a linear characteristic, although some designs allow for an equal-percentage characteristic. When the disks or flow areas of the trim are identical throughout the entire stack, the trim follows a linear characteristic. An equal-percentage characteristic is generally obtained by using different disks or passageways that increase the flow as the stroke continues. Another method of modifying an anticavitation linear characteristic is by using a shaped cam in the actuator positioner.

11.2.10   Anticavitation-Trim Sizing

Although methods of sizing a valve with anticavitation trim vary according to different valve manufacturers, the following procedure utilizes σ values and provides a general idea of the steps involved. The first step is to calculate the required C_v for the given application (see

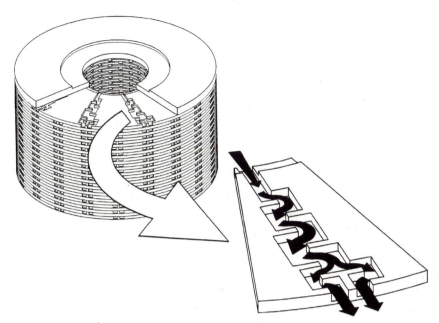

Figure 11.16 Etched tortuous-path trim for horizontal flow. (*Courtesy of Control Components Inc., an IMI company*)

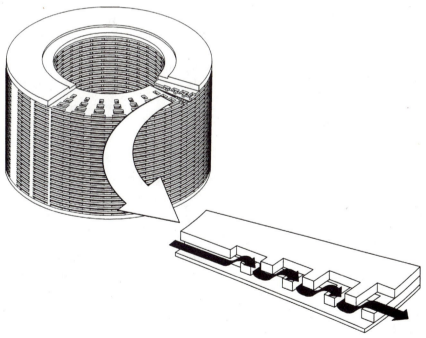

Figure 11.17 Punched tortuous-path trim for vertical and horizontal flows. (*Courtesy of Control Components Inc., an IMI company*)

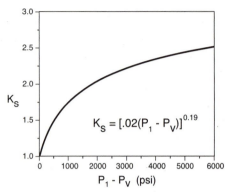

Figure 11.18 Pressure scale for factor K_S (water service), where $K_S = [0.02 (P_1 - P_V)]^{0.19}$. (*Courtesy of Valtek International*)

Chap. 9). Choked-flow conditions should be considered, and the F_L factor should be adjusted to compensate for the use of the anticavitation trim. In this case, the required C_v for a valve with anticavitation trim will most likely be smaller than a conventional valve. Using the o formula, the operating o can be calculated from the flow conditions for the required C_v. The difference between the upstream pressure and the vapor pressure should be calculated ($P_1 - P_V$). Following this calculation, the K_S factor can be determined by referring to Fig. 11.18 or by using the following calculation:

$$K_S = [0.02(P_1 - P_V)]^{0.19}$$

The service σ can now be calculated for each pressure:

$$\sigma_{service} = \frac{\sigma}{K_S}$$

The manufacturer of the anticavitation trim provides tables (Table 11.3) that provide C_v and σ values. If the requirement of the service σ is less than the minimum requirement of the calculated σ, then a larger valve with more stages will mostly likely be needed. The velocity of the flow should also be considered to ensure that it does not approach the maximum velocity capacity of the trim.

Table 11.3 Cavitation Trim Sizing Table*

Body Size	Trim No. (Seat Dia.)	Stages	Stroke	C_v	$\sigma_{min.}$	Bore Area
1½	1.38	2	1.50	17	.170	1.77
	1.25	3	1.50	11	.070	1.48
	1.12	4	1.50	6	.020	.99
2	1.38	2	1.50	18	.170	1.77
	1.25	3	1.50	12	.070	1.49
	1.12	4	1.50	7	.020	.99
3	2.50	2	2.50	50	.200	5.41
	2.38	3	2.50	34	.080	4.91
	2.00	4	2.50	20	.025	3.55
	1.62	5	2.50	12	.007	2.41
	1.25	6	2.50	7	.002	1.49
4	3.50	2	3.00	85	.200	10.3
	3.12	3	3.00	54	.080	8.3
	2.75	4	3.00	33	.025	6.51
	2.38	5	3.00	21	.007	4.91
	1.88	6	3.00	12	.002	3.14
6	5.25	2	4.00	175	.200	22.7
	4.75	3	4.00	105	.080	18.7
	4.25	4	4.00	65	.025	15.1
	3.50	5	4.00	40	.007	10.3
	3.00	6	4.00	25	.002	7.67
8	6.50	2	6.00	320	.200	34.5
	6.00	3	6.00	200	.080	29.5
	5.50	4	6.00	130	.025	24.8
	5.00	5	6.00	85	.007	20.6
	4.50	6	6.00	55	.002	16.8
10	8.75	2	7.50	530	.230	61.9
	8.38	3	7.50	350	.090	56.7
	7.88	4	7.50	230	.028	50.3
	7.38	5	7.50	155	.008	44.2
	6.88	6	7.50	105	.002	38.5
12	9.75	2	8.00	640	.230	76.6
	9.00	3	8.00	400	.090	65.4
	8.38	4	8.00	260	.028	56.7
	7.88	5	8.00	180	.008	50.3
	7.38	6	8.00	125	.002	44.2
14	11.00	2	8.00	720	.240	97.2
	10.25	3	8.00	460	.095	84.5
	9.50	4	8.00	300	.030	72.8
	8.75	5	8.00	200	.008	61.9
	8.00	6	8.00	135	.002	51.8

Courtesy of Valtek International.

11.2.11 Anticavitation-Trim Sizing Example

The following service conditions apply to this example:

Fluid	Water
Maximum flow	515 gal/min
Inlet temperature	208°F
Inlet pressure	287 psia
Outlet pressure	24 psia
Vapor pressure	13.57
Specific gravity	0.92

Using the flow capacity calculations in Chap. 9, the required C_v is calculated at $C_v = 30$. σ is calculated as follows:

$$\sigma = \frac{P_2 - P_V}{P_1 - P_2} = \frac{287 - 13.57}{287 - 24} = 0.04$$

Using the K_S chart (Fig. 11.18), the K_S is 1.38. Knowing K_S, the allowable σ can be calculated as follows:

$$\sigma_{\text{service}} = \frac{0.04}{1.38} = 0.029$$

Using an anticavitation trim table from the manufacturer (Table 11.3) for an application requiring a C_v of 33 and a σ value of 0.029, the required valve would be a 4-in (DN 100) valve with a four-stage anticavitation trim.

11.2.12 Other Cavitation-Control Solutions

A number of other solutions to cavitation control or elimination exist, such as characterized cages or separation of the valve's seat and the throttling mechanism. In applications where the pressure drop decreases as the plug travel and flow rate increase, a characterizable cage can be used. For example, a typical characterizable cage would have two stages of pressure reduction, the middle portion would have one stage of pressure reduction, and the top portion would have straight-through flow. With this design, cavitation control is provided at the early stages of plug travel, when it is needed most. As the travel

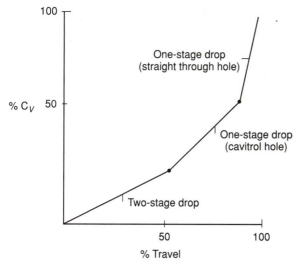

Figure 11.19 Flow curve showing effects of two-stage characterized cage. (*Courtesy of Fisher Controls International, Inc.*)

continues and the pressure drop and chance of cavitation decrease, the mechanism allows greater flow with less restriction. Figure 11.19 plots how a characterizable cage works in relation with flow versus travel.

In most flow-over-the-plug applications, the pressure reduction device is located above the seat in the body gallery. However, in some cavitating applications where tight shutoff is important, the body can be designed with the seat separate from the pressure-reduction mechanism. As shown in Fig. 11.20, the seat is located above the anticavitation trim, which is contained in the downstream portion of the valve. The trim area above the seat is designed to take a large flow, hence a lower pressure drop. This design keeps the velocities at a minimum through the seat, which improves the stability of the valve plug close to the seat and makes for easier shutoff.

Traditionally, anticavitation trim is associated with linear throttling valves, although some designs exist for quarter-turn valves. For example, a plug valve can utilize a special plug (Fig. 11.21) to take an additional stage of pressure reduction for those applications where the pressure drop falls just below the vapor pressure. As the plug closes, the grid turns into the flow, taking a pressure drop and preventing cavitation from forming. The grid prevents severe cavitation from forming and channels remaining cavitation away from metal bound-

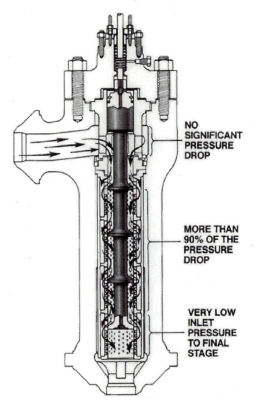

NO
SIGNIFICANT
PRESSURE
DROP

MORE THAN
90% OF THE
PRESSURE
DROP

VERY LOW
INLET
PRESSURE
TO FINAL
STAGE

Figure 11.20 Anticavitation trim located downstream from the seating surface. (*Courtesy of
Fisher Controls International, Inc.*)

aries. This design also allows large particulates to bounce off the grid
and be flushed downstream.

11.3   Flashing

11.3.1   Introduction to Flashing

In liquid applications, when the downstream pressure is equal to or
less than the vapor pressure, the vapor bubbles generated at the vena
contracta stay intact and do not collapse. This happens because the
pressure recovery is high enough for this to happen. As shown in Fig.
11.22, this phenomenon is known as *flashing*. When flashing occurs, the
fluid downstream is a mixture of vapor and liquid moving at very

Figure 11.21 Anticavitation plug for quarter-turn plug valves. (*Courtesy of The Duriron Company, Valve Division*)

high velocities, resulting in erosion in the valve and in the downstream piping (Fig. 11.23).

11.3.2 Controlling Flashing

Unfortunately, eliminating flashing completely involves modifying the system itself, in particular the downstream pressure or the vapor pressure. However, not all systems are easily modified and this may not be an option. The location of the valve may be considered—especially if the valve empties the downstream flashing flow into a large vessel, tank, or condenser. Placing the valve closer to the larger vessel will allow the flow to impinge into the larger volume of the vessel and away from any critical surfaces. When flashing occurs, no solution can be designed into the valve, such as is the case with anticavitation or cavitation-control trim, except to offer hardened trim materials.

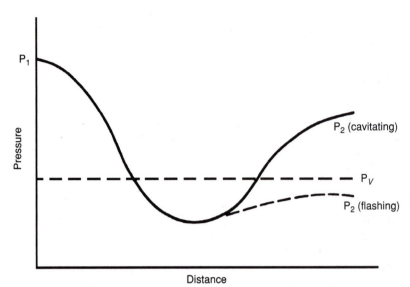

Figure 11.22 Pressure curve showing outlet pressure below the vapor pressure, resulting in flashing. (*Courtesy of Fisher Controls International, Inc.*)

11.4   Choked Flow

11.4.1   Introduction to Choked Flow

Choked flow occurs in gases and vapors when the velocity of a process fluid achieves sonic speeds in the valve or the downstream piping. As the fluid in the valve reaches the valve restriction, the pressure

Figure 11.23 Plug damaged by flashing. (*Courtesy of Fisher Controls International, Inc.*)

decreases and the specific volume increases until sonic velocities are achieved. When choked flow occurs, the flow rate is limited to the amount of flow that can pass through the valve at that point and cannot be increased unless the service conditions are changed.

In liquid applications, the presence of vapor bubbles caused by cavitation or flashing significantly increases the specific volume of the fluid. This increase rises at a faster rate than that generated by the pressure differential. In liquid choked flow conditions, if upstream pressure remains constant, decreasing the downstream pressure will not increase the flow rate. In gas applications, the velocity at any portion of the valve or downstream piping is limited to Mach 1 (sonic speed). Hence, the gaseous flow rate is limited to the flow that is achieved at sonic velocity in the valve's trim or the downstream piping.

As noted in Sec. 9.2, choked flow must be considered when sizing a valve, especially when considering $\Delta P_{\text{allowable}}$ and the valve recovery coefficient K_M.

11.5 High Velocities

11.5.1 Introduction to High Velocities

In general, large pressure differentials create high velocities through a valve and in downstream piping. This in turn creates turbulence and vibration in liquid applications and high noise levels in gas applications. The velocity is inversely related to the pressure losses and gains as the flow moves through the vena contracta (Fig. 11.24). The velocity

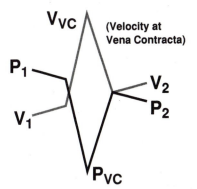

Figure 11.24 Velocity and pressure profiles as flow travels through an orifice restriction. (*Courtesy of Valtek International*)

reaches its maximum peak just slightly after the vena contracta, which is when the pressure is at its lowest point.

11.5.2 Velocity Limits

The following general rules apply to velocities: Liquids should generally not exceed 50 ft/s (15.2 m/s) (or 30 ft/s or 9 m/s in cavitating services). Gases should not exceed sonic speeds (Mach 1.0). And, mixtures of gases and liquids (such as flashing applications) should not exceed 500 ft/s (152 m/s). These general rules can vary, however, according to the size of the valve. For example, smaller-sized valves can normally handle higher velocities, while larger valves only handle lower velocities.

Generally, process liquids that have temperatures close to the saturation point must keep the velocity at or under 30 ft/s (9 m/s) to avoid the fluid pressure from falling below the vapor pressure and creating cavitation. The 30-ft/s rule also applies to cavitating services, where higher velocities result in greater cavitation damage to downstream piping. Lower velocities will also reduce problems associated with flashing and erosion.

11.6 Water-Hammer Effects

11.6.1 Definition of the Water-Hammer Effect

In liquid applications, whenever flow suddenly stops, shock waves of a large magnitude are generated both upstream and downstream. This phenomena is known as the *water-hammer effect*. It is typically caused by a sudden pump shutoff or a valve slamming shut when the closure element is suddenly sucked into the seat ("bathtub stopper effect") as the valve nears shutoff. In control valves, the bathtub stopper effect is caused by a low-thrust actuator that does not have the stiffness to hold a position close to the seat. In some cases, valves with a quick-open or an installed linear flow characteristic can also cause water-hammer effects.

Although water hammer generates considerable noise, the real damage occurs through mechanical failure. Because of the drastic changes from kinetic energy to the static pipe pressure, water hammer has been known to burst piping or damage piping supports as well as damage piping connections. In valves, water hammer can create severe shock through the trim, which can cause trim, gasket, or packing failure.

11.6.2 Water-Hammer Control

With valves, the best defense against water hammer is to prevent any sudden pressure changes to the system. This may involve slowing the closure of the valve itself or providing a greater degree of stiffness as the closure element approaches the seat. To avoid pressure surges, the valve should be closed with a uniform rate of change. In some cases, when a quick-open or installed linear characteristic (which approaches the quick-open characteristic) is used, a change to an equal-percentage characteristic may be required. With control valves that must throttle close to the seat, using an exceptionally stiff actuator—such as a spring cylinder pneumatic actuator or a hydraulic actuator—or a special notch in the stroke collar of a manual quarter-turn operator will minimize or prevent the bathtub stopper effect. Adding some type of surge protection to the piping system can also reduce water hammer. This may be accomplished with a pressure-relief valve or a rubber hose containing a gas, which can be run down the length of the piping. In addition, gas may be injected into the system. Gas injection reduces the density of the fluid and provides some compressibility to handle any unexpected surge.

11.7 High Noise Levels

11.7.1 Introduction to Noise

One of the most noticeable and uncomfortable problems associated with valves is noise. To the human, not only can noise be annoying, but it can also cause permanent hearing loss and unsafe working conditions. Extensive studies have shown that human hearing is damaged by long exposures to high noise levels. Hearing damage is cumulative and irreversible and begins with the loss of high frequencies. As hearing loss continues, lower frequencies are eventually lost, which affects the ability to understand normal speech patterns. When subjected to noise at lower frequencies, the performance of human organs, such as the heart or the liver, can also be affected. In addition, noise and the accompanying vibration can affect the valve's performance and cause fatigue in the valve, piping, and nearby process equipment.

In essence, noise is génerated when vibration produces wide variations in atmospheric pressure, which are then transferred to the eardrums as noise. Noise spreads at the speed of sound [which is 1100 ft/s (335 m/s) or 750 mi/h (1200 km/h)]. Noise in valves can be created in a number of different ways; however, the most common cause is

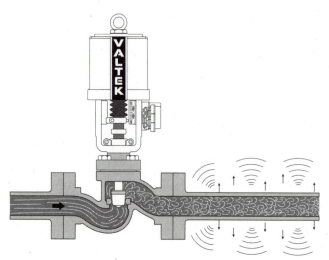

Figure 11.25 Downstream pipeline vibration caused by valve turbulence. (*Courtesy of Valtek International*)

turbulence generated by the geometry of the valve, which is radiated by the downstream piping (Fig. 11.25). In many cases, noise does not radiate from the valve itself, because the body itself is stiff and unyielding.

Process turbulence can create mechanical vibration of the valve or valve components. Such noise is caused by vibrations created by random pressure fluctuations within the body assembly or the fluid impinging on obstacles in the fluid steam, such as the plug, disk, or other closure element. This often causes a rattling noise, as the closure element impacts continually against its guides. Because the frequency level is less than 1500 hertz (Hz), it is normally not annoying to the hearer. However, this rattling of the stem or shaft with the guides can damage critical guiding or seating surfaces. One side benefit of a rattling noise associated with valve parts is that such secondary noise provides a warning signal that turbulence is taking place inside the valve and that corrections may be necessary before failure occurs. Vibration can also be caused by certain valve parts or accessories that resonate at their natural frequency, which is often found in lower noise levels—less than 100 dBA. This type of noise is characterized by a single tone or hum (with a frequency between 3000 and 7000 Hz). Although this noise is not an annoyance, it does produce high levels of stress in the material, which may fatigue the material of the component and cause it to weaken. Noise can also be generated by hydrodynamic and aerodynamic fluid sound. With liquid applications, hydro-dynamic noise is caused by the turbulence of the flow, cavitation, flashing, or the high velocities that occur as the flow moves through the vena contracta. Generally, however, the noise generated by the liquid flow does not occur at high levels and can be tolerated by workers. In severe cavitating or flashing applications, noise levels can reach higher levels and must be dealt with by changing the process or installing anticavitation components in the valve.

When cavitation occurs in liquid services, the noise generated by the implosion of the bubbles occurs just slightly downstream from the valve and sounds similar to rocks flowing down the pipe. Overall, this noise is simply irritating and does not reach levels that cause harm. On the other hand, aerodynamic noise is often a problem for nearby workers when dealing with gaseous services. It generates frequencies in the range between 1000 and 8000 Hz, the range that is most sensitive to the human ear. In many cases, gaseous noise levels rise above 100 dBA (decibels for human hearing) and in some extreme cases, above 150 dBA.

In general, the noise level is a function of the velocity of the flow stream. As the pressure profile indicates, when pressure drops at the vena contracta, the velocity increases proportionately. Because of the vena contracta, high noise levels can be generated as velocity increases through the restriction, even though the velocity decreases as low as Mach 0.4 as the flow reaches the downstream side of the valve.

The mechanisms used in cavitation control—tortuous paths, staged pressure drops, and expanding flow areas—can also be applied in order to lower sound levels in gas services. In addition, the mechanism in providing a flow path with sudden expansions and contractions is also used to lower aerodynamic noise.

11.7.2 Sound Pressure Level

Vibrations or atmospheric pressure changes are based upon the number of cycles per second (hertz). A young hearer has a hearing range of 20 to 20,000 cycles per second (20 kilocycles or 20 kHz). The intensity of sound that is heard by a hearer is expressed as in units as *decibels*. In order to understand decibels, the relationship of microbars to 1 Newton per square meter must be understood. One µbar is one-millionth of a normal atmospheric pressure and 10 µbar equal 1 N/m². Zero decibel (dB) is defined as 0.00002 N/m², which is considered the absolute threshold of hearing for a young hearer. Decibels are applied to three common weighted sound levels; dBA for human hearing, dBB

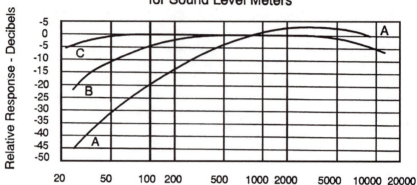

Figure 11.26 Decibel curves for *A*, *B*, and *C* scales. (*Courtesy of Valtek International*)

for an intermediate range, and dBC for equipment (Fig. 11.26). In nearly all cases, dBA is the most commonly applied sound level because it is weighted to account for the sensitivity of human hearing. With the dBA-weighted scale, the loudness of a particular noise at a certain frequency is compared to the loudness given for a 1000-Hz level. In other words, at 1000 Hz, the dBA value is zero. With the 1000-Hz scale, the sound pressure level is equal to the actual dB level. However, if a different hertz level is applied, the noise may sound less loud. For example, with 200 Hz, a sound measured at approximately 120 dB is lower in loudness (110 dB). Or in other words, the correction for dBA at 200 Hz is −10, as shown in Fig. 11.26. Table 11.4 indicates a number of common sounds measured in dBA levels.

Valve noise is calculated as a *sound pressure level,* which is defined as

$$\text{SPL} = 20 \log_{10} \frac{P}{0.0002 \ \mu\text{bar}} \ \text{dB}$$

where SPL = sound pressure level
P = root-mean-square sound pressure (N/m^2)

Approximately 90 dB equals one sound pressure level, and this level doubles every 6 dB. Therefore, 96 dB is two times the sound pressure level and 102 dB is four times the sound pressure level. To illustrate the magnitude of this change, the range between 80 and 120 dB is

Table 11.4 Typical dBA Sound
Levels*

Sound	dBA
Threshold of hearing	0
Soft whisper at 5 feet (1.3 meters)	10
Average home residence	50
Busy highway	57
Freight train at 100 feet (25.4 meters)	67
Subway train at 20 feet (5.1 meters)	80
Textile weaving plant	83
Electric furnace	90
Pneumatic peen hammer	94
Riveting machine	100
Discomfort threshold	110
Jet take-off at 200 feet (50.8 meters)	123
50 HP siren at 100 feet (25.4 meters)	135
Pain threshold	140

Courtesy of Valtek International.

100 times the sound pressure level. Noise radiating from a single point decays at a rate of 6 dB for every doubling of distance. However, if the noise is radiating from a radial line source—such as noise radiating from a pipeline—the noise decays at half that rate or 3 dB for every doubling of distance. Conversely, hard surfaces close to the noise source can increase the noise by reflecting sound. A single hard surface, such as a floor, increases the noise level by 3 dBA. Two hard surfaces, such as a floor and wall, reflect an additional 6 dBA and three hard surfaces (a corner) add 9 dBA. Theoretically, if the noise was enclosed in a completely sealed room with hard surfaces, noise levels would approach infinity—although this is highly unlikely with atmospheric friction. However, the possibility exists that a loud valve installed in a small metal building could easily achieve the pain threshold of 140 dBA.

Sound pressure levels are measured by a sound-level meter, which is normally held 1 m downstream from the valve's outlet and 1 m away from the pipe itself. Because of the effect reflective surfaces can have on the sound pressure levels, the measurement must be taken in a free-field area without any reflective surfaces. In some cases, sound intensity levels may be preferred for measuring or comparing sound intensities. This is calculated as

$$\text{sound intensity level} = 10 \log_{10} \frac{\dfrac{P_S^2}{\rho C}}{10^{-16}} \text{ dB}$$

where P_S = amplitude of sound pressure
ρ = density
C = sonic velocity

   In some cases, two noise sources may be occurring at the same time, which will increase the overall sound pressure level. The energy of the two sources is logarithmically combined as one noise source. Table 11.5 represents a simple method of determining the increase in noise when two noise sources are combined. After sound pressure levels are taken at each source, the difference between the two readings is used to find the correct dB factor, which is then added to the loudest noise source. As Table 11.5 shows, as the difference in the sound pressure level between two sources widens, the overall noise increase lessens. Therefore, the obvious solution is to concentrate on correcting the source with the loudest noise.

11.7.3 Turbulence

To achieve an understanding of how to decrease valve noise, the causes of turbulence must be examined. As the flow moves through the valve, the flow stream is interrupted by the valve geometry, such as the presence of a seat, disk, plug, or a sharp contour of the body. Turbulence causes pressure fluctuations in a variety of ways; however, in simple terms the pressures work against the wall of the downstream piping and cause wall fluctuations, which radiates the noise frequencies to the atmosphere. Figure 11.27 shows the pressure profile of a throttling valve as the upstream pressure is released to atmosphere. The profile shows a wide range of fluctuations in the downstream pressure that can vary by more than 15 psi (1.0 bar). As the upstream pressure decreases, the pressure drop decreases, and the variations of downstream pressure and resultant noise are less. Using the same test data, Fig. 11.28 shows a downstream test plot of the sound pressure

Table 11.5 dB Factors for Two Noise Sources†

Difference in dB between two sources	dB Factor*
0	3.01
1	2.54
2	2.12
3	1.76
4	1.46
5	1.20
6	0.97
7	0.79
8	0.64
9	0.52
10	0.42
11	0.33
12	0.27
13	0.22
14	0.17
15	0.14
16	0.11
17	0.09
18	0.07
19	0.06
20	0.05

†*Data courtesy of Fisher Controls International, Inc.*
*Added to loudest source to provide overall sound pressure level.

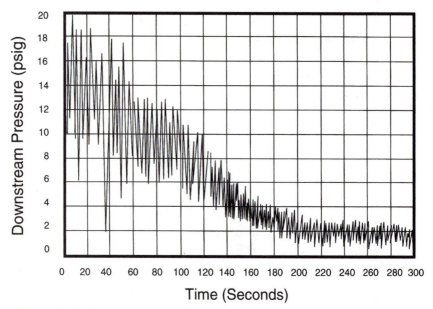

Figure 11.27 Pressure vs. time profile—downstream from valve. (*Courtesy of Valtek International*)

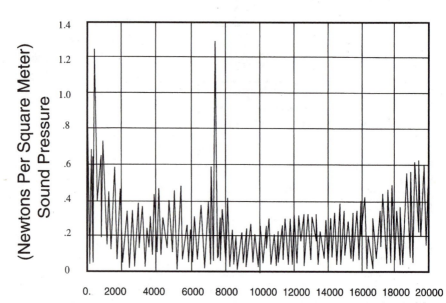

Figure 11.28 Plot of sound pressure level—downstream from valve. (*Courtesy of Valtek International*)

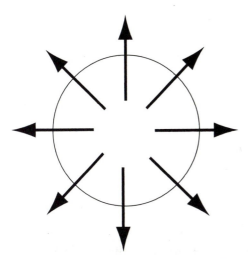

Figure 11.29 Monopole noise source. (*Courtesy of Valtek International*)

level (in hertz). The plot shows one discrete frequency peak occurring at 7500 Hz. Such peaks in that range are commonplace with valves that experience prolonged high noise levels. Although the test data indicate the presence of a wide range of subharmonics, the discrete peak frequency is principally responsible for the valve noise.

Turbulence is designated as one of three classifications: monopole, dipole, and quadrupole. *Monopole turbulence* is often described as an expanding and contracting source of noise (Fig. 11.29). The energy generated by a monopole-turbulent source is directly proportional to the flowing energy of the process fluid times the Mach number of the fluid, or in equation form:

$$(\text{turbulent energy}) \; \alpha \; (\text{flowing energy}) \times (\text{Mach number})$$

The formula of monopole turbulence indicates that the greater speed of the flow stream will convert to more turbulent energy. Monopole energy can be easily illustrated by using a Hartmann generator (Fig. 11.30). Air flows through the nozzle (d_0) into the bore (d), causing shock waves to form inside the bore and attach to the flat surface at the bottom of the bore. As these shock waves resonate back and forth, they create discrete peak frequencies, resulting in noise that can increase by as much as 24 dBA. The importance of the Hartmann gen-

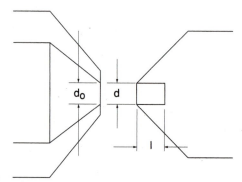

Figure 11.30 Hartmann generator (monopole noise source). (*Courtesy of Valtek International*)

erator can be seen, for example, if one envisions the shapes inside a globe valve that is made from barstock. The fluid follows through a small opening (seat ring) to a flat surface in the cavity leading to the outlet port (the bottom of the valve body). In an open situation, flow moves past the seat ring into the flat bottom portion of the body, where shock waves can attach and resonate. The position of the plug plays a large role in how much flow and velocity occur as well as the resulting noise (Fig. 11.31). However, studies have shown that if the seat-ring design is modified to a very narrow surface on the inside

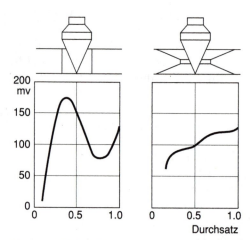

Figure 11.31 Effect of monopole noise with conventional globe valve's closure element. (*Courtesy of Valtek International*)

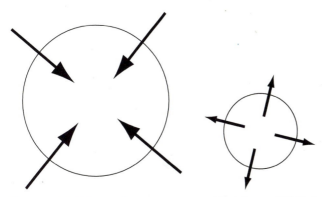

Figure 11.32 Dipole noise source. (*Courtesy of Valtek International*)

diameter, the shock waves seemed to attach themselves to the sharp point and do not resonate in the body cavity.

Dipole turbulence is defined by two energy sources, one contracting in size as the other expands inversely (Fig. 11.32). With dipole turbulence, the energy of the turbulence is proportional to the Mach number cubed or in equation form:

$$\text{(turbulent energy)} \; \alpha \; \text{(flowing energy)} \times \text{(Mach number)}^3$$

Because of the cubed Mach number, higher velocities are much more critical in dipole turbulence than monopole turbulence. A common example of dipole turbulence is the "singing" telephone line (Fig. 11.33), in which alternate vortices are generated from both the top and bottom of the wire. These alternate vortices produce a discrete frequency, which can vary in pitch as the velocity changes. Dipole turbu-

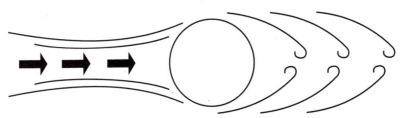

Figure 11.33 Karmen vortex street (dipole noise source). (*Courtesy of Valtek International*)

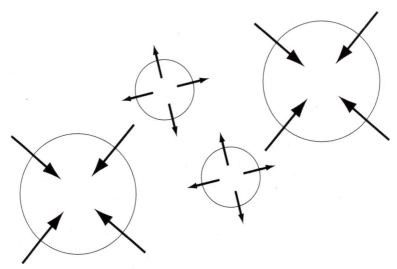

Figure 11.34 Quadrupole noise source. (*Courtesy of Valtek International*)

lence can be created in valves, such as with the sharp edges of a butterfly valve body and the disk. In addition, some trim-hole designs can generate dipole noise—a good example is the large flow characteristic holes designed into cage-guided trim.

Quadrupole turbulence is related to dipole noise; however, it involves two pairs of dipole turbulent energy. Although each pair is in phase (contracting and expanding inversely) with each turbulent source, the two pairs are out of phase with each other (Fig. 11.34). In this case, the turbulent energy varies according to the Mach number to the fifth power, or in equation form:

$$(\text{turbulent energy}) \; \alpha \; (\text{flowing energy}) \times (\text{Mach number})^5$$

Even more than in dipole turbulence, velocity is critical to the formation of quadrupole turbulence. One important difference with quadrupole turbulence is that it involves a number of random peak frequencies rather than one discrete frequency. Nearly all noise radiating from a downstream pipe is related to quadrupole turbulence. As shown in Fig. 11.25, as turbulence generated by a valve travels downstream inside the pipe, the turbulence has a tendency to move to the outer wall while smoother portions of the flow stay in the center of the pipe.

11.7.4 Noise Regulations

A growing number of organizations monitor the amount of noise workers can be safely exposed to. For example, in the United States, the Occupational Safety and Health Act (OSHA) and the Environmental Protection Agency (EPA) both regulate noise as it affects workers and the surrounding community. Initially, the Occupational Safety and Health Act (1970) stipulated that workers could be exposed to no more than 90 dBA for an 8-hour work day. Later, the Walsh Healy Public Contracts Act was enacted to further protect workers. It regulates the exact amount of time workers may work around noise. According to this legislation, the higher the dBA level, the less time workers can spend in that area, as outlined in Table 11.6.

Table 11.6 Permissible Noise Levels*
Walsh Healy Public Contracts Act

Duration Per Day (hours)	dBA
0.25 or less	115
0.5	110
1.0	100
2.0	97
4.0	95
6.0	92
8.0	90

Data courtesy of Valtek International.

11.7.5 Hydrodynamic Noise Prediction

Similar in some aspects to calculating the advent of cavitation and flashing, the prediction of noise levels in liquid services is based upon a number of common factors, including the pressure drop and flow capacity. In addition, the factors associated with pipe attenuation and distance from hearers are also considered. Using these factors, the following empirical equation can be used to predict hydrodynamic noise:

$$dBA = DP_S + C_S + R_S + K_S + D_S$$

where dBA = sound pressure level
DP_S = pressure-drop factor
C_S = flow capacity factor
R_S = ratio factor
K_S = pipe attenuation factor
D_S = distance factor

To calculate R_S and DP_S the pressure-drop ratio (DP_F) must be determined, which involves the following equation:

$$DP_F = \frac{\Delta P}{P_1 - P_v}$$

where DP_F = pressure-drop ratio
ΔP = pressure drop
P_1 = upstream pressure
P_v = vapor pressure

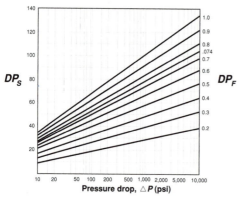

Figure 11.35 Pressure-drop factor. (*Courtesy of Valtek International*)

If DP_F is calculated to be 1 or greater, a flashing situation is occurring in the valve. Because flashing is indicative of a system problem, no modification to the valve will abate flashing and the resultant noise.

Once the pressure-drop ratio DP_F is determined, the pressure-drop factor DP_S can be determined using Fig. 11.35 and the ratio factor R_S can then be found using Fig. 11.36. Figure 11.37 provides a typical representation of the flow-capacity factor C_S. Table 11.7 provides typical distance factors D_S. Pipe attenuation factors K_S are found in Table 11.8.

11.7.6 Hydrodynamic Noise Example

The following service conditions apply for this example:

Fluid	Water
Upstream pressure	300 psig
Downstream pressure	90 psig
Vapor pressure	29.89 psia
Required C_v	34.8
Pipe size	2 in
Pipe schedule	Schedule 40
Distance of hearer	3 ft

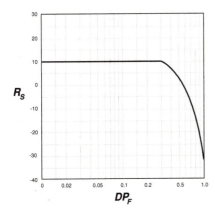

Figure 11.36 Ratio factor. (*Courtesy of Valtek International*)

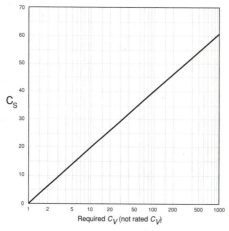

Figure 11.37 Flow-capacity factor. (*Courtesy of Valtek International*)

Table 11.7 Distance Factors*

Distance of hearer from noise source (feet/meters)	D_s
3 feet 0.9 meters	0 dBA
6 feet 1.8 meters	-5 dBA
12 feet 3.6 meters	-10 dBA
24 feet 7.2 meters	-15 dBA
48 feet 14.4 meters	-20 dBA
96 feet 28.8 meters	-25 dBA

*Data courtesy of Valtek International.

Note: Factors are affected by type of noise source, as well as any reflecting surfaces close to the valve.

Table 11.8 Pipe Attenuation Factors for Liquids*

	Pipe Schedule												
	10	20	30	40	60	80	100	120	140	160	STD.	XS	XXS
0.5				0		-5				-11	0	-5	-15
0.75				0		-5				-11	0	-5	-15
1.0				0		-6				-12	0	-6	-15
1.5				0		-6				-12	0	-6	-14
2				0		-6				-12	0	-6	-14
3				0		-7				-13	0	-7	-16
4				0		-7		-9		-13	0	-7	-14
6				0		-8		-10		-14	0	-8	
8		4	3	0	-3	-9	-8	-12	-13	-18	0	-9	
10		5	3	0	-5	-9	-9	-13	-14	-19	0	-7	
12		6	2	-1	-6	-10	-11	-14	-15	-20	0	-6	
14	6	3	0	-2	-6	-11	-12	-15	-16	-22	0	-4	
16	6	3	0	-4	-8	-12	-13	-16	-18	-24	0	-4	
18	5	3	-2	-6	-9	-13	-15	-18	-19	-25	0	-4	
20	5	0	-4	-6	-10	-14	-16	-19	-21	-26	0	-4	
24	5	0	-6	-8	-12	-15	-19	-21	-23	-27	0	-4	
30	3	-4	-7	-8		-15				-27	0	-4	
36	3	-4	-7	-9		-15				-27	0	-4	
42		-4	-7			-15							

Courtesy of Valtek International.

By using the pressure-drop ratio equation, DP_F is calculated as 0.74:

$$DP_F = \frac{DP}{P_1 - P_V} = \frac{314.7 - 104.7}{314.7 - 29.89} = 0.74$$

From Figs. 11.35 to 11.37 and Tables 11.7 and 11.8, the following factors apply: $DP_S = 60$, $R_S = -10$, $C_S = 31$, $D_S = 0$, and $K_S = 0$. Therefore, the hydrodynamic noise equation can be used to predict the noise from this application:

$$dBA = DP_S + R_S + C_S + D_S + K_S = 60 + (-10) + 31 + 0 + 0 = 81 \text{ dBA}$$

With a predicted sound pressure level at 81 dB, hearers could safely work in the vicinity of the valve for 8 h per day (as outlined by the Walsh Healy Act).

11.7.7 Aerodynamic Noise Prediction

Because aerodynamic noise is the most irritating type of noise to nearby hearers and communities, predicting the noise level emitted from a valve is critical to the sizing and selection process. The noise prediction for gas services varies from the hydrodynamic noise equation in that factors relating to pressure, temperature, and gas properties must also be considered. The following empirical equation is used:

$$\text{dBA} = V_S + P_S + E_S + T_S + G_S + A_S + D_S$$

where V_S = flow factor
 P_S = pressure factor
 E_S = pressure ratio factor
 T_S = temperature correction factor
 G_S = gas property factor
 A_S = attenuation factor

The flow factor V_S is determined by using the valve's required C_v, as shown in Fig. 11.38. The pressure factor P_S is found by using the valve's upstream pressure (Fig. 11.39). To determine the pressure ratio factor E_S, the ratio between the upstream and downstream pressures must be calculated (Fig. 11.40). The temperature correction factor T_S is determined by Table 11.9. The gas property factor G_S is found by applying the molecular weight of the gas against Fig. 11.41. The attenuation factor A_S is found for a given pipe size and schedule in Table 11.10. The same distance factor table (Table 11.7) that was used in the hydrodynamic calculations still applies for D_S.

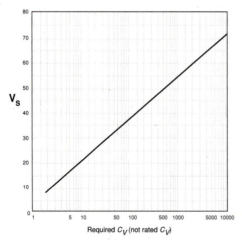

Figure 11.38 Flow factor. (*Courtesy of Valtek International*)

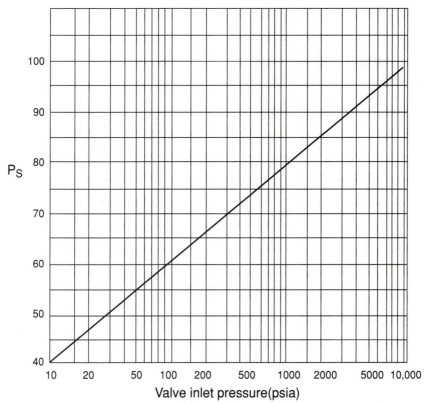

Figure 11.39 Pressure factor. (*Courtesy of Valtek International*)

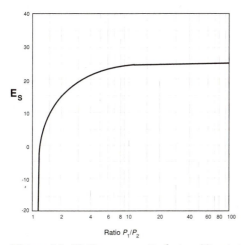

Figure 11.40 Pressure ratio factor. (*Courtesy of Valtek International*)

Table 11.9 Temperature Correction
Factors*

Flowing Temperature of Gas	T_s
70° F / 21° C	0.0
100° F / 38° C	0.0
200° F / 93° C	-1.0
300° F / 150° C	-1.5
500° F / 260° C	-2.0
700° F / 370° C	-3.0
1000° F / 540 ° C	-3.5
1200° F / 490° C	-4.0

Courtesy of Valtek International.

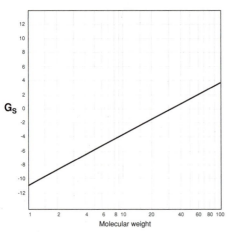

Figure 11.41 Gas property factor. (*Courtesy of Valtek International*)

Table 11.10 Pipe Attenuation Factors for Gases*

Pipe Size	Pipe Schedule												
	10	20	30	40	60	80	100	120	140	160	Std.	XS	XXS
½				-8.0		-11.5				-13.1	-8.0	-11.5	-18.5
¾				-10.0		-13.7				-17.0	-10.0	-13.7	-21.0
1				-11.4		-15.0				-18.8	-11.4	-15.0	-22.2
1½				-13.6		-17.6				-21.6	-13.6	-17.6	-25.5
2				-14.8		-19.2				-24.6	-14.8	-19.2	-27.4
3				-16.6		-20.8				-25.4	-16.8	-20.8	-29.1
4				-18.0		-22.4		-25.7		-28.0	-18.0	-22.4	-30.9
6				-20.0		-25.5		-28.8		-31.8	-20.0	-25.5	-34.1
8		-18.1	-19.4	-21.4	-24.3	-27.0	-29.1	-31.5	-33.1	-34.4	-21.4	-27.0	-34.0
10		-17.6	-20.3	-22.5	-26.6	-28.7	-31.2	-33.2	-35.3	-36.8	-22.5	-26.6	-35.3
12		-18.2	-21.8	-24.5	-28.5	-31.2	-33.8	-36.0	-37.4	-39.3	-24.5	-28.7	-36.0
14	-18.8	-21.6	-24.0	-26.0	-29.9	-32.9	-35.5	-37.7	-39.3	-40.8	-24.0	-28.5	
16	-19.5	-22.4	-24.8	-28.5	-32.0	-35.2	-37.7	-39.8	-41.9	-43.2	-24.8	-28.5	
18	-20.2	-23.1	-27.4	-30.7	-35.4	-37.2	-39.9	-42.1	-43.7	-45.3	-25.4	-29.0	
20	-20.8	-26.1	-29.8	-32.0	-36.0	-39.1	-41.8	-43.8	-45.7	-47.2	-26.1	-29.8	
24	-21.9	-27.1	-32.3	-34.9	-39.3	-42.3	-45.2	-47.3	-48.9	-50.5	-27.1	-29.5	
30	-26.1	-32.2	-35.1	-38.5	-42.7	-45.5	-48.3	-50.5			-26.0	-32.2	
36	-27.2	-33.3	-36.2	-42.0	-45.5	-48.5	-51.2				-26.4	-33.3	
42	-28.7	-37.0	-40.3	-44.5	-48.0	-50.7	-53.7				-26.7	-30.4	
48	-29.8	-39.0	-42.5	-46.5	-50.3	-53.0					-27.0	-30.5	
54	-30.5	-41.0	-44.3	-48.5	-52.2								
60	-31.2	-42.5	-45.7	-50.3	-53.5								

Courtesy of Valtek International.

11.7.8 Aerodynamic Noise Example

The following service conditions apply to this example:

Fluid	Steam
Upstream pressure	139.7 psig
Downstream pressure	29.7 psig
Required C_v	46.2
Pipe size	2 in
Pipe schedule	Schedule 40
Distance of hearer	3 ft
Molecular weight	18.02

Using the upstream and downstream pressures, the ratio P_1/P_2 is:

$$\frac{P_1}{P_2} = \frac{139.7}{29.7} = 4.70$$

From Figs. 11.38 to 11.41, and Tables 11.7 and 11.8, the following factors are applied: $V_S = 31$, $P_S = 61$, $E_S = 22.5$, $T_S = -2$, $G_S = -1.0$, $D_S =$

0, and $A_S = -18.0$. With these factors, the aerodynamic noise equation can be used to predict the noise from this application:

$$dBA = V_S + P_S + E_S + T_S + G_S + A_S + D_S$$

$$= 31 + 61 + 22.5 + (-2) + (-1.0) + (-18.0) + 0$$

$$= 93.5 \text{ dBA}$$

According to the Walsh Healy Act, at 93.5 dB, hearers could remain in the vicinity of the valve for 4 h per day.

11.8 Noise Attenuation

11.8.1 Introduction to Attenuation

Because hydrodynamic noise is often associated with cavitating services, it can be controlled with anticavitation measures. Generally, however, noise is associated with gas applications. This section emphasizes methods to lower noise levels in gaseous applications, although some methods may be applicable to liquid applications also. The process of lowering noise or sound pressure levels is called *attenuation*. Noise pollution is a primary environmental concern, for both plant and community environments. In many cases, the sound pressure levels must be reduced by noise attenuation of the source itself (the valve) or the path (downstream piping). Correcting the offending source is the ideal situation, but this involves sophisticated attenuation devices that reduce sound pressure levels to comfortable levels. Unfortunately, the costs associated with these special attenuation devices are high. Depending on the size of the valve, the cost could increase anywhere from 40 to 200 percent. If material fatigue or diminished performance is not a concern, path attenuation may be a less expensive, easier option, although it is only treating the symptom rather than the root cause.

Valve manufacturers, especially those that offer sizing and selection software programs, routinely predict noise as part of the valve selection process. However, the user should be aware that these predicted sound pressure levels assume that the valve is installed in a completely nonreflective environment and do not consider the additional noise levels associated with walls, floors, and ceilings. For example, a valve installed in a natural-gas pressure-reduction application is predicted to produce 85 dBA, which is within the safety standards of most regulations. However, because the valve is installed in a metal building, which is highly sound-reflective, the sound pressure level rises to 115 dBA. Therefore, the location of the valve should always be considered before determining that noise-attenuation devices or preventative measures are not necessary.

As this section outlines, a great deal of options are available to either reduce or eliminate noise. Some are more expensive than others, while some present additional problems, such as increased maintenance or added potential leak paths. Because of the costs and safety factors involved, the user should examine all options before deciding on installing an expensive antinoise valve.

11.8.2 Valve Attenuation Options

Although many users consider expensive valve trims the only solution to valve noise, a number of less expensive options exist that should be explored prior to specifying a specially engineered valve. The most simple, but overlooked option would be to restrict the access of workers to a high noise level or to provide ear protection while in that area. If equipment damage is not an issue, the main benefit of reducing noise levels is to protect the hearing of nearby workers. If workers do not need access to the affected area, then safety warnings and requirements for ear protection can be mandated and the process left alone.

Changes to the process may also be an option. The velocity may be varied by slightly changing the upstream or downstream pressures. In many cases, a discrete signal, which is within the range of hearers, is often prevented by a slight pressure variation to either side of the valve. The valve's position can also be slightly increased or decreased, allowing a minor change in flow that may disrupt the retention of shock waves on a given surface.

An interesting aspect of noise is that some linear valve styles, such as a globe valve, produce a discrete signal at 30 percent lift, despite the valve size or length of stroke. One way to deal with this phenomenon is to use a special diverting seat ring that has a special lip built into the bottom of the seat ring and breaks up the formation of shock waves.

As discussed in Sec. 11.7, velocities are directly related to turbulence and noise and can be controlled through right-angle turns. Frictional losses associated with 60 ft (18 m) of straight pipe are equal to the frictional forces produced by one 90° elbow, which will slow the velocity. Designing the system with several elbows can produce attenuation. In addition, placing two or more globe valves in series will produce a staged pressure drop and also add two or more right-angle turns per valve.

When a gas process is vented to atmosphere, high noise levels above 100 dBA can be generated. Such noise can be channeled in the opposite direction using shields or shrouds, which may lower the sound pressure level within acceptable limits. Moving the vent to a distant location may be an option in some cases, although the additional piping may be cost prohibitive.

The valve style can have some bearing on the type of noise that is generated. As explained in the preceding section, rotary valves are more apt to produce sharp dipole vortices as the flow travels past the sharp edges of the body. In addition, valve bodies produced from barstock commonly cause monopole noise as the flow moves through the seat and attaches to the flat surface of the outlet port. On the other hand, the conventional casting design of a globe body would avoid sharp edges associated with rotary valves and the flat surfaces of barstock bodies. The user should remember that different valve styles and internal geometries react differently to the same process. As a last resort, trial and error may be required to discover the one valve style that is able to handle the service without producing turbulence and subsequent pressure fluctuations that lead to noise.

When the flow direction is not a critical element of the application, the valve can be installed backwards (inlet port is installed downstream, and the downstream port is installed upstream), so that the flow direction is opposite the normal operation. (For example, a flow-over-the-plug linear valve will become a flow-under-the-plug valve.) When this is done, the fluid will then flow through a different valve geometry, in which monopole, dipole, or quadrupole noise is less likely to be created. For example, changing from flow-over-the-plug to flow-under-the-plug may avoid monopole noise that would be created from the Hartmann generator effect (Sec. 11.7.2). (The process stream flows up through the seat into the upper gallery, where the geometry provides no flat surface perpendicular to the flow where shock waves can form.)

In some cases, modifications can be made to the existing valve trim to attenuate the noise without installing expensive trims or downstream attenuation equipment. As discussed earlier, monopole noise will attach itself to a very narrow landing on the seating surface. Reducing this landing through machining may be possible, as long as the seat's seating surface and overall strength is not affected.

The location of the valve is vitally critical to the amount of noise generated by turbulence. Often noise is generated by turbulence in the valve and is then carried to downstream piping. The noise radiates the pressure fluctuations through the downstream pipe wall to the environment as sound waves. This phenomenon occurs with long, straight sections of thin-walled piping that are more apt to flex. Conversely, piping elbows and other nonlinear piping configurations are stiffer and are not apt to allow wall fluctuations. If a valve is included in a long stretch of piping, the preferred arrangement would be a long length of pipe on the upstream side of the valve and on the downstream side an elbow or a shorter length of pipe. The longer the pipe, the more sound radiation is possible. Piping supports can also be used to stiffen long lengths of piping, preventing the flex of the pipe wall.

In some process services in which the valve discharges fluid into a large vessel, the valve can be located next to the vessel without a long expanse of pipe. This will allow the valve to discharge the fluid into the vessel and the noise to be absorbed in a larger area.

If the valve and downstream piping are located in a room or protective shed with a number of close-by hard reflective surfaces, the sound pressure levels may increase significantly, upwards of 30 to 40 dB. However, by moving the location of the valve to the wall, the downstream side of the pipe can be placed outside of the room. Not only will the noise be eliminated from the room, but the noise radiated to the environment outside of the room will also be less.

Another option is to specify a thicker wall schedule in the downstream piping, which provides greater stiffness. For example, using a schedule 80 pipe instead of a schedule 40 pipe will lower the sound pressure level by approximately 4 dB. Table 11.11 provides a correction factor for noise attenuation for piping that has a heavier wall schedule (assuming schedule 40 pipe wall thickness is standard.)

One of the more common methods of dealing with high sound pressure levels is to absorb the noise with thermal or acoustic insulation, which can be wrapped around the valve or downstream piping. This is the best solution only when high sound pressure levels offer no threat of fatigue to materials or substandard performance of instrumentation. Generally, 1 in (2.5 cm) of normal thermal insulation will provide a reduction in sound pressure level of between 3 and 5 dB. Acoustic insulation is manufactured to absorb more sound energy and can provide a reduction of 8 to 10 dBA per inch of insulation. Depending on the R value of the insulation, a 3-in insulation will provide the maximum attenuation anywhere from 15 to 24 dB. (Additional insulation does not attenuate the noise any further.) Table 11.12 outlines typical insulation factors.

One caution should be noted, however. As previously explained, noise levels close to a valve and its immediate downstream piping may be reduced with an elbow, thick schedule pipe, or insulation.

Table 11.11 Pipe-Wall Attenuation*

Pipe Size	Schedule 40	Schedule 80	Schedule 120	Schedule 160
2-inch (DN 50)	0 dBA	- 6 dBA	-8 dBA	-12 dBA
3-inch (DN 75)	0 dBA	-7 dBA	-9 dBA	-13 dBA
4-inch (DN 100)	0 dBA	-7 dBA	-10 dBA	-13 dBA
6-inch (DN 150)	0 dBA	-8 dBA	-12 dBA	-15 dBA
8-inch (DN 200)	0 dBA	-9 dBA	-14 dBA	-18 dBA
10-inch (DN 250)	0 dBA	-10 dBA	-14 dBA	-19 dBA
12-inch (DN 300)	0 dBA	-11 dBA	-16 dBA	-20 dBA

Data courtesy of Fisher Controls International, Inc.

However, these methods only protect the hearer in the immediate vicinity of the valve. Since these methods do not attenuate the source of the noise, sound will continue in the downstream piping and may surface at an unprotected point further downstream (Fig. 11.42). At that point, either the noise must be tolerated or additional corrective action must be taken.

11.8.3 Downstream Antinoise Equipment

In some applications, adding an antinoise element immediately downstream from the valve may be effective in attenuating the noise to reasonable levels. In addition, these elements can absorb energy or straighten turbulent flow so that noise is not carried downstream. The cost associated with these supplemental devices is less than or equal to special valve trim. Access to a downstream element is much easier for

Table 11.12 Insulation Factors*

Depth of Insulation	dBA Reduction
1 in / 2.5 cm	-5 dBA
2 in / 5.1 cm	-10 dBA
3 in / 7.6 cm	-15 dBA

*Data courtesy of Valtek International.

maintenance purposes than gaining access to special trim. Common antinoise elements include attenuator plates, diffusers, silencers, and external stacks. Because these devices all utilize small holes or flow paths, they are susceptible to plugging if the process fluid contains particulate matter, which may require additional maintenance.

Placed downstream in series with the valve, the *attenuator plate* is a downstream antinoise element (Fig. 11.43) that provides anywhere from single to multiple stages of pressure reduction (Fig. 11.44). Attenuator plates typically reduce the overall sound pressure level by up to 15 dB. Using a pattern of holes, each stage of the attenuator plate

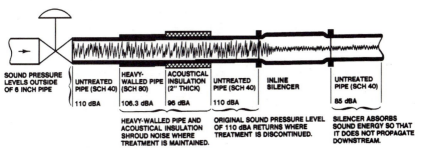

Figure 11.42 Multiple methods of path treatment of noise. (*Courtesy of Fisher Controls International, Inc.*)

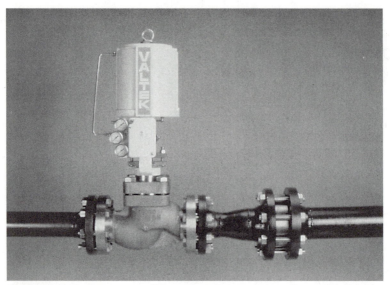

Figure 11.43 Attenuator plate mounted downstream from a globe control valve. (*Courtesy of Valtek International*)

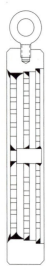

Figure 11.44 Three-stage attenuator plate.
(*Courtesy of Valtek International*)

has its own individual flow capacity. As Fig. 11.44 illustrates, each suc-ceeding stage has a larger flow area, which provides the staged pres-sure reduction and maintains velocities at lower levels. The multiple holes act as a straightening device for the turbulent fluid, providing a series of controlled, smaller fluid streams instead of a large turbulent eddy. Although these smaller fluid streams still have some turbulence, they are more easily dissipated throughout the overall process stream because of their size. Since the area of the plate is limited to the inside diameter of the pipe, as well as the hole pattern, only so much flow can pass through the first stage. The maximum flow capacity through the attenuator plate is achieved with a pressure ratio of 4.5 to 1 (or less). High rangeability is highly unlikely with attenuator plates; therefore, they should be installed only in moderate to low rangeabili-ty applications. Because the flow capacity is limited, attenuator plates should be considered only for those applications that can handle such a reduction in flow. In some applications where additional flow is needed, a larger plate can be specified with more flow area, but pipe expanders or reducers must be used to allow the installation of the larger plate in a smaller pipeline. Not only does this raise costs, but it also adds a number of line penetrations that could leak.

For applications that require greater flow than offered by an attenu-ator plate, a *diffuser* is often specified, which also offers reductions of up to 15 dB. As shown in Fig. 11.45, a diffuser is a long cylinder tube with a closed end that can vary in length according to the flow needed. As with attenuator plates, the diffuser is installed downstream in series with the valve. The diffuser is designed to fit inside the pipeline, allowing for a specific clearance between the inside diameter of the pipe and outside diameter of the diffuser. The diffuser is held in place between the raised face flanges of the valve and pipeline, or it can be welded in place. A diffuser can also be directly bolted or welded to the valve and be used to vent to atmosphere. When venting to atmos-phere, a diffuser can be equipped with shrouds to direct the noise away from hearers. Although the diffuser shares the overall pressure drop with the valve, its flow capacity can be expanded by making the diffuser longer and adding more holes. These holes control the sound pressure level by passing the flow through the holes to absorb sound energy and minimize turbulence. The major disadvantage of a diffuser is the maintenance problems associated with the small holes, which can become plugged if the process contains oversized particulates. Because the holes in the diffuser are perpendicular to the piping, they have a tendency to impinge condensates and particulates directly on the piping wall, which may lead to erosion.

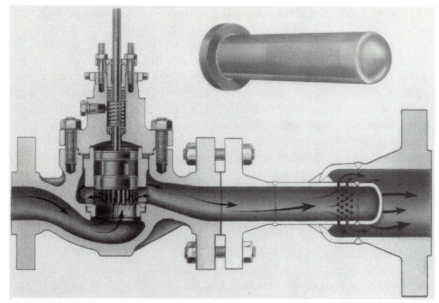

Figure 11.45 Downstream diffuser. (*Courtesy of Fisher Controls International, Inc.*)

A *silencer* is used when reductions of sound pressure levels of more than 15 dB are required, which are beyond the capability of attenuator plates or diffusers. Depending on the design and process service conditions, attenuations as high as 35 dB can be achieved with a silencer. Similar to a diffuser in that it shares the pressure drop with the valve, a silencer also lowers the sound pressure level by absorbing noise. As shown in Fig. 11.46, a common silencer incorporates a series of compartments that use tubes with holes, much like minidiffusers. Acoustic material is used throughout the silencer to absorb sound and process energy. The primary disadvantage of silencers is that they are designed to attenuate a particular frequency. Overall, silencers are good for applications with a constant flow. However, if the application is such that the flow varies routinely, the frequency will also vary and may render the silencer ineffective. While a silencer is less expensive than other antinoise options, it requires some piping modifications, including piping supports. Depending on the application, the size of the silencer can be quite large. This may become a factor where space is limited. Silencers are normally flanged and bolted to the pipeline, although they can also be used to vent to atmosphere.

Some valves use an *external stack* (also known as an *atmospheric resistor*) as a downstream element to reduce noise in venting or blowdown

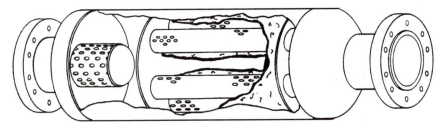

Figure 11.46 In-line silencer. (*Courtesy of Fisher Controls International, Inc.*)

applications (Fig. 11.47). Instead of installing antinoise trim inside the valve, the stack is placed immediately downstream from the valve's outlet port. This design provides several benefits. First, the physical characteristics of the stack can be larger, allowing a greater outside diameter and stack height. This allows greater flow and increased

Figure 11.47 External stack mounted on outlet of angle body control valve. (*Courtesy of Control Components Inc., an IMI company*)

attenuation than a stack inside the valve—which is limited by the body's gallery height. Second, the antinoise mechanisms built into the stack—such as expanding flow areas, tortuous paths, etc.—can lower the exit velocity and share the pressure drop with the valve. This design provides greater attenuation while not affecting the overall flow rate.

11.8.4 Downstream Antinoise Equipment Sizing

Sizing for the flow capacity of downstream equipment is based on the number of stages of pressure drop taken. These stages can be taken through one element (such as an attenuator plate) or a number of single-stage elements in series (such as two diffusers). A common equation for attenuation plates follows:

$$C_v = \frac{1}{\sqrt{\left(\dfrac{1}{C_{v_1}}\right)^2 + \left(\dfrac{1}{C_{v_2}}\right)^2 + \left(\dfrac{1}{C_{v_3}}\right)^2 + \left(\dfrac{1}{C_{v_N}}\right)^2}}$$

where C_v = total flow capacity
C_{v_1} = flow coefficient of the first control element (or first stage)
C_{v_2} = flow coefficient of the second control element (or second stage)
C_{v_3} = flow coefficient of the third control element (or third stage)
$C_v N$ = flow coefficients of any additional control elements

11.8.5 Downstream Antinoise Equipment Sound-Pressure-Level Prediction

Predicting the overall sound pressure level is determined by the following two equations:

$$\text{SPL} = 22 + 12 \log_{10}\left(\frac{P_1}{P_O} - 1.05\right) + 10 \log_{10}(C_v F_L) + 10 \log_{10}(P_1 P_2)$$

$$+ 30 \log_{10}\left(\frac{t_{40}}{t}\right) - G_S + T_L$$

$$\frac{P_1}{P_O} = \frac{\dfrac{P_1}{P_2} + Z}{Z}$$

where SPL = sound pressure level
Z = number of elements (or stages)
P_1 = inlet pressure
P_2 = outlet pressure
P_O = outlet pressure for each stage
t_{40} = schedule 40 pipe wall thickness
t = wall thickness for given wall pipe
G_S = gas property correction factor (Table 11.12)
T_L = SPL velocity-correction factor for gas discharges above Mach 0.15 ($T_L = 20 \log_{10} [1/(1.1 - M)]$)
M = Mach number of outlet pipe

When two elements are combined in series, up to 3 dB should be added to the total sound pressure level to compensate for having two separate noise sources. Figure 11.48 provides this data. If two noise sources have identical sound pressure levels, the overall intensity will not be equal to that level, but will be greater than either noise source. A 6-dB insertion loss factor should be included in the overall sound pressure level to compensate for a close connection between the element and the valve. *Close connection* is defined as one pipe reducer length. The sound pressure level of venting applications can also be

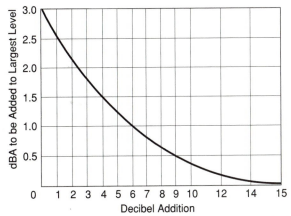

Figure 11.48 dBA addition for two elements installed in series. (*Courtesy of Valtek International*)

determined. Although the general rule is that spherical radiation of noise reduces the sound pressure level by 6 dB for every doubling of distance, noise emitted over long distances can absorb even more sound because of atmospheric absorption, and the attenuating effects of nearby objects and the ground.

The following equation can be used to calculate the sound pressure level for all gaseous venting applications, except steam:

$$SPL_{intermediate} = SPL_{elements} - 10 \log_{10} \left(\frac{(3.2)(10^{-11}) \, P_2 \, D^2 \, T^{0.5} \, R^2}{t^3} \right) - \frac{G_S}{2}$$

where $SPL_{intermediate}$ = uncorrected sound pressure level from vent
$SPL_{elements}$ = sound pressure level emitted from control elements
D = downstream nominal pipe diameter
P_2 = valve downstream pressure
R = distance from vent
T = absolute temperature

Although this equation is used to find the total sound pressure level emitting from the vent, the sound pressure level can also be lowered by the direction of the noise.

The following equation is used to calculate the sound pressure level for steam-venting applications:

$$SPL_{intermediate}$$

$$= SPL_{elements} - 10 \log_{10} \left(\frac{(3.2)(10^{-10}) \, P_2 \, D^2 \, R^2 \, (1 + 0.00126 T_{SN})^3}{t^3} \right)$$

where T_{SN} = superheated steam temperature

Sound pressure levels are also reduced if the noise radiates in a directional nature rather than spherical. In other words, the farther the noise is pointed away from the hearer, the less noise is heard. This phenomena is called *directivity*. This concept is illustrated in Fig. 11.49 and Table 11.13. With particular venting applications, directivity can occur if the vent is pointed away from workers or nearby communities or if a resistor shroud is used to direct the sound upward (or away from the hearer). In venting applications where directivity occurs, the reduction of the sound pressure level can be determined by the following equation:

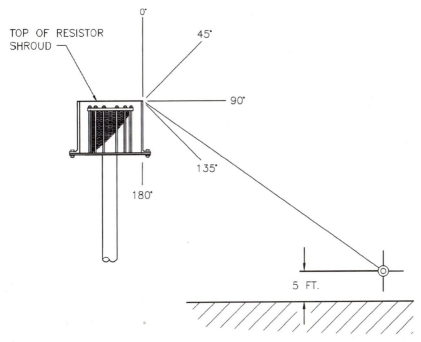

Figure 11.49 Angle location of hearer from noise source as associated with the directivity index. (*Courtesy of Control Components Inc., an IMI company*)

$$SPL_{total} = SPL_{intermediate} + DI$$

where SPL_{total} = total noise emitting from the vent
  DI = directivity index (Table 11.13)

Atmospheric noise can be divided into *near field* and *far field*. The near-field noise is the noise that is generated within 3 to 10 ft (1 to 3 m) of the source, while the far-field noise is that generated beyond 10 ft from the source. In far-field situations where sound spreads in a radial, homogeneous pattern, the noise intensity is attenuated by the distance—intensity decreases inversely to the (distance)^2 from the vent. This relationship is found in the following equation:

$$I = \frac{W}{4\pi r^2}$$

Table 11.13 Typical
Directivity Index*

Angle Away from Axis of Resister Shroud	dBA Addition or Subtraction
0°	0
20°	+1
40°	+8
60°	+2
80°	-4
100°	-8
120°	-11
140°	-13
160°	-15
180°	-17

Data courtesy of Control Components, Inc.

where I = sound intensity (W/m²)
    W = sound power (W)
    r = distance from sound source

This calculation applies only to far-field situations, in which large distances are involved and can be affected by a number of different factors, including humidity, wind, presence of trees, etc.

11.8.6 Antinoise Valve Trims

In difficult gaseous applications, noise must be treated at the source rather than treating the symptom with insulation, heavier or nonlinear piping, or ear protection. This means that modifications must be made to the valve to minimize or eliminate the high sound pressure levels and resultant vibration that can fatigue metal or affect the performance of nearby instrumentation. The antinoise trim must reduce the pressure drop, so that the resultant high velocities do not approach sonic levels. The most common approach to this problem is to install special trims in globe valves. In principle, these trims channel the fluid through a series of turns, which affects the velocities and pressures involved. Each turn is typically called a *stage*. Antinoise trims can include anywhere from 1 to 40 or even 50 stages, based on the design. While anticavitation trims are designed to flow over the plug in linear valves, antinoise trims are designed to flow under the plug. This direction allows an expanding flow area in the later stages of the antinoise device, which slows the velocity to subsonic levels.

A number of different antinoise trim devices are in existence, but for the most part they can be categorized into four different styles: slotted, multihole, tortuous path, and expanding teeth. *Slotted trim* use is a single-stage cage or retainer that contains long, narrow slots around the entire diameter (Fig. 11.50). As the fluid passes through the slots, tur-

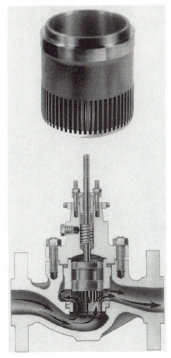

Figure 11.50 Single-stage multiple-slotted cage. (*Courtesy of Fisher Controls International, Inc.*)

bulence is broken up into smaller eddies, and the velocity is distributed evenly throughout the gallery of the globe body. This design works best when the pressure drop to inlet pressure ratio ($\Delta P/P_1$) is equal to or less than 0.65 and when the maximum downstream pressure (P_2) is less than half of the fluid's sonic velocity. If the $\Delta P/P_1$ ratio is higher than 0.65, the pressure drop may be handled by adding a second device (such as a downstream element) to share the pressure drop. Slotted cages or retainers offer noise attenuation up to 15 dB and are relatively inexpensive when compared to other antinoise trims. Outlet velocity is limited to below Mach 0.5. Additional dB reduction can be handled by adding an attenuation plate or diffuser downstream to the valve, which can also be cost effective when compared to other antinoise trims.

Multihole trim utilizes a number of cylinders, also known as stages, with drilled or punched holes that control turbulence in the flow stream (Fig. 11.51). This device also has a secondary use as a seat-ring retainer, which allows a clearance between the plug and the inside diameter of the retainer (Fig. 11.52). This device can also be designed as a cage, where the plug guides on the inside diameter (Fig. 11.53).

Figure 11.51 Single- and multiple-stage attenuators. (*Courtesy of Valtek International*)

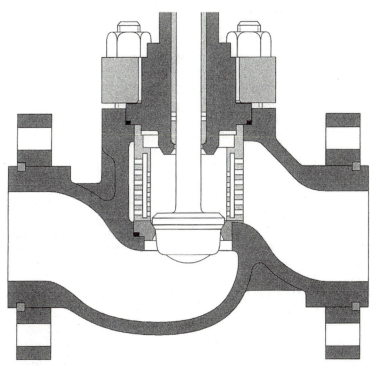

Figure 11.52 Globe valve equipped with two-stage attenuator. (*Courtesy of Valtek International.*)

Figure 11.53 Globe valve equipped with two-stage attenuation cage. (*Courtesy of Fisher Controls International, Inc.*)

One of the key engineering elements of multihole trim is its utilization of sudden expansions and contractions. With flow under the plug, the pressure drop occurs as the flow moves through the seat to the inside diameter (an expansion), through the first stage cylinder (a contraction), through the area between cylinders (an expansion), through the second-stage cylinder (a contraction), and so forth. With this method, a portion of the pressure drop is taken at each stage. As the pressure drop is taken in stages, velocity is maintained at acceptable and reasonable levels of around Mach 0.33. The number of stages, flow areas, and flow-area ratios are determined by the velocity control required to avoid high sound pressure levels. In other words, the greater the control, the more stages and flow area that are required. The only limitations to the number of stages are the inside dimensions of the globe-body gallery and the amount of flow required to pass through the valves.

As the flow moves through the valve's vena contracta, the increased velocity, along with the geometry of the seat, creates turbulence. If untreated, this may create pressure fluctuations and eventually noise as the flow carries down the pipe. With multihole devices, the large turbulent eddy is broken up into smaller eddies. As the flow moves through the entire trim, the resulting small eddies are easily dissipated into the overall flow stream, which is illustrated in Fig. 11.54. The use

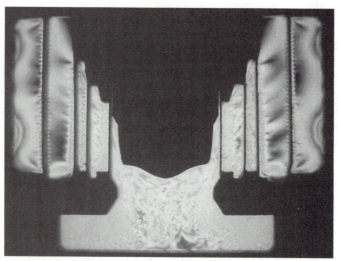

Figure 11.54 Schlieren display showing dissipation of turbulent eddies with attenuation trim. (*Courtesy of Valtek International*)

of smaller holes also decreases the noise energy significantly. If one hole in a cage generates 90 dB, studies have shown that two smaller holes (which add up to the total area of the original hole) will generate less noise—in this case 84 dB. This is due, in part, to the principle that the energy generated by noise is proportional to the square of the hole area. Therefore, using one hole instead of two will provide the desired flow, but will also double the sound pressure level. Each succeeding cylinder is designed with more or larger holes. Not only does this provide an increased flow area, but it also handles the increased gas volume that results from the pressure drop. In addition, the materials and overall design of multistage devices are selected to provide maximum acoustic impedance, avoiding any geometry that may create monopole, dipole, or quadrupole noise. This is especially important when the plug is throttled close to the seating surface where noise is most likely to occur.

When used as part of the valve's trim, multihole devices can achieve attenuation of sound pressure levels up to 15 dB for one- and two-stage devices, while multistage devices can achieve up to 30 dB. When high-pressure ratios ($\Delta P/P_1$) are greater than 0.8, the addition of a downstream element (in conjunction with the valve) can divide the pressure drop between the two. However, both should be engineered to produce the same noise level so as to not increase the overall sound pressure level. As discussed in Sec. 11.2, quarter-turn plug valves can be equipped with severe service grids, similar to multihole devices, which take an additional pressure drop and control turbulence.

As detailed extensively in Sec. 11.2.7, a tortuous-path device uses a series of 90° turns etched or machined into a stack of metal disks to slow velocity to acceptable levels. For gas service, this same device can be used, although the flow direction is opposite that of liquid applications. The flow direction moves from the inside diameter of the stack to the outside diameter. The tortuous path becomes wider and/or deeper as it progresses, widening the flow area. Each turn in the tortuous-path device is considered to be one stage. With some mazelike paths, upwards of 40 right-hand turns are possible, achieving the same number of stages and providing extremely high attenuation. Tortuous-path devices typically provide attenuation up to 30 dB.

Like the tortuous-path trim, *expanding-teeth trim* uses a stack of disks. Instead of a tortuous path, however, expanding-teeth trim uses a series of concentric grooves (referred to as *teeth*) that are machined onto both sides (face and backside) of the disk (refer again to Fig. 11.13). Flow arrives from under the plug to the inside diameter where it passes through the wavelike teeth in a radial manner. As shown in that figure, the spacing between the teeth grows significantly larger as the flow moves to the outside diameter, permitting flow expansion, increasing pressure, and decreasing velocity. In addition, as the flow moves over the grooves, the phenomenon of sudden expansions and contractions takes place, which provides staged pressure-drop reduction and increased frictional losses. Figure 11.55 shows where the fluid expansions and contractions take place as the flow moves over the grooves. One advantage of the expanding tooth design over the tortuous-path or multiple-hole trims is that its passages are wider than the beginning of a tortuous path or a hole, which allows for particulates to flow through the stack without clogging the inlet passages. Each groove (or tooth) in the stack is considered to be a stage, and in most cases this trim can have up to seven grooves, providing seven stages of pressure drop. Depending on the number of teeth in the design, this trim can provide up to 30 dB attenuation.

An antinoise trim can often be used in series with a downstream element to attenuate noise to acceptable levels. For example, when noise is close to the threshold of pain (140 dBA) and the valve cannot be removed from a reverberate chamber, such as a metal building, installing antinoise trim may make a significant reduction of up to 30 dB. However, to reduce the sound pressure level down to 85 dBA (allowing employees to work an entire 8-h shift), a downstream element must be installed to reduce the noise by another 15 dBA, which

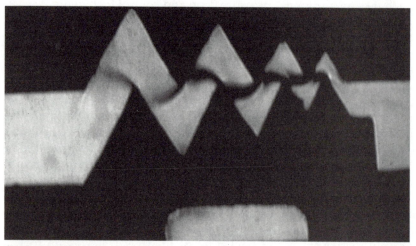

Figure 11.55 Schlieren display showing pressure reduction through sudden expansions and contractions with expanding-teeth trim. (*Courtesy of Valtek International*)

brings the noise level to 95 dBA. Insulation will most likely be needed to bring the level close to the desired 85 dBA. Although this is an extreme application, it does indicate the multiple options necessary to bring extremely high noise down to acceptable levels.

Antinoise trims with multiple stages that provide attenuation up to 30 dB are often the most expensive method of noise control. One- and two-stage devices are often less expensive but only provide attenuation up to 15 dB. The addition of a diffuser, for example, with a two-stage trim may provide the same attenuation as the more expensive multistage trim. When the required noise attenuation is 15 dB or less, in many cases a downstream element will accomplish what a two-stage trim can but at a lower cost. And in some cases, simple modifications to the process system or the orientation of the valve or changing of the pipe configuration may be even better cost-effective options—as long as the noise is only a hearing concern and is not destructive to the equipment.

11.9 Fugitive Emissions

11.9.1 Introduction to Fugitive Emissions

In many industrial regions of the world, increasing levels of environmental pollution have led to enactment of strict antipollution laws, which target emissions from automobiles, home heating systems, and industry. In particular, process industries have been under legislative mandate to reduce or eliminated fugitive emissions from their process systems. These antipollution laws target all devices that penetrate a process line, such as valves, sensors, regulators, flow meters, etc. Although many users see such legislation as costly and labor-consuming, a side benefit to tighter fugitive-emissions control is a more efficient system, with less lost product and greater efficiency. Even if a user is not under legislative mandate to reduce emissions, maintaining a strict antifugitive emissions program can provide greater production savings than the actual cost of the program. A case in point is the power-generation industry that, in the past, has accepted leakage of steam applications as standard operating procedure. Although steam (being water-based) is not a fugitive emission, power plants have discovered that using high-temperature seals prevents significant steam losses, which in turn lowers operating costs. In addition, power plants are operating more in the range of high-pressure superheated steam to improve energy efficiencies, which requires new sealing systems for safety reasons.

11.9.2 Clean Air Legislation

In the United States, the Clean Air Act was amended in 1990 to include some of the strictest laws regarding industrial pollution. In general terms, it mandates lower fugitive emissions from process equipment, including valves. Because most valves in today's chemical plants were installed prior to the new standards, maintenance personnel face a choice of retrofitting existing valves to the new standard or replacing them with new valves equipped with packing-box designs that comply with the Environmental Protection Agency (EPA). The Clean Air Act mandates a 500-parts-per-million (ppm) standard on all valves. As compared to past leakage standards, this new standard is 20 times more stringent. The Clean Air Act lists 189 hazardous materials that must be monitored by the law; 149 of these hazardous materials are volatile organic compounds (VOC), which can be easily monitored using an organic sniffer (Fig. 11.56). The Clean Air Act provides an

Figure 11.56 Organic sniffer used to detect fugitive emissions. (*Courtesy of Valtek International*)

incentive of fewer inspections if the valves are tested below the mandated 500 ppm. On the other hand, process systems with fugitive emissions higher than 500 ppm must increase the number of inspections and/or implement programs designed to improve the quality of the system.

The final phase of the Clean Air Act began in April 1997. A 500-ppm standard applies, but quarterly testing is permitted if less than 2 percent of all valves fail to meet the standard. If the failure rate is higher, monthly testing is mandatory unless a quality-improvement program is instituted. A plant can earn semiannual testing status if less than 1 percent of the valves fail to meet the standard. And finally, if less than 0.5 percent of the valves do not meet the standard, the plant can earn an annual test status. With the number of valves in a typical plant numbering in the hundreds and even thousands, achieving the higher semiannual or annual test status is important in order for the plant to avoid additional paperwork, testing, and maintenance. A graph indicating the program as outlined by the Clean Air Act is shown in Fig. 11.57.

11.9.3 Detection Standards

Clean air legislation calls for field monitoring of all line penetrations. Static seals at the flanges or body gaskets retain their seals for some

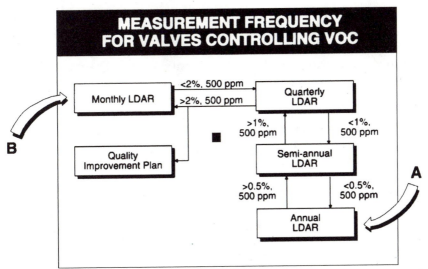

Figure 11.57 Monitoring frequency required by the Clean Air Act (United States). (*Courtesy of Fisher Controls International, Inc.*)

time. However, the sliding seal at the stem or shaft is more apt to leak because packing damage occurs over time due to friction. Because of the potential leak paths, valve packing boxes attract the most attention when fugitive emissions testing is performed.

The leak detection and repair (LDAR) procedure outlines the procedures for inspections and leak repairs. In addition to the LDAR procedure, a related regulation is "Method 21: Determination of Volatile Organic Compound Leaks" (included in App. B). In general terms, Method 21 provides leakage definitions, as well as the proper procedures for using an organic sniffer to detect a leak or measure the leakage from the valve's static seals and the dynamic seal at the packing box. With linear-motion valves, the leakage reading is taken where the rising-stem slides out of the bonnet. With rotary valves, the reading is taken where the shaft penetrates the body. Measurements are also taken at all static seals around the body, bonnet, and flange gaskets.

When metal bellows-sealed valves are used (Sec. 11.9.6), they can be equipped with a leak-detection port, which can be used to monitor any fluid leakage between the bellows and the packing box. Although a negative (no emissions) measurement can be taken at the seal, the user can also read a leak-detection gauge for visual verification that the bellows has remained pressurized.

11.9.4 Packing-Box Upgrades

The user may replace an existing valve with one that has EPA-compliant designs. However, before the valve is replaced, its design should be reviewed to determine if the valve packing box can be upgraded to an improved packing or a live-loaded configuration. Overall, upgrades are more cost effective than purchasing a newer design. However, the upgrade may affect valve performance with more stem friction that can create sticking or erratic stroking. Upgrading the valve also means that continual monitoring is required during a period of break-in. Maintenance costs will also increase.

Because the packing box is the valve's primary dynamic seal it usually receives the most attention rather than the static seals (body, bonnet, and flange gaskets). One criteria for the new packing-box design should be its ability to compensate for packing consolidation, which occurs when the packing volume is reduced by wear, cold flow, plastic deformation, or extrusion. When packing is first installed, a certain amount of space can be found between the rings. As the packing is compressed to form a seal, these gaps slowly collapse. As packing loses its seal through friction, more force is applied to once again provide a seal. After several tightenings, all available space between the rings is exhausted. The packing is now one solid block and is incapable of further compression. When continued force is applied to consolidated packing, if the packing is soft and fluid, it may extrude up or down the stem or shaft. A photograph of packing that has extruded is found in Chap. 2 (Fig. 2.42).

Since 1990 when the Clean Air Act was amended, valve manufacturers introduced a number of packing-box designs that comply with the EPA requirements, many of which can be upgraded or retrofitted into existing valves. In nearly all cases, the costs associated with upgrading an existing valve are far less than installing a new valve. The following criteria should be evaluated before determining if a valve can be upgraded to an EPA-compliant packing. First, the user should ensure that the upgrade can be accomplished easily, safely, and economically. In some cases, the valve can remain in the line while the retrofit takes place—although the line should be drained and decontaminated, if necessary, for safety reasons. In some cases, the retrofit procedure may be so complicated that the valve must be sent to the factory or an authorized repair center for the conversion. This may present a problem if the valve is a critical final element of the system or if a replacement valve is not available. Second, the user should ensure that the upgraded packing box will meet the 500-ppm standard without continual packing readjustments. In addition, the packing box should continue to perform under the 500-ppm standard for long periods of time. Third, some consideration should be given to whether the upgraded packing-box design requires new maintenance procedures or installation equipment (which may require additional training for maintenance personnel). The best solution, and the least costly, is an upgrade that permits using the original bonnet, body, stem, or shaft. If live-loading is necessary, space for the fasteners and live-loading mechanism must be available above the bonnet or body. In some cases, an upgrade requires a new bonnet for linear valves or a new body for rotary valves. Unfortunately, the introduction of these expensive new parts often increases costs so much that an overall new valve is the best option.

A careful review of an existing valve's packing-box features should be conducted to reveal upgrade possibilities and the probability of success. Some packing-box designs have features that are better suited for upgrading, while others have features that may result in leakage or premature failure. A number of design features improve the likelihood of success in upgrading packing boxes. Bonnets manufactured from forgings or barstock inherently seal better than bonnets made from castings. Although less expensive, bonnets made from castings may have minuscule cracks or porosity, which sometimes cannot be detected during manufacture and inspection without use of a dye penetrate test. The problem with these minute cracks or porosity is that leakage from these avenues cannot be halted by tightening the packing. Double-top stem guiding is commonly used in linear-motion valves to contain the packing with both the top and bottom guides. This arrangement provides a concentric and constant alignment between the plug stem and the bonnet bore. The lower guide also acts as a barrier against particulates or other impurities, which may affect the integrity of the packing. Double-top stem guiding also avoids the problems inherent to caged-guided trim, which may lead to increased fugitive emissions. The longer distance between the two guiding elements (the upper guide and the cage) allows column loading and stem flex. Plug stems with small diameters can create side loading in the packing box and possible leakage. Because the packing box itself lacks a bottom guide, particulates in the fluid can damage the "wiper" set of packing.

Deep packing boxes are designed to allow for a wider separation of upper and lower guides in the double-top stem guiding design, which provides accurate guiding of the plug head into the seat. Regarding fugitive emissions, a side benefit of a deep packing box is that it allows the upper set of packing to be completely separated from the lower set, which is designed to protect and "wipe" the fluid medium from the plug stem. This wide spacing of the packing sets avoids contact with any part of the plug stem exposed to the flowing medium. Shallow packing-box designs permit the exposed plug stem to contaminate the upper seal. A buildup of process material could also damage the dynamic seal between the stem and packing.

Packing works best with a highly polished plug stem or shaft. A typical plug stem or shaft will be approximately 8 μin root mean squared (rms). On the other hand, a static seal (such as a bonnet bore in a linear valve or a body bore in a rotary valve) would be designed with a surface finish of 32 μin rms.

If the application requiring low fugitive emissions can utilize either a linear or rotary valve, a rotary valve may be the best choice. Because of the circular action of the ball or disk, the seal between the packing and the shaft travels around the shaft circumference instead of linearly up the shaft. This shorter action produces less friction and wear and in the long term promotes packing life. Additionally, consolidation of the packing is far less because the individual rings are stressed in a tangential direction rather than an axial direction.

11.9.5 Live-Loading

Live-loading is often installed to apply a constant packing load without requiring continual retightening of the packing bolting. Live-loading is designed to compensate for packing load losses due to consolidation as well as thermal contraction and expansion. If space exists between the gland flange and the top-works of the valve, live-loading can be retrofitted on most linear and rotary valves. As illustrated in Fig. 11.58, a typical live-loading design uses disk springs above the packing flange to provide a constant load to the packing when properly torqued. The typical disk spring is a metal washer, with the inside diameter formed so that it rises higher than the outside diameter. Two disk springs are placed from inside diameter to inside diameter and stacked with other sets, allowing for a springlike configuration. Disk springs are normally made from corrosion-resistant stainless steel, although Inconel is sometimes used for highly corrosive environments.

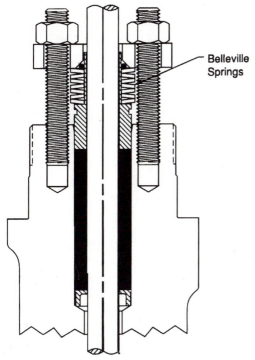

Figure 11.58 Conventional live-loading design with single stack of disk springs (*Courtesy of Fisher Controls International, Inc.*)

In live-loading, the disk springs are compressed by the gland-flange bolting, allowing a certain percentage of possible travel (typically 80 to 85 percent). As the packing volume decreases due to extrusion or friction, the disk spring's action continues to provide a load to the packing without retorquing. This is especially important since most packings can lose at least 0.02 in (0.5 mm) during the early stage of compression. Without live-loading this height loss would result in the relaxation of the packing and eventual leakage, unless the user retightens the packing. The use of live-loading compensates for this first initial loss in height. As packing settles over time, causing the springs to return to their natural position, the spring force will decrease slightly. However, the overall loss is so low that the seal is not normally affected. The amount of force applied by the live-loading can be controlled by the type of disk spring as well as the compression of the disk spring.

In addition to the reduced need for retorquing, live-loading is ideal for applications in which thermal cycling is a problem. With normal packing configurations, if the packing is tightened when the temperature is high, the packing will leak when the temperature lowers. If the packing is tightened when the temperature is low, the stem or shaft may grab or stick due to thermal expansion when the temperature increases.

Live-loading has other disadvantages than the initial cost as well as the acquisition and installation of new parts. With some valves, little or no room exists between the packing box and the top-works of the valve for upgrading to live-loading, although some manufacturers provide special live-loading configurations for limited space applications, as shown in Fig. 11.59. This design uses an upper plate as the gland flange and a lower plate as the packing compressor with stacks of disk springs located on the outside fringes of the two plates.

The torque values provided by the manufacturer to maintain the proper spring compression of the washers may be affected by the condition of the bolting. If the bolting is new and lubricated, the resulting torque value will be much different than if the threads are corroded and nonlubricated. Some packings may not respond to live-loading as well. For example, because of its high density, graphite packing requires a greater load than the manufacturer specifies for normal packings. If the live-loading is placed in a corrosive atmosphere, the disk springs can also lose strength through corrosion or even bond together, restricting free movement of the disk springs.

Some users argue that the use of live-loading actually contributes to early failure of packing through extrusion by applying more force to the packing than is required to achieve an adequate seal. If extrusion

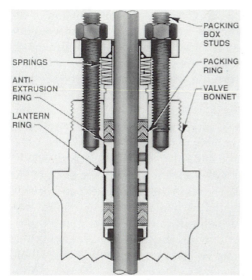

Figure 11.59 Live-loading design used for limited space applications. (*Courtesy of Fisher Controls International, Inc.*)

occurs, after some time the packing box will lose so much packing material that a seal will not be possible. Consequently, if live-loading is desired, antiextrusion rings should be included inside the packing box, especially if a soft packing is used. Too much compression may also be the problem. In that case, a thinner disk spring (which will apply less force) can be specified.

Another argument against live-loading is that, unless the live-loading provides equal amounts of force on the packing, it can cause stem-alignment problems with linear valves. This can occur if tolerance buildup occurs on some disk-spring stacks and not others, causing an unbalanced packing load and slightly affecting the stem alignment (especially with extremely thin stems or shafts which can flex). Such misalignment can affect both the shutoff and packing seal. This may be remedied, however, by using stem guides that have close-fitting guide liners or by using linear valves with oversized stems.

11.9.6 Metal-Bellows Seals

As a safety measure to workers and the general community, hazardous and corrosive applications must not be allowed to leak any fugitive emissions. In some toxic or lethal processes, however, the

Environmental Protection Agency (EPA) can designate a portion of the plant as a nonattainment area where a small amount of fugitive emissions are allowed. If a process is expanded to include more line penetrations, the parameters of the nonattainment area are often not easily expanded by regulations; therefore the user must not introduce new fugitive emissions. In this case, valves that are incapable of leaking are often required.

Linear valves equipped with a standard packing box always present a risk of leakage. When zero leakage is required, a *metal-bellows seal* is usually specified. A typical metal-bellows seal design contains the fluid with a specially formed metal-bellows welded to the stem of the plug. As shown in Fig. 11.60, the bellows is designed to expand or contract with the linear stroke of the valve, while providing a solid, permanent barrier between the fluid medium in the body and any potential leak paths to atmosphere. A metal bellows presents the best solution against fugitive emissions, as long as the body gaskets hold

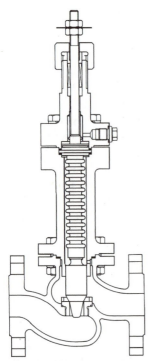

Figure 11.60 Hydroformed bellows with extended bonnet design. (*Courtesy of Kammer Valves*)

their static seals and the metal-bellows seal remain intact. Valves equipped with bellows seals do have some limitations, such as shorter strokes, decreased stroking life, and greater height. Valves with bellows seals can also cost 20 to 40 percent more, although that cost is offset by less monitoring and packing maintenance.

In throttling applications, the bellows is welded to the stem in the middle of the stroke. In the middle of the stroke the bellows is in a "relaxed" state and is equally stretched at the full-open and full-closed positions. This maximizes the life of the bellows. In applications in which a majority of the throttling is done between the 25 and 75 percent range, a bellows life of up to 200,000 strokes is possible. If a full stroke is required (0 to 100 percent), the life drops dramatically—up to 60,000 strokes. On the other hand, in applications in which the valve remains shut (or wide open) for a good portion of the time, the bellows can be welded at different locations in the plug. The bellows stays in the relaxed position for a majority of the service, prolonging its life. A metal-bellows cycle life is expressed as the number of times that the bellows can be stretched to its full limit and then compressed without failure. Because a full cycle involves a complete expansion and contraction, a bellows rated at 10,000 cycles actually translates into 20,000 full valve strokes. Because throttling service may not require a full-open or full-closed position, the bellows may be stretched or compressed less than a full stroke, which will further prolong the bellows life. The bellows life can also be prolonged by changing the tuning setting on the process controller. Process controllers can be so highly tuned that they continually search for the correct signal, sending minute signals to the valve that varies in position with each signal. Although minimal, this continual movement of the valve will shorten the overall life of the bellows. The rated bellows life number is determined by the minimum number of cycles that a bellows can withstand at the maximum operating temperature and pressure. Although a bellows is designed for the operating services, the actual operating conditions are usually less than the maximum temperature and pressure, which further prolongs bellows life. This means that the bellows life can be many more times than expected. Some applications require minimal stroke travel in a service with lower-than-rated service conditions. For example, a bellows rated at 10,000 cycles can provide beyond 100,000 strokes, given the right conditions. Table 11.14 shows how reducing the stroke by half significantly prolongs the life of the bellows.

Bellows life is also dependent on the process pressures that act on the bellows. Bellows can be designed to allow the process fluid to be

Table 11.14 Bellows Cycle Life*

Valve Size	Maximum Bellows Travel	Half Stroke (cycles)	Full Stroke (cycles)
0.5-inch	0.56 inches	1,400,000	150,000
DN 12	1.42 cm		
2-inch	0.84 inches	1,400,000	150,000
DN 50	2.13 cm		
3-inch	1.12 inches	700,000	300,000
DN 80	2.84 cm		
4-inch	1.50 inches	450,000	100,000
DN 100	3.81 cm		

Data courtesy of Fisher Controls International, Inc.

Note: Data based on single-wall formed bellows, Inconel 625 material, 100°F (39°C), and 150 psig (10.3 bar).

contained in the inside or on the outside of the bellows. However, because a bellows is harder to compress externally than to expand internally, external pressure can double the life of the bellows. A bellows typically handles process pressures from 250 to 550 psi (17.2 to 37.9 bar). It can also be designed with up to four walls, ranging in wall thickness from 0.004 to 0.006 in (0.1 to 0.15 mm)—depending on the pressure and temperature ratings. Multiwall designs provide longer cycle life, because the multiple walls all share the stress of the process pressure instead of a single wall bearing the entire stress of the pressure. Multiple walls also allow for higher pressures over single-wall designs, as shown in Table 11.15.

Although many standard bellows are designed for pressures between 250 and 550 psi (between 17.2 and 37.9 bar), severe service bellows can be designed for pressures up to 3800 psi (262 bar) and temperature ranges from −320 to 1000°F (−195 to 535°C). Both high temperatures and pressures can affect the cycle life of the bellows, as is shown in Fig. 11.61. As a safety measure, bellows-seal valve manufac-

Table 11.15 Pressure Ratings for Single- and Double-Wall Bellows*

Valve Size	Bellows Walls	100° F 38° C	300° F 149° C	500° F 260° C	800° F 427° C
0.5 - 2-inch DN 12 - 50	Single	550 psig 38 bar	497 psig 34 bar	429 psig 30 bar	396 psig 27 bar
0.5 - 2-inch DN 12 - 50	Double	1,000 psig 69 bar	870 psig 60 bar	780 psig 54 bar	720 psig 50 bar
3 - 4-inch DN 80 - 100	Single	346 psig 24 bar	296 psig 20 bar	265 psig 18 bar	245 psig 17 bar
3 - 4-inch DN 80 - 100	Double	625 psig 43 bar	544 psig 37 bar	488 psig 34 bar	450 psig 31 bar

Data courtesy of Fisher Controls International, Inc.

turers usually pressure-test each bellows seal at or over the rated service pressure.

Because of corrosion or erosion problems, the bellows is not normally placed in direct contact with the fluid; instead, it is placed just outside the flow stream, usually above the plug. A hole or a number of holes are used to allow the process fluid and pressure to bleed either to the outside or inside of the bellows. One problem that can occur with bellows pressurization is that process fluid leaving the flow stream may enter the area next to the bellows, where it cools and thickens. This can cause maintenance problems or undue bellows fatigue. In this case, external pressurization is preferred (Fig. 11.62), since cleaning the outside surfaces of a bellows during maintenance is much easier. Larger bleed holes can allow more liquid to circulate around the bellows and prevent the fluid from cooling.

Two types of metal bellows are in general use today and each is classified by its method of manufacture. *Welded bellows* (Fig. 11.63), also referred to as *diaphragm bellows*, are fabricated using a series of flat

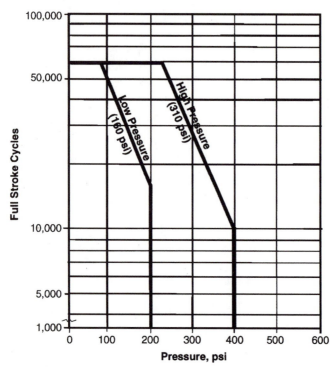

Figure 11.61 Full-stroke cycle life according to pressure. (*Courtesy of Valtek International*)

rings that are joined at the outside diameter and inside diameter by a fillerless tungsten inert gas (TIG) weld, creating a series of uniform convolutions. These convolutions have the general appearance of an accordion. Because welded bellows are made from flat rings, the overall height is quite compact and therefore can be contained in a relatively small area, adding only minimal height to the valve. For those applications requiring a small stroke, bellows can be contained inside the body (Fig. 11.64). This is particularly important where space consideration is critical or where seismic requirements limit the height of the valve's top-works. A primary disadvantage of the welded bellows is the welded edges of each convolution, which are easily stressed during expansion or contraction and are usually the first area to fail. Another problem can occur when particulates or solid matter becomes caught in the tight crevices of the convolutions. When this happens, these solids can create stress points in the convolutions and can cause premature failure. Welded bellows are also susceptible to corrosion

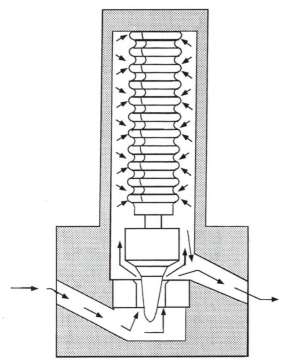

Figure 11.62 External pressurization of bellows.
(*Courtesy of Kammer USA*)

because of the thin plate used in manufacture, especially when process fluid is continually trapped in the crevices. In addition, due to the difficulties associated with welding some alloys, material selection is limited. Because of the welded edges of the convolutions, the outside diameter of the bellows may restrict their use with some valve styles.

Hydroformed bellows (again refer to Fig. 11.63) is made from a flat metal sheet, which is rolled and fusion welded for solid construction. This tube is then mechanically or hydraulically pressed to create a series of uniform corrugations. More space is required for a complete corrugation—up to three times longer than a single convolution of a welded bellows. For this reason, hydroformed bellows are much longer than welded bellows for the same stroke length. They are encased inside an extended bonnet and have a greater height than normal valves (refer again to Fig. 11.60). One important advantage of the rolled construction is that process matter does not become entrapped in the folds, as is the case with welded bellows. Generally, formed bellows last longer than welded bellows because of the minimal welding,

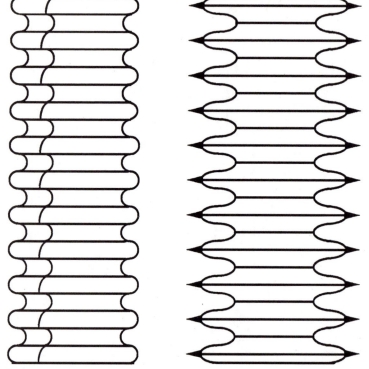

Figure 11.63 Hydroformed (left) and welded bellows. (*Courtesy of Kammer USA*)

the overall strength of the corrugations, and the limited travel of each fold (as compare to welded flat rings). They also handle higher pressures because of their greater strength. The main disadvantage is that formed bellows must be three times longer than welded bellows to handle the same stroke. The longer length may present problems with upgrading if space restrictions or seismic limitations exist.

In most designs, a packing box is placed above the bellows as a backup, in case the bellows ruptures from mechanical failure. To provide a warning of a bellows failure, a "telltale tap" can be installed in the bonnet, which is connected to an alarm system. Although not failproof, a metal-bellows seal provides the most reliable seal against leakage to atmosphere. Bellows can be made from a number of different materials, depending on the application, but 300 series stainless steels, Inconel, or Hastelloy C are standard materials because of their ability to resist stress fatigue and corrosion. Bellows can also be made from titanium, nickel, or Monel.

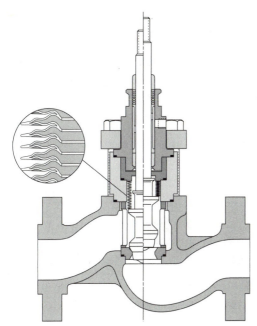

Figure 11.64 Body-contained welded bellows. (*Courtesy of Valtek International*)

A metal-bellows sealed valve may be seal-welded between the bonnet and the body as a precautionary measure with lethal or highly toxic services.

Because bellows seals are highly complex, retrofitting a linear valve is equally complex as well as very costly. In most cases, a new bonnet, plug or bellows assembly, and housing must be acquired, which can cost more than a new metal-bellows sealed valve.

11.9.7 Packing-Box Issues

When valves are initially installed in service, their packing boxes normally meet fugitive-emissions requirements. However, over time with continual operation, the packing will consolidate somewhat and begin to leak, requiring retightening of the gland-flange bolting. Most packing boxes will require retightening over time, until the packing reaches full compression. Further retightening only results in crushing the packing, rendering it useless. Manufacturers often provide suggested torque rates for given packing-box designs. This torque is applied to the gland-flange bolting, which in turn compresses the gland flange against the packing guide, finally resulting in full compression of the packing. Because of the problems associated with exact torque measurements, some designs have been simplified with the packing bolting tightened to just a flat or two past finger-tight. A manufacturer's recommended torque value can be affected by environmental corrosion or a lack of adequate lubrication, which can cause increased thread resistance and a false torque reading. Ideally, correct packing compression can be determined by measuring the packing's height when uncompressed and then applying torque until the manufacturer's ideal packing height is reached. Normally, the manufacturer's recommended packing height requires a 15 to 30 percent compression.

Maintenance technicians will sometimes overtorque the gland-flange bolting, believing that overcompression is better than under-compression. Unfortunately, too much torque can crush the packing, creating even greater leak paths. Because the packing will be compressed against the stem or shaft, high torques will boost the breakout force, causing an uneven (jerky) stroking motion. Due to the severe nature of the process or wide temperature swings, some applications require retightening often. For example, superheated steam applications may require retightening every few days. If this is not done, the packing box may develop a serious leak and be destroyed quickly by the high temperatures and pressures of the superheated steam.

The issue of torque is related directly to balancing leakage rates versus stem friction. As compression is applied to the packing by an axial load, packing deforms radially, pushing against two surfaces: the wall of the packing box and the stem or shaft. With greater compression, the greater the stress will be applied against these surfaces. As the packing deforms against the wall, any voids are closed off, permitting an effective seal. However, more compression also increases stem friction as the inside diameter of the packing grips the plug or shaft stem. This leads to erratic stem movement. Conversely, if the force to the packing is decreased to allow for smoother stroking, the packing may not fully grip the stem or shaft and leakage can occur.

Another factor that plays an important part in packing-box friction is the amount of contact between the packing and the stem or shaft. As the surface area of the packing touching the sliding stem or rotating shaft increases, more friction is produced that must be overcome to produce movement. High levels of friction will require greater force by the actuator, or a longer lever or larger diameter handwheel with manual valves. Some packings are V-shaped, which provides a very narrow point of contact and generates minimal friction. On the other hand, square packing (such as graphite) provides full contact and increased packing friction.

Most valve manufacturers today provide highly polished valve plug stems or shafts to accommodate the dynamic seal between the inside diameter of the packing and the stem or shaft. However, valve stroking or the service conditions of the process itself can deform, pit, or corrode the stem or shaft. Such wear can significantly increase the stem friction, while decreasing valve performance. This problem is often compounded when the seal is leaking and additional torque is applied to the packing to stop the leak. During routine maintenance, the stem or shaft should be carefully examined to ensure a smooth surface finish. If the finish is not smooth, that part should be replaced or repaired if the scratches or pits are not too deep.

11.9.8 Packings Specified for Fugitive-Emission Control

Today's packing materials are well suited to control fugitive emissions and can be adapted for retrofitting. Although no packing material or design is universal, many different packings exist that have broader applications than in years past. Choosing the correct packing is critical to the successful performance of the packing box. The packing should be compatible with the process fluid and service temperature and pressures as well as provide the desired seal between maintenance checks, without excessively high torque of the gland-flange bolting. The proper packing should also withstand consolidation and should minimize the friction on the stem or shaft, avoiding poor stroking performance.

A number of packing materials are commonly applied to anti-fugitive-emission packing boxes. Recently introduced in the past several years, perfluoroelastomer (PFE) packing is generally regarded by valve users as the best packing for complying with fugitive-emission standards. PFE provides an excellent seal with even the most difficult application. It resists degradation and chemical attack and is very resilient and elastic. PFE is rated to handle service temperatures from 20 to 550°F. A special low-temperature PFE has been developed that handles temperatures down to −40°F. As shown in Fig. 11.65, PFE requires a rigid backup V-ring system to support the packing. A property unique to PFE is that it wears well and compensates for any consolidation that takes place—although PFE can consolidate and eventually extrude if not supported by backup rings. Live-loading is not normally required, since PFE has an ability to return to its precom-

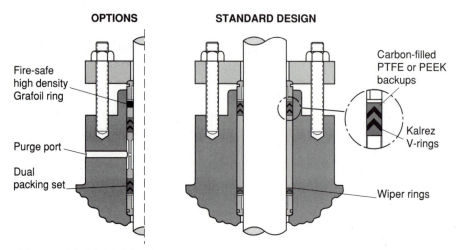

Figure 11.65 PFE backup ring packing configuration. (*Courtesy of Valtek International*)

pressed position. However, in applications with large thermal gradients, live-loading should be considered. The chief disadvantage of PFE is its high cost, although the initial cost of the packing is easily offset by reduced maintenance and increased up-time.

One of the most common and least expensive packings, "virgin" polytetrafluoroethylene (PTFE) packing is typically applied in a V-ring design (see Fig. 2.33 in Chap. 2). Virgin PTFE is often chosen because it has numerous advantages. Due to its pressure-energized design, coupled with "feather" edges, little compression is required to create a strong seal. Overall, virgin PTFE has good elasticity, which minimizes packing consolidation while responding well to a live-loading option. Because it is inert to many chemicals, virgin PTFE is found in a wide range of process services. The surfaces of virgin PTFE are extremely smooth; therefore, little breakout force is required to begin stroking the valve. Despite its wide application, virgin PTFE has some disadvantages. Its performance is limited to temperatures between −20 and 350°F. If the packing bolting is overtorqued—providing an excess load on the packing—the voids between the male and female rings can compress and result in consolidation. In addition, the spaces between packing spacers and the plug stem can result in extrusion, although antiextrusion rings or close-fitting spacers can be installed to prevent extrusion. Because of its tendency to cold flow and consolidate over time, virgin PTFE does require retorque on occasion.

The composition of "filled" PTFE contains 15 to 20 percent glass or carbon, which creates a more rigid V-ring design that is less likely to produce consolidation (which is common to virgin PTFE). Because its elasticity is less, filled PTFE does not seal as well as virgin PTFE. It also produces greater friction and is slightly abrasive to the stem or shaft. And, it is more expensive than virgin PTFE. Sometimes, as a compromise between virgin and filled PTFE, rings of both materials are alternated in the packing configuration to provide a good seal with reduced consolidation. Live-loading can also be used with filled PTFE to minimize retorquing.

Graphite and other carbon-based packings are commonly manufactured in die-formed or straight braided carbon-ring sets. As a measure against graphite migration, braided rings are often included in die-formed packing sets. This feature also protects the graphite rings from foreign particles. Braided rings are known to cause additional friction and leakage in high-compression applications. The main advantage of graphite packing is its ability to handle high temperatures (up to 800°F with a standard-length bonnet in an oxidizing environment). Graphite packings are usually offered in low-density or high-density graphite. Low-density graphite seals well and has lower friction, but must be retorqued often. High-density graphite has higher friction and provides a marginal seal but allows for a longer retorque cycle. To convert low-density to high-density packing, the packing can be torqued several times over a period of time. Compared to other packings, graphite packings are more expensive and do not respond well to live-loading systems. Also, the higher friction can affect the performance of the valve, requiring high breakout forces that may result in unstable stem movement. Typically, torque requirements for graphite ring packings can be eight to 10 times higher than those of PTFE or PFE packings. This usually requires the use of a torque wrench to ensure that overcompression does not occur. Overcompression will crush the graphite, causing it to extrude from the packing box.

11.9.9 Other Packing Considerations

Some users believe that if using the standard number of packing rings provides a good seal, using more rings should provide an even better seal against fugitive emissions. If a packing box is exceptionally deep, a user may be tempted to double the number of rings during routine maintenance. However, the use of extra-ring compounds several problems. First, multiple rings maximize the adverse affects of thermal expansion of the packing. Second, they increase stem friction substantially. Third, the manufacturer's recommend torque values will now be incorrect, providing far less compression than required. This may necessitate a trial-and-error approach to determining the correct torque value, which could shorten the life of the packing. Fourth, with more soft packing material in the packing box, unnecessary consolidation and extrusion can take place.

With rotary valves, the closure or regulating element or the actuation unit can apply stresses to the shaft, causing an incorrect center alignment. If a small-diameter stem (linear valves) is used, the force applied by the actuation to the closure element in the seated position can actually flex the stem. Whenever the stem or shaft are off-center with the packing box, a leak path for fugitive emissions can occur on one side. This problem can usually be avoided by using valves that feature oversized stems or shafts. Oversized stems or shafts present a large contact area between the stem or shaft and the packing, which will result in higher friction—although this is not an issue with high-thrust actuation units, such as piston cylinder actuators.

12

Valve Purchasing Issues

12.1 Life-Cycle Costs

12.1.1 Introduction to Installed Cost

Some manually operated on–off valves are considered to be an off-the-shelf commodity—valves that are purchased, rarely serviced, used until they fail, and then discarded. On the other hand, more complicated valves—such as throttling valves—have a number of cost factors that should be considered in addition to the initial cost of the valve. These critical cost factors are considered the *true installed cost over time.* This terminology is commonly used to describe all the costs related to the valve from the point of purchase to final replacement. The true installed cost of a valve includes the following factors: the purchase price, installation and startup costs, training of technical and maintenance staff, maintenance frequency, the complexity and length of time required to perform routine service on the valve, and the cost and availability of spare parts. Another factor that can affect the true installed cost is the overall *reliability* of the valve or the valve's ability to stay in service between routine maintenance without unplanned failure.

12.1.2 Maintenance and Reliability Issues

Less time spent servicing a valve means money saved over the life of the valve. Compared to overall maintenance costs associated with a valve over its life, the initial purchase cost may not be very significant. That is a primary reason why buyers should confer with the technicians who service and repair the valves. Often the valves with the lowest purchase price are those that take longest to repair and require more frequent repairs. What is saved in the initial purchase is quickly surpassed by increased maintenance costs. The valve purchaser should always consider the degree of difficulty and length of time it takes to service or overhaul a valve. Factors affecting the ease of maintenance include accurate and easily understood literature, designs that permit easy maintenance and calibration, accessibility of technical assistance, frequency of maintenance, cost and availability of spare parts, effect that downtime has on production, and emergency repairs.

Maintenance and service instructions should be provided with the delivered valve. They should be written in a simple and straightforward style, using a thorough, detailed step-by-step process, and should be illustrated with drawings and photographs. The literature should also be written and illustrated for the specific model being serviced, and include instructions for installation, disassembly, and reassembly as well as periodic maintenance schedules, troubleshooting charts, torque values, recommended spare parts, warning or caution notes, and special tool requirements. Some manufacturers offer videotapes or CD-ROMs that complement the written instructions and are helpful for training maintenance technicians.

The basic valve design can influence maintenance costs. Several valve-body designs, such as the split-body or butterfly design, must be totally removed from the line for disassembly, which adds to the time required for servicing. Top-entry valves allow access to the trim while allowing the body to remain in-line, reducing the labor hours and process downtime. Complex valves (such as special service control valves) have a higher number of parts and accessories, which can lengthen the disassembly and reassembly time process. Occasionally, disassembly and/or reassembly of complex valves requires special tools, which may be purchased from the manufacturer (adding to the true installed cost of the valve) or may be included with the initial valve purchase.

Several valve designs have been redesigned over the years to simplify maintenance. For example, many older linear valve designs were built with fixed threaded seats, which require lapping of the plug and seat to reach the required shutoff. Today, recent trim designs permit the seat to be self-centering, which is faster and easier to reach without lapping. Another time-saving feature offered in some designs is the metal-to-metal (bonnet to body) gasket compression, which eliminates the need for torque values and wrenches to ensure proper gasket sealing.

The actuator or actuator system should be reviewed carefully before purchase. Some positioners and actuators are designed to be calibrated once and left to function. Others require a periodic follow-up calibration, which adds to the maintenance function. Process problems (such as vibration) can occasionally disrupt the calibration of positioners and actuator accessories that are extremely sensitive. The process engineer or maintenance technician should review the calibration process for simplicity. If calibration is complex, instructions should be available that are thorough and fully illustrated, or training should be available.

The costs and requirements of normal maintenance—which requires the use of spare parts—may vary significantly between manufacturers. The costs of two types of spare parts should be considered: first, the cost of those parts replaced during normal maintenance (gaskets, packing, O-rings, diaphragms, etc.), and second, the cost of less frequently replaced parts (valve trim, clamps, bonnets, etc.). Because of the high cost of spare parts associated with original equipment manufacturer (OEM) parts, many users have resorted to purchasing imitation parts. (For a detailed discussion of the merits of using imitation parts, refer to Sec. 12.2.) The buyer should ensure that spare parts needed for maintenance, as well as operational spares, are easily available from the closest OEM source. Quick accessibility or delivery of spare parts can eliminate lengthy downtime or the need for carrying a large inventory of spare parts.

Maintenance is normally scheduled during routine downtimes—assuming the valve is reliable. Buyers should look at reliability as an essential element when purchasing control valves. An unreliable control valve, with its associated unexpected and higher maintenance costs, can interrupt the production process and be extremely costly, especially if the spare parts are not readily available or the overhaul time is quite lengthy. Although valves are designed to provide reliable service between scheduled maintenance, an unexpected problem will occasionally occur. A valve's failure could be attributed to a damaging process, a flaw in material or workmanship, a loss of signal or power, or recalibration. Whatever the cause, a particular valve model with a reputation for unreliability can significantly affect the total installed cost of the valve. An emergency repair results in the additional cost of the loss of production, cost of spare parts, and the labor required. In some systems, downtime can be avoided if a bypass system is provided to circumvent the critical valve. This option is best when minimal work, such as adjusting a packing box, is required on the valve. Another valve may be available to temporarily replace the one requiring service. This allows for minimal downtime of the system, especially if the repair process is lengthy, such as replacement of the trim. Unfortunately, the cost of having a spare replacement valve available is usually only cost-effective if the process requires a number of identical valves or if the valve in question is an inexpensive commodity valve.

Another option is to have the valve serviced by an outside source. Oftentimes, when the user has limited internal resources to repair the valve, the original manufacturer, a manufacturer's authorized repair facility, or an independent repair facility is available to perform the necessary work. Outside sources can repair or recondition the valve in the required timeframe at labor costs similar to or even less than those incurred by using internal resources. Outside sources are typically well versed in proper repair procedures, have the tools available for repair, and have wide inventories to draw from. The manufacturer's local representative or distributor should be available to provide additional service or knowledge. In start-up or shutdown situations, factory-trained assistance is often necessary, requiring the valve manufacturer to provide on-site support, such as application engineers or field technicians. Some valve manufacturers also offer on-site repair and/or inventory management, relieving the user of that responsibility.

12.1.3 Valve Life-Cycle Cost: Example A

Process plant A overhauls roughly 15 percent of its installed control valves on an annual basis. If the plant is designed for a 40-year life, a 15 percent frequency interval indicates that each valve would be overhauled six times during the life of the plant. Plant A's cost of maintenance spare parts is 5 percent of the purchase price of the valve. The designated valve is an off-the-shelf (nonspecial) 2-in, carbon steel globe control valve, with an average price of U.S.\$2200. For such a valve, the assumption is made that 12 hours are required to remove, overhaul, calibrate, and reinstall the control valve. The plant's hourly labor burden rate is U.S.\$45/hour. Over the first year's life of the valve, the total direct costs associated with the valve are

Purchase price of valve	U.S.\$2200
Spare parts (\$2200 × 5% × 6)	660
Maintenance (12 × \$45 × 6)	3240
True installed cost	U.S.\$6100

This example represents the net present value because the cost of inflation is not added into the spare-part or maintenance costs. In addition, the example does not take into account the loss of product if the system must be shut down or cannot be bypassed for the required maintenance. According to this example, the valve's original purchase price represents only 35 percent of the total installed cost. Note that over the full 40-year life of the valve, the spare-part cost would be $26,400 and the maintenance labor would be $129,600.

When compared against the life of the valve, the cost of the valve in this example is insignificant. Yet today, many plants purchase valves solely on the initial price without taking into account other nonpurchase cost issues. The second example illustrates the impact an inexpensive, high-maintenance valve has on the true installed cost.

12.1.4 Valve Life-Cycle Cost: Example B

This example examines the impact on the "total installed cost" if a less expensive valve, which requires a more frequent maintenance schedule that the valve used in Example A, is purchased. This is illustrated by assuming a 15 percent price decrease on the purchase price of the valve, with 25 percent of the plant's valves requiring maintenance annually (10 times during the life of each valve). This less-expensive valve changes the numbers drastically:

Purchase price of valve	U.S.$1870
Spare parts ($1870 × 5% × 10)	935
Maintenance (12 × $45 × 10)	5400
True installed cost	U.S.$8205

This example illustrates the point that the purchase of a less-expensive, high-maintenance valve, actually resulted in a 35 percent increase to the installed cost over the life of the valve—even after paying 15 percent less for the valve.

When examined over the entire 40-year life, the savings of using the more expensive and reliable valve would be nearly $100,000. Multiplied by hundreds of valves in a typical process plant, the savings can be significant. As illustrated by the two examples, the total spare-part and maintenance costs of a particular valve design should always be considered before the final valve purchase decision is made.

12.2 Spare Parts

12.2.1 Introduction to Spare Parts

The cost of spare parts, and the associated quality of parts, should be an important calculation into the overall life-cycle cost of the valve.

Consideration should be given to the number of recommended spare parts, availability, and parts interchangeability. Valve manufacturers typically supply a list of recommended spare parts as part of the paper work included with the valve shipment. This list includes soft goods (especially dynamic soft goods, such as the packing, a soft seat insert, or an actuator diaphragm) and hard goods that are subject to friction wear, impact, or fluid dynamics (such as a closure element or guides).

With escalating costs associated with keeping large spare-part inventories, users may look to the local supplier, distributor, representative, or the original equipment manufacturer to supply the needed parts, using a just-in-time delivery system. Unfortunately, this system relies heavily upon the inventory of the local supplier and is no guarantee of immediate delivery. Most users keep a necessary supply of critical soft goods and anticipate the requirements of hard goods through a periodical maintenance program. Users who maintain in-house parts inventories benefit through those valve designs that offer a wide interchangeability of spare parts between sizes or models. When a spare part can be used in more than one size or model, inventory costs can be reduced.

12.2.2 Spare-Part Concerns

Since the late 1960s, some valve users have achieved cost savings on escalating spare-part costs by using imitation valve parts (commonly called "pirate" parts). Unfortunately, the user generally finds that money saved by using imitation parts is lost through increased downtime, marginal performance, and lost production. As this debate of imitation versus OEM parts has grown, three main issues have surfaced for valve users.

First, by virtue of appearance, no significant difference seems to exist between OEM and imitation parts. In many cases, the user simply cannot see any difference between an OEM part and an imitation part—visually, both appear identical. The assumption is made that if both appear similar, then their performance must be identical. Unfortunately, performance cannot be judged entirely by overall dimensions, surface finish, or general appearance. Unless an imitation parts producer pays a licensee fee to the manufacturer, it will not have access to the original design drawings and material codes. Therefore, the imitation-part producer must make its drawings and material selection based upon field examination. This may take place when the user allows the imitation producer access to the user's spare-part inventory. However, in most cases, the imitation producer merely collects discarded or obsolete OEM parts and makes the necessary measurements and material selection from those parts.

Because of their non-OEM origins, imitation parts, and their dimensions, materials, flow characteristics, and tolerances can vary significantly from the OEM parts. In addition to performance differences, these discrepancies can also create mechanical difficulties with the existing OEM parts. Because OEM parts are often engineered and manufactured in tandem with each other, performance and reliability are dependent on the use of OEM spare parts. These differences can result in inaccurate process control, decreased safety, and increased service costs.

The second issue is one of cost. The purchase price of imitation parts is seen as inexpensive when compared with OEM parts. Some valve users claim that OEM parts cost too much—that by using imitation parts, savings of up to 30 percent or more are possible. In reality, the cost of a product is more than the cost of the material and time it takes to make it. OEM companies have larger costs associated with doing business because they typically provide a wider range of services to the user—the cost of which is included in the cost of OEM products, including spare parts. For example, the OEM performed the original application engineering function. In this age of complex processes, fewer applications are considered standard—pressures, temperatures, and fluids can vary widely according to the user's process. In order to match equipment with the need, the OEM company must examine the application thoroughly to specify the correct equipment or spare parts for that particular application. Occasionally, the user's application will surpass the normal limits of the OEM equipment, such as with high velocities, extreme temperatures or pressures, severe cavitation, or high noise levels. In these situations, the OEM must use additional engineering time and effort to engineer, design, and manufacture the special equipment to meet these applications. The cost of certified dimensional drawings is also included in the OEM product cost. Prior to installation, the user often requires the OEM to provide certified dimensional drawings to plan piping schemes and room requirements. Most OEM manufacturers provide training at the factory or on-site to acquaint the user with the operation and service of the equipment. With the complexities of today's systems, the user often requires extensive testing to prove the performance and reliability of the equipment. The time and effort involved with a plant start-up are often minimized with OEM engineers available on-site, who make application recommendations and changes as process requirements change. As a general rule, an imitation-part producer does not offer these services because the user has already accepted them from the OEM user. Therefore, its overall cost of doing business—as well as the product cost—is far less than the OEM.

Because today's valve market is very competitive, the OEM expends a portion of its business costs with product promotions and visits to user plants. Although some users see these costs as unnecessary, good business decisions can be based only on strong communication between the vendor and the user. An imitation-part company does not normally promote its business because its day-to-day business depends solely on the success of the OEM. It cannot sell imitation parts until the user purchases the OEM valves.

The third issue is that imitation parts are seen as more readily available and accessible since most are manufactured by a producer located near the plant. Overall, the range of imitation parts is limited. The imitation-part producer cannot supply the user with all spare-part needs. It only manufactures parts that it can build easily without costly and extensive tooling or engineering expertise. It normally does not provide small or inexpensive nonprofitable parts, such as positioner parts, O-rings, nuts, bolts, springs, stroke plates, tag plates, and warning stickers. It is not interested in providing parts requiring extensive tooling, such as yokes, cylinders, diaphragm cases, bodies, rotary balls, and rotary disks. It is especially not interested in providing highly technical parts, such as trims that handle high pressure drop, cavitation, or high noise levels. Such parts require extensive engineering expertise, which is unavailable to the imitation part manufacturer.

12.2.3 Manufacturing and Quality Issues

The method of manufacture and overall quality is what widely separates the OEM spare parts from imitation parts. Because the OEM deals with large volumes of parts, it normally uses computer-aided manufacturing techniques for high production runs. Using these state-of-the-art manufacturing techniques allow the OEM to produce higher quality parts at lower costs than manual manufacturing methods. On the other hand, the imitation-part producer deals with lower volumes and cannot justify the use of more expensive manufacturing equipment. Because manual machining is used, the same parts may vary widely from the OEM part. For example, an OEM may typically offer full penetration welds between the plug head and the stem, as well as a solid hardened alloy plug head. Although computer-controlled welding equipment is required for such manufacture, because of low volume, imitation producers can only manually weld a hardened alloy overlay. The imitation-part producer also uses more basic technology, such as simple fillet welds, which may cause premature failure.

Figure 12.1 graphically illustrates this difference in two plug and seat sets. Both sets were used in the same angle-style valve in a high-velocity bleaching clay application. Because of the severe service, the OEM parts (shown on the left) were removed during periodic maintenance and replaced with imitation parts (shown on the right). Because of bleaching clay's erosive and corrosive characteristics, the OEM plug was manufactured from 316 stainless steel with a full-bore overlay of a hardened alloy. The OEM plug and seat were placed in service for six months and were functioning as expected when pulled during routine maintenance. They were replaced by an imitation plug and seat. The imitation trim began to deteriorate the next day and was replaced on the third day. A second imitation trim set was then installed. The failure occurred again within three days. An examination by the user indicated that the hardened alloy application was extremely thin in some places on the plug due to inexact machining. Further tests indicated that the hardness was a soft RC-5, showing that the hard-facing had worn off and exposed the core material to the severe process condi-

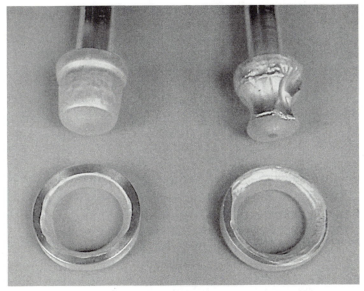

Figure 12.1 OEM (left) and imitation trim parts following service in a bleaching clay application. (*Courtesy of Valtek International*)

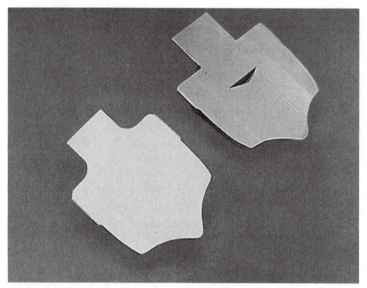

Figure 12.2 Internal construction of OEM (lower left) and imitation plug heads. (*Courtesy of Valtek International*)

tions. Conversely, the OEM plug that was replaced was tested to RC-35—even after six months of service.

Figure 12.2 shows another example of poor quality of imitation parts. Shown on the left is a globe valve's plug manufactured by the OEM. The stem and plug head were friction-welded to bond the stem and the plug head together—a technique that provides a bond similar to one-piece construction. The OEM plug design also provided a radius between the stem and plug head to allow a lower stress concentration across the entire area of the plug stem and the radius. In comparison, the imitation plug (shown on the right) was produced in a different manner. The plug head was drilled with the stem inserted into the hole. The two pieces were then welded together using a simple fillet weld, resulting in some porosity. (This construction is illustrated in Fig. 12.3.) If the fillet weld does not meet industry standards for full penetration welds, the weld will most likely fail in service eventually. The imitation-part manufacturer also failed to include the radius between the plug head and stem. This oversight increases the stress loads by up to a factor of 3, which will cause stress cracks and eventual failure.

From a quality standpoint, an OEM part is manufactured using repeatable industry standards, allowing the same part to be used uni-

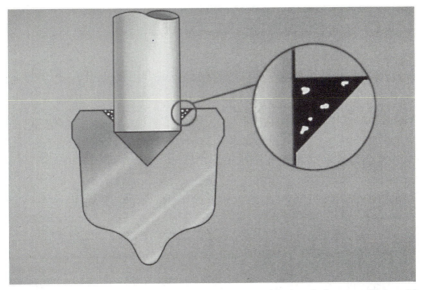

Figure 12.3 Internal construction of imitation plug head showing fillet weld between head and stem. (*Courtesy of Valtek International*)

versally in any identical valve. For example, in globe valves, plug heads can be used to provide the flow characteristic for the valve. The shape of the plug head is critical to the exactness of the flow characteristic and is typically produced using computer-generated templates to ensure repeatability. Unless the imitation-part manufacturer has a license for the OEM technology, it can only trace an existing plug to develop its own manufacturing template. This inexact method can create discrepancies, especially if the plug that was used is obsolete or taken from service in a worn condition. Even if the template was hand-traced from a new OEM part, a 0.01-in difference—the width of a pencil point—can change the flow characteristic significantly in small C_v applications. The difficulty in manufacturing an accurate flow characteristic is illustrated in Fig. 12.4, which shows three plugs. The bottom plug was produced by the OEM, while the other two were imitations. The imitation plugs have plug contours significantly different from the OEM plug, which will most likely result in a different flow characteristic.

Another important consideration is that an OEM may upgrade the performance of a particular valve design over time. This improvement may affect the material selection, design parameters, or function of a particular part. Unaware of this change, the imitation-part producer

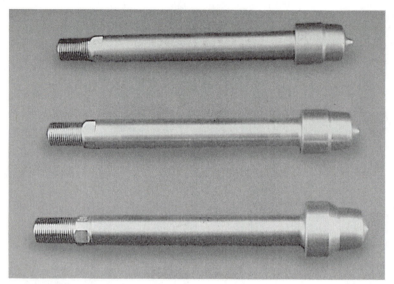

Figure 12.4 Plug contour variations between OEM (bottom) and imitation plugs. (*Courtesy of Valtek International*)

will continue to offer the previous, obsolete part. This may cause undue performance and maintenance problems that were previously solved by the upgraded OEM part. Worse yet, the obsolete part may not perform correctly in the new design. An example of this situation is shown in Fig. 12.5. The improved OEM plug head design (shown on the left) has a new flow-characteristic contour, as well as a new back angle and contour, and a smaller outside diameter. However, the imitation part (right), which was manufactured after the OEM improvement, shows continued use of the obsolete design. Although both the OEM and the imitation parts had identical part numbers, they had far different flow characteristics when flow-tested.

12.2.4 Warranty Issues

Most valves built by established OEMs are supplied with some type of manufacturer's warranty. A typical valve warranty will guarantee the performance and life of the valve for a certain period of time, as long as it is used in the service it was designed or specified for. If the valve fails while the warranty is in effect, the OEM normally agrees to repair or replace the defective parts or the valve itself. To protect itself against the use of imitation parts, the OEM may invalidate portions of

Figure 12.5 OEM (right) and imitation plug heads showing variations in back angle, characteristic contour, and head diameter. (*Courtesy of Valtek International*

a valve's warranty—or sometimes the entire warranty—if imitation parts are used. The position of the OEM is that it is not obligated to honor a warranty or a portion of it if an OEM valve is maintained with another manufacturer's part that can affect performance or lead to damage. As an example, the user replaces an OEM gasket with a generic gasket provided by a local gasket supplier. Because the generic gasket has a ± 0.01-in tolerance on the gasket thickness (as compared to the ± 0.005-in tolerance required by the OEM design), gasket compression is not fully achieved using the recommended torque value. This results in the gasket failing, causing a process leak between the bonnet and body, which may erode the metal surfaces on the gasket step. In this case, the OEM warranty may not cover the damaged OEM parts—the body and bonnet—because the non-OEM gasket caused the failure. Overall, the use of imitation parts in OEM equipment may cost the user more in terms of subsequent replacement parts than the added cost of OEM parts, as well as a loss of the warranty.

12.2.5 Material Issues

Valves built for special services may have a material selection that includes special alloys. To ensure that the correct materials are used with the valve, most OEM procedures include material identification (see Fig. 2.28 in Chap. 2) to verify materials according to the user's original specifications. Imitation-part manufacturers must ask the user for the material data, which can be wrong or subject to supposition. The imitation-parts producer may not have an exact alloy in stock, but may choose to supply a close material. In some cavitating or corrosive services, even a slight material variation may cause valve failure.

Bibliography

Books

Hutchinson, J. W., *ISA Handbook of Control Valves*, Pittsburgh: Instrument Society of America, 1976.

Kane, Les, *Handbook of Advanced Process Control Systems and Instrumentation*, Houston: Gulf Publishing, 1987.

Knapp, Robert T., Dailey, James W., and Hammit, Frederick G., *Cavitation*, Iowa City: Institute of Hydraulic Research, University of Iowa, 1970.

Lyons, Jerry L., *Lyon's Valve Designers Handbook*, New York: Van Nostrand Reinhold, 1982.

Merrick, Ronald C., *Valve Selection and Specification Guide*, New York: Van Nostrand Reinhold, 1991.

Pearson, G. H., *Valve Design*, London: Mechanical Engineering Publications, 1978.

Smith, E., and Vivian, B. E., *An Introductory Guide to Valve Selection*, London: Mechanical Engineering Publications, 1995.

Tullis, J. Paul, *Hydraulics of Pipelines: Pumps, Valves, Cavitation, Transients*, New York: Wiley, 1989.

Ulanski, Wayne, *Valve and Actuator Technology*, New York: McGraw-Hill, 1991.

Zappe, R. W., *Valve Selection Handbook*, 3d ed, Houston: Gulf Publishing, 1991.

Encyclopedia References

"Irrigation," in Grolier Multimedia Encyclopedia, Grolier Inc., 1995.

"Newcomen, Thomas," Grolier Multimedia Encyclopedia, Grolier Inc., 1995.

"Steam Engine," Grolier Multimedia Encyclopedia, Grolier Inc., 1995.

"Valves," Grolier Multimedia Encyclopedia, Grolier Inc., 1995.

Periodicals

Abboud, Jason S., and Steinke, Joseph H., "The Dangers of Counterfeit Valve Parts," *Valve Magazine*, Spring 1995, pp. 18–20, 50–51.

Adams, James W., "Clean Air Act Update: Fugitive Emissions and Valve Standards," *Valve Magazine*, Summer 1995, pp. 10–12.

Arant, James B., "Control Valve Notes and Positioner Myths," *Control,* November 1993.

Arant, James B., "The Pneumatic Actuator Argument: Piston or Diaphragm?" *Control,* October 1990.

Beasley, Marvin E., "Choosing Valves for Slurry Service," *Chemical Engineering,* September 1992.

Beasley, Marvin E., "Look Hard at Ball Valve Materials for Abrasive Chemical Processes," *Chemical Engineering Progress,* December 1992, pp. 73–76.

Beasley, Marvin E., "Transient Effects on Valve Performance," *Valve Magazine,* Spring 1995.

Blickley, George J., "Valves Unite with DCSs to Get Smarter," *Control Engineering,* September 1994, pp. 103–104.

Brailey, Edwin J., and Miller, Herbert, "Control Valves Limit Turbine Temperature Swings," *Power Engineering,* April 1991, pp. 47–50.

Bravenic, Edward, "Metallurgical Failure Analysis of Valves & Valve Components," *Valve Magazine,* Fall 1995, pp. 10–16.

Brockbank, Grant, and Glenn, Alan, "Compressible Flow Noise," in *Valve Magazine,* Fall 1992, pp. 12–16.

Burns, Ralph, "Four Industry Buzzwords: Are They Meaningful?" *Chemical Processing,* October 1992.

Butcher, Charles, "Applying Intelligence," *Chemical Processing Technology International,* 1994, pp. 179–182.

Dana, Christopher A., "Consider Quarter-turn Gates for Gate Valve Applications," *Hydrocarbon Processing,* June 1993.

"Fieldbus Emerges as the Key Driver in New Alliances of DCS Vendors and `Smart' Control Valve Vendors," *Control Engineering,* March 1994.

Fitzgerald, Bill, "Keep Valve Top Works From Becoming a Bottom Priority," *Chemical Engineering,* December 1995, pp. 74–81.

Fusaro, Dave, "Lack of Fieldbus Standard Makes Digitizing a Question," *Control,* August 1992, pp. 28–30.

Hart, David, "Pitfalls of Gasket Selection," *Valve Magazine,* Summer 1995, pp. 18–23.

Katz, Bob, "How Drag Technology Contributes to Reducing O&M Costs," *Nuclear Engineering International,* July 1993, pp. 50–52.

Laing, David E., Miller, Herbert L., and McCaskill, John W., "Redesigned Recycle Valves Abate Compressor Vibration," *Oil & Gas Journal,* June 5, 1995, pp. 46–51.

Larsen, Keith, "Time is Finally Ripe for Accepting Smart Control Valves," *Control,* March 1994.

Lueders, Wallace M., "Safety Valve Maintenance: A Key to Plant Safety," *Valve Magazine,* Fall 1995, pp. 26–38.

Miller, Herbert L., "Controlling Valve Fluid Velocity." *INTECH,* May 1993, pp. 32–34.

Paul, Brayton, "Fire-safe Stem Packing Meets Emissions Regulations," *Chemical Processing,* February 1995.

Paul, Brayton, "Safety Valves Provide Zero Leakage," *Chemical Processing,* June 1995, pp. 40–41.

Paul, Brayton, Winkel, Laren, and Wolz, Darren, "Controlling Fugitive Emissions by Retrofitting Valves," *Chemical Processing,* April 1995.

Ritz, George, "Advances in Control Valve Technology," *Control,* March 1993, pp. 32–35.

Ritz, George, "Control Valve Actuator Options," *Control,* June 1994.

Ritz, George, "Exploring Control Valve Actuator Options," *Control,* July 1994, pp. 44–47.

Ritz, George, "Tracking Down Fugitive Emissions," *Control,* August 1995, pp. 32–38.

Skousen, Philip L., "Imitation Valve Parts: You Get What You Pay For," *Valve Magazine,* Winter 1995.

"Valve-Actuator Assemblies Key to Improved Gas Turbine Operation," *Diesel & Gas Turbine Worldwide,* April 1994.

Warnett, Chris, "Automating Valve Operation," *Plant Engineering,* July 10, 1995, pp. 72–74.

White, Barry A., "Rethink the Role of Control Valves," *Chemical Engineering Progress,* December 1993, pp. 31–34.

Wolz, Darren, "How to Evaluate Actuators for Automatic Control Valves," *Power,* August 1989, pp. 47–53.

Wolz, Darren, "Understanding Control Valve Seat Leakage Classes," *Power Engineering,* March 1990.

Privately Issued Publications

Control Valve Handbook, Marshalltown, Iowa: Fisher Controls International, 1977.

Control Valve Sourcebook, Marshalltown, Iowa: Fisher Controls International, 1990.

Guidelines for Controlling Fugitive Emissions with Valve Stem Sealing Systems, Park Ridge, Illinois: Society of Tribologists and Lubrication Engineers, October 1992.

1994 Industrial Valve and Actuator Data Book, Washington D.C.: Valve Manufacturers Association of America.

Sizing and Selection Manual, Springville, Utah: Valtek International, 1996.

White Papers

Beatty, Ken, "Benefits of Process Measurement and Intelligence in the Final Control Element," Springville, Utah: Valtek International.

Cain, Fred M., and Nelson, Michael P., "Valtek Guide to Cavitation in Control Valves," Springville, Utah: Valtek International.

Johnson, Bruce A., "Digital Communications with...Control Valves," Marshalltown, Iowa: Fisher Controls International.

Latty, Cyril X., "Controlling Fugitive Emissions From Valves, Pumps and Flanges," European Sealing Association e. V, October 1995.

Miller, Herbert, "Tortuous Path Control Valves for Vibration and Noise Control," Rancho Santa Margarita, California: Control Components Inc.

Miller, Herbert, and Sterud, Curtis, "A High Pressure Pump Recirculation Valve," Rancho Santa Margarita, California: Control Components Inc.

Appendix A

Common Conversion Factors

Although the Imperial units used by process industry in the United States are somewhat recognized and even accepted in some portions of the industrial world, the fact remains that a majority of the process users in the world use the metric system. One complication is that many of the large valve manufacturers are based inside North America and use the Imperial system daily in design and marketing efforts.

Conversion factors allow the manipulation of data between different units, as long as the units used are consistent within any given equation. This section provides a list of common conversion factors that are used in flow calculations and valve sizing. To understand how conversion factors work, one set of units can be divided into another to derive the required units. To illustrate this principle, the length of one foot is also equal to 12 in. This length can be written in one of two ways:

$$L = 12 \text{ in} \quad \text{or} \quad L = 1 \text{ ft}$$

Although the numerical values are different, the actual dimension remains the same. By dividing one set of units into the other, the conversion factor is always 1:

$$\frac{12 \text{ in}}{1 \text{ ft}} = 1.0$$

Thus, a conversion factor (which is always 1.0) can be used in any equation to change the units without changing the numerical values:

$$4 \text{ ft} \times \frac{12 \text{ in}}{1 \text{ ft}} = 48 \text{ in}$$

The following example uses a conversion factor to convert bars and square inches to pounds force:

$$f = 8.5 \text{ bars} \times 2.07 \text{ in}^2 \times \frac{14.5 \text{ lb}}{\text{in}^2 \times \text{bars}} = 225.13 \text{ lb}$$

Another example takes a flow rate of 4500 ft³/h. Using two conversion factors, the flow can be converted to gal/min:

$$Q = \frac{4500 \text{ ft}^3}{h} \times \frac{7.481 \text{ gal}}{1 \text{ ft}} \times \frac{1 \text{ h}}{60 \text{ min}} = 561.1 \text{ gal}$$

Some conversion factors are quite simple and easy to factor, such as

$$\frac{1.8 \text{ degree Rankine}}{1 \text{ degree Kelvin}}$$

or

$$\frac{1000 \text{ grams}}{1 \text{ kilogram}}$$

Other conversion factors contain more than just one set of units:

$$\frac{778 \text{ foot-pounds}}{1 \text{ British thermal unit}}$$

or

$$\frac{32.174 \text{ foot-pounds}_m}{1 \text{ pound-second}^2}$$

where m = minute

Only a small number of conversion factors are required for calculations, because two or more factors can be combined to produce a more complicated factor.

Unit A (multiply)	Conversion factor (by)	Unit B (to obtain)
abcoulombs	2.998×10^{10}	statcoulombs
acres	10	square chain (gunters)
acres	160	rods
acres	10^5	square links (gunters)
acres	0.4047	hectares (square hectometer)
acres	43,560	square feet
acres	4,047	square meters
acres	0.001562	square miles
acres	4,840	square yards
acre-feet	43,560	cubic feet
acre-feet	0.00003259	gallons (U.S.)
amperes/square centimeter	6.452	amperes/square inch
amperes/square centimeter	10,000	amperes/square meter
amperes/square inch	0.155	amperes/square centimeter
amperes/square inch	1,550	amperes/square meter
amperes/square meter	0.0001	amperes/square centimeter
amperes/square meter	0.0006452	amperes/square inch
ampere-hours	3,600	coulombs
ampere-hours	0.03731	faradays
ampere-turns	1.257	gilberts
ampere-turns/centimeter	2.540	ampere-turns/inch
ampere-turns/centimeter	100	ampere-turns/meter
ampere-turns/centimeter	1.257	gilberts/centimeter
ampere-turns/inch	0.3937	ampere-turns/centimeter
ampere-turns/inch	39.37	ampere-turns/meter
ampere-turns/inch	0.4950	gilberts/centimeter
ampere-turns/meter	0.01	ampere-turns/centimeter
ampere-turns/meter	0.0254	ampere-turns/inch
ampere-turns/meter	0.01257	gilberts/centimeter
angstrom units	3937×10^{-9}	inches
angstrom units	10^{-10}	meters
angstrom units	0.0001	microns
ares	0.02471	acres
ares	119.6	square yards
ares	0.02471	acres
ares	100	square meters
astronomical units	14,950	kilometers
atmospheres	0.007348	tons/square inches
atmospheres	76	centimeters mercury
atmospheres	33.9	feet of water (39°F or 4°C)
atmospheres	29.92	inches mercury (32°F or 0°C)
atmospheres	1.0333	kilograms/square centimeter

Unit A (multiply)	Conversion factor (by)	Unit B (to obtain)
atmospheres	10,333	kilograms/square meter
atmospheres	14.7	pounds/square inch
atmospheres	1.058	tons/square foot
barrels (U.S. dry)	7,056	cubic inches
barrels (U.S. dry)	105	quarts (dry)
barrels (U.S. liquid)	31.5	gallons (U.S.)
barrels (oil)	42	gallons (oil)
barrels (oil)/day	0.029	gallons (oil)/minute
bars	0.9869	atmospheres
bars	1,000,000	dynes/square centimeter
bars	10,200	kilograms/square meter
bars	2,089	pounds/square foot
bars	14.5	pounds/square inch
baryls	1	dynes/square centimeter
bolts (U.S.)	36.576	meters
British thermal unit (Btu)	10.409	liters-atmosphere
Btu	1.055×10^{10}	ergs
Btu	778.3	foot-pounds
Btu	252	grams-calorie
Btu	0.0003931	horsepower-hour
Btu	1,054.8	joules
Btu	0.252	kilograms-calorie
Btu	107.5	kilograms-meter
Btu	0.0002928	kilowatts-hour
Btu/hour	0.2162	foot-pounds/second
Btu/hour	0.07	gram-calories/second
Btu/hour	0.0003929	horsepower-hour
Btu/hour	0.2931	watt
Btu/minute	12.96	foot-pounds/second
Btu/minute	0.02356	horsepower
Btu/minute	0.01757	kilowatts
Btu/minute	17.57	watts
Btu square feet/minute	0.1221	watts/square inch
buckets (British dry)	18,180	cubic centimeters
bushels	1.2445	cubic feet
bushels	2,150.4	cubic inches
bushels	0.03524	cubic meters
bushels	35.24	liters
bushels	4	pecks
bushels	64	pints (dry)
bushels	32	quarts (dry)
calories/gram (mean)	0.0039685	Btu (mean)

Unit A (multiply)	Conversion factor (by)	Unit B (to obtain)
candle/square centimeter	3.142	lamberts
candle/square inch	0.4870	lamberts
centares (centiares)	1	square meters
centigrams	0.01	grams
centiliters	0.3382	fluid ounces (U.S.)
centiliters	0.6103	cubic inches
centiliters	2.705	drams
centiliter	0.01	liters
centimeters	0.03281	feet
centimeters	0.3937	inches
centimeters	0.00001	kilometers
centimeters	0.01	meters
centimeters	0.000006214	miles
centimeters	10	millimeters
centimeters	393.7	mils
centimeters	0.01094	yards
centimeter-dynes	0.00102	centimeter-gram
centimeter-dynes	1.02×10^{-8}	meter-kilograms
centimeter-dynes	7.376×10^{-8}	pound-feet
centimeter-grams	980.7	centimeter-dynes
centimeter-grams	0.00001	meter-kilograms
centimeter-grams	0.00007233	pound-feet
centimeters mercury	0.01316	atmospheres
centimeters mercury	0.4461	feet of water
centimeters mercury	136	kilograms/square-meter
centimeters mercury	27.85	pounds/square-foot
centimeters mercury	0.1934	pounds/square-inch
centimeters/second	1.1969	feet/minute
centimeters/second	0.03281	feet/second
centimeters/second	0.036	kilometers/hour
centimeters/second	0.1943	knots
centimeters/second	0.6	meters/minute
centimeters/second	0.02237	miles/hour
centimeters/second	0.0003728	miles/minute
(centimeters/second)/second	0.03281	(feet/second)/second
(centimeters/second)/second	0.036	(kilometers/hour)/second
(centimeters/second)/second	0.01	(meters/second)/second
(centimeters/second)/second	0.02237	(miles/hour)/second
chain	792	inches
chain	20.12	meters
chain (gunter)	22	yards
circular mils	0.000005067	square centimeters
circular mils	0.7854	square mils

Unit A (multiply)	Conversion factor (by)	Unit B (to obtain)
circumference	6.283	radians
circular mils	7.854×10^{-7}	square inches
cords	8	cord feet
cord feet	16	cubic feet
coulombs	2.998×10^9	statcoulombs
coulombs	0.00001036	faradays
coulombs/square centimeter	64.52	coulombs/square inch
coulombs/square centimeter	10,000	coulombs/square meter
coulombs/square inch	0.155	coulombs/square centimeter
coulombs/square inch	1,550	coulombs/square meter
coulombs/square meter	0.0001	coulombs/square centimeter
coulombs/square meter	0.0006452	coulombs/square inch
cubic centimeters	0.00003531	cubic feet
cubic centimeters	0.06102	cubic inches
cubic centimeters	0.000001	cubic meters
cubic centimeters	0.000001308	cubic yards
cubic centimeters	0.0002642	gallons (U.S. liquid)
cubic centimeters	0.001	liters
cubic centimeters	0.002113	pints (U.S. liquid)
cubic centimeters	0.001057	quarts (U.S. liquid)
cubic feet	0.8036	bushels (dry)
cubic feet	28,320	cubic centimeters
cubic feet	1,728	cubic inches
cubic feet	0.02832	cubic meters
cubic feet	0.03704	cubic yards
cubic feet	7.48052	gallons (U.S. liquid)
cubic feet	28.32	liters
cubic feet	59.84	pints (U.S. liquid)
cubic feet	29.92	quarts (U.S. liquid)
cubic feet/minute	472.0	cubic centimeters/second
cubic feet/minute	0.1247	gallons (U.S.)/second
cubic feet/minute	0.472	liters/second
cubic feet/minute	62.43	pounds of water/minute
cubic feet/second	0.646317	million gallons (U.S.)/day
cubic feet/second	448.831	gallons (U.S.)/minute
cubic inches	16.39	cubic centimeters
cubic inches	0.0005787	cubic feet
cubic inches	0.00001639	cubic meters
cubic inches	0.00002143	cubic yards
cubic inches	0.004329	gallons (U.S.)
cubic inches	0.01639	liters
cubic inches	106,100	mil-feet

Unit A (multiply)	Conversion factor (by)	Unit B (to obtain)
cubic inches	0.03463	pints (U.S. liquid)
cubic inches	0.01732	quarts (U.S. liquid)
cubic meters	28.38	bushels (dry)
cubic meters	1,000,000	cubic centimeters
cubic meters	35.31	cubic feet
cubic meters	61,023	cubic inches
cubic meters	1.308	cubic yards
cubic meters	264.2	gallons (U.S. liquid)
cubic meters	1,000	liters
cubic meters	2,113	pints (U.S. liquid)
cubic meters	1,057	quarts (U.S. liquid)
cubic meters/hour	4.4	gallons (U.S.)/minute
cubic yards	764,600	cubic centimeters
cubic yards	27	cubic feet
cubic yards	46,656	cubic inches
cubic yards	0.7646	cubic meters
cubic yards	202	gallons (U.S. liquid)
cubic yards	764.6	liters
cubic yards	1,615.9	pints (U.S. liquid)
cubic yards	807.9	quarts (U.S. liquid)
cubic yards/minute	0.45	cubic feet/second
cubic yards/minute	3.367	gallons (U.S.)/second
cubic yards/minute	12.74	liters/second
dalton	1.65×10^{-24}	grams
days	24	hours
days	1,440	minutes
days	86,400	seconds
decigrams	0.1	grams
deciliters	0.1	liters
decimeters	0.1	meters
degrees (angle)	0.01111	quadrants
degrees (angle)	0.01745	radians
degrees (angle)	3,600	seconds
degrees Centigrade	$(X\,°C \times \frac{9}{5}) + 32$	degrees Fahrenheit
degrees Fahrenheit	$\frac{5}{9}\,(X\,°F - 32)$	degrees Celsius
degrees/second	0.01745	radians/second
degrees/second	0.1667	revolutions/minute
degrees/second	0.002778	revolutions/second
decagrams	10	grams
deciliters	10	liters
decameters	10	meters
drams (apothecaries or troy)	0.1371429	ounces (avoidupois)

Unit A (multiply)	Conversion factor (by)	Unit B (to obtain)
drams (apothecaries or troy)	0.125	ounces (troy)
drams (U.S. fluid or apothecaries)	3.6967	cubic centimeters
drams	1.7718	grams
drams	27.3437	grains
drams	0.0625	ounces
dynes/centimeter	0.01	erg/square millimeter
dynes/square centimeter	9.869×10^{-7}	atmospheres
dynes/square centimeter	0.000001	bars
dynes/square centimeter	0.00002953	inches of mercury (32°F or 0°C)
dynes/square centimeter	0.0004015	inches of water (39°F or 4°C)
dynes	0.00102	grams
dynes	10^{-7}	joules/centimeter
dynes	0.00001	joules/meter (Newtons)
dynes	0.00000102	kilograms
dynes	0.00007233	poundals
dynes	0.000002248	pounds
ell	114.3	centimeters
ell	45	inches
em/pica	0.167	inches
em/pica	0.4233	centimeters
ergs	9.480×10^{-11}	Btu
ergs	1	dyne-centimeters
ergs	7.367×10^{-8}	foot-pounds
ergs	0.2389×10^{-7}	gram-calories
ergs	0.00102	gram-centimeters
ergs	3.725×10^{-14}	horsepower-hours
ergs	10^{-7}	joules
ergs	2.389×10^{-11}	kilogram-calories
ergs	1.02×10^{-8}	kilogram-meters
ergs	0.2778×10^{-13}	kilowatt-hours
ergs	0.2778×10^{-10}	watt-hours
ergs/second	$5,688 \times 10^{-9}$	Btu/minute
ergs/second	1	dyne-centimeter/second
ergs/second	0.000004427	foot-pounds/minute
ergs/second	7.3756×10^{-8}	foot-pounds/second
ergs/second	1.341×10^{-10}	horsepower
ergs/second	1.433×10^{-9}	kilogram-calories/minute
ergs/second	10^{-10}	kilowatts
farads	1,000,000	microfarads
faradays/second	96,500	amperes (absolute)
faradays	26.8	ampere-hours

Unit A (multiply)	Conversion factor (by)	Unit B (to obtain)
faradays	96,490	coulombs
fathoms	1.828804	meter
fathoms	6	feet
feet	30.48	centimeters
feet	0.0003048	kilometers
feet	0.3048	meters
feet	0.0001645	miles (nautical)
feet	0.0001894	miles (statute)
feet	304.8	millimeters
feet	1,200	mils
feet of water	0.02950	atmospheres
feet of water	0.8826	inches of mercury
feet of water	0.03048	kilograms/square meter
feet of water	304.8	kilograms/square meter
feet of water	62.43	pounds/square foot
feet of water	0.4335	pounds/square inch
feet/minute	0.5080	centimeters/second
feet/minute	0.01667	feet/second
feet/minute	0.01829	kilometers/hour
feet/minute	0.3048	meters/minute
feet/minute	0.01136	miles/hour
feet/second	30.48	centimeters/second
feet/second	1.097	kilometers/hour
feet/second	0.5921	knots
feet/second	18.29	meters/minute
feet/second	0.6818	miles/hour
feet/second	0.01136	miles/minute
(feet/second)/second	30.48	(centimeters/second)/second
(feet/second)/second	1.097	(kilometers/hour)/second
(feet/second)/second	0.3048	(meters/second)/second
(feet/second)/second	0.6818	(miles/hour)/second
feet/100 feet	1.0	percent grade
foot-candle	10.764	lumen/square meter
foot-pounds	0.001286	Btu
foot-pounds	1.356×10^7	ergs
foot-pounds	0.3238	gram-calories
foot-pounds	5.050×10^{-7}	horsepower-hours
foot-pounds	1.356	joules
foot-pounds	0.000324	kilogram-calories
foot-pounds	0.1383	kilogram-meters
foot-pounds	3.766×10^{-7}	kilowatt-hours
foot-pounds/minute	0.001286	Btu/minute

Unit A (multiply)	Conversion factor (by)	Unit B (to obtain)
foot-pounds/minute	0.01667	foot-pounds/second
foot-pounds/minute	0.0000303	horsepower
foot-pounds/minute	0.000324	kilogram-calories/minute
foot-pounds/minute	0.0000226	kilowatts
foot-pounds/second	4.6263	Btu/hour
foot-pounds/second	0.07717	Btu/minute
foot-pounds/second	0.001818	horsepower
foot-pounds/second	0.01945	kilograms-calories/minute
foot-pounds/second	0.001356	kilowatts
furlongs	0.125	miles (U.S.)
furlongs	40	rods
furlongs	660	feet
gallons (British Imperial)	1.20095	gallons (U.S.)
gallons (U.S.)	3,785	cubic centimeters
gallons (U.S.)	0.1337	cubic feet
gallons (U.S.)	231	cubic inches
gallons (U.S.)	0.003785	cubic meters
gallons (U.S.)	0.004951	cubic yards
gallons (U.S.)	0.83267	gallons (British Imperial)
gallons (U.S.)	3.785	liters
gallons (U.S.)	8.3453	pounds water
gallons (U.S.)/minute	0.002228	cubic feet/second
gallons (U.S.)/minute	0.06308	liters/second
gallons (U.S.)/minute	8.0208	cubic feet/hour
gausses	6.452	lines/square inch
gausses	0.00000001	webers/square centimeter
gausses	6.452×10^{-8}	webers/square inch
gausses	0.0001	webers/square meter
gilberts	0.7958	ampere-turns
gilberts/centimeter	0.7958	ampere-turns/centimeter
gilberts/centimeter	2.021	ampere-turns/inch
gills (British)	142.07	cubic centimeter
gills	0.1183	liters
gills	0.25	pints (liquid)
grade	0.01571	radian
grains	0.03657143	drams (avoirdupois)
grains (troy)	1	grains (avoirdupois)
grains (troy)	0.06480	grams
grains (troy)	0.0020833	ounces (avoirdupois)
grains (troy)	0.04167	pennyweight (troy)
grains/gallon (U.S.)	17.118	parts/million
grains/gallon (U.S.)	142.86	pounds/million gallons (U.S.)
grains/gallon (British Imperial)	14.286	parts/million

Unit A (multiply)	Conversion factor (by)	Unit B (to obtain)
grams	980.7	dynes
grams	15.43	grains
grams	0.00009807	joules/centimeter
grams	0.009807	joules/meter (newtons)
grams	0.001	kilograms
grams	1,000	milligrams
grams	0.03527	ounces (avoirdupois)
grams	0.03215	ounces (troy)
grams	0.07093	poundals
grams	2,205	pounds
grams/centimeter	0.0056	pounds/inch
grams/cubic centimeter	62.43	pounds/cubic foot
grams/cubic centimeter	0.03613	pounds/cubic inch
grams/cubic centimeter	0.0000003405	pounds/mil-foot
grams/liter	58.417	grains/gallon (U.S.)
grams/liter	8.345	pounds/1,000 gallons (U.S.)
grams/liter	0.062427	pounds/cubic feet
grams/liter	1,000	parts/million
grams/liter	0.134	average ounce/gallon (U.S.)
grams/square centimeter	2.0481	pounds/square foot
gram-calories	0.0039683	Btu
gram-calories	4.1868×10^7	ergs
gram-calories	3.088	foot-pounds
gram-calories	0.0000015596	horsepower-hours
gram-calories	0.0000011630	kilowatt-hours
gram-calories	0.0011630	watt-hours
gram-calories/second	14.286	Btu/hour
gram-centimeters	9.297×10^{-8}	Btu
gram-centimeters	980.7	ergs
gram-centimeters	0.00009807	joules
gram-centimeters	2.343×10^{-8}	kilogram-calories
gram-centimeters	0.00001	kilogram-meters
hand	10.16	centimeters
hectares	2.471	acres
hectares	107,600	square feet
hectograms	100	grams
hectoliters	100	liters
hectometers	100	meters
hectowatts	100	watts
henries	1,000	millihenries
hogsheads (British)	10.114	cubic feet
hogsheads (U.S.)	8.42184	cubic feet
hogsheads (U.S.)	63	gallons (U.S.)

Unit A (multiply)	Conversion factor (by)	Unit B (to obtain)
horsepower	42.44	Btu/minute
horsepower	33,000	foot-pounds/minute
horsepower	550	foot-pounds/second
horsepower	1.014	horsepower (metric)
horsepower	10.68	kilogram-calories/minute
horsepower	0.7457	kilowatts
horsepower	745.7	watts
horsepower (boiler)	33,479	Btu/hour
horsepower (boiler)	9.803	kilowatts
horsepower (boiler)	34.5	pounds water evaporated/ hour (212°F or 100°C)
horsepower-hours	2,547	Btu
horsepower-hours	2.6845×10^{13}	ergs
horsepower-hours	1,980,000	foot-pounds
horsepower-hours	641,190	gram-calories
horsepower-hours	2,684,000	joules
horsepower-hours	641.1	kilogram-calories
horsepower-hours	273,700	kilogram-meters
horsepower-hours	0.7457	kilowatt-hours
horsepower (metric)	542.5	foot-pounds/second
horsepower (metric)	0.9863	horsepower
hours	0.04167	days
hours	60	minutes
hours	3600	seconds
hours	0.005952	weeks
hundredweights (long)	112	pounds
hundredweights (long)	0.05	tons (long)
hundredweights (short)	1600	ounces (avoirdupois)
hundredweights (short)	100	pounds
hundredweights (short)	0.0453592	tons (metric)
hundredweights (short)	0.0446429	tons (long)
inches	2.54	centimeters
inches	0.0254	meters
inches	0.00001578	miles
inches	25.4	millimeters
inches	1,000	mils
inches	0.02778	yards
inches mercury	0.03342	atmospheres
inches mercury	1.133	feet water
inches mercury	0.03453	kilograms/square centimeter
inches mercury	345.3	kilograms/square meter
inches mercury	70.73	pounds/square foot

Unit A (multiply)	Conversion factor (by)	Unit B (to obtain)
inches of mercury	0.4912	pounds/square inch
inches water (39°F or 4°C)	0.002458	atmospheres
inches water (39°F or 4°C)	0.07355	inches mercury
inches water (39°F or 4°C)	0.00254	kilograms/square centimeter
inches water (39°F or 4°C)	0.5781	ounces/square inch
inches water (39°F or 4°C)	5.204	pounds/square feet
inches water (39°F or 4°C)	0.03613	pounds/square inch
international ampere	0.9998	ampere (absolute)
international volt	1.0003	volts (absolute)
international volt	1.593×10^{-19}	joules (absolute)
international volt	96,540	joules
joules	0.000948	Btu
joules	10^7	ergs
joules	0.7376	foot-pounds
joules	0.0002389	kilogram-calories
joules	0.1020	kilogram-meters
joules	0.0002778	watt-hours
joules/centimeter	1.02×10^4	grams
joules/centimeter	10^7	dynes
joules/centimeter	100	joules/meter (newtons)
joules/centimeter	723.3	poundals
joules/centimeter	22.48	pounds
kilograms	980,655	dynes
kilograms	1,000	grams
kilograms	0.09807	joules/centimeter
kilograms	9.807	joules/meter (newtons)
kilograms	70.93	poundals
kilograms	2.205	pounds
kilograms	0.0009842	tons (long)
kilograms	0.001102	tons (short)
kilograms/cubic meter	0.001	grams/cubic centimeter
kilograms/cubic meter	0.06243	pounds/cubic foot
kilograms/cubic meter	0.00003613	pounds/cubic inches
kilograms/meter	0.6720	pounds/foot
kilograms/cubic meter	3.405×10^{-10}	pounds/mil-foot
kilograms/square centimeter	980,655	dynes
kilograms/square centimeter	0.9678	atmospheres
kilograms/square centimeter	32.81	feet water
kilograms/square centimeter	28.96	inches mercury
kilograms/square centimeter	2,048	pounds/square foot
kilograms/square centimeter	14.22	pounds/square inch

Unit A (multiply)	Conversion factor (by)	Unit B (to obtain)
kilograms/square meter	0.00009678	atmospheres
kilograms/square meter	0.00009807	bars
kilograms/square meter	0.003281	feet water
kilograms/square meter	0.002896	inches mercury
kilograms/square meter	0.2048	pounds/square foot
kilograms/square meter	0.001422	pounds/square inch
kilograms/square millimeter	1,000,000	kilograms/square meter
kilogram-calories	3.968	Btu
kilogram-calories	3,088	foot-pounds
kilogram-calories	0.001560	horsepower-hours
kilogram-calories	4,186	joules
kilogram-calories	426.9	kilogram-meters
kilogram-calories	4.186	kilojoules
kilogram-calories	0.001163	kilowatt-hours
kilogram-meters	0.009294	Btu
kilogram-meters	9.804×10^7	ergs
kilogram-meters	7.233	foot-pounds
kilogram-meters	9.804	joules
kilogram-meters	0.002342	kilogram-calories
kilogram-meters	0.000002723	kilowatt-hours
kilolines	1,000	maxwells
kiloliters	1,000	liters
kilometers	100,000	centimeters
kilometers	3,281	feet
kilometers	39,370	inches
kilometers	1,000	meters
kilometers	0.6214	miles
kilometers	1,000,000	millimeters
kilometers	1,094	yards
kilometers/hour	27.78	centimeters/second
kilometers/hour	54.68	feet/minute
kilometers/hour	0.9113	feet/second
kilometers/hour	0.5396	knots
kilometers/hour	16.67	meters/minute
kilometers/hour	0.6214	miles/hour
(kilometers/hour)/second	27.78	(centimeters/second)/second
(kilometers/hour)/second	0.9113	(feet/second)/second
(kilometers/hour)/second	0.2778	(meters/second)/second
(kilometers/hour)/second	0.6214	(miles/hour)/second
kilowatts	56.92	Btu/minute
kilowatts	44,260	foot-pounds/minute
kilowatts	737.6	foot-pounds/second

Unit A (multiply)	Conversion factor (by)	Unit B (to obtain)
kilowatts	1.341	horsepower
kilowatts	14.34	kilogram-calories/minute
kilowatts	1,000	watts
kilowatt-hours	3,413	Btu
kilowatt-hours	3.600×10^{13}	ergs
kilowatt-hours	26,550	foot-pounds
kilowatt-hours	859,850	gram-calories
kilowatt-hours	1.341	horsepower-hours
kilowatt-hours	3,600,000	joules
kilowatt-hours	860.5	kilogram-calories
kilowatt-hours	367,100	kilogram-meters
kilowatt-hours	3.53	pounds of water evaporated from 212°F (100°C)
kilowatt-hours	22.75	pounds of water raised from 62 to 212°F (17 to 100°C)
knots	6,080	feet/hour
knots	1.8532	kilometers/hour
knots	1	nautical miles/hour
knots	1.151	statute miles/hour
knots	2.027	yards/hour
knots	1.689	feet/second
league	3	miles (approximately)
light year	5.9×10^{12}	miles
light year	5.56	milometers
lines/square centimeter	1	gausses
lines/square inch	0.155	gausses
lines/square inch	1.55×10^{-9}	webers/square centimeter
lines/square inch	10^{-8}	webers/square inch
lines/square inch	0.0000155	webers/square meter
links (engineer's)	12	inches
links (surveyor's)	7.92	inches
liters	0.02838	bushels (U.S. dry)
liters	1,000	cubic centimeters
liters	0.03531	cubic feet
liters	61.02	cubic inches
liters	0.001	cubic meter
liters	0.001308	cubic yards
liters	0.2642	gallons (U.S. liquid)
liters	2.113	pints (U.S. liquid)
liters	1.057	quarts (U.S. liquid)
liters/second	15.85	gallons (U.S.)/minute
liters/minute	0.0005886	cubic feet/second

Unit A (multiply)	Conversion factor (by)	Unit B (to obtain)
liters/minute	0.004403	gallons (U.S.)/second
liters/hour	0.0044	gallons (U.S.)/minute
lumens/square foot	1	foot-candles
lumen	0.07958	spherical candle
lumen	0.001496	watt
lumen/square foot	10.76	lumen/square meter
lux	0.0929	foot-candles
maxwells	0.001	kilolines
maxwells	10^{-8}	webers
megalines	10,000	maxwells
megohms	10^{12}	microhms
megohms	1,000,000	ohms
meters	100	centimeters
meters	3.281	feet
meters	39.37	inches
meters	0.001	kilometers
meters	0.0005396	miles (nautical)
meters	0.0006214	miles (statute)
meters	1,000	millimeters
meters	1.094	yards
meters	1.179	varas
meters/minute	1.667	centimeters/second
meters/minute	3.281	feet/minute
meters/minute	0.05468	feet/second
meters/minute	0.06	kilometers/hour
meters/minute	0.03238	knots
meters/minute	0.03728	miles/hour
meters/second	196.8	feet/minute
meters/second	3.281	feet/second
meters/second	3.6	kilometers/hour
meters/second	0.06	kilometers/hour
meters/second	2.237	miles/hour
meters/second	0.03728	miles/minute
(meters/second)/second	100	(centimeters/second)/second
(meters/second)/second	3.281	(feet/second)/second
(meters/second)/second	3.6	(kilometers/hour)/second
(meters/second)/second	2.237	(miles/hour)/second
meter-kilograms	9.807×10^{7}	centimeter-dynes
meter-kilograms	100,000	centimeter-grams
meter-kilograms	7.233	pound-feet
microfarads	0.000001	farads

Unit A (multiply)	Conversion factor (by)	Unit B (to obtain)
micrograms	0.000001	grams
microhms	10^{-12}	megohms
microhms	0.000001	ohms
microliters	0.000001	liters
microns	0.000001	meters
miles (nautical)	6,080.27	feet
miles (nautical)	1.853	kilometers
miles (nautical)	1,853	meters
miles (nautical)	1.15166	miles (statute)
miles (nautical)	2,027	yards
miles (statute)	160,900	centimeters
miles (statute)	5,280	feet
miles (statute)	63,360	inches
miles (statute)	1.609	kilometers
miles (statute)	1,609	meters
miles (statute)	0.8684	miles (nautical)
miles (statute)	1,760	yards
miles/hour	44.7	centimeters/second
miles/hour	88	feet/minute
miles/hour	1.467	feet/second
miles/hour	1,609	kilometers/hour
miles/hour	0.02682	kilometers/minute
miles/hour	0.8684	knots
miles/hour	26.82	meters/minute
miles/hour	0.1667	miles/minute
(miles/hour)/second	44.7	(centimeters/second)/second
(miles/hour)/second	1.467	(feet/second)/second
(miles/hour)/second	1.609	(kilometers/hour)/second
(miles/hour)/second	0.447	(meters/second)/second
miles/minute	2,682	centimeters/second
miles/minute	88	feet/second
miles/minute	1.609	kilometers/minute
miles/minute	0.8684	knots/minute
miles/minute	60	miles/hour
mil-feet	0.000009425	cubic inches
milliers	1,000	kilograms
millimicrons	10^{-9}	meters
milligrams	0.01543236	grains
milligrams	0.001	grams
milligrams/liter	1	parts/million
millihenries	0.001	henries
milliliters	0.001	liters

Unit A (multiply)	Conversion factor (by)	Unit B (to obtain)
millimeters	0.1	centimeters
millimeters	0.003281	feet
millimeters	0.03937	inches
millimeters	0.000001	kilometers
millimeters	0.001	meters
millimeters	6.214×10^{-7}	miles
millimeters	39.37	mils
millimeters	0.001094	yards
million gallons (U.S.)/day	1.54723	cubic feet/second
mils	0.00254	centimeters
mils	0.00008333	feet
mils	0.001	inches
mils	2.54×10^{-8}	kilometers
mils	0.00002778	yards
miner's inches	1.5	cubic feet/minute
minims (British)	0.059192	cubic centimeters
minims (U.S. fluid)	0.061612	cubic centimeters
minutes	60	seconds
minutes	0.1667	hours
minutes	0.000694	days
minutes (angles)	0.01667	degrees
minutes (angles)	0.0001852	quadrants
minutes (angles)	0.0002909	radians
minutes (angles)	60	seconds
myriagrams	10	kilograms
myriagrams	10	kilograms
myriawatts	10	kilowatts
nepers	8.686	decibels
newtons	105	dynes
ohms (international)	1.0005	ohms (absolute)
ohms	0.000001	megohms
ohms	1,000,000	microhms
ounces	16	drams
ounces	437.5	grains
ounces	28.349527	grams
ounces	0.0625	pounds
ounces	0.9115	ounces (troy)
ounces	0.0000279	tons (long)
ounces	0.00002835	tons (metric)
ounces (fluid)	1.805	cubic inches
ounces (fluid)	0.02957	liters
ounces (troy)	480	grains
ounces (troy)	31.103481	grams
ounces (troy)	1.09714	ounces (avoirdupois)

Unit A (multiply)	Conversion factor (by)	Unit B (to obtain)
ounces (troy)	20	pennyweights (troy)
ounces (troy)	0.08333	pounds (troy)
ounce/square inch	4,309	dynes/square centimeter
ounces/square inch	0.0625	pounds/square inch
parsec	19×10^{12}	miles
parsec	3.084×10^{13}	kilometers
parts/million	0.0584	grains/gallon (U.S.)
parts/million	0.07016	grains/gallon (imperial)
parts/million	8.345	pounds/million gallons (U.S.)
pecks (British)	554.6	cubic inches
pecks (British)	9.091901	liters
pecks (U.S.)	0.25	bushels
pecks (U.S.)	537.605	cubic inches
pecks (U.S.)	8.809582	liters
pecks (U.S.)	8	quarts (dry)
pennyweights (troy)	24	grains
pennyweights (troy)	0.05	ounces (troy)
pennyweights (troy)	1.55517	grams
pennyweights (troy)	0.0041667	pounds (troy)
pints (dry)	33.6	cubic inches
pints (liquid)	473.2	cubic centimeters
pints (liquid)	0.01671	cubic feet
pints (liquid)	28.87	cubic inches
pints (liquid)	0.0004732	cubic meters
pints (liquid)	0.0006189	cubic yards
pints (liquid)	0.125	gallons (U.S.)
pints (liquid)	0.4732	liters
pints (liquid)	0.5	quarts (liquid)
planck's quantum	6.624×10^{-27}	erg-second
poise	1	grams/centimeter-second
pounds (avoirdupois)	14.5833	ounces (troy)
poundals	13,826	dynes
poundals	14.1	grams
poundals	0.001383	joules/centimeter
poundals	0.1383	joules/meter (newtons)
poundals	0.0141	kilograms
poundals	0.03108	pounds
pounds	256	drams
pounds	444,823	dynes
pounds	7,000	grains
pounds	453.5924	grams
pounds	0.04448	joules/centimeter
pounds	4.448	joules/meter (newtons)
pounds	0.4536	kilograms

Unit A (multiply)	Conversion factor (by)	Unit B (to obtain)
pounds	16	ounces
pounds	14.5833	ounces (troy)
pounds	32.17	poundals
pounds	1.21528	pounds (troy)
pounds	0.0005	tons (short)
pounds (troy)	5,760	grains
pounds (troy)	373.24177	grams
pounds (troy)	13.1657	ounces (avoirdupois)
pounds (troy)	12	ounces (troy)
pounds (troy)	240	pennyweights (troy)
pounds (troy)	0.822857	pounds (avoirdupois)
pounds (troy)	0.00036735	tons (long)
pounds (troy)	0.00037324	tons (metric)
pounds (troy)	0.00041143	tons (short)
pounds water	0.01602	cubic feet
pounds water	27.68	cubic inches
pounds water	0.1198	gallons (U.S.)
pounds water/minute	0.000267	cubic feet/second
pound-feet	1.356×10^7	centimeter-dynes
pound-feet	13,825	centimeter-grams
pound-feet	0.1383	meter-kilograms
pounds/cubic feet	0.01602	grams/cubic centimeter
pounds/cubic feet	16.02	kilograms/cubic meter
pounds/cubic feet	0.0005787	pounds/cubic inch
pounds/cubic feet	5.456×10^{-9}	pounds/mil-foot
pounds/cubic inch	27.68	grams/cubic centimeter
pounds/cubic inch	27,680	kilograms/cubic meter
pounds/cubic inch	1,728	pounds/cubic foot
pounds/cubic inch	0.000009425	pounds/mil-foot
pounds/foot	1.488	kilograms/meter
pounds/inch	178.6	grams/centimeter
pounds/mil-foot	23,060	grams/cubic centimeter
pounds/square foot	0.0004725	atmospheres
pounds/square foot	0.01602	feet water
pounds/square foot	0.01414	inches mercury
pounds/square foot	4.882	kilograms/square meter
pounds/square foot	0.006944	pounds/square inch
pounds/square inch	0.06804	atmospheres
pounds/square inch	2.307	feet water
pounds/square inch	2.036	inches mercury
pounds/square inch	703.1	kilograms/square meter
pounds/square inch	144	pounds/square foot
quadrants (angle)	90	degrees
quadrants (angle)	5,400	minutes
quadrants (angle)	1.571	radians

Unit A (multiply)	Conversion factor (by)	Unit B (to obtain)
quadrants (angle)	324,000	seconds
quarts (dry)	67.2	cubic inches
quarts (liquid)	946.4	cubic centimeters
quarts (liquid)	0.03342	cubic feet
quarts (liquid)	57.75	cubic inches
quarts (liquid)	0.0009464	cubic meters
quarts (liquid)	0.001238	cubic yards
quarts (liquid)	0.25	gallons (U.S.)
quarts (liquid)	0.9463	liters
radians	57.3	degrees
radians	3,438	minutes
radians	0.6366	quadrants
radians	2.063	seconds
radians/second	57.3	degrees/second
radians/second	9.549	revolutions/minute
radians/second	0.1592	revolutions/second
(radians/second)/second	573	(revolutions/minute)/minute
(radians/second)/second	9.549	(revolutions/minute)/minute
(radians/second)/second	0.1592	(revolutions/second)/second
revolutions	360	degrees
revolutions	4	quadrants
revolutions	6.283	radians
revolutions	6	degree/second
revolutions/minute	0.1047	radians/second
revolutions/minute	0.01667	revolutions/second
(revolutions/minute)/minute	0.001745	(radians/second)/second
(revolutions/minute)/minute	0.01667	(revolutions/minute)/second
(revolutions/minute)/minute	0.0002778	(revolutions/second)/second
revolutions/second	360	degrees/second
revolutions/second	6.283	radians/second
revolutions/second	60	revolutions/minute
(revolutions/second)/second	6.283	(radians/second)/second
(revolutions/second)/second	3,600	(revolutions/minute)/minute
(revolutions/second)/second	60	(revolutions/minute)/second
rods	0.25	chain (gunters)
rods	5.029	meters
rods	16.5	feet
rods (surveyor)	5.5	yards

Unit A (multiply)	Conversion factor (by)	Unit B (to obtain)
scruples	20	grains
seconds	0.000012	days
seconds	0.000278	hours
seconds	0.1667	minutes
seconds	0.00000165	weeks
seconds (angle)	0.0002778	degrees
seconds (angle)	0.01667	minutes
seconds (angles)	0.000003087	quadrants
seconds (angles)	0.000004848	radians
slug	14.59	kilograms
slug	32.17	pounds
sphere	12.57	steradians
square centimeters	197,300	circular mils
square centimeters	0.001076	square feet
square centimeters	0.155	square inches
square centimeters	0.0001	square meters
square centimeters	3.861×10^{-11}	square miles
square centimeters	100	square millimeters
square centimeters	0.0001196	square yards
square feet	0.00002296	acres
square feet	1.833×10^{8}	circular mils
square feet	929	square centimeters
square feet	144	square inches
square feet	0.09290	square meters
square feet	3.587×10^{-8}	square miles
square feet	92,900	square millimeters
square feet	0.1111	square yards
square inches	1,273,000	circular mils
square inches	6.452	square centimeters
square inches	0.006944	square feet
square inches	645.2	square millimeters
square inches	1,000,000	square mils
square inches	0.0007716	square yards
square kilometers	247.1	acres
square kilometers	10^{10}	square centimeters
square kilometers	10,760,000	square feet
square kilometers	1.55×10^{9}	square inches
square kilometers	1,000,000	square meters
square kilometers	1,196,000	square yards
square meters	0.0002471	acres
square meters	10,000	square centimeters
square meters	10.76	square feet
square meters	1,550	square inches
square meters	3.861×10^{-7}	square miles
square meters	1,000,000	square millimeters
square yards	1.196	square yards

Unit A (multiply)	Conversion factor (by)	Unit B (to obtain)
square miles	640	acres
square miles	27,880,000	square feet
square miles	2.59	square kilometers
square miles	2,590,000	square meters
square miles	3,098,000	square yards
square millimeters	1,973	circular mils
square millimeters	0.01	square centimeters
square millimeters	0.00001076	square feet
square millimeters	1,550	square inches
square mils	1.273	circular mils
square mils	0.000006452	square centimeters
square mils	0.000001	square inches
square yards	0.0002066	acres
square yards	8,361	square centimeters
square yards	9	square feet
square yards	1,296	square inches
square yards	0.8361	square meters
square yards	3.228×10^{-7}	square miles
square yards	836,100	square millimeters
temperature, °C	(°C + 273)	absolute temperature, °C
temperature, °C	(°C + 17.78)(1.8)	temperature, °F
temperature, °F	(°F + 460)	absolute temperature, °F
temperature, °F	(°F − 32)(0.555)	temperature, °C
tons (long)	1,016	kilograms
tons (long)	2,240	pounds
tons (long)	1.12	tons (short)
tons (metric)	1,000	kilograms
tons (metric)	2,205	pounds
tons (short)	907.1848	kilograms
tons (short)	32,000	ounces
tons (short)	29,166.66	ounces (troy)
tons (short)	2,000	pounds
tons (short)	2,430.56	pounds (troy)
tons (short)	0.89287	tons (long)
tons (short)	0.9078	tons (metric)
tons (short)/square foot	9,765	kilograms/square meter
tons (short)/square foot	2,000	pounds/square inch
tons (short)/hour	4/Sp. Gr.	gallons (U.S.)/minute
tons refrigerant	12,000	Btu/hour
tons water/day	83.333	pounds water/hour
tons water/day	0.16643	gallons (U.S.)/minute
tons water/day	1.3349	cubic feet/hour
volt/inch	0.3937	volt/centimeter

Unit A (multiply)	Conversion factor (by)	Unit B (to obtain)
volt (absolute)	0.003336	statvolts
watts	3.4192	Btu/hour
watts	0.05688	Btu/minute
watts	10^7	ergs/second
watts	44.27	foot-pounds/minute
watts	0.7378	foot-pounds/second
watts	0.001341	horsepower
watts	0.0036	horsepower (metric)
watts	0.01433	kilogram-calories/minute
watts	0.001	kilowatts
watts (absolute)	0.056884	Btu (mean)/minute
watts (absolute)	1	joules/second
watt-hours	3.413	Btu
watt-hours	3.6×10^{10}	ergs
watt-hours	2.656	foot-pounds
watt-hours	859.85	gram-calories
watt-hours	0.001341	horsepower-hours
watt-hours	0.8605	kilogram-calories
watt-hours	367.2	kilogram-meters
watt-hours	0.001	kilogram-hours
watt (international)	1.0002	watt (absolute)
webers	10^8	maxwells
webers	100,000	kilolines
webers/square inch	1.55×10^7	gausses
webers/square inch	10^8	lines/square inch
webers/square inch	0.155	webers/square centimeter
webers/square meter	10,000	gausses
webers/square meter	64,520	lines/square inch
webers/square meter	0.0001	webers/square centimeter
webers/square meter	0.0006452	webers/square inch
weeks	7	days
weeks	168	hours
weeks	10,080	minutes
weeks	604,800	seconds
yards	91.44	centimeters
yards	0.0009144	kilometers
yards	0.9144	meters
yards	0.0004934	miles (nautical)
yards	0.0005682	miles (statute)
yards	914.4	millimeters

Appendix B

Fluid Data

Formula	Fluid	Molecular weight	Critical temperature, °F (°C)	Critical pressure, psia (bar)	Specific-heat ratio
Ar	Argon	39.95	−189 (−121)	707 (48.8)	1.67
BCl$_3$	Boron trichloride	117.17	354 (179)	562 (38.8)	0.00
BF$_3$	Boron trifluoride	67.81	9 (−13)	723 (49.9)	0.00
Br$_2$	Bromine	159.81	591 (311)	1499 (103.4)	1.30
CBrF$_3$	Trifluorobromomethane	148.91	152 (67)	576 (39.7)	0.00
CClF$_3$	Chlorotrifluoromethane	104.46	84 (29)	569 (39.2)	1.14
CCl$_2$F$_2$	Dichlorodifluoromethane	120.91	233 (112)	598 (41.2)	1.13
CCl$_2$O	Phosgene	98.92	359 (182)	823 (56.8)	1.17
CCl$_3$F	Trichlorofluoromethane	137.37	388 (198)	639 (44.1)	1.12
CCl$_4$	Carbon tetrachloride	153.82	542 (283)	662 (45.7)	1.11
CF$_4$	Carbon tetrafluoride	88.01	−50 (−46)	542 (37.4)	1.16
CHClF$_2$	Chlorodifluoromethane	86.47	205 (96)	722 (49.8)	1.18
CHCl$_2$F	Dichloromonofluoromethane	102.92	353 (178)	750 (51.7)	1.16
CHCl$_3$	Chloroform	119.38	506 (263)	794 (54.8)	1.15
CHN	Hydrogen cyanide	27.03	362 (183)	782 (53.9)	1.30
CH$_2$Br$_2$	Dibromomethane	173.84	589 (308)	1044 (72.0)	0.00
CH$_2$Cl$_2$	Dichloromethane	84.93	458 (237)	882 (60.1)	1.20
CH$_2$O	Formaldehyde	30.03	274 (134)	956 (65.9)	1.32
CH$_2$O$_2$	Formic acid	46.03	584 (307)	0 (0.0)	1.23
CH$_3$Br	Methyl bromide	94.94	375 (191)	1250 (86.2)	1.25
CH$_3$Cl	Methyl chloride	50.49	289 (143)	969 (66.8)	1.26
CH$_3$F	Methyl fluoride	34.03	112 (44)	853 (58.8)	1.29
CH$_3$I	Methyl iodine	141.94	490 (254)	955 (65.9)	1.24
CH$_3$NO$_2$	Nitromethane	61.04	598 (314)	916 (63.2)	1.17

Formula	Name				
CH_4	Methane	16.04	−117 (−83)	667 (46.0)	1.31
CH_4O	Methanol	32.04	463 (239)	1174 (81.0)	1.24
CH_4S	Methyl mercaptan	48.11	386 (197)	1050 (72.4)	1.20
CH_5N	Methyl amine	31.06	314 (157)	1081 (74.6)	1.21
CH_6N_2	Methyl hydrazine	46.07	561 (294)	1166 (80.4)	0.00
$ClNO$	Nitrosyl chloride	65.46	332 (167)	1323 (91.2)	1.33
Cl_2	Chlorine	70.91	291 (144)	1117 (77.0)	1.33
Cl_3P	Phosphorus trichloride	137.33	553 (290)	0 (0.0)	0.00
Cl_4Si	Silicon tetrachloride	169.90	453 (234)	544 (37.5)	0.00
CO	Carbon monoxide	28.01	−221 (−141)	507 (35.0)	1.40
CO_2	Carbon dioxide	44.01	88 (31)	1070 (73.8)	1.29
COS	Carbonyl sulfide	60.07	215 (102)	853 (58.8)	1.25
CS_2	Carbon disulfide	76.13	534 (279)	1147 (79.1)	1.22
C_2ClF_5	Chloropentafluoroethane	154.47	176 (80)	459 (31.7)	1.18
$C_2Cl_2F_4$	1,1-Dichloro-1,2,2,2	170.92	293 (145)	479 (33.0)	1.08
$C_2Cl_2F_4$	1,2-Dichloro-1,1,2,2	170.92	294 (146)	473 (32.6)	1.08
$C_2Cl_3F_3$	1,2,2-Trichloro-1,1,2	187.38	417 (214)	495 (34.1)	1.07
C_2Cl_4	Tetrachloroethylene	165.83	656 (347)	647 (44.6)	1.10
$C_2Cl_4F_2$	1,1,2,2-Tetrachloro-1	203.83	532 (278)	0 (0.0)	0.00
C_2F_4	Perfluoroethene	100.02	92 (33)	572 (39.5)	1.12
C_2F_6	Perfluoroethane	138.01	67 (19)	0 (0.0)	1.09
C_2HCl_3	Trichloroethylene	131.39	568 (298)	713 (49.2)	1.12
$C_2HF_3O_2$	Trifluoroacetic acid	114.02	424 (218)	473 (32.6)	0.00
C_2H_2	Acetylene	26.04	95 (35)	891 (61.5)	1.23
$C_2H_2F_2$	1,1-Difluoroethylene	64.04	85 (29)	647 (44.6)	1.17
C_2H_2O	Ketene	42.04	224 (106)	941 (64.9)	1.22
C_2H_3Cl	Vinyl chloride	62.50	313 (156)	813 (56.1)	1.19
$C_2H_3ClF_2$	1-Chloro-1,1-difluoro	100.47	278 (137)	598 (41.2)	1.11
C_2H_3ClO	Acetyl chloride	78.50	454 (234)	852 (58.8)	1.14

Formula	Fluid	Molecular weight	Critical temperature, °F (°C)		Critical pressure, psia (bar)		Specific-heat ratio
$C_2H_3Cl_3$	1,1,2-Trichloroethane	133.41	624	(329)	603	(41.6)	1.11
C_2H_3F	Vinyl fluoride	46.04	130	(54)	760	(52.4)	0.00
$C_2H_3F_3$	1,1,1-Trifluoroethane	84.04	163	(73)	545	(37.6)	1.12
C_2H_3N	Acetonitrile	41.05	526	(274)	701	(48.3)	1.19
C_2H_3NO	Methyl isocyanate	57.05	424	(218)	809	(55.8)	1.15
C_2H_4	Ethylene	28.05	48	(9)	731	(50.4)	1.24
$C_2H_4Cl_2$	1,1-Dichloroethane	94.96	481	(249)	735	(50.7)	1.12
$C_2H_4Cl_2$	1,2-Dichloroethane	98.96	550	(288)	779	(53.7)	1.12
$C_2H_4F_2$	1,1-Difluoroethane	66.05	236	(113)	653	(45.0)	1.14
C_2H_4O	Acetaldehyde	44.05	370	(188)	809	(55.8)	1.19
C_2H_4O	Ethylene oxide	44.05	384	(196)	1044	(72.0)	1.21
$C_2H_4O_2$	Acetic acid	60.05	610	(321)	839	(57.9)	1.14
$C_2H_4O_2$	Methyl formate	60.05	417	(214)	870	(60.0)	1.15
C_2H_5Br	Ethyl bromide	108.97	447	(231)	904	(62.3)	1.15
C_2H_5Cl	Ethyl chloride	64.52	369	(187)	764	(52.7)	1.16
C_2H_5F	Ethyl fluoride	48.06	216	(102)	729	(50.3)	1.16
C_2H_5N	Ethylene imine	43.07	−460	(−273)	0	(0.0)	1.19
C_2H_6	Ethane	30.07	90	(32)	709	(48.9)	1.19
C_2H_6O	Dimethyl ether	46.07	260	(127)	779	(53.7)	1.15
C_2H_6O	Ethanol	46.07	469	(242)	926	(63.9)	1.15
$C_2H_6O_2$	Ethylene glycol	62.07	701	(371)	1117	(77.0)	1.09
C_2H_6S	Ethyl mercaptan	62.13	438	(226)	797	(55.0)	1.13
C_2H_6S	Dimethyl sulfide	62.13	445	(229)	803	(55.4)	1.13
C_2H_7N	Ethyl amine	45.09	361	(183)	816	(56.3)	1.13

Formula	Name	MW			Ratio
C_2H_7N	Dimethyl amine	45.09	328 (164)	770 (53.1)	1.14
C_2H_7NO	Monoethanolamine	61.08	645 (341)	647 (44.6)	1.11
$C_2H_8N_2$	Ethylenediamine	60.10	607 (319)	911 (62.8)	1.09
C_2N_2	Cyanogen	52.04	260 (127)	867 (59.8)	1.17
C_3H_3N	Acrylonitrile	53.06	505 (263)	662 (45.7)	1.15
C_3H_4	Propadiene	40.07	247 (119)	794 (54.8)	1.17
C_3H_4	Methyl acetylene	40.07	264 (129)	816 (56.3)	1.16
C_3H_4O	Acrolein	56.06	451 (233)	750 (51.7)	1.15
$C_3H_4O_2$	Acrylic acid	72.06	647 (342)	823 (56.8)	1.12
$C_3H_4O_2$	Vinyl formate	72.06	395 (202)	838 (57.8)	1.13
C_3H_5Cl	Allyl chloride	76.53	465 (241)	691 (47.7)	1.13
$C_3H_5Cl_3$	1,2,3-Trichloropropane	147.43	712 (378)	573 (39.5)	1.08
C_3H_5N	Propionitrile	55.08	556 (291)	607 (41.9)	1.13
C_3H_6	Cyclopropane	42.08	256 (124)	797 (55.0)	1.18
C_3H_6	Propylene	42.08	197 (92)	670 (46.2)	1.15
$C_3H_6Cl_2$	1,2-Dichloropropane	112.99	579 (304)	647 (44.6)	1.09
C_3H_6O	Acetone	58.08	455 (235)	682 (47.0)	1.13
C_3H_6O	Allyl alcohol	58.08	521 (272)	829 (57.2)	1.12
C_3H_6O	Propionaldehyde	58.08	433 (223)	691 (47.7)	1.12
C_3H_6O	Propylene oxide	58.08	408 (209)	714 (49.2)	1.13
C_3H_6O	Vinyl methyl ether	58.08	325 (163)	691 (47.7)	1.12
$C_3H_6O_2$	Propionic acid	74.08	642 (339)	779 (53.7)	1.10
$C_3H_6O_2$	Ethyl formate	74.08	455 (235)	688 (47.5)	1.10
$C_3H_6O_2$	Methyl acetate	74.08	452 (233)	681 (47.0)	1.12
C_3H_7Cl	Propyl chloride	78.54	445 (229)	664 (45.8)	1.11
C_3H_7Cl	Isopropyl chloride	78.54	413 (212)	685 (47.2)	1.11
C_3H_8	Propane	44.10	205 (96)	616 (42.5)	1.13
C_3H_8O	1-Propanol	60.10	506 (263)	750 (51.7)	1.11

Formula	Fluid	Molecular weight	Critical temperature, °F (°C)		Critical pressure, psia (bar)		Specific-heat ratio
C_3H_8O	Isopropyl alcohol	60.10	455	(235)	691	(47.7)	1.10
C_3H_8O	Methyl ethyl ether	60.10	328	(164)	638	(44.0)	1.10
$C_3H_8O_2$	Methylal	76.10	435	(224)	0	(0.0)	0.00
$C_3H_8O_2$	1,2-Propanediol	76.10	665	(352)	882	(60.8)	1.09
$C_3H_8O_2$	1,3-Propanediol	76.10	724	(384)	867	(59.8)	1.09
$C_3H_8O_3$	Glycerol	92.10	847	(452)	970	(66.9)	1.08
C_3H_8S	Methyl ethyl sulfide	76.16	499	(259)	617	(42.6)	1.10
C_3H_9N	N-Propyl amine	59.11	435	(224)	688	(47.5)	1.10
C_3H_9N	Isopropyl amine	59.11	397	(203)	735	(50.7)	1.10
C_3H_9N	Trimethyl amine	59.11	320	(160)	591	(40.8)	1.10
$C_4H_2O_3$	Maleic anhydride	98.06	-460	(-273)	0	(0.0)	1.13
C_4H_4	Vinylacetylene	52.08	359	(182)	720	(49.7)	1.13
C_4H_4O	Furan	68.08	422	(217)	798	(55.0)	1.15
C_4H_4S	Thiophene	84.14	583	(306)	826	(57.0)	1.13
C_4H_5N	Allyl cyanide	67.09	593	(312)	573	(39.5)	1.11
C_4H_5N	Pyrrole	67.09	692	(367)	0	(0.0)	0.00
C_4H_6	1-Butyne	54.09	374	(190)	684	(47.2)	1.12
C_4H_6	2-Butyne	54.09	419	(215)	738	(50.9)	1.12
C_4H_6	1,2-Butadiene	54.09	339	(171)	653	(45.0)	1.12
C_4H_6	1,3-Butadiene	54.09	305	(152)	628	(43.3)	1.12
$C_4H_6O_2$	Vinyl acetate	86.09	485	(252)	632	(43.6)	1.10
$C_4H_6O_3$	Acetic anhydride	102.09	564	(296)	679	(46.8)	1.09
$C_4H_6O_4$	Dimethyl oxalate	118.09	670	(354)	578	(39.9)	0.00
$C_4H_6O_4$	Succinic acid	118.09	-460	(273)	0	(0.0)	0.00

Formula	Name	MW			
C_4H_7N	Butyronitrile	69.11	588 (309)	550 (37.9)	1.10
$C_4H_7O_2$	Methyl acrylate	86.09	505 (263)	617 (42.6)	1.10
C_4H_8	1-Butene	56.11	295 (146)	583 (40.2)	1.11
C_4H_8	cis-2-Butene	56.11	324 (162)	610 (42.1)	1.12
C_4H_8	trans-1-Butene	56.11	311 (155)	595 (41.0)	1.11
C_4H_8	Cyclobutane	56.11	368 (187)	723 (49.9)	1.14
C_4H_8	Isobutylene	56.11	292 (144)	580 (40.0)	1.10
C_4H_8O	N-Butyraldehyde	72.11	483 (251)	588 (40.6)	1.09
C_4H_8O	Isobutyraldehyde	72.11	463 (239)	603 (41.6)	1.08
C_4H_8O	Methyl ethyl ketone	72.11	504 (262)	603 (41.6)	1.09
C_4H_8O	Tetrahydrofuran	72.11	512 (267)	753 (51.9)	1.09
C_4H_8O	Vinyl ethyl ether	72.11	395 (202)	591 (40.8)	1.09
$C_4H_8O_2$	N-Butyric acid	88.11	670 (354)	764 (52.7)	1.08
$C_4H_8O_2$	1,4-Dioxane	88.11	597 (314)	756 (52.1)	1.10
$C_4H_8O_2$	Ethyl acetate	88.11	481 (249)	556 (38.3)	1.08
$C_4H_8O_2$	Isobutyric acid	88.11	636 (336)	588 (40.6)	1.08
$C_4H_8O_2$	Methyl propionate	88.11	495 (257)	581 (40.1)	1.09
$C_4H_8O_2$	N-Propyl formate	88.11	508 (264)	589 (40.4)	0.00
C_4H_9Cl	1-Chlorobutane	92.57	516 (269)	535 (36.9)	1.09
C_4H_9Cl	2-Chlorobutane	92.57	477 (247)	573 (39.5)	1.08
C_4H_9Cl	Tert-butyl chloride	92.57	453 (234)	573 (39.5)	1.08
C_4H_9N	Pyrrolidine	71.12	563 (295)	814 (56.1)	1.12
C_4H_9NO	Morpholine	87.12	652 (344)	794 (54.8)	1.10
C_4H_{10}	N-Butane	58.12	305 (152)	551 (38.0)	1.09
C_4H_{10}	Isobutane	58.12	275 (135)	529 (36.5)	1.09
$C_4H_{10}O$	N-Butanol	74.12	553 (289)	641 (44.2)	1.08
$C_4H_{10}O$	2-Butanol	74.12	505 (263)	609 (42.0)	1.08
$C_4H_{10}O$	Isobutanol	74.12	526 (274)	623 (43.0)	1.08
$C_4H_{10}O$	Tert-butanol	74.12	451 (233)	576 (39.7)	1.08

Formula	Fluid	Molecular weight	Critical temperature, °F (°C)		Critical pressure, psia (bar)		Specific-heat ratio
$C_4H_{10}O$	Ethyl ether	74.12	380	(193)	528	(36.4)	1.08
$C_4H_{10}O_2$	1,2-Dimethoxyethane	90.12	505	(263)	562	(36.7)	1.07
$C_4H_{10}O_3$	Diethylene glycol	106.12	766	(408)	676	(46.6)	1.05
$C_4H_{10}S$	Diethyl sulfide	90.18	543	(284)	575	(39.7)	1.08
$C_4H_{10}S_2$	Diethyl disulfide	122.24	696	(369)	0	(0.0)	1.06
$C_4H_{11}N$	N-Butyl amine	73.14	483	(251)	603	(41.6)	1.08
$C_4H_{11}N$	Isobutyl amine	73.14	469	(243)	617	(42.6)	1.07
$C_4H_{11}N$	Diethyl amine	73.14	434	(223)	538	(37.1)	1.08
C_5H_5N	Pyridine	79.10	656	(347)	817	(56.3)	1.06
C_5H_8	Cyclopentene	68.12	451	(233)	0	(0.0)	1.13
C_5H_8	1,2-Pentadiene	68.12	445	(229)	591	(40.8)	1.09
C_5H_8	1-*trans*-3-Pentadiene	68.12	433	(223)	579	(39.9)	1.08
C_5H_8	1,4-Pentadiene	68.12	400	(204)	550	(37.9)	1.09
C_5H_8	1-Pentyne	68.12	428	(220)	588	(40.6)	1.09
C_5H_8	2-Methyl-1,3-butadiene	68.12	411	(211)	559	(38.6)	1.09
C_5H_8	3-Methyl-1,2-butadiene	68.12	433	(223)	597	(41.2)	1.09
C_5H_8O	Cyclopentanone	84.12	667	(353)	779	(53.7)	1.11
$C_5H_8O_2$	Ethyl acrylate	100.12	534	(279)	544	(37.5)	1.01
C_5H_{10}	Cyclopentane	70.14	461	(238)	654	(45.1)	1.11
C_5H_{10}	1-Pentene	70.14	376	(191)	588	(40.6)	1.08
C_5H_{10}	*cis*-2-Pentene	70.14	397	(203)	529	(36.5)	1.09
C_5H_{10}	*trans*-2-Pentene	70.14	395	(202)	531	(36.6)	1.08
C_5H_{10}	2-Methyl-1-butene	70.14	377	(192)	500	(34.5)	1.08

C$_5$H$_{10}$	2-Methyl-2-butene	70.14	386	(197)	500	(34.5)	1.09
C$_5$H$_{10}$	3-Methyl-1-butene	70.14	350	(177)	510	(35.2)	1.07
C$_5$H$_{10}$O	Valeraldehyde	86.13	537	(281)	515	(35.5)	1.07
C$_5$H$_{10}$O	Methyl N-propyl ketone	86.13	555	(291)	564	(38.9)	1.07
C$_5$H$_{10}$O	Methyl isopropyl ketone	86.13	536	(280)	559	(38.6)	1.07
C$_5$H$_{10}$O	Diethyl ketone	86.13	550	(288)	542	(37.4)	1.07
C$_5$H$_{10}$O$_2$	N-Valeric acid	102.13	712	(378)	559	(38.6)	1.06
C$_5$H$_{10}$O$_2$	Isobutyl formate	102.13	532	(278)	563	(38.8)	1.07
C$_5$H$_{10}$O$_2$	N-Propyl acetate	102.13	529	(276)	484	(33.4)	1.07
C$_5$H$_{10}$O$_2$	Ethyl propionate	102.13	523	(273)	488	(33.7)	1.07
C$_5$H$_{10}$O$_2$	Methyl butyrate	102.13	538	(281)	504	(34.8)	0.00
C$_5$H$_{10}$O$_2$	Methyl isobutyrate	102.13	513	(267)	498	(34.3)	0.00
C$_5$H$_{11}$N	Piperidine	85.15	609	(321)	691	(47.7)	1.09
C$_5$H$_{12}$	N-Pentane	72.15	385	(196)	490	(33.8)	1.08
C$_5$H$_{12}$	2-Methyl butane	72.15	369	(187)	491	(33.9)	1.08
C$_5$H$_{12}$	2,2-Dimethyl propane	72.15	321	(161)	465	(32.1)	1.07
C$_5$H$_{12}$O	1-Pentanol	88.15	595	(313)	559	(38.6)	1.07
C$_5$H$_{12}$O	2-Methyl-1-butanol	88.15	568	(298)	559	(38.6)	1.07
C$_5$H$_{12}$O	3-Methyl-1-butanol	88.15	583	(306)	559	(38.6)	1.07
C$_5$H$_{12}$O	2-Methyl-2-butanol	88.15	521	(272)	573	(39.5)	1.07
C$_5$H$_{12}$O	2,2-Dimethyl-1-propane	88.15	528	(276)	573	(39.5)	1.06
C$_5$H$_{12}$O	Ethyl propyl ether	88.15	441	(227)	472	(32.6)	0.00
C$_6$F$_6$	Perfluorobenzene	186.06	470	(243)	479	(33.0)	1.06
C$_6$F$_{12}$	Perfluorocyclohexane	300.05	363	(184)	353	(24.3)	1.06
C$_6$F$_{14}$	Perfluoro-N-hexane	338.04	353	(178)	276	(19.0)	0.00
C$_6$H$_4$Cl$_2$	O-Dichlorobenzene	147.00	795	(424)	595	(41.0)	1.08
C$_6$H$_4$Cl$_2$	M-Dichlorobenzene	147.00	771	(411)	559	(38.6)	1.08
C$_6$H$_4$Cl$_2$	P-Dichlorobenzene	147.00	773	(412)	573	(39.5)	1.08
C$_6$H$_5$Br	Bromobenzene	157.01	746	(397)	656	(45.2)	1.09

Formula	Fluid	Molecular weight	Critical temperature, °F (°C)	Critical pressure, psia (bar)	Specific-heat ratio
C_6H_5Cl	Chlorobenzene	112.56	678 (359)	656 (45.2)	1.10
C_6H_5F	Fluorobenzene	96.10	548 (287)	660 (45.5)	1.10
C_6H_5I	Iodobenzene	204.01	838 (448)	656 (45.2)	1.09
C_6H_6	Benzene	78.11	552 (289)	710 (49.0)	1.11
C_6H_6O	Phenol	94.11	790 (421)	889 (61.3)	1.09
C_6H_7N	Aniline	93.13	798 (426)	770 (53.1)	1.08
C_6H_7N	4-Methyl pyridine	93.13	703 (373)	647 (44.6)	1.09
C_6H_{10}	1,5-Hexadiene	82.15	453 (234)	500 (34.5)	0.00
C_6H_{10}	Cyclohexene	82.15	549 (287)	631 (43.5)	1.09
$C_6H_{10}O$	Cyclohexanone	98.15	672 (356)	559 (38.6)	1.08
C_6H_{12}	Cyclohexane	84.16	536 (280)	591 (40.1)	1.09
C_6H_{12}	Methycyclopentane	84.16	499 (259)	550 (37.9)	1.08
C_6H_{12}	1-Hexene	84.16	447 (231)	460 (31.7)	1.07
C_6H_{12}	cis-2-Hexene	84.16	472 (244)	476 (32.8)	1.07
C_6H_{12}	trans-2-Hexene	84.16	469 (243)	475 (32.8)	1.07
C_6H_{12}	cis-3-Hexene	84.16	471 (244)	476 (32.8)	1.07
C_6H_{12}	trans-3-Hexene	84.16	476 (247)	472 (32.6)	1.07
C_6H_{12}	2-Methyl-2-pentene	84.16	472 (244)	476 (32.8)	1.07
C_6H_{12}	3-Methyl-cis-2-pentene	84.16	472 (244)	476 (32.8)	1.07
C_6H_{12}	3-Methyl-trans-2-pentene	84.16	478 (247)	478 (33.0)	1.07
C_6H_{12}	4-Methyl-cis-2-pentene	84.16	422 (217)	441 (30.4)	1.07
C_6H_{12}	4-Methyl-trans-2-pentene	84.16	427 (219)	441 (30.4)	1.06
C_6H_{12}	2,3-Dimethyl-1-butene	84.16	442 (228)	470 (32.4)	1.06
C_6H_{12}	3,3-Dimethyl-1-butene	84.16	422 (217)	472 (32.6)	1.07

Formula	Name	MW					
$C_6H_{12}O$	Cyclohexanol	100.16	665	(352)	544	(37.5)	1.07
$C_6H_{12}O$	Methyl isobutyl ketone	100.16	568	(298)	475	(32.8)	1.07
$C_6H_{12}O_2$	N-Butyl acetate	116.16	582	(306)	456	(31.5)	1.06
$C_6H_{12}O_2$	Isobutyl acetate	116.16	550	(288)	441	(30.4)	1.06
$C_6H_{12}O_2$	Ethyl butyrate	116.16	559	(293)	456	(31.5)	1.06
$C_6H_{12}O_2$	Ethyl isobutyrate	116.16	535	(279)	441	(30.4)	0.00
$C_6H_{12}O_2$	N-Propyl propionate	116.16	580	(304)	0	(0.0)	0.00
C_6H_{14}	N-Hexane	86.18	453	(234)	431	(29.7)	1.06
C_6H_{14}	2-Methyl pentane	86.18	436	(224)	437	(30.1)	1.06
C_6H_{14}	3-Methyl pentane	86.18	448	(231)	453	(31.2)	1.06
C_6H_{14}	2,2-Dimethyl butane	86.18	420	(216)	447	(30.8)	1.06
C_6H_{14}	2,3-Dimethyl butane	86.18	440	(227)	454	(31.3)	1.06
$C_6H_{14}O$	1-Hexanol	102.18	638	(337)	588	(40.6)	1.06
$C_6H_{14}O$	Ethyl butyl ether	102.18	496	(258)	441	(30.4)	1.05
$C_6H_{14}O$	Diisopropyl ether	102.18	440	(227)	417	(28.8)	1.06
$C_6H_{15}N$	Dipropylamine	101.19	530	(277)	456	(31.5)	1.05
$C_6H_{15}N$	Triethylamine	101.19	503	(262)	441	(30.4)	1.06
C_7F_{14}	Perfluoromethylcyclohexane	350.06	416	(213)	338	(23.3)	0.00
C_7F_{16}	Perfluoro-N-heptane	388.06	395	(202)	235	(16.2)	0.00
C_7H_5N	Benzonitrile	103.12	799	(426)	612	(42.2)	1.08
C_7H_6O	Benzaldehyde	106.12	791	(422)	676	(46.6)	1.08
$C_7H_6O_2$	Benzoic acid	122.12	894	(479)	662	(45.7)	1.09
C_7H_8	Toluene	92.14	605	(318)	597	(41.2)	1.09
C_7H_8O	Methyl phenyl ether	108.14	694	(368)	606	(41.8)	0.00
C_7H_8O	Benzyl alcohol	108.14	759	(404)	676	(46.6)	1.07
C_7H_8O	O-Cresol	108.14	796	(424)	726	(50.1)	1.07
C_7H_8O	M-Cresol	108.14	810	(432)	662	(45.7)	1.07
C_7H_8O	P-Cresol	108.14	808	(431)	747	(51.5)	1.07

Formula	Fluid	Molecular weight	Critical temperature, °F (°C)	Critical pressure, psia (bar)	Specific-heat ratio
C_7H_9N	2,3-Dimethylpyridine	107.16	720 (383)	0 (0.0)	0.00
C_7H_9N	2,5-Dimethylpyridine	107.16	700 (371)	0 (0.0)	0.00
C_7H_9N	3,4-Dimethylpyridine	107.16	771 (411)	0 (0.0)	0.00
C_7H_9N	3,5-Dimethylpyridine	107.16	741 (394)	0 (0.0)	0.00
C_7H_9N	Methylphenylamine	107.16	802 (428)	754 (52)	0.00
C_7H_9N	O-Toluidine	107.16	789 (421)	544 (37.5)	0.00
C_7H_9N	M-Toluidine	107.16	816 (436)	603 (41.6)	1.07
C_7H_9N	P-Toluidine	107.16	741 (394)	0 (0.0)	0.00
C_7H_{14}	Cycloheptane	98.19	600 (316)	539 (37.2)	1.07
C_7H_{14}	1,1-Dimethylcyclopentane	98.19	525 (273)	500 (34.5)	1.07
C_7H_{14}	cis-1,2-Dimethylcyclopentane	98.19	557 (292)	500 (34.5)	1.07
C_7H_{14}	trans-1,2-Dimethylcyclopentane	98.19	536 (280)	500 (34.5)	1.07
C_7H_{14}	Ethylcyclopentane	98.19	565 (296)	492 (33.9)	1.07
C_7H_{14}	Methylcyclohexane	98.19	570 (299)	504 (34.8)	1.07
C_7H_{14}	1-Heptene	98.19	507 (264)	412 (28.4)	1.06
C_7H_{14}	2,3,3-Trimethyl-1-butane	98.19	499 (259)	420 (29.0)	0.00
C_7H_{16}	N-Heptane	100.21	512 (267)	397 (27.4)	1.05
C_7H_{16}	2-Methylhexane	100.21	495 (257)	397 (27.4)	1.05
C_7H_{16}	3-Methylhexane	100.21	503 (262)	409 (28.0)	1.05
C_7H_{16}	2,2-Dimethylpentane	100.21	477 (247)	403 (27.8)	1.05
C_7H_{16}	2,3-Dimethylpentane	100.21	507 (264)	422 (29.1)	1.05
C_7H_{16}	2,4-Dimethylpentane	100.21	475 (246)	397 (27.4)	1.05
C_7H_{16}	3,3-Dimethylpentane	100.21	505 (263)	428 (29.5)	1.05
C_7H_{16}	3-Ethylpentane	100.21	513 (267)	419 (28.9)	1.05

Formula	Compound	Mol. wt.					
C_7H_{16}	2,2,3-Trimethylbutane	100.21	496	(258)	429	(29.6)	1.05
$C_7H_{16}O$	1-Heptanol	116.20	679	(359)	441	(30.4)	1.05
$C_8H_4O_3$	Phthalic anhydride	148.12	998	(537)	691	(47.7)	1.06
C_8H_8	Styrene	104.15	705	(374)	579	(39.9)	1.07
C_8H_8O	Methyl phenyl ketone	120.15	802	(428)	559	(38.6)	1.07
$C_8H_8O_2$	Methyl benzoate	136.15	786	(419)	529	(36.5)	1.07
C_8H_{10}	O-Xylene	106.17	674	(357)	541	(37.3)	1.07
C_8H_{10}	M-Xylene	106.17	651	(344)	515	(35.5)	1.07
C_8H_{10}	P-Xylene	106.17	649	(343)	510	(35.2)	1.07
C_8H_{10}	Ethylbenzene	106.17	651	(344)	523	(36.1)	1.07
$C_8H_{10}O$	O-Ethylphenol	122.17	805	(429)	0	(0.0)	0.00
$C_8H_{10}O$	M-Ethylphenol	122.17	830	(443)	0	(0.0)	0.00
$C_8H_{10}O$	P-Ethylphenol	122.17	830	(443)	0	(0.0)	0.00
$C_8H_{10}O$	Phenetole	122.17	705	(374)	497	(34.3)	0.00
$C_8H_{10}O$	2,3-Xylenol	122.17	841	(449)	0	(0.0)	0.00
$C_8H_{10}O$	2,4-Xylenol	122.17	814	(434)	0	(0.0)	0.00
$C_8H_{10}O$	2,5-Xylenol	122.17	841	(449)	0	(0.0)	0.00
$C_8H_{10}O$	2,6-Xylenol	122.17	802	(428)	0	(0.0)	0.00
$C_8H_{10}O$	3,4-Xylenol	122.17	854	(457)	0	(0.0)	0.00
$C_8H_{10}O$	3,5-Xylenol	122.17	828	(442)	0	(0.0)	0.00
$C_8H_{11}N$	N,N-Dimethylaniline	121.18	777	(414)	526	(36.3)	1.06
C_8H_{16}	1,1-Dimethylcyclohexane	112.22	604	(318)	430	(29.7)	1.06
C_8H_{16}	cis-1,2-Dimethylcyclohexane	112.22	631	(333)	431	(29.7)	1.06
C_8H_{16}	trans-1,2-Dimethylcyclohexane	112.22	613	(323)	431	(29.7)	1.06
C_8H_{16}	cis-1,3-Dimethylcyclohexane	112.22	604	(318)	431	(29.7)	1.06
C_8H_{16}	trans-1,3-Dimethylcyclohexane	112.22	616	(324)	431	(29.7)	1.06
C_8H_{16}	cis-1,4-Dimethylcyclohexane	112.22	616	(324)	431	(29.7)	1.06
C_8H_{16}	trans-1,4-Dimethylcyclohexane	112.22	602	(317)	431	(29.7)	1.06
C_8H_{16}	Ethylcyclohexane	112.22	636	(336)	440	(30.3)	1.06

Formula	Fluid	Molecular weight	Critical temperature, °F (°C)	Critical pressure, psia (bar)	Specific-heat ratio
C_8H_{16}	1,1,2-Trimethylcyclopentane	112.22	583 (306)	426 (29.4)	0.00
C_8H_{16}	1,1,3-Trimethylcyclopentane	112.22	565 (296)	410 (28.3)	0.00
C_8H_{16}	cis,cis,trans-1,2,4-Trimethyl	112.22	582 (306)	417 (28.8)	0.00
C_8H_{16}	cis,trans-cis-1,2,4-Trimethyl	112.22	568 (298)	407 (28.1)	0.00
C_8H_{16}	1-Methyl-1-ethylcyclohexane	112.22	606 (319)	434 (29.9)	0.00
C_8H_{16}	N-Propylcyclopentane	112.22	625 (329)	435 (30.0)	1.06
C_8H_{16}	Isopropylcyclopentane	112.22	622 (328)	435 (30.0)	0.00
C_8H_{16}	1-Octene	112.22	560 (293)	381 (26.3)	1.05
C_8H_{16}	trans-2-Octene	112.22	584 (307)	401 (27.7)	1.05
C_8H_{16}	N-Octane	114.23	564 (296)	360 (24.8)	1.05
C_8H_{18}	2-Methylheptane	114.23	547 (286)	360 (24.8)	1.05
C_8H_{18}	3-Methylheptane	114.23	554 (290)	369 (25.5)	1.05
C_8H_{18}	4-Methylheptane	114.23	551 (288)	369 (25.5)	1.05
C_8H_{18}	2,2-Dimethylhexane	114.23	530 (277)	368 (25.4)	1.05
C_8H_{18}	2,3-Dimethylhexane	114.23	554 (290)	381 (26.3)	1.05
C_8H_{18}	2,4-Dimethylhexane	114.23	536 (280)	370 (25.5)	1.05
C_8H_{18}	2,5-Dimethylhexane	114.23	530 (277)	360 (24.8)	1.05
C_8H_{18}	3,3-Dimethylhexane	114.23	552 (289)	385 (26.6)	1.05
C_8H_{18}	3,4-Dimethylhexane	114.23	564 (296)	391 (27.0)	1.05
C_8H_{18}	3-Ethylhexane	114.23	558 (292)	378 (26.1)	1.05
C_8H_{18}	2,2,3-Trimethylpentane	114.23	554 (290)	395 (27.2)	1.05
C_8H_{18}	2,2,4-Trimethylpentane	114.23	519 (270)	372 (25.7)	1.05
C_8H_{18}	2,3,3-Trimethylpentane	114.23	572 (300)	408 (28.1)	1.05
C_8H_{18}	2,3,4-Trimethylpentane	114.23	559 (293)	395 (27.2)	1.05

Formula	Name						
C_8H_{18}	2-Methyl-3-ethylpentane	114.23	561	(294)	392	(27.0)	1.05
C_8H_{18}	3-Methyl-3-ethylpentane	114.23	578	(303)	407	(28.1)	1.05
$C_8H_{18}O$	1-Octanol	130.23	724	(384)	500	(34.5)	1.04
$C_8H_{18}O$	2-Octanol	130.23	687	(364)	397	(27.4)	1.04
$C_8H_{18}O$	2-Ethylhexanol	130.23	643	(339)	400	(27.6)	1.04
$C_8H_{18}O$	Butyl ether	130.23	584	(307)	368	(25.4)	1.04
$C_8H_{19}N$	Dibutylamine	129.25	613	(323)	368	(25.4)	1.04
C_9H_{10}	α-Methyl styrene	118.18	717	(381)	494	(34.1)	1.06
$C_9H_{10}O_2$	Ethyl benzoate	150.18	795	(424)	470	(32.4)	1.05
C_9H_{12}	N-Propylbenzene	120.20	689	(365)	465	(32.1)	1.06
C_9H_{12}	Isopropylbenzene	120.20	676	(358)	466	(32.1)	1.06
C_9H_{12}	1-Methyl-2-ethylbenzene	120.20	712	(378)	441	(30.4)	1.06
C_9H_{12}	1-Methyl-2-ethylbenzene	120.20	687	(364)	412	(28.4)	1.06
C_9H_{12}	1-Methyl-4-ethylbenzene	120.20	692	(367)	426	(29.4)	1.06
C_9H_{12}	1,2,3-Trimethylbenzene	120.20	736	(391)	501	(34.6)	1.06
C_9H_{12}	1,2,4-Trimethylbenzene	120.20	708	(376)	469	(32.3)	1.06
C_9H_{12}	1,2,5-Trimethylbenzene	120.20	687	(364)	454	(31.3)	1.06
C_9H_{18}	N-Propylcyclohexane	126.24	690	(366)	407	(28.1)	1.05
C_9H_{18}	Isopropylcylcohexane	126.24	692	(367)	412	(28.4)	0.00
C_9H_{18}	1-Nonene	126.24	606	(319)	340	(23.5)	1.04
C_9H_{20}	N-Nonane	128.26	610	(321)	335	(23.1)	1.05
C_9H_{20}	2,2,3-Trimethylhexane	128.26	598	(314)	362	(25.0)	1.04
C_9H_{20}	2,2,4-Trimethylhexane	128.26	573	(301)	344	(23.7)	1.04
C_9H_{20}	2,2,5-Trimethylhexane	128.26	562	(278)	338	(23.3)	1.04
C_9H_{20}	3,3-Diethylpentane	128.26	638	(337)	388	(26.8)	1.04
C_9H_{20}	2,2,3,3-Tetramethylpentane	128.26	634	(334)	397	(27.4)	1.04
C_9H_{20}	2,2,3,4-Tetramethylpentane	128.26	607	(319)	378	(26.1)	1.04
C_9H_{20}	2,2,4,4-Tetramethylpentane	128.26	574	(301)	360	(24.8)	1.04

Formula	Fluid	Molecular weight	Critical temperature, °F (°C)	Critical pressure, psia (bar)	Specific-heat ratio
C_9H_{20}	2,3,3,4-Tetramethylpentane	128.26	634 (334)	394 (27.2)	1.04
$C_{10}H_8$	Napthalene	128.17	887 (475)	588 (40.6)	1.07
$C_{10}H_{12}$	1,2,3,4-Tetrahydronapthalene	132.21	834 (446)	510 (35.2)	0.00
$C_{10}H_{14}$	N-Butylbenzene	134.22	728 (387)	419 (29.9)	1.05
$C_{10}H_{14}$	Isobutylbenzene	134.22	710 (378)	456 (31.5)	0.00
$C_{10}H_{14}$	sec-Butylbenzene	134.22	735 (391)	427 (29.5)	1.05
$C_{10}H_{14}$	tert-Butylbenzene	134.22	728 (387)	431 (29.7)	1.05
$C_{10}H_{14}$	1-Methyl-2-isopropylbenzene	134.22	746 (397)	420 (29.0)	0.00
$C_{10}H_{14}$	1-Methyl-3-isopropylbenzene	134.22	739 (393)	426 (29.4)	1.05
$C_{10}H_{14}$	1-Methyl-4-isopropylbenzene	134.22	715 (379)	410 (28.3)	0.00
$C_{10}H_{14}$	1,4-Diethylbenzene	134.22	724 (384)	407 (28.1)	1.05
$C_{10}H_{14}$	1,2,4,5-Tetramethylbenzene	134.22	755 (402)	426 (29.4)	1.05
$C_{10}H_{15}N$	N-Butylaniline	149.24	838 (448)	412 (28.4)	1.05
$C_{10}H_{18}$	cis-Decalin	138.25	804 (429)	456 (31.5)	1.05
$C_{10}H_{18}$	trans-Decalin	138.25	782 (417)	456 (31.5)	1.05
$C_{10}H_{18}N$	Caprylonitrile	153.27	660 (349)	472 (32.6)	0.00
$C_{10}H_{20}$	N-Butylcyclohexane	140.27	741 (394)	457 (31.5)	1.04
$C_{10}H_{20}$	Isobutylcyclohexane	140.27	726 (386)	453 (31.2)	0.00
$C_{10}H_{20}$	sec-Butylcyclohexane	140.27	744 (396)	388 (26.8)	0.00
$C_{10}H_{20}$	tert-Butylcyclohexane	140.27	726 (386)	387 (26.7)	0.00
$C_{10}H_{20}$	1-Decene	140.27	647 (342)	320 (22.1)	1.04
$C_{10}H_{22}$	N-Decane	142.29	652 (344)	306 (21.1)	1.04
$C_{10}H_{22}$	3,3,5-Trimethylheptane	142.29	637 (336)	337 (23.2)	1.04
$C_{10}H_{22}$	2,2,3,3-Tetramethylheptane	142.29	662 (350)	365 (25.2)	1.04

Formula	Name				
$C_{10}H_{22}$	2,2,5,5-Tetramethylheptane	142.29	587 (308)	318 (21.9)	1.04
$C_{10}H_{22}O$	1-Decanol	158.29	800 (427)	323 (22.3)	1.04
$C_{11}H_{10}$	1-Methylnaphthalene	142.20	930 (499)	517 (35.7)	1.06
$C_{11}H_{10}$	2-Methylnaphthalene	142.20	910 (488)	509 (35.1)	1.06
$C_{11}H_{14}O_2$	Butyl benzoate	178.23	841 (449)	382 (26.3)	1.04
$C_{11}H_{22}$	N-Hexylcyclopentane	154.30	728 (387)	310 (21.4)	1.04
$C_{11}H_{22}$	1-Undecene	154.30	687 (364)	290 (20.0)	1.04
$C_{11}H_{24}$	N-Undecane	156.31	690 (366)	285 (19.7)	1.03
$C_{12}H_{10}$	Diphenyl	154.21	960 (516)	559 (38.6)	1.06
$C_{12}H_{10}O$	Diphenyl ether	170.21	919 (493)	456 (31.5)	1.05
$C_{12}H_{24}$	N-Heptylcyclopentane	168.32	762 (406)	282 (19.5)	1.04
$C_{12}H_{24}$	1-Dodecene	168.32	723 (384)	269 (18.6)	1.03
$C_{12}H_{26}$	N-Dodecane	170.34	725 (385)	265 (18.3)	1.03
$C_{12}H_{26}O$	Dehexyl ether	186.340	723 (384)	265 (18.3)	1.03
$C_{12}H_{26}O$	Dodecanol	186.340	762 (406)	279 (19.2)	1.03
$C_{12}H_{27}N$	Tributylamine	185.36	697 (369)	265 (18.3)	1.03
$C_{13}H_{12}$	Diphenylmethane	168.240	921 (494)	432 (29.8)	0.00
$C_{13}H_{26}$	N-Octylcyclopentane	182.35	789 (421)	260 (17.9)	1.03
$C_{13}H_{26}$	1-Tridecene	182.35	753 (401)	247 (17.0)	1.03
$C_{13}H_{28}$	N-Tridecene	184.37	756 (402)	250 (17.2)	1.03
$C_{14}H_{10}$	Anthracene	178.23	1129 (609)	0 (0.0)	1.05
$C_{14}H_{10}$	Phenanthrene	178.23	1120 (604)	0 (0.0)	1.05
$C_{14}H_{28}$	N-Nonylcyclopentane	196.38	819 (437)	240 (16.6)	1.03
$C_{14}H_{28}$	1-Tetradecene	196.38	780 (416)	226 (15.6)	1.03
$C_{14}H_{30}$	N-Tetradecane	198.39	789 (421)	235 (16.2)	1.03
$C_{15}H_{30}$	N-Decylcyclopentane	210.41	843 (451)	221 (15.2)	1.03
$C_{15}H_{30}$	1-Pentadecene	210.41	807 (431)	212 (14.6)	1.03
$C_{15}H_{32}$	N-Pentadecane	212.42	813 (434)	221 (15.2)	1.02

Formula	Fluid	Molecular weight	Critical temperature, °F (°C)	Critical pressure, psia (bar)	Specific-heat ratio
$C_{16}H_{22}O_4$	Dibutyl-O-phthalate	278.35	−460 (−273)	0 (0.0)	1.03
$C_{16}H_{32}$	N-Decylcyclohexane	224.43	890 (477)	197 (13.6)	1.03
$C_{16}H_{32}$	1-Hexadecene	224.43	831 (444)	194 (13.4)	1.02
$C_{16}H_{34}$	N-Hexadecane	226.45	831 (444)	206 (14.2)	1.02
$C_{17}H_{34}$	N-Dodecylcyclopentane	238.46	890 (477)	188 (13.0)	1.02
$C_{17}H_{36}$	N-Heptadecane	240.48	859 (309)	191 (13.2)	1.02
$C_{17}H_{36}O$	Heptadecanol	256.47	865 (463)	206 (14.2)	1.02
$C_{18}H_{14}$	O-Terphenyl	230.31	1144 (618)	566 (39.0)	0.00
$C_{18}H_{14}$	M-Terphenyl	230.31	1205 (652)	509 (35.1)	0.00
$C_{18}H_{14}$	P-Terphenyl	230.31	1207 (653)	482 (33.2)	0.00
$C_{18}H_{36}$	1-Octadecene	252.49	870 (466)	165 (11.4)	1.02
$C_{18}H_{36}$	N-Tridecylcyclopentane	252.49	910 (488)	175 (12.1)	1.02
$C_{18}H_{38}$	N-Octadecane	254.50	881 (472)	175 (12.1)	1.02
$C_{18}H_{38}O$	1-Octadecanol	270.50	885 (474)	206 (14.2)	1.02
$C_{19}H_{38}$	N-Tetradecylcyclopentane	266.51	930 (499)	163 (11.2)	1.02
$C_{19}H_{40}$	N-Nonadecane	268.53	901 (483)	162 (11.2)	1.02
$C_{20}H_{40}$	N-Pentadecylcyclopentane	280.54	944 (507)	148 (10.2)	1.02
$C_{20}H_{42}$	N-Eicosane	282.56	921 (494)	162 (11.21)	1.02
$C_{20}H_{42}O$	1-Eicosanol	298.56	926 (497)	176 (12.1)	1.02
$C_{21}H_{42}$	N-Hexadecylcyclopentane	294.57	964 (518)	141 (9.7)	1.02
CO	Carbon monoxide	28.01	−221 (−141)	507 (35.0)	1.40
COS	Carbonyl sulfide	60.07	215 (102)	853 (58.8)	1.25
CO_2	Carbon dioxide	44.01	88 (31)	1070 (73.8)	1.29
CS_2	Carbon sulfide	76.13	534 (279)	1147 (79.1)	1.22

Formula	Name	Molar mass			
D_2	Deuterium	4.03	-391 (-235)	241 (16.6)	1.40
D_2O	Deuterium oxide	20.03	699 (371)	3143 (216.8)	0.00
F_2	Fluorine	38.00	-200 (-129)	757 (52.2)	1.36
F_3N	Nitrogen trifluoride	71.00	-39 (-39)	657 (45.3)	0.00
F_4Si	Silicon tetrafluoride	104.08	6 (-14)	539 (37.2)	0.00
F_6S	Sulfur hexafluoride	146.05	114 (45)	545 (37.6)	0.00
HBr	Hydrogen bromide	80.91	194 (90)	1241 (85.6)	1.40
HCl	Hydrogen chloride	36.46	124 (51)	1205 (83.1)	1.40
HF	Hydrogen fluoride	20.01	370 (188)	941 (64.9)	1.40
HI	Hydrogen iodide	127.91	303 (151)	1205 (83.1)	1.40
H_2	Hydrogen	2.02	-400 (-240)	188 (13.0)	1.40
H_2O	Water	18.02	705 (374)	3199 (220.6)	1.33
H_2S	Hydrogen sulfide	34.08	212 (100)	1297 (89.5)	1.32
H_3N	Ammonia	17.03	270 (132)	1636 (112.8)	1.31
H_4N_2	Hydrazine	32.05	715 (379)	2132 (147.0)	1.19
He	Helium	4.00	-451 (-268)	34 (2.3)	0.00
I_2	Iodine	253.81	1014 (546)	1691 (116.6)	1.29
Kr	Krypton	83.80	-83 (-64)	798 (55.0)	0.00
NO	Nitric oxide	30.01	-136 (-93)	940 (64.8)	1.39
NO_2	Nitrogen oxide	46.01	317 (158)	1470 (101.4)	1.29
N_2	Nitrogen	28.01	-233 (-147)	492 (33.9)	1.40
N_2O	Nitrous oxide	44.01	97 (36)	1051 (72.5)	1.28
Ne	Neon	20.18	-380 (-229)	400 (27.6)	0.00
O_2	Oxygen	32.00	-182 (-119)	732 (50.5)	1.40
O_2S	Sulfur dioxide	64.06	315 (157)	1144 (78.9)	1.27
O_3	Ozone	48.00	10 (-12)	808 (55.7)	1.27
O_3S	Sulfur trioxide	80.06	424 (218)	1191 (82.1)	0.00
Xe	Xenon	131.30	61 (16)	847 (58.4)	0.00

Appendix C

Method 21: Determination of Volatile Organic Compound Leaks*

1. *Applicability and Principle*
   1.1 *Applicability.* This method applies to determination of volatile organic compound (VOC) leaks from process equipment. These sources include, but are not limited to, valves, flanges and other connections, pumps and compressors, pressure relief devices, process drains, open-ended valves, pump and compressor seal system degassing vents, accumulator vessel vents, agitator seals, and access door seals.
   1.2 *Principle.* A portable instrument is used to detect VOC leaks from individual sources. The instrument detector type is not specified, but it must meet the specifications and performance criteria contained in Section 3. A leak definition concentration based on a reference compound is specified in each applicable regulation. This procedure is intended to locate and classify leaks only, and is not to be used as a direct measure of mass emission rates from individual sources.
2. *Definitions*
   2.1 *Leak Definition Concentration.* The local VOC concentration at the surface of a leak source that indicates that a VOC emission (leak)

*From Code of Federal Regulations, Title 40, Part 60, Appendix A.

is present. The leak definition is an instrument meter reading based on a reference compound.

2.2 *Reference Compound.* The VOC specifies selected as an instrument calibration basis for specification of the leak definition concentration. (For example: If a leak definition concentration is 10,000 ppmv* as methane, then any source emission that results in a local concentration that yields a meter reading of 10,000 on an instrument calibrated with methane would be classified as a leak. In this example, the leak definition is 10,000 ppmv, and the reference compound is methane.)

2.3 *Calibration Gas.* The VOC compound used to adjust the instrument meter reading to a known value. The calibration gas is usually the reference compound at a concentration approximately equal to the leak definition concentration.

2.4 *No Detectable Emission.* The local VOC concentration at the surface of a leak source that indicates that a VOC emission (leak) is not present. Since background VOC concentrations may exist, and to account for instrument drift and imperfect reproducibility, a difference between the source surface concentration and the local ambient concentration is determined. A difference based on meter readings of less than a concentration corresponding to a minimum readability specification indicates that a VOC emission (leak) is not present. (For example, if the leak definition in a regulation is 10,000 ppmv, then the allowable increase in surface concentration versus local ambient concentration would be 500 ppmv based on the instrument meter readings.)

2.5 *Response Factor.* The ratio of the known concentration of a VOC compound to the observed meter reading when measured using an instrument calibrated with the reference compound specified in the application regulation.

2.6 *Calibration Precision.* The degree of agreement between measurements of the same known value, expressed as the relative percentage of the average difference between the meter readings and the known concentration to the know concentration.

2.7 *Response Time.* The time interval from a step change in VOC concentration at the input of the sampling system to the time at which 90 percent of the corresponding final value is reached as displayed on the instrument readout meter.

---

*ppmv = parts per million by volume.

3. *Apparatus*
   3.1 *Monitoring Instrument*
      3.1.1 *Specifications*
         a. The VOC instrument detector shall respond to the compounds being processed. Detector types which may meet this requirement include, but are not limited to, catalytic oxidation, flame ionization, infrared absorption, and photoionization.
         b. The instrument shall be capable of measuring a leak definition concentration specified in the regulation.
         c. The scale of the instrument meter shall be readable to ± 5 percent of the specified leak definition concentration.
         d. The instrument shall be equipped with a pump so that a continuous sample is provided to the detector. The nominal sample flow rate shall be $\frac{1}{2}$ to 3 liters per minute.
         e. The instrument shall be intrinsically safe for operation in explosive atmospheres as defined by the applicable U.S.A. standards (e.g., National Electrical Code by the National Fire Prevention Association).
      3.1.2 *Performance Criteria*
         a. The instrument response factors for the individual compounds to be measured must be less than 10.
         b. The instrument response time must be equal to or less than 30 seconds. The response time must be determined for the instrument configuration to be used during testing.
         c. The calibration precision must be equal to or less than 10 percent of the calibration gas value.
         d. The evaluation procedure for each parameter is given in Section 4.4.
      3.1.3 *Performance Evaluation Requirements*
         a. A response factor must be determined for each compound that is to be measured, either by testing or from reference sources. The response factor tests are required before placing the analyzer into service, but do not have to be repeated at subsequent intervals.
         b. The calibration precision test must be completed prior to placing the analyzer into service, and at subsequent 3-month intervals or at the next used, whichever is later.
         c. The response time test is required prior to placing the instrument into service. If a modification to the sample pumping system or flow concentration is made that

would change the response time, a new test is required prior to further use.

3.2 *Calibration Gases.* The monitoring instrument is calibrated in terms of parts per million by volume (ppmv) of the reference compound specified in the applicable regulation. The calibration gases required for monitoring and instrument performance evaluation are a zero gas (air, less than 10 ppmv VOC) and a calibration gas in air mixture approximately equal to the leak definition specified in the regulation. If the cylinder calibration gas mixture are used, they must be analyzed and certified by the manufacturer to be within ± 2 percent accuracy, and a shelf life must be specified. Cylinder standards must be either reanalyzed or replaced at the end of the specified shelf life. Alternately, calibration gases may be prepared by the user according to any accepted gaseous standards preparation procedure that will yield a mixture accurate to within ± percent. Prepared standards must be replaced each day of use unless it can be demonstrated that degradation does not occur during storage.

Calibrations may be performed using a compound other than the reference compound if a conversion factor is determined for that alternative compound so that the resulting meter readings during source surveys can be converted to reference compound results.

4. *Procedures*

4.1 *Pretest Preparations.* Perform the instrument evaluation procedures given in Section 4.4 if the evaluation requirements of Section 3.1.3 have not been met.

4.2 *Calibration Procedures.* Assemble and start up the VOC analyzer according to the manufacturer's instructions. After the appropriate warm-up period and zero internal calibration procedure, introduce the calibration gas into the instrument sample probe. Adjust the instrument meter readout to correspond to the calibration gas value.

Note: If the meter readout cannot be adjusted to the proper value, a malfunction of the analyzer is indicated and corrective actions are necessary before use.

4.3 *Individual Source Surveys*

4.3.1 *Type I—Leak Definition Based on Concentration.* Place the probe inlet at the surface of the component interface when leakage could occur. Move the probe along the interface periphery while observing the instrument readout. If an increased meter reading is observed, slowly sample the interface where leakage is indicated until the maximum meter reading is obtained. Leave the probe inlet at this maximum reading location for approximately two times the instrument response time. If the maximum observed meter reading is greater than the leak definition in the applicable regulation, record and report the results as specified in the regulation reporting requirements. Examples of the application of this general technique to specific equipment types are:

a. Valves—The most common source of leaks from valves is at the seal between the stem and housing. Place the probe at the interface where the stem exists the packing gland and sample the stem circumference. Also, place the probe at the interface of the packing gland take-up flange seat and sample the periphery. In addition, survey valve housing of multipara assembly at the surface of all interfaces where a leak could occur.

b. Flanges and Other Connections—For welded flanges, place the probe at the outer edge of the flange–gasket interface and sample the circumference of the flange. Sample other types of non-permanent joints (such as threaded connections) with a similar traverse.

c. Pumps and Compressors—Conduct a circumferential traverse at the outer surface of the pump or compressor shaft and seal interface. If the source is a rotating shaft, position the probe inlet within 1 cm of the shaft–seal interface for the survey. If the housing configuration prevents a complete traverse of the shaft periphery, sample all accessible portions. Sample all other joints on pump or compressor housing where leakage could occur.

d. Pressure Relief Devices—The configuration of most pressure relief devices prevents sampling at the sealing seat interface. For those devices equipped with an enclosed extension, or horn, place the probe inlet at approximately the center of the exhaust area to the atmosphere.

e. Process Drains—For open drains, place the probe inlet at approximately the center of the area open to the atmosphere. For covered drains, place the probe at the surface of the cover interface and conduct a peripheral traverse.

f. Open-Ended Lines or Valves—Place the probe inlet at approximately the center of the opening to the atmosphere.

g. Seal System Degassing Vents and Accumulator Vents—Place the probe inlet at approximately the center of the opening to the atmosphere.

h. Access Door Seals—Place the probe inlet at the surface of the door seal interface and conduct a peripheral traverse.

4.3.2 *Type II—"No Detectable Emission."*   Determine the local ambient concentration around the source by moving the probe inlet randomly upwind and downwind at a distance of one to two meters from the source. If an interference exists with this determination due to a nearby emission or leak, the local ambient concentration may be determined at distances closer to the source, but in no case shall the distance be less than centimeters. Then move the probe inlet to the surface of the source and determine the concentration described in 4.3.1. The difference between these concentrations determines whether there are no detectable emissions. Record and report the results as specified by the regulation.

For those cases where the regulation requires a specific device installation, or that specified vents be ducted or piped according to a control device, the existence of these conditions shall be visually confirmed. When the regulation also requires that no detectable emissions exist, visual observations and samples are required. Examples of this technique are:

(a) Pump or Compressor Seals—If applicable, determine the type of shaft seal. Perform a survey of the local area ambient VOC concentration and determine if detectable emissions exist as described above.

(b) Seal System Degassing Vents, Accumulator Vessel Vents, Pressure Relief Devices—If applicable, observe whether or not the applicable ducting or piping exists. Also, determine if any sources exist in the ducting or piping where emissions could occur prior to the control device. If the required ducting or piping exists and there are no sources where the emissions could be vented to the atmosphere prior to the control device, then it is presumed that no detectable emissions are present. If there are sources in the ducting or piping where emissions could be vented or sources where leaks could occur, the sampling surveys described in this paragraph shall be used to determine if detectable emissions exist.

4.3.3 *Alternative Screening Procedure.*   A screening procedure based on the formation of bubbles in a soap solution that is sprayed on a potential leak source may be used for those sources that do not have continuously moving parts, that do not have surface temperatures greater than the boiling point or less than the freezing point of the soap solution, that do not have open areas to the atmosphere that the soap solution cannot bridge or that do not exhibit evidence of liquid leakage. Sources that have these conditions present must be surveyed using the instrument techniques of 4.3.1 or 4.3.2.

Spray a soap solution over all potential leak sources. The soap solution may be a commercially available leak detection solution or may be prepared using concentrated detergent and water. A pressure sprayer or a squeeze bottle may be used to dispense the solution. Observe the potential leak sites to determine if any bubbles are formed. If no bubbles are observed, the source is presumed to have no detectable emissions of leaks as applicable. If any bubbles are observed, the instrument techniques of 4.3.1 or 4.3.2 shall be used to determine if a leak exists, or if the source had detectable emissions, as applicable.

4.4 *Instrument Evaluation Procedures.* At the beginning of the instrument performance evaluation test, assemble and start up the instrument according to the manufacturer's instructions for recommended warm-up period and preliminary adjustments.

4.4.1 *Response Factor.* Calibrate the instrument with the reference compound as specified in the applicable regulation. For each organic species that is to be measured during individual source surveys, obtain or prepare a known standard in air at a concentration of approximately 80 percent of the applicable leak definition unless limited by volatility or explosivity. In these cases, prepare a standard at 90 percent of the saturation concentration, or 70 percent of the lower explosive limit, respectively. Introduce this mixture to the analyzer and record the observed meter reading. Introduce zero air until a stable reading is obtained. Make a total of three measurements by alternating between the known mixture and zero air. Calculate the response factor for each repetition and the average response factor.

Alternatively, if response factors have been published for the compounds of interest for the instrument or detector type, the response factor determination is not required and existing results may be referenced. Examples of published response factors may be referenced. Examples of published response factors for flame ionization and catalytic oxidation detectors are included in Section 5.

4.4.2 *Calibration Precision.* Make a total of three measurements by alternately using zero gas and the specified calibration gas. Record the meter readings. Calculate the average algebraic difference between the meter readings and the known value. Divide this average difference by the know calibration value and multiply by 100 to express the resulting calibration precision as a percentage.

4.4.3 *Response Time.* Introduce zero gas into the instrument sample probe. When the meter reading has stabilized, switch quickly to the specified calibration gas. Measure the time from switching to when 90 percent of the final stable reading is attained. Perform this test sequence three times and record the results. Calculate the average response time.

5. *Bibliography*

5.1 *DuBose, D. A., and G. E. Harris.* Response Factors of VOC Analyzers at a meter reading of 10,000 ppmv for Selected Organic Compounds. U.S. Environmental Protection Agency, Research Triangle Park, N.C. Publication No. EPA 600/2-81-051. September 1981.

5.2 *Brown, G. E., et al.* Response Factors of VOC Analyzers Calibrated with Methane for Selected Organic Compounds. U.S. Environmental Protection Agency, Research Triangle Park, N.C. Publication No. EPA 600/2-81-022. May 1981.

5.3 *DuBose, D. A., et al.* Response of Portable VOC Analyzers to Chemical Mixtures. U.S. Environmental Protection Agency, Research Triangle Park, N.C. Publication No. EPA 600/2-81-110. September 1981.

Abbreviations of Related Organizations and Standards

Abbreviation	Organization Name
ACI	Alloy Castings Institute
AFNOR	Association Française de Norme (France)
AIChE	American Institute of Chemical Engineers
AISI	American Iron and Steel Institute
AMS	Aerospace Materials System
ANSI	American National Standards Institute
API	American Petroleum Institute
ASM	American Society of Metals
ASME	American Society of Mechanical Engineers
ASTM	American Society for Testing and Materials
AWS	American Welding Society
BASEEFA	Health and Safety Executive (Great Britain)
BSI	British Standards Institution
CENELEC	European Committee for Electrotechnical Standardization
CGA	Compressed Gas Association
CSA	Canadian Standards Association
DIN	Deutsche Industrie Norme (Germany)
DOT	Department of Transportation (United States)
EPA	Environmental Protection Agency (United States)
FDA	Food and Drug Administration (United States)
FM	Factory Mutual Association
IEC	International Electrotechnical Commission
ISA	Instrument Society of America
ISO	International Standards Organization
JIS	Japan Industrial Standard

MSS	Manufacturers Standardization Society of the Valve and Fitting Industry
NACE	National Association of Corrosion Engineers
NEMA	National Electrical Manufacturers Association
NFPA	National Fire Prevention Association
OCMA	Oil Companies Materials Association
OSHA	Occupational Safety and Health Act (United States)
PFI	Pipe Fabrication Institute
PIMA	Paper Industry Management Association
PLCA	Pipe Line Contractors Association
PTB	Physikalisch-Technische Bundesanstalt (Germany)
SAA	Standards Association of Australia
SAE	Society of Automotive Engineers
SCS	Sira Certification Service (Great Britain)
SEIA	Solar Energy Industries Association
SPE	Society of Petroleum Engineers
TAPPI	Technical Association of the Pulp and Paper Industry
TEMA	Thermal Exchanger Manufacturers Association
3A	International Association of Milk, Food and Environmental Sanitarians
UL	Underwriters Laboratories
UNI	Unificazione Italiana (Italy)
USDA	United States Department of Agriculture
VMA	Valve Manufacturers Association

Glossary

Accessory: A device attached to an actuator to provide a special function not normally handled by the actuator.

Acme threads: A common thread pattern used to thread plug stems to actuator stems. Acme threads feature 29° angles and flattened tops, with the thread height identical to the thread width.

Actual pressure drop: The difference in pressure between the inlet pressure and the outlet pressure of a valve.

Actuation system: A separate actuator assembly that can be installed on a manual valve and used for either on–off or throttling applications.

Actuator: A power-driven device that provides the force to open, close, or throttle a valve.

Actuator barrier: A dynamic actuator part that separates the actuator chamber from the atmosphere or another actuator chamber. When the actuator barrier is acted upon by the medium in the chamber(s), the actuator moves.

Actuator stem: A rod used in linear valves to connect the actuator barrier with the stem of the valve.

Actuator stem force: The amount of force that an actuator must generate to move the actuator stem (as well as the closure element in the valve).

Adjusting screw: A bolt used to compress the spring in a cylinder actuator.

Advanced cavitation: The point when cavitation is at its maximum level.

Air consumption: The amount of air used by pneumatic devices that bleed air constantly.

Air filter: An accessory used to prevent dirt, oil, or water in the air or signal supplies from reaching the actuator or other air-driven devices.

Airset: A device used to limit the air supply to an actuator. Also known as a *pressure regulator.*

Air spring: A fail-safe system for actuators that employs a locked-up volume of air to drive the actuator to the failure position.

Air usage: The amount of air used to stroke a pneumatic actuator.

Allowable pressure drop: The pressure drop that is required for calculating a valve size. In applications where choked flow occurs, the allowable pressure drop may be less than the actual pressure drop in some valves.

Alloy steels: A special formula of iron combined with particular elements (nickel, vanadium, chromium, or manganese) to produce unique characteristics. Alloys are used to avoid corrosion or erosion attack by certain processes.

Angle valve: A valve-body style where the first port is perpendicular to the second port.

Arithmetic average roughness height: A measurement of the smoothness of a particular surface. Usually expressed in microinches, AARH is sometimes designated as "Ra." As AARH or Ra measurement decreases, the surface becomes smoother.

Atmospheric resistor: An attenuator trim installed outside the valve in a venting application. Also referred to as an *external stack*.

Attenuation: The capability of lowering the sound pressure level radiating from a valve.

Attenuation plate: An antinoise device installed downstream from a valve in a gaseous service. Attenuation plates normally use staged pressure reduction to reduce sound levels.

Attenuation trim: A trim installed inside a valve that uses a tortuous path, multiple holes, expanding flow area, or sudden expansions and contractions to lower the sound pressure level generated by a valve.

Auto-ignition temperature: The surface temperature of a device that will cause a flammable material or atmosphere to ignite.

Automatic control system: Any system that can regulate a process without user involvement.

Automatic control valve: A final control element that receives a signal from a process controller to regulate the pressure, flow, or temperature of a liquid or gas process. Also known as a *control valve*.

Auxiliary handwheel: A special handwheel attached to an actuator to allow for manual operation of the valve.

Average wall thickness: The dimension of the ideal wall thickness, which includes a tolerance below and above that dimension. For example, an average wall thickness of 0.625 in with a $\pm$ 0.01-in tolerance would be between 0.615 to 0.635 in.

Backflow: Process flow direction that occurs opposite the normal or expected flow direction.

Back-seat: In linear valves, a special design that permits a portion of the stem or plug head to seat against the bottom of the bonnet to help prevent process flow from migrating into the packing box.

Ball: A spherical part in quarter-turn rotary valves that rotates through the body seal, allowing flow to pass when the opening in the ball is exposed.

Ball valve: A quarter-turn valve that features a spherical closure device. As the ball moves radially across the seal, the opening in the ball is exposed, which allows the flow to move through the valve.

Bearings: Cylindrical supports located on both sides of a butterfly disk to stabilize the disk and shaft during stroking.

Belleville disk spring: A cone-shaped metal washer used to produce a constant load to the packing box without retightening.

Block valves: Simple manual valves that are used to start or stop the process flow. Common block valves include gate, quarter-turn plug, ball, pressure-relief, and tank-bottom valves. Also referred to as *on–off valves*.

Blowdown: The discharge of process fluid in a pressure-relief valve when the upstream process pressure exceeds the preset designation.

Body: The major pressure-retaining component of a valve that houses the closure device, as well as the inlet and outlet ports.

Body end connection: The part of a valve that matches and joins with the mating piping. Also referred to as *end connection*.

Body subassembly: The portion of the valve that includes the body, flanges, and associated bolting, bonnet, packing box, cage or seat retainers, and closure element.

Bolt circle: The diameter of a circle where the centerlines of bolt holes interconnect—usually refers to the design of end and bonnet flanges.

Bonnet: A pressure-retaining part that houses the packing box and guides, as well as seals the top-works of a valve body.

Bonnet cap: A pressure-retaining part that seals the top-works of a valve body and usually does not contain a packing box. Also known as a *top cap*.

Bonnet flange: The flange used to retain the bonnet to the body.

Bonnet flange bolting: Fasteners used to secure the bonnet flange and bonnet to the valve body.

Bottom flange: In reverse-acting valves, a special cap mounted on the bottom of the valve body that allows access to the trim.

Bottom port: In angle valves, the port concentric with the rising stem.

Breakout torque: The amount of torque needed to open a rotary valve.

Brinell hardness number: A number between 111 and 745 that indicates the relative hardness of a material. As the number increases, the material is designated as harder.

Brinell hardness test: A standard hardness test that uses an indentor to apply a standard load to a material. The resulting indentation determines the Brinell hardness number.

Bubble-tight: A condition where no measurable seat leakage occurs through the closure element of a valve during a certain amount of time. The typical bubble-tight test involves air-under-water testing.

Bushing: Another term for the guide(s) found in the packing box, which is used for guiding the stem or shaft.

Butterfly valve: A quarter-turn rotary valve design that has a narrow body face-to-face and a circular disk closure element.

Buttweld end connection: A special end connection used to prepare a valve for welding into the line. A buttweld connection has a series of angles that match up with the similar angles on the pipe end connection. When matched up, the two ends present a V-shaped gap that is filled with a weld.

Bypass valve: A manual valve used in a bypass line that circumvents a larger valve (usually a control valve). A bypass line usually involves using one or more manual valves, which block the flow upstream of the main valve and detour the flow on the downstream side.

Cage: A cylinder contained inside the flow gallery of a globe valve body that is used to guide the plug and possibly retain the seat ring. Special designed holes in the cage can be used to provide the flow characteristic or control the pressure drop in order to manage high velocities, cavitation, flashing, or high sound pressure levels.

Calibration: The correct adjustment of a mechanical device to ensure the preferred operating parameters.

Capacity: The amount of flow that can pass through a valve under certain conditions, without the valve choking. Also referred to as *flow capacity* or *valve capacity*.

Carbon steel: Iron-based metal that contains 0.1 to 0.3 percent carbon. Carbon steel is a base steel from which alloys are created by mixing other metals with it.

Cartridge: An anticavitation retainer or cage used inside a linear-motion valve.

Cavitation: A situation in liquid services when the pressure at the vena contracta falls below the vapor pressure, followed by a pressure recovery above the vapor pressure. The pressure reduction below the vapor pressure at the vena contracta causes vapor bubbles to form, which then collapse as the pressure recovers. This implosion of the vapor bubbles can erode metal surfaces in both the valve body and downstream piping, as well as cause noise.

Certified dimensional drawing: A drawing that guarantees the overall critical dimensions of a valve for installation purposes.

Certified material test report: A history of a particular metal traced back to the heat number and batch number from the foundry.

Chain wheel: A handwheel design that is chain-driven for hard-to-reach applications. The handwheel portion usually has a series of teeth or grooves, as well as a chain guide, to accommodate the use of a chain.

Characterizability: The ability to vary the flow characteristic of a valve by designing a special shape into a closure element, such as holes in a cage, a plug contour, the orifice shape of a ball, etc.

Charpy impact test: A quality test that measures a material's ability to resist fracture on a V-shaped notch cut into the material when an impact load is applied to the side opposite the notch. Charpy tests are normally conducted on materials that are subject to extremely cold temperature. (In cold applications, some metals can become more brittle and apt to fracture.)

Check valve: A valve that prevents the process flow from reversing. The closure element of a check valve normally uses gravity, fluid, and/or spring force to close the valve and does not require any type of manual operation or actuation.

Chevron packing: A packing style characterized by V-shaped (in cross-section) seamless rings. When axial force is applied to the ring, the radial forces cause the thin edges of the ring to press tightly against the wall of the packing box as well as the stem or shaft. Chevron packing typically provides a strong seal with minimal friction in moderate temperatures. Also referred to as *V-ring packing.*

Choked flow: A condition where the flow rate cannot be increased even if the downstream pressure is lowered. In liquid applications, choked flow occurs when cavitation or flashing causes vapor bubbles to form in the vena contracta, which consequently crowd the flow passage and will not allow a further increase in flow. In gas applications, choked flow occurs when the velocity reaches sonic proportions and a reduction in downstream pressure can no longer increase the gas flow.

Choked pressure drop: The point at which the pressure drop causes choked flow to occur in a valve.

Choke valve: A special angle valve used for well-head applications.

Class: A term used in conjunction with a pressure class, i.e., ANSI Class 2500.

Clearance flow: Any flow that occurs below the lower end of a valve's rangeability and the actual closing of the closure element.

Closure element: A device or combination of devices used to close or open the flow passageway of a valve. Typical closure elements include a quarter-turn plug or sleeve, linear plug or seat ring, butterfly disk or seat, ball or seat, etc.

Cock valve: Simple on–off valve (similar to a faucet design) used for low-pressure fluids, commonly used with beverage containers.

Cold box: An extension used in conjunction with a bonnet (or integral to the bonnet) that is used to allow a stagnated gas to form in moderate temperatures. This extension protects the packing box from the cryogenic process flowing through the valve.

Cold flow: The ability of a soft material to undergo plastic deformation under sustained pressure, regardless of the temperature.

Collar: The portion of the top-works of a quarter-turn manual valve that limits the motion of the closure element or operator.

Compressor: The part used in pinch valves to squeeze the walls of the elastomeric body together.

Concentric butterfly valve: A butterfly-valve design with the disk installed in the center of the body.

Concentric disk: A butterfly valve's disk that is positioned exactly at the center of the body.

Consolidation: A reduction in a packing's volume due to wear, cold flow, plastic deformation, or extrusion. Consolidation usually occurs with soft valve materials, such as packing and gaskets.

Continuously connected handwheel: A handwheel design that allows for continual retraction or extension of the stem, while also acting as a low- or high-limit stop. The handwheel can also be placed in a neutral position to allow for automatic operation of the control valve.

Controller: A microprocessor dedicated to monitoring and correcting the actual pressures, temperatures, or flow levels of a process. The controller constantly monitors the set point of the condition (pressure, temperature, or flow). When the actual measurement varies significantly from the set point, the controller sends an electronic or pneumatic signal to the control valve. The valve, in turn, corrects the imbalanced service conditions until the process achieves the conditions established at the set point.

Control loop: A process feedback system that consists of a regulator or control valve, sensing element or transmitter (for flow, pressure, or temperature), and a controller. The controller receives the input from the transmitter and compares it to a set point. By comparing the actual input against the set point, the controller can correct the process by sending a signal to the control valve until the set point is reached.

Control valve: A throttling valve equipped with an actuator or actuation system to respond to an input signal from a controller. Control valves are used to regulate the flow, temperature, or pressure of a process system. Also known as an *automatic control valve*.

Corrosion: Any deterioration of metal that is created by a chemical reaction with the metal.

Corrosion-resistant: Any material that does not react with the chemical it is exposed to, such as stainless steel.

Cracking pressure: In check valves, the positive line pressure that allows the closure element to begin opening and allow flow through the valve.

Critical temperature: The temperature at which a metal's crystal structure becomes austenite.

Cryogenic valve: A special valve used in services with temperatures below $-50°F$ ($-45°C$). A cryogenic valve has a protective cold box as part of the body to allow a vapor barrier to form between the liquefied gas and the packing box.

C_v: The term used to measure flow through a valve. The C_v of valve is used to calculate the ideal valve size to pass the required flow rate, while providing overall stability to the process. C_v is defined as 1 U.S. gallon (3.8 liters) of 60°F (16°C) water during 1 min with a 1 psi (6.9 kPa) pressure drop. Also referred to as *valve coefficient* or *flow coefficient*.

Cylinder: A pressure-retaining device used in an actuator to house the actuator barrier (a piston) as well as to contain the power (pneumatic or hydraulic) supply in a cylinder actuator.

Cylinder actuator: A double-acting actuator that uses a piston to separate pressure chambers on both sides of the piston. The piston moves by varying the power supply pressures on either side (using a positioner), thereby moving the position of the valve's closure device. Also referred to as *piston cylinder actuator.*

DCS: Abbreviation for *distributive control system.*

Dead band: The maximum input change needed to reverse an observable movement of a valve stem.

Delta P: The pressure difference between the upstream and downstream pressures. Also referred to as ΔP or *pressure drop,* or *differential pressure.*

Design pressure: The pressure used to determine overall design criteria for a valve, including flange rating, bolting torques and threads, wall thicknesses, packing-box configurations, bellows seals design, etc. The design pressure includes certain allowances and safety factors to compensate for pressure surges, water-hammer effects, or other unexpected phenomena to the process line. For this reason, the design pressure is always a greater value than the actual operating pressure.

Design temperature: The temperature used to determine overall design criteria for a valve, including flange rating, bolting, wall thicknesses, packing-box configurations, bellows seals design, etc. The design temperature includes certain allowances and safety factors to compensate for unexpected temperatures at or beyond the operating temperature. For this reason, the design temperature is always a greater value than the actual operating temperature.

Destructive test: A special test that uses mechanical or chemical methods to destroy a part in order to discover its properties.

Diaphragm: A flexible elastomer used in an actuator that responds to varying air pressures in order to transfer force to the diaphragm plate and ultimately to the valve's closure or regulating element.

Diaphragm actuator: Single-acting actuator that functions on low operating air pressure and uses a diaphragm to separate the air chamber from the atmosphere. The force used to load the pressurized air chamber must overcome an opposing spring on the nonpressurized side. As the diaphragm moves when acted on by the air pressure, the diaphragm plate moves accordingly, which varies the closure or throttling device.

Diaphragm case: A pressurized housing that contains the diaphragm and diaphragm plate, consisting of two sections joined together by bolting.

Diaphragm check valve: A check valve that uses a preformed elastomeric closure element, which opens upon positive flow and reverts back to its preformed closed position upon reverse flow.

Diaphragm plate: An actuator barrier that is concentric with the diaphragm and is used to transmit force from the diaphragm to the actuator stem.

Diaphragm pressure span: The range of pressure over which a diaphragm can operate. Diaphragm pressure span is noted as the maximum and minimum pressure.

Diaphragm valve: A manual valve related to a pinch valve that compresses an elastomeric diaphragm against the bottom of a metal body to shut off the flow.

Differential pressure: The pressure difference between the upstream and downstream pressures. Also referred to as ΔP, *delta P*, or *pressure drop.*

Diffuser: A single-stage downstream element used to attenuate high sound pressure levels. Diffusers are designed with multiple holes, which divide the turbulent flow into smaller eddies and manage the turbulent energy so that it does not create noise.

Digital positioner: A positioner that uses a microprocessor to position an actuator and to monitor and record certain data.

Direct-acting actuator: A term for a diaphragm actuator that allows the actuator stem to extend (closing the closure element), with air pressure placed in the chamber above the diaphragm.

Direct-acting pressure-relief valve: A pressure-relief valve that allows line pressure to act on one side of the closure element, while a predetermined spring applies a mechanical load to the other side. When the line pressure reaches its maximum limit, the line pressure overcomes the spring load and the valve opens until the line pressure falls below the preset level.

Directivity: The reduction in noise as the source of the noise is pointed away from the hearer.

Disassembly clearance: The amount of space needed between the valve or actuator and its surroundings for removal of the valve, or to gain access to the internal structure of the valve.

Disk: The closure component in a butterfly valve used to swing through the body to open, close, or throttle the flow. Also, the closure element of a pressure-relief valve (also known as a *pallet*).

Disk stop: A portion of the butterfly body that prevents the disk from overstroking.

Distributive control system: A process plant's overall data management system that takes input from management requirements and transfers that data into process management, utilizing a number of control loops.

DN: The ISO standard abbreviation for the nominal diameter of a pipe size, i.e., DN100.

Double-acting actuator: An actuator that can supply and exhaust air to both sides of a piston or diaphragm at the same time. Double-acting designs require the use of a positioner.

Double-acting positioner: A positioner that has the ability to supply and exhaust air to both sides of an actuator piston or diaphragm at the same time.

Double-disk check valve: A check valve with two half-circle disks hinged together that fold together upon positive flow and retract to a full-circle to close against reverse flow. Also known as a *split-disk check valve*.

Double-ported trim: A trim with two closure elements that work in unison for reducing the unbalanced forces as the valve opens or closes.

Double-top stem guiding: A packing-box configuration that includes two guides at each end of the packing box to guide a linear-motion closure device, such as a plug.

Downstream: The process portion of a system following a valve.

Downstream back-pressure device: A pressure-limiting device installed after a valve to take an additional pressure drop.

Drain: A special outlet (usually involving a threaded port with a plug) that allows process fluids to flow out of a pipe or a cavity, such as the bottom of a globe valve.

Drop-tight: A bubble-tight test that involves water-under-air testing.

Dry lubricant: A solid or powdered lubricant used to coat mating parts, such as perfluoroelastomer.

Dual springs: A heavy-duty spring actuator design that permits one small spring to fit inside another, which allows for shorter actuators with similar thrust to a longer heavy-duty spring.

Ductile iron: A special cast iron with 18 percent ductility.

Ductility: The capability of a metal to deform when placed under pressure or when acted upon by a force. Ductility is measured by the percentage increase of a stretched test specimen just before fracture.

Dye penetrant: A quality test that uses bright red or fluorescent dye to detect surface cracks, pits, or porosity on a nonporous surface. A dye-penetrant test involves spraying the special dye on the part. When the excess dye on the surface is wiped away, surface flaws are detected when natural or fluorescent light highlights the remaining dye in the crack or pit.

Dynamic seal: A seal that involves two or more elements that physically move as they act upon each other. Usually such seals include those associated with the closure element (such as between a plug and a seat, butterfly disk and a seal, ball and a seal, etc.), packing box (where a stem or a shaft move against the packing), or the actuator (where a piston may move between two pressure chambers).

Dynamic torque: The torque required to throttle a valve in midstroke.

Dynamic unbalance: The net force of the process fluid acting on the plug of a globe valve in the open position.

Eccentric butterfly valve: A butterfly-valve design in which the valve shaft is slightly offset from the center of the disk, allowing the disk to move in an elliptical motion as it leaves the sealing surface. This unique motion permits minimal friction and wear to the closure element.

Eccentric cammed disk: A butterfly valve's disk that is offset both vertically and horizontally from the center of the valve.

Eccentric plug valve: A quarter-turn rotary valve that uses an offset plug to swing into a seat as the closure or regulating element.

Effective area: In a diaphragm actuator, the area of a diaphragm that can be acted upon by the air pressure that results in stem or shaft movement. Because the shape of the diaphragm changes as air pressure builds, the effective area can change over a given signal.

Elastomer: A polymer part that is flexible and resilient, which is used to seal joints or provide a moving barrier.

Elbow: A special pipe fitting that permits an angle turn in the pipeline (typically a 45° or 90° turn). The standard elbow fitting has a matching end connection on both ends for mating with the upstream and downstream piping.

Electrohydraulic actuator: A hydraulic actuator with a self-contained hydraulic source that is a physical part of the actuator and is electrically driven.

Enclosed-body pinch valve: A pinch valve in which the elastomeric body is protected by a solid metal housing.

End connection: The part of a valve that matches and joins with the mating piping. Also referred to as a *body end connection.*

End-to-end: The dimension from one end connection to the opposite end connection on a valve. End-to-end is similar to *face-to-face*, except that it is used with valves that do no have flat-faced surfaces on the end connection, such as buttweld ends.

Equal-percentage characteristic: A flow characteristic that permits a change in flow per unit of valve stroke, which is directly proportional to the flow occurring just before the change is made.

Erosion: Material wear inside a valve or pipeline caused by the flowing action of the process fluid. Erosion deterioration is often hastened when entrained solids are present in the flow.

Examination: The physical review of a part or finished valve to ensure that it is in full compliance with user requirements. Examinations are typically conducted by the user, while *inspections* are conducted by the user or a designated third-party agent.

Expanded outlet valve: A body configuration that has oversized inlet and outlet ports for a given valve size, for example, a 2-in valve with 4-in ports.

Expanding-teeth trim: An antinoise trim that uses a stack of grooved disks to provide a series of sudden expansions and contractions of the fluid.

Explosion-proof: The assurance that an electrical device can be placed in a potentially explosive atmosphere. An explosion-proof device must separate from any electrical device which may arc from the atmosphere.

Extension bonnet: A bonnet that is longer than normal that is used to protect the packing box and actuator from the effects of severe temperatures. Also referred to as *extended bonnet*.

External stack: An attenuator trim installed outside the valve in a venting application. Also referred to as an *atmospheric resistor*.

Extrusion: A condition that occurs when force is applied to a soft material, causing it to deform and eventually fill empty spaces or migrate through openings.

Face: In the closed position, the side of a butterfly disk that faces the seat.

Face-to-face: The body dimension between the face of one end connection and the opposite face.

Fail-closed: A planned design of a valve and/or actuator that allows the valve to move to the full-closed position upon loss of power to the actuator.

Fail-lock-in-place: An actuation system that uses lock-up valves to allow the valve to remain in its last position upon loss of power to the actuator.

Fail-open: A planned design of a valve and/or actuator that allows the valve to move to the full-open position upon loss of power to the actuator.

Fail-safe: An actuator system that allows a valve to move to a certain position (open or shut) or to retain the current position, should the actuator power supply fail.

Far field: The noise that is generated beyond 10 ft (3 m) from the source.

Feedback system: In control valves using a positioner, the system in which a return signal from the stem position is mechanically fed to a positioner, allowing for verification and/or correction of the valve's position.

FEP: Abbreviation for *fluorinated ethylene propylene* copolymer.

Fieldbus: A standardized digital communications language that allows field devices to communicate directly with a controller or DCS as well as other field devices.

Field-reversible: The ability for actuators and positioners to be modified from air-to-open mode to air-to-close mode (or vice versa) with no additional parts or special procedures requiring removal from the line.

Filter: A device used to screen the actuator's power supply medium of impurities.

Final control element: High-performance process equipment that provides power and accuracy to control the flowing medium to the desired service conditions.

Fire-resistant: The ability of a valve to withstand and survive a fire in terms of reaching and maintaining the failure position of the closure element. A valve developed for fire-resistant service can be designed to allow the flow direction to assist with the failure action of the valve. It can also be equipped with backup metal seats to soft seats, external fire-resistant insulation, external fire-resistant enclosures, and other design features that intrinsically resist fire damage.

Fire-safe: The ability of a valve to maintain certain standards after being subjected to a fire test. A fire-safe valve is expected to minimize the amount of process fluid lost downstream or to the atmosphere during a fire. Also referred to as *fire-tested.*

Flange: A flat round portion of the valve's end connection that has a greater outside diameter than the valve hub. The flange has a number of bolt holes for connecting a valve to a similar end connection on the end of a length of pipe. A flange that is a solid part of the body is called an *integral flange*. A flange that slides over the body hub and is held in place with half-rings is called a *separable flange*.

Flangeless body: A body of a rotary valve with a short face-to-face that can be sandwiched between two flanged piping end connections, using long studs and nuts. Also referred to as a *wafer-style body*.

Flashing: A common valve problem in which the pressure at the vena contracta falls below the vapor pressure, followed by a pressure recovery that remains below the vapor pressure. This pressure reduction below the vapor pressure at the vena contracta causes vapor bubbles to form and to continue downstream. This liquid–gas mixture downstream causes the overall velocity to accelerate, which can lead to excessive noise and eventual erosion.

Flat face: An integral flanged end connection that has no raised face or serrations for the gasket to adhere to. The face is continually smooth from the outside diameter of the inlet or outlet port to the outside diameter of the integral flange. The gasket used with flat-face end connections extends to the outside diameter of the entire flange.

Flat gasket: A gasket produced with simple inside and outside diameters with even, flat surfaces.

Flexible gate: A closure element in a gate valve that uses a solid gate and a flexible seat. Also known as a *split wedge gate*.

Flexible valve: A manual valve with an elastomeric closure element, such as a pinch or diaphragm valve design.

Floating ball: A ball- or check-valve design that features a ball that is not fixed to the valve body.

Floating seat: A seat ring that is not fixed to the valve body. Floating seats can shift position to conform better to the shape of the closure element, providing better shutoff.

Flow booster: Actuator accessories used to provide a quick stroking action when large input signal changes are made.

Flow capacity: The amount of flow that can pass through a valve under certain flow conditions, without the valve choking. Also, referred to as *capacity* or *valve capacity*.

Flow characteristic: A valve's relationship between the flow coefficient (C_v) and the valve stroke, from 0 to 100 percent. A flow characteristic is usually differentiated as either an inherent flow characteristic or an installed flow characteristic. The three most common types of flow characteristics are equal percentage, linear, and quick open.

Flow coefficient: The measurement of flow that is commonly applied to valves. The flow coefficient is used to determine the best valve size to pass the required flow rate while providing overall stability to the process. The flow coefficient is expressed by the term C_V, which is defined as 1 U.S. gallon (3.8 liters) of 60°F (16°C) water during 1 min with a 1 psi (6.9 kPa) pressure drop. Also referred to as *valve coefficient*.

Fluid: Any material that can flow given a particular set of circumstances. Fluids include gases, liquids, powders, pellets, and slurries.

Fluorinated ethylene propylene copolymer: A common fluoropolymer used to provide linings for bodies as well as seals for soft seats and soft seals. Also abbreviated as *FEP*.

Fluoroplastic: A polymer with a molecular structure similar to hydrocarbons, except that fluorine atoms take the place of carbon atoms. Also referred to as *fluoropolymer, polytetrafluoroethylene,* or *PTFE*.

Fluoropolymer: A polymer with a molecular structure similar to hydrocarbons, except that fluorine atoms take the place of carbon atoms. Also referred to as *fluoroplastic, polytetrafluoroethylene,* or *PTFE*.

Four-way positioner: A positioner that sends and exhausts air to both sides of an actuator.

Four-way solenoid: A solenoid used to operate on–off actuators, providing two-way direction.

Fracture toughness: A measurement of a metal's ability to resist fracture using a Charpy test. Fracture toughness is a major concern with metals placed in services in which cryogenic temperatures can exist.

Free air: The flow or volume rate of air at standard atmospheric temperature (70°F or 21°C) and pressure (14.7 psia or 1 bar).

Frequency response: The measurement of how a system or actuation device responds to a constant-amplitude sinusoidal input signal. In other words, frequency response determines the overall speed of the system by measuring how well the system keeps up with changing input signal. With frequency response, the phase shift and the output amplitude are measured at different frequencies and are plotted as amplitude ratio and phase shift versus frequency.

Full-bore valve: Any valve where the opening of the closure element (such as the seat) has the same area as the inside diameter of the inlet and outlet ports. Also referred to as *full-port*.

Full-closed: The valve's position when the valve's closure element is fully seated.

Full lift: A pressure-relief valve design in which the valve opens to the full-open position immediately upon overpressurization.

Full-open: The valve's position when the valve's closure element allows for maximum flow through the valve.

Full-port valve: Any valve where the opening of the closure element (such as a seat or a full ball) has the same area as the inside diameter of the inlet and outlet ports. Also referred to as a *full-bore valve*.

Full trim: The area of a valve's seat that can pass the maximum amount of flow for that particular size of valve.

Gage: A device that measures a particular function and reports the results through a display. Gages can display information through a dial or through a digital display. Also spelled *gauge* outside North America.

Gain: The ratio of actuator pressure unbalance to instrument pressure change when the stem is locked in place.

Galling: Damage to two mating parts when microscopic portions interact and bond together, which, when each part moves against the other, results in tearing of the two surfaces. Galling typically happens when two parts of the same material are used together without lubrication. In the case of threaded nuts, bolts, or studs, galling can destroy both male and female threads. Galling is also common between guides and stems or shafts, which not only damages smooth surfaces, but can also destroy packing or other soft parts.

Gasket: A hard or soft material used to seal a joint from the process fluid. Gaskets are used with flanged end connections. They are also used between the bonnet and the body, and the body and seat ring.

Gate: A flat and broad closure element that intersects the flow steam.

Gate valve: A linear-motion valve in which the closure element (a gate) is flat and broad. The gate slides up and down seating surfaces found on both sides of the body. Gate valves are normally used for on–off service.

Gear operator: A manual operator that uses gearing to produce high output thrust.

General-service valve: A versatile valve design that can be used in several applications without modification to the design.

Gland bushing: A part found at the top of a packing box that protects the packing box from atmospheric elements and transfers force from the gland-flange bolting to the packing. Also referred to as a *packing follower*.

Gland flange: Valve part used to retain and compress the guides, packing, antiextrusion rings, and packing spacers in the packing box.

The gland flange is usually attached to the bonnet in linear and quarter-turn valves and the body in rotary valves.

Globe valve: A valve with a rounded valve body made from a casting with in-line ports. Normally the inlet and outlet passages feature constant areas and are streamlined to ensure smooth flow through the valve. The closure element is usually a round member called the plug, which fits into a seat, and is perpendicular to the inlet and outlet passages.

Graphite: A carbon-based packing or gasket used in high-temperature applications. Graphite is normally produced as die-formed rings or braided rings.

Grease fitting: A small check mechanism that allows grease to be injected into an area containing a bearing.

Guide: A cylindrical part that aligns the valve stem or shaft with the closure element. In most cases, the guide or guides are found in the packing box or close to the closure element. In some designs, the *upper guide* is also used to transmit axial force from the gland-flange bolting to the packing. Also referred to as *guide bushing.*

Guide bushing: Another term for a *guide.*

Guiding: Mechanism used to maintain the correct position of the stem or shaft of the closure device. In the case of linear valves, the guiding ensures that the plug aligns with the seat ring. With rotary valves, the guiding ensures that the disk or ball aligns with the body seal.

Half-ring: A ring divided in half that is used to retain a separable flange in place on a valve body.

Handle: A simple lever used to manually open, close, or position a quarter-turn, butterfly, or ball valve, which may or may not include a latching device. Also referred to as a *handlever* or *actuator lever.*

Handlever: A manual operator with two spring-loaded levers used to operate a quarter-turn rotary valve. In a static position, it is locked in place. To operate the valve, the two levers must be squeezed together to disengage the handlever.

Handwheel: A manual operator with a wheel that turns clockwise or counterclockwise to manually operate a valve.

Hardened trim: Valve trim that is overlaid with a special material designed to withstand the effects of cavitation, corrosion, or erosion.

Hardfacing: The welding of a harder alloy over a softer base material (such as steel) to produce additional resistance to the effects of severe services, such as cavitation and flashing.

Hardness: A material's ability to resist indentation.

Hardness Rockwell test: Measurement of a material's hardness depending on the size of an indentation. Higher numbers indicate greater hardness of the material. Also referred to *Rockwell hardness,* or specifically as *HRB* or *HRC* (depending on the scale used).

Hazardous location: Any location where an explosion or fire may result because of flammable vapors or materials in the atmosphere.

Header: A large process line that is designed to feed several smaller process lines.

Heat treating: The process of heating and cooling a metal in order to create or enhance the properties of the metal. In some cases, a record of the temperature cycle can be generated by temperature recorders, which are called *heat treat charts.*

High-performance valve: A valve designed specifically for exceptional throttling performance.

High recovery: The inherent ability of a particular valve design to pass flow without taking a large pressure drop, allowing the downstream pressure to recover close to the upstream pressure. Valves with high-recovery factors have similar design characteristics, such as straight-through or streamlined internal passageways, and closure elements that, when open, are outside the flow stream. Gate, quarter-turn plug, and ball valves are good examples of valves with high recovery factors.

HRB: Abbreviation for B scale of the *Rockwell hardness* test.

HRC: Abbreviation for C scale of the *Rockwell hardness* test.

Hydraulic actuator: A highly accurate and fast actuator that uses a plant's hydraulic fluid (or other external hydraulic supply) to provide the thrust to open, close, or position a valve.

Hydrostatic test: A valve test using water pressure to detect the presence of leaks through sealed joints, vessel walls, or the closure element. As a safety measure, a hyrdotest typically generates 1.5 times the pressure called for by the design.

Hysteresis: In actuation devices, the amount of position error caused when an identical input signal is approached from opposite directions.

Impact test: A test used to determine the toughness of a material by measuring the force required to fracture a test specimen.

Incipient cavitation: The point when vapor bubbles begin to form in cavitating services.

Inclusion: A foreign object or particle found in a casting, forging, or weld that may weaken the material or create a potential leak path.

Increaser: A special pipe fitting designed to expand one pipe size to another pipe size. If the increaser has universal end connections, such as flanged or buttweld, it may be used to either reduce or expand the pipeline. If the expander has one-way connections (male-to-female connection, for example), it may only be used to expand.

Independent linearity: The amount of deviation of an actuator stem from a true straight line.

Indicator: A pointing mechanism (usually attached to the stem or shaft) that provides visual verification of valve position.

Indicator plate: A plate with increments of travel that shows the entire length of the valve's stroke. The valve's indicator moves along the indicator plate to show current position.

Inherent flow characteristic: A flow characteristic of a valve that operates with a constant pressure drop. A valve's inherent flow characteristic does not account for the effects of piping or other process equipment, such as pumps.

Inherent rangeability: The ratio of maximum to minimum flow that can be recognized and acted upon by a throttling valve after receiving a signal from a controller, taking into account any significant deviation from the inherent flow characteristic.

Inlet: The port where the process fluid enters the valve.

Inspection: An examination of a valve, part, or system by the user or by an authorized inspector (not associated with the manufacturer) to ensure that it was manufactured to the specifications approved by the user.

Installed flow characteristic: A flow characteristic of a valve that changes from the inherent flow characteristic as it takes into account the system effects of valves, pumps, piping configurations, etc.

Installed set: In throttling valves, the high- and low-pressure values applied to an actuator that indicates the range that the valve will stroke—after installation in a process system.

Instrument pressure: The output pressure from a controller to an actuator (usually without a positioner).

Instrument signal: A signal (electric or pneumatic) between a controller and a valve that communicates the desired valve position.

Integral flange: An end connection that features a solid flange either cast or fabricated as part of the valve body.

Integral seat: A seating surface machined into the body itself. Integral seats are usually used with oversized closure elements to provide greater flow than is normally provided in a given valve size.

Intelligent system: A microprocessor-based controller installed on a control valve that provides local process control, diagnostics, and safety management functions.

Intrinsically safe: An electrical device that is not capable of producing enough heat to cause an ignition of a flammable material or atmosphere.

Isolating valve: A valve placed between the packing box and the packing lubricator used to control or stop the flow of lubricant to the packing box.

Lapping: A process in which an abrasive compound is placed on the seating surfaces of a plug and a seat; the plug is turned in the seat until a full-contact seal is created.

Leakage: The measured amount of process fluid that continues to flow through the closure element of a valve (at a certain temperature, pressure, and pressure drop) when in the full-closed position.

Lever actuator: Another name for a *lever operator.*

Lever operator: Manual operator with a pivot handle used to open or close, or adjust the position of the closure device of a valve. Lever operators may also be used to position other flow devices, such as louvers or dampers. Also referred to as a *lever actuator.*

Lift check valve: A check valve that uses a free-floating closure element, consisting of a piston or poppet and a seat ring.

Limit-stop: A device in an actuator that restricts or limits the linear or rotary motion of an actuator. Limit-stops may be adjustable or fixed.

Limit switch: An electromechanical accessory attached to an actuator used to indicate or verify the valve's position (open, closed, or an intermediate position). Usually the signal (or lack of a signal) generates some action in the process, such as operating a motor, pump, or other device.

Line: A long container used to move fluid or pressure from one point to another in a process plant. Also referred to as *piping.*

Linear flow characteristic: The inherent flow characteristic that produces equal changes in flow per unit of valve stroke at a constant pressure drop.

Linear-motion valve: A valve with a sliding stem design that pushes the closure or regulating element into an open, closed, or throttling position. Also known as a *linear valve.*

Linear valve: Another name for a *linear-motion valve.*

Liquid-penetrant examination: A test that is used on nonporous surfaces to detect small flaws, such as porosity or cracks. Liquid penetrant is usually a red or fluorescent dye that is applied to the surface and all excess removed, revealing the flaws in the surface.

Loading pressure: In positioners installed on actuators, the output pressure from the positioner to position an actuator.

Locking device: A mechanism attached to a valve or an actuator that prevents accidental operation or use by unauthorized individuals.

Lock-up system: System used with an actuator to hold the actuator in position if the power supply fails.

Lower valve body: In split-body valves, the half of the valve body that houses the flow area under the seat ring.

Low recovery: The inherent ability of a particular valve design to take a large pressure drop, resulting in the downstream pressure only slightly recovering in relationship to the upstream pressure. Valves with low recovery factors have similar design characteristics, such as highly contoured internal passageways and closure elements that remain in the flow stream when open. Globe and butterfly valves are good examples of valves with low recovery factors.

Lubricator: A mechanical device that feeds lubricant to a packing box to minimize friction between the packing and the stem or shaft. Also referred to as the *packing lubricator.*

Lug body: In butterfly valves, a body with an integral flange that has a threaded hole pattern identical to the hole pattern of the piping flanges.

Magnetic particle examination: A method used to discover small cracks or porosity in the surface of a material that cannot be detected by visual examination. Magnetic particle examination involves spreading iron filings over a metal surface and passing an electric current through the metal. Flaws are then revealed as the iron filings cluster around imperfections in the metal.

Maintenance repair operations: The function of servicing a valve, either through valve failure or periodic servicing. Also referred to as *MRO.*

Manual handwheel: A handwheel design that is not used in conjunction with an actuator but is the sole means of moving the valve's closure element.

Manual operator: Any device that requires the presence of a human being to provide the energy to operate the valve.

Manual valve: Any valve that operates through a manual operator, such as a handwheel or handlever. Most manual valves are used for on–off service, although some can be used for throttling service.

Maximum allowable operating pressure: The maximum pressure a pressure-retaining vessel, such as a valve, can safely hold on a continuous basis. The maximum allowable operating pressure is expressed in psi or bar (or kilopascal) and is determined by the vessel's material, maximum temperature, and given pressure class. Also referred to as *maximum allowable working pressure*.

Maximum flow capacity: The volume of air that can flow into an actuator during a particular time period.

Mechanical preload effect: A butterfly seat design that allows the inside diameter of the seat to interfere slightly with the outside diameter of the disk.

Mechanical tubing: Hollow piping that is used for non-pressure-retaining purposes, such as structural supports or to hold wiring, etc. Typically, the piping is round, but other shapes are possible.

Metal bellows seal: An accordion-shaped, flexible device that is welded to a stem or a shaft to provide a solid barrier between the process fluid and the atmosphere.

Metal seat: A seat design in which the mating surfaces in the closure element are both made from metal, lacking an elastomer surface. Metal-to-metal seats have greater leakage rates than soft seats but can handle higher temperatures and pressures.

Milliampere: The electrical current measurement used in process instrumentation. Also designated as *mA*.

Mill test report: A document that indicates the results of chemical testing and physical testing performed on a base material. Mill test reports are normally provided by the producer of the material to the valve manufacturer. Some users require access to mill test reports to ensure compliance to user specifications and requirements.

Minimum wall thickness: Specifying a wall thickness so that the wall measurement never falls below a particular dimension. Tolerance is usually added to the minimum wall thickness; for example, a minimum wall thickness of 0.625 in would have a 0.01-in tolerance (0.625 to 0.635).

Modulating lift: In pressure-relief valves, a design that permits the valve to open only enough to relieve the overpressurization and not the entire line pressure.

Multihole trim: An antinoise trim that uses a number of cylinders with drilled or punched holes.

National pipe thread: A slightly tapered thread used with pressure connections in piping. As the male and female ends are threaded together, the tapered threads have a tendency to seal the connection (especially when a tape thread or sealant is used in conjunction with connection).

National straight thread: A straight thread used with piping that requires a gasket to seal the joint.

Near field The noise that is measured or heard within 3 to 10 ft (1 to 3 m) of the source.

Needle-valve trim: A low-flow trim set in which the plug head is designed to be very narrow and sharp, with the general appearance of a needle. Needle-valve trims are used for extremely low C_vs and are normally required for research applications.

Nominal diameter: A dimensionless numerical designation for a valve or a pipe. Also designated as DN with a number, such as DN 50.

Nominal pipe size: A reference to the size of a pipe or valve. Nominal pipe size is expressed in inches measured across the inside diameter of the pipe or the inlet or outlet of the valve.

Nominal pressure: The pressure rating used with ISO standards. Also, designated as PN with a number, such as PN 100.

Nondestructive examination: Any test that determines the characteristics of a material without requiring destruction of the material.

Nondestructive test: Any test that determines the function, performance, or reliability of a part, assembly, or entire valve without destroying it or damaging its operation.

Nonreturn valves: Valves that allow the flow of the process fluid in only one direction. Nonreturn valves are designed so that any flow or pressure in the opposite direction is mechanically restricted. All *check valves* are nonreturn valves.

Nonrising stem: In manual handwheels, in which the closure element is threaded and the turning of a stationary operator causes the closure element to lift or close.

Normally closed: A valve design in which the valve remains closed unless it receives a control signal or is manually operated.

Normally open: A valve design in which the valve remains open unless it receives a control signal or is manually operated.

Nozzle: In pressure-relief valves, the seating portion of the closure element that the disk interacts with.

Off-balance area: The difference in the surface areas between the two sides of a closure element, usually the upstream side and the downstream side. The greater the off-balance area, the greater the process forces act against or with the closure element.

Offset globe body: A special body configuration that features an inlet port and an outlet port that are parallel but not in-line.

On–off valves: Simple manual valves used to start or stop the flow of the process fluid. Also referred to as *block valves*.

Open-body pinch valve: A pinch valve with no metal casing surrounding the elastomeric body, which is supported by a metal skeleton structure.

Open bonnet: In safety valves, a bonnet that exposes the spring to the atmosphere and is used in high-temperature applications in which an enclosed spring would normally lose some strength.

Opening pressure: In full-lift pressure-relief valves, the overpressurization required before the valve moves to the full-open position.

Open-loop gain: The ratio of unbalance that occurs when an instrument signal change is made when the actuator is locked in position.

Operating medium: The power source used to operate an actuation system or actuator, which is generally pressurized air, electricity, or hydraulics.

Operating pressure: The pressure that a valve normally operates under during everyday service. The operating pressure should not be confused with the design pressure, which is defined as the operating pressure plus outside factors.

Operating temperature: The temperature that a valve normally operates under during everyday service. The operating temperature should not be confused with the design temperature, which is defined as the operating temperature plus outside factors.

Operative limits: The range that a device can operate under without damage to the device or impaired operation.

Operator: Any device that is used to provide force, leverage, mechanical action, or torque to open, close, or regulate the closure element.

O-ring: An elastomer ring that is used to seal a joint or to seal two dynamic pressure chambers, such as is the case with a piston actuator.

Outlet: The port where the fluid exits the valve.

Outside diameter: The measurement of a round structure from one side to its opposite side.

Oversized actuator: The selection of a larger actuator than is normally required for a given service in order to provide more thrust.

Oxidation: A chemical reaction between iron-based metals and oxygen, resulting in rust or scale.

Packing: A soft material used to prevent the leakage of process fluid from around a valve's stem or shaft and the bore of the bonnet.

Packing box: The configuration of packing, packing spacer(s), lantern ring, packing spring, extrusion rings, wiper rings, guides, gland flange, live-loading, gland-flange followers, etc. grouped together in a bonnet. The packing box is designed to prevent process fluid from escaping through the stem or shaft. The gland flange is located above the packing box and is used to apply axial force to the packing.

Packing follower: Similar in appearance to an upper guide, a part in which the basic purpose is to transfer the axial load from the gland flange to the packing, as well as protecting the packing from the outside environment. Also referred to as a *gland bushing*.

Packing lubricator: Another term for *lubricator*.

Pallet: The closure element of a pressure-relief valve. Also known as a *disk*.

Parallel gate valve: A gate valve that uses a flat disk gate, which fits between two parallel free-floating seats.

PEEK: Common abbreviation for polyetheretherketone.

Perfluoroalkoxy: A fluoropolymer similar to polytetrafluoroethylene (PTFE), with some different properties that make it ideal for valve linings, seals, or soft seats. Also abbreviated as *PFA*.

Permanent pressure drop: The difference between the upstream and downstream pressures, which is caused by frictional losses as the fluid moves through the valve.

PFA: Common abbreviation for perfluoroalkoxy.

PID control: A loop-tuning process where the proportional, integral, and derivative settings are adjusted continuously.

Pig: A device the same shape as the inside diameter of a pipe, which is pushed down the pipeline by fluid pressure. A pig has two purposes: first, to act as a barrier between an earlier process fluid and a new fluid, and second, to clean the inside walls of the pipe.

Pilot actuation: The process in which a pilot valve mechanism inside a pressure-relief valve monitors the system pressure and triggers the opening of the main valve when the line pressure exceeds the limit.

Pinch valve: A flexible valve with an elastomeric body that can be closed by using a mechanism or fluid pressure to push the walls of the body together.

Pinion: A small gear that works in conjunction with a larger gear or a flat-tooth rack. Pinion gears are usually used in actuation systems.

Piping: A hollow cylinder of a particular length and diameter used to retain pressure and fluid. Standard pipe lengths have uniform wall thickness and outside diameter. Also referred to as a *line*.

Piping and instrument diagram: A schematic indicating the design of a process system, including the piping, valves, pumps, controllers, sensors, and related instrumentation. The piping and instrument diagram is not to scale and only shows the general order of the equipment and does not address equipment orientation or specific location.

Piping schedule: In pressure pipe, a number that describes the thickness of the pipe wall, such as schedule 40 pipe. The larger the number, the larger the pipe wall and the more pressure the pipe can handle.

Piping tee: A common pipe fitting that has three end connections and has the general appearance of the letter T. In piping configurations, tees are generally used for joining or separating flows.

Piston: A round, flat actuator barrier that is used inside a piston actuator to separate and seal between two air chambers. The piston remains stationary when the air pressures on both sides are equal. When the air pressures change, the piston moves toward the air chamber with less pressure, until both sides are equal. In check valves, a free-floating closure element that seats against a fixed seat ring.

Piston cylinder actuator: A pneumatic actuator that features a piston inside a cylinder. Because piston actuators are double-acting, positioners are usually required for operation. They typically provide higher thrust than other types of pneumatic actuators. Also referred to as a *cylinder actuator*.

Pitting corrosion: A specific form of corrosion with the appearance of small holes or cavities. Over time, pitting corrosion will cause the cavities to combine and create larger cavities.

Plug: In linear globe valves, the part that extends into the seat as a closure device and, in some cases, provides the flow characteristic. In quarter-turn plug valves, a cone-shaped part with a flow passage that turns to open, close, or regulate the flow.

Plug head: In linear globe valves, the portion of the plug that fits into a seat ring.

Plug stem: In linear globe valves, the portion of the plug that connects the plug head with the operator.

Plug valve: A manually operated quarter-turn valve featuring a cylinder-shaped or cone-shaped closure element with an internal flow passage. Plug valves can be turned to allow flow through valve or turned 90° to block the flow.

Pneumatic actuator: An actuation system that is powered by air pressure.

Poisson effect: The assistance of process pressure to cause deformation of a soft material against a seating surface for shutoff of a closure element.

Polyetheretherketone: A rigid elastomer that is used in high-temperature services. Also abbreviated as *PEEK*.

Polyethylene: A common thermoplastic used for piping and some valve parts, such as the seat. Polyethylene is known for its flexibility.

Polypropylene: A common thermoplastic used for piping as well as pipe liners. Polypropylene is not as flexible as polyethylene.

Polytetrafluoroethylene: A common and inexpensive gasket and packing material. Polytetrafluoroethylene is widely used in general services because it is highly elastic, generates little friction, and requires minimal compression to achieve a strong seal. Abbreviated as *PTFE*.

Poppet: A closure element in a check valve that uses a spring to keep the element closed unless positive line pressure is applied.

Porosity: Small air bubbles in the molten metal during the casting process. After the metal has cooled, these bubbles leave small pits or holes in the casting. Porosity can cause a failure in a pressure-retaining vessel.

Port: Fixed-area holes in the valve body where the process fluid enters and exits after passing through the closure element.

Positioner: A feedback device that receives a signal (pneumatic or electric) from a controller and compares that signal to the actual position of the valve. If the valve position and the signal are not the same, the positioner sends or bleeds air pressure to or from the actuator until the correct valve position is achieved.

Position indicator: A device on the valve or the actuator that shows the position of the closure element.

Position transmitter: An accessory normally mounted to an actuator that continuously relays the valve's open, closed, or intermediate position, using an electric or pneumatic signal.

Positive material identification: An independent process in which the identity of a material is verified. A nuclear analyzer is typically used for verification.

Potentiometer: An actuator accessory that uses a coil and an output wire to produce a variable electrical output that corresponds to the position of the actuator.

Pressure: The force that fluids exerts on the containment walls.

Pressure-assisted pinch valve: A pinch valve that uses an outside pressure source rather than a mechanical device to close the valve.

Pressure-balanced trim: A special trim modification that allows the upstream pressure to act on both sides of the plug, significantly reducing the off-balance forces and requiring less thrust to close the valve. Pressure-balanced trim is often used when high pressure drops exist or when smaller or lower-thrust actuators are used.

Pressure class: The amount of pressure that a valve's wall thickness and connections are designed to handle without failure.

Pressure drop: The difference between the upstream and downstream pressures of a valve.

Pressure recovery: The difference between the pressure at the vena contracta and the downstream pressure.

Pressure regulator: A small valve used to limit the air supply to an actuator.

Pressure relief valve: A self-actuated valve designed to move to the wide-open position when the upstream pressure reaches a preset pressure, relieving pressure from the line. When the pressure is under the preset value, a spring or poppet keeps the closure element closed. When the pressure builds, the spring or poppet tension is overcome, allowing the closure element to open. Also referred to as a *relief valve* for liquid applications and a *safety valve* for gas applications.

Pressurized bonnet: In relief valves, bonnets that totally enclose the spring and do not allow leakage of the process to atmosphere.

Process flow diagram: A schematic that outlines the overall processes in a plant, including the major control instrumentation and equipment. The process flow diagram generally does not include specific piping sizes or detailed instrumentation.

Proximity switch: A special limit switch that indicates valve position without using a mechanical contact. Rather, a proximity switch relies on a magnetic or electronic sensor to determine valve position.

psi: Standard abbreviation of *pounds per square inch.*

psia: Standard abbreviation of *pounds per square inch, absolute.* The unit psia is used when pressure is expressed without considering ambient pressures.

psig: Standard abbreviation of *pounds per square inch, gage (gauge).* The unit psig is used when pressure is expressed to standard atmospheric pressure (noted as 14.7 psia).

PT: Abbreviation for *penetrant test.*

PTFE: Standard abbreviation for *polytetrafluoroethylene.*

Push-only handwheel: A handwheel design part of an actuator that can extend the actuator stem by pushing against the actuator stem. It can also act as a high-limit stop.

Quarter-turn motion: The movement of a valve's closure element from 0° (full-closed) to 90° (full-open).

Quick exhaust valve: An actuator accessory that is designed to quickly vent one side of a double-acting actuator when a full-open or full-closed position is required immediately.

Quick-open flow characteristic: An inherent flow characteristic that produces a maximum amount of flow with minimal lift or turn of the closure element.

Rack: In rack-and-pinion operators, a flat plate with a series of linear teeth that move linearly with the piston and rotates the pinion.

Rack-and-pinion actuator: An actuation system used with quarter-turn valves. The actuator uses a pneumatic or hydraulic force to move a flat-toothed rack, turning a gear to move the closure element.

Radiography: An examination method that uses γ rays to produce an internal image of a material. Radiography is used to reveal porosity, cracks, and other inclusions under the surface of the material.

Raised-face flange: A valve and piping end connection that includes a flange (either integral or separable). The flat portion of the valve hub extends farther than the flange. This flat portion, which is called the *raised face,* permits a greater load on the gasket between the valve and pipe than comparable flat-faced end connections. Raised faces are usually spiral serrated or concentric grooved for better gasket sealing.

Ram valve: A valve placed at the bottom of a tank that allows for drainage. Usually the actuator or handlever faces down with the closure element facing up into the tank. Also known as a *tank bottom valve.*

Range: The region between the lower and upper limits. The range in valves usually applies to temperatures (such as −50 to 100°C) or instrument signals (such as 3 to 15 psi).

Rangeability: The ratio of maximum to minimum flow that can be recognized and acted upon by a throttling control valve after receiving a signal from a controller.

Range spring: In single-acting diaphragm actuators, the internal spring that opposes the air chamber. In positioners, the internal spring that opposes the incoming pneumatic signal.

Rated C_v: The flow coefficient C_v of a valve when the valve is in the full-open position.

Rated travel: The measured linear movement of the valve closure element from the full-closed position to the full-open position.

Reduced-port valve: A valve inlet or outlet which is smaller than the pipe size.

Reduced trim: In throttling valves, a smaller diameter seat that is expected to pass a smaller amount of flow than that valve's rated flow capacity.

Reducer: A special pipe fitting designed to decrease (or increase) one pipe size to another pipe size. If the reducer has universal end connections, such as flanged or buttweld, it may be used to reduce or expand the pipeline. If the reducer has one-way connections (male to female connection, for example), it may only be used to reduce.

Regulating element: A closure element that can be used for throttling control of the valve. Also known as a *throttling element.*

Regulator valve: A valve designed to ensure process output is constant, regardless of pressure fluctuations.

Reliability: A valve's ability to stay in service between routine maintenance without any unplanned failures.

Relief valve: In liquid services, a self-actuated valve designed to move to the wide-open position when the upstream pressure reaches a preset pressure, thus relieving pressure from the line. When the pressure is under the preset value, a spring or poppet keeps the closure element closed. However, when the pressure builds, the spring or poppet tension is overcome, allowing the closure element to open.

Repeatability: The maximum amount of change in the input signal that requires a change in valve stem or shaft position when approached from the same direction.

Reseating pressure: In pressure-relief valves, the point at which the line depressurization allows the valve to return to the closed position.

Resilient seat: A seat in a closure element that is made wholly or partially with a flexible or semiflexible elastomer.

Resolution: The minimum amount of change in valve stem or shaft position when an input signal is received.

Response level: The maximum amount of input signal required by the valve in order for the valve stem or shaft position to move in one direction.

Reverse-acting actuator: In a diaphragm actuator, a design in which the actuator stem retracts (or opens the closure element) as the air pressure is applied to the chamber below the diaphragm.

Reverse flow: When a valve is capable of being installed backwards, with the upstream flow passing into the outlet port and the inlet port discharging flow to downstream. Only some valves are capable of reverse flow (for example, some globe and butterfly valves), although the flow coefficient and leakage limits may vary. In addition, some actuators may not be able to provide the correction failure mode with reverse flow (especially if the process flow direction assists with the failure mode).

Ring-type joint: Flanged end connections with a special groove on each face, where a soft piece of metal is placed before the flange is tightened. Also abbreviated *RTJ*.

Rising stem: A valve design in which the stem of the manual operator lifts with the opening of the closure element.

rms: Abbreviation for *root mean squared*.

Rockwell hardness: Specific hardness of a material measured by comparing the size of an indentation against a standard scale. Also abbreviated as *HRB* for the Rockwell B scale and *HRC* for the Rockwell C scale. Also referred to as a *hardness Rockwell test*.

Root mean squared: A measurement in microinches of how rough a surface is. Although it is not exactly equivalent to the *arithmetic average roughness height,* the two values for the same measurement are very close. Also abbreviated as *rms.*

Rotary-motion valve: Any valve design that involves a quarter-turn rotation to open or close the valve's closure element. Also known as a *rotary valve.*

Rotary valve: Another term for a *rotary-motion valve.*

RTJ: Common abbreviation for *ring-type joint.*

Safety barrier: An electrical restriction device located outside a hazardous location that is designed to restrict the voltage and current being sent to a device inside a hazardous location.

Safety valve: A pressure-relief valve designed to reduce overpressurization in gas or steam services. Also, a valve used with volume tanks to prevent overpressurization of the tank.

Sanitary trim: Special self-draining, stainless-steel trim used for the food and beverage industry.

Scraper: A device in the shape of the inside diameter of a pipeline and operated by line pressure that is used to clean the inside walls of a pipeline.

Screwed bonnet: A bonnet designed with male threads at the joint with the valve body (which is threaded with female threads).

Screwed end connection: Body end connection that features a female National Pipe Thread (NPT), which mates with a male NPT on a pipe.

Seal: The static or fixed portion of a closure element in rotary valves. A seal is normally associated with quarter-turn manual plug or ball valves.

Seal load: In rotary valves, the force that must be generated by an actuator to overcome the static, shaft, and dynamic unbalanced forces acting on the closure element.

Seal weld: Following assembly of two parts, the welding of a joint to prevent any leakage to atmosphere.

Seal-welded bonnet: A special bonnet that can be welded to the valve body after assembly.

Seal-welded cover: A special cap that can be welded to the valve body following assembly.

Seat: The static or fixed portion of a closure element in a valve. A seat is normally associated with globe, gate, butterfly, and eccentric plug valves.

Seating pressure differential: In pressure-relief valves, the difference between the operating and set pressures.

Seating torque: The amount of torque that must be produced by a rotary actuator to close or open the valve.

Seat load: In linear valves, the force that must be generated by an actuator to overcome the static, shaft, and dynamic unbalanced forces acting on the closure element.

Seat retainer: In linear globe valves, a trim component that transmits force from the tightened bonnet to the seat ring, when a clamped-in seat-ring design is provided. Unlike cages, seat retainers are not used to guide the plug or to provide a flow characteristic.

Seat ring: In linear globe valves, a round orifice in which the plug fits to achieve shutoff. Seat rings may be either solid metal or include a soft seal material.

Segmented ball: A style of ball in which only a portion of the sphere is used instead of the entire sphere.

Sensitivity: The specific change in the flow area opening produced by a given change in a regulating element when compared to the previous position.

Separable flange: A flange that is not integrally connected to the body, but rather is held in place with half-rings placed within a groove in the body hub. Because the separable flange is not part of the body, it can be made from less costly materials, such as carbon or stainless steel, when alloy bodies are specified.

Set point: The input value that determines the best desired position of a controllable device.

Set pressure: In pressure-relief valves, the point at which the pressure of the system overcomes the spring force holding the disk to the nozzle and the valve begins to open.

Severe-service valve: A valve that is highly engineered to handle difficult applications, such as those with cavitation, high noise levels, flashing, a high pressure drop, etc.

Shaft: In rotary valves, the component attached to both the closure device and the actuation device or handlever.

Shim: In ball valves, extremely thin gaskets used in a series to adjust the deflection of the ball to the seal.

Shutoff: The point at which the valve's closure element is in the closed position and flow ceases through the valve.

Shutoff valve: A valve that is used to block (or shut off) the flow. Shutoff valves are usually used for emergency shutdown situations.

Side-mounted handwheel: An auxiliary handwheel mounted on the side of the actuator.

Side port: In angle valves, the port perpendicular to the rising stem.

Sigma (σ): The cavitation index that describes the ratio of the potential for preventing cavitation to the potential for causing cavitation.

Signal: An electronic or pneumatic piece of information sent from one controlling device to another.

Silencer: A downstream antinoise device that offers large noise reduction with a series of compartments of acoustic material to absorb noise.

Single-acting: A term used to describe the action of diaphragm actuators in which air pressure is applied to one air chamber. This pressure pushes a plate, which is opposed by a range spring on the opposite side of the plate.

Single-acting positioner: A positioner that can only send air to one side of a piston or diaphragm.

Single-loop control: The process of an input sensor sending information to a controller, which sends a correcting signal to a control valve until the correct process condition is achieved.

Single-seated trim: A closure or regulating element with a sole point of closure.

Sleeve: In globe valves, the part that provides a sliding seal surface for a pressure-balanced plug. In quarter-turn plug valves, an elastomeric cylinder used to seal and retain the plug.

Sliding gate valve: A gate valve that uses a flat, rectangular gate for irrigation or waterway services. Also known as a *sluice valve*.

Sliding seal: In actuators used with rotary valves, a special seal located where the actuator stem exits the lower pressure chamber of the actuator. Because of the rotary shaft connection (usually a rod end bearing), the motion of the actuator stem must move in a nonlinear fashion, thereby needing a moving seal.

Sliding valve: A manual valve that uses a flat perpendicular closure element that intersects the flow, such as a gate-valve design.

Slotted trim: A single-stage cage or seat retainer that uses long, narrow slots to reduce noise.

Sluice valve: A gate valve that uses a flat, rectangular gate for irrigation or waterway services. Also known as a *sliding gate valve.*

Slurry: A process fluid characterized by a mixture of undissolved solids and liquids.

Smart valve: A control valve equipped with an onboard microprocessor or a digital positioner.

Socketweld end connection: An end connection that allows the inside diameter bore of the body hub to mate with the outside diameter of the pipe. The body and the pipe are then connected using a weld between the face of the body hub and pipe's outside diameter.

Soft-seat plug: In globe valves, a linear plug in which an elastomer insert is placed in the seating area of the plug and is used with a metal seat ring. This design usually requires the plug to be disassembled to install or replace the elastomer. Soft seat plugs are used to provide bubble-tight shutoff.

Soft-seat ring: A seat ring made up of two metal pieces, with an elastomer insert sandwiched between the two, that is used with a metal plug. Soft seats are usually required to achieve bubble-tight shutoff. Also referred to as a *resilient seat.*

Solenoid: An actuator accessory that acts as a control device to supply air to an actuator in on–off applications or to control signal pressure in throttling applications.

Sound pressure level: The standard measurement of noise in a valve. One sound pressure level is equal to 90 dB and doubles every 6 dB.

Span: The difference between the high and low limits of a range. For example, a 3- to 15-psi (0.2- to 1.0-bar) signal has a span of 12 psi (0.8 bar).

Special-service valve: A valve designed only for a specific application.

Specification: Any requirement that a valve must conform to in order to meet the user's expectations. Specifications can vary significantly, outlining requirements for performance, materials, design tolerances, quality, delivery, and method of shipment.

Speed control valve: An actuator accessory used to regulate or limit the stroking speed for a control valve by restricting the air supply to the actuator.

Speed of response: The rate of travel provided by a particular actuator, given certain conditions. Also referred to as *stroking speed.*

Spiral-wound gasket: A gasket consisting of alternate layers of metallic and nonmetallic materials wound together.

Spline: A series of equal-sized grooves cut into a shaft clamped to a matching coupling.

Split body: A body that is separated into parts with the closure or regulating element found at the joint.

Split-disk check valve: A check valve with two half-circle disks hinged together, which fold together upon positive flow and retract to a full circle to close against reverse flow. Also known as a *double-disk check valve.*

Split range: In positioners, the partial use of an available signal range.

Split wedge gate: A closure element with a gate valve that uses a solid gate and a flexible seat. Also known as a *flexible gate.*

Spring: In diaphragm actuators, the part that provides force to act against the opposing air chamber. In piston cylinder actuators, the part that provides force for the actuator to move to the correct failure position.

Spring button: A flat, round part found in actuators that is responsible for the holding the spring in place.

Spring rate: The amount of force generated by a spring when the spring is compressed to a certain measurement.

Staged pressure reduction: A measure to reduce effects of a high pressure drop by taking a series of small pressure drops, rather than one large pressure drop.

Stainless steel: An iron alloy that may be mixed with different elements to produce a metal that resists oxidation in the open atmosphere. The capabilities of stainless steels vary depending on the percentage of chrome, nickel, and other elements added to the iron.

Standard: Any procedure or requirement that is part of a written document.

Standard flow: In butterfly and ball valves, where the flow enters into the port of the body that is closest to the seat and exits from the port that is farthest from the seat.

Static seal: Any sealing design that occurs between two parts that do not move after being secured into place. Because friction is limited, static seals have a longer life and are generally more reliable than dynamic seals. Common static seals include joints between the body and bonnet, body and the seat ring, yoke and the actuator casing, etc.

Static unbalance: The difference between the forces of the process fluid pressure that act on both sides of the closure element when in the closed position.

Steady-state air consumption: In double-acting actuators and four-way positioners, the air consumption required to maintain a required position.

Steam jacket: A metal covering that is welded or attached to a valve body to provide space for a separate fluid. A steam jacket is used to provide external cooling or heating of the body or the process.

Stem: The rod portion of a closure or regulating element that is attached to the actuator or a hand-operated device. The stem is sealed by the packing box and transmits the force applied by the actuator to the closure or regulating element.

Stem clamp: A component used to secure or clamp the actuator stem to the plug stem of a linear valve. A stem clamp also has secondary purposes, such as providing feedback to a positioner, indicating the position of the valve, preventing plug rotation, and tripping limit switches.

Stop: A mechanical device attached to a rotary-valve shaft or closure element that prevents further motion. In particular, rotary valves have stops to prevent excessive actuator force from driving through the seat and destroying it.

Stopper valve: A manual valve that uses a linear-motion, round closure element perpendicular to the centerline of the piping, such as a globe- or piston-valve design.

Stroke: The amount of travel a valve is capable of. In linear valves, the stroke is usually a linear measurement. In quarter-turn rotary valves, the stroke is usually a measurement of an angle between 0 and 90°.

Stroking: The act of moving the position of a closure element.

Stroking speed: The rate of travel provided by a particular actuator given certain conditions. Also referred to as *speed of response.*

Stud: A threaded rod in which one end is threaded into a drilled and tapped hole and the other end is secured with a nut.

Stuffing box: Another term for packing box, commonly used in Great Britain.

Supply pressure: In positioners and actuators, the air pressure that drives the actuation system.

Supply-pressure effect: The effect on a valve's position when the actuator's air supply is changed by 10 psi (0.7 bar).

Swing check valve: A check valve designed with the closure element attached to the top of the cap. The closure element can be pushed aside by the flow, but swings back into the closed position if the flow reverses.

Switching valve: An actuator accessory that senses a loss of air-supply pressure and then activates a fail-safe system.

System actuation: In pressure relief valves, the process in which line pressure acts on one side of the closure element and a predetermined spring applies a mechanical load to the other side. When the line pressure reaches its maximum limit, the line pressure overcomes the spring load and the valve opens until the line pressure falls below the preset level.

Tack weld: A small weld designed to hold two parts together during major welding or to allow handling without misalignment of the parts.

Tank-bottom valve: A valve placed at the bottom of a tank to allow for drainage. The actuator or handlever usually faces down (underneath the tank) with the closure element faces up into the tank. Also known as a *ram valve.*

T body: In linear-motion valves, a body style designed to allow the valve to be installed in a straight piping configuration with the rising-stem action perpendicular to the piping.

Tensile strength: The maximum amount of force that can be applied to a part before failure occurs. Also referred to as *ultimate strength.*

Test certificate: Documentation that outlines the results of a physical or chemical test.

Thermoplastic: A common plastic material used in plastic piping and valve parts. Thermoplastics are sensitive to temperature variances, losing their strength as the temperature increases, while gaining strength as the temperature decreases.

Threaded end connection: A valve's end connection that is designed with female NPT threads, which mates with the piping end connection designed with male NPT threads.

Three-way adapter: An upper-body extension that adds a third port to a conventional T-style globe or angle body.

Three-way positioner: A positioner that sends and exhausts air to only one side of an actuator.

Three-way solenoid: An actuator accessory used to interrupt a signal to a pneumatic positioner or to operate a spring diaphragm valve.

Three-way valve: A valve-body configuration that has three ports, allowing for the flow to be diverted to one of two paths or to combine two separate flows.

Throttling: Regulating the position of a valve between the full-open and full-closed positions so that pressure or fluid are adjusted to meet the requirements of the process system.

Throttling element: A closure element that can be used for throttling control of the valve. Also known as a *regulating element.*

Throttling valve: A high-performance valve that is used to regulate the flow, temperature, or pressure of a process. Although a throttling valve can fully open or close, its main purpose is to provide an exact measurement somewhere between those two points.

Through-bolt connection: A piping-to-valve connection that involves a narrow face-to-face, flangeless valve body installed between two piping flanges with longer bolting.

Through-conduit gate valve: A full-area gate valve that has a body the same shape as the pipe. Not only does a through conduit provide for continuous flow without interruption, but it also allows the unrestricted passage of pigs or scrapers.

Thrust: The force generated by an actuator, actuation system or a manually operated actuation device.

Tilting disk check valve: A check valve with a round closure element with two pivot points located on each side of the element.

Top cap: A pressure-retaining part that seals the top-works of a valve body and does not contain a packing box. Also known as a *bonnet cap*.

Top-mounted handwheel: An auxiliary handwheel mounted above the actuator.

Top-works: Any equipment or part mounted on a valve that is located above the bonnet. Top-works can include any number of parts depending on the design, but typically the term refers to the operator, yoke, positioner, position indicator, etc.

Torque: Rotational force applied to a shaft.

Torque switch: A switch that opens or closes when a certain amount of rotational force is applied to a shaft.

Tortuous-path trim: A trim designed with a series of holes, channels, teeth, etc., to increase flow resistance, lower velocities, and reduce the pressure drop.

Toughness: A material's ability to remain intact when force is applied. A tough material will deform under the force, rather than fracture.

Transducer: A positioner accessory used to convert an electrical signal to a pneumatic signal.

Transfer case: A mechanism used in rotary valves that generates rotary shaft motion from linear actuator motion.

Transflow valve: A valve with three or more ports, in which the flow from the inlet port is always flowing through one of the remaining outlet ports. In a transflow valve, if flow is switched from one outlet to another, the increasing flow in one outlet is inversely related to the decreasing flow in the other outlet. Therefore, flow is continual, although being channeled to another line.

Trim: A term for wetted parts and related parts of the closure element in a linear globe valve. The trim usually includes the following parts: the cage (or seat retainer or severe service cartridge or stack), plug, and seat ring.

True installed cost over time: Term used to describe all the costs related to the valve from the point of purchase to final replacement.

Trunnion-mounted ball: A ball that is supported by both a shaft and a post opposite the shaft.

Tubing: Any length of metal pipe that is purposely manufactured according to nonstandard piping measurements and materials.

Turbulence: Flow that is characterized by high velocities and a series of eddies created by obstructions or odd angles in the flow stream, such as the closure element on a valve.

Turndown ratio: A ratio of the rangeability that can be used by a valve. If a valve has a rangeability of 100:1 and the maximum controlled flow is equivalent to 75 percent of the valve's capacity, the turndown ratio is 75:1.

Ultimate strength: The maximum amount of force that can be applied to a part before failure occurs. Also referred to as *tensile strength*.

Ultrahigh-molecular-weight polyethylene: A higher grade of polyethylene used in closure seals of ball and butterfly valves.

Ultrasonic testing: A testing process that involves bombarding a surface of a material with high frequencies to find hidden inclusions, cracks, pits, etc. As the sound waves make contact with a flaw, the reflected signal can be evaluated to determine the depth of the flaw.

Union end connection: A piping end connection found on some smaller valves that involves an external nut is held in place on one end connection and the mating end is threaded. The end with the external nut is then engaged with the threaded end and tightened.

Upper guide: The guide that is furthest from the closure element, usually at the entrance to the packing box. The upper guide not only provides correct alignment of the stem or shaft with the closure element, but it is also used to transmit axial force from the gland-flange bolting to the packing and to protect the packing from the outside environment.

Upper valve body: In split-body valves, the half of the valve body that houses the flow area over the seat ring.

Upstream: The process preceding a valve, including the fluid, pressure, temperature, piping, tank, process equipment, etc.

Valve: A mechanical device designed to divert, start, stop, mix, or regulate a process fluid, which may be either a liquid or a gas. Specifically, a valve is used to handle the flow, pressure, or temperature of a process fluid.

Valve capacity: The amount of flow that can pass through a valve under certain circumstances, without the valve choking. Also referred to as *capacity* or *flow capacity*.

Valve coefficient: The measurement of flow that is commonly applied to valves. The valve coefficient is used to determine the best valve size to pass the required flow rate, while providing overall stability to the process. The flow coefficient is expressed by the term C_v, which is defined as 1 U.S. gallon (3.8 liters) of 60°F (16°C) water during 1 min with a 1 psi (6.9 kPa) pressure drop. Also referred to as *flow coefficient*.

Velocity: The speed at which a fluid moves through a valve.

Vena contracta: The narrowest constriction in the fluid stream as process flow moves through a valve. The vena contracta can be caused by a small orifice or a valve seat or seal. At that point, the flow velocity is at its highest rate, while pressure is at its lowest point.

Vent: A portion of a pipeline where process fluid (usually a gas, such as air) is allowed to escape to atmosphere.

Vent element: Similar to a severe-service trim stack, except that it is mounted downstream from the outlet of the valve as an antinoise measure when venting to atmosphere. Vent elements allow for the use of larger stacks than physically possible inside the valve, thus providing greater staged pressure reduction and reduced velocity control.

Venturi seat ring: A special extended seat ring used in an angle valve to protect the outlet portion of the valve body from damage caused by gas-born particulates, cavitation, flashing, process erosion, etc. In some cases, the Venturi seat ring can be extended through the entire length of the downstream portion of the valve and even into the piping itself.

Viscosity: A measurement of how thick or thin a fluid is. Highly viscous fluids require more energy in the form of pressure or heat to move through the process system.

Visual examination: A sight review of the quality of a casting or manufacturer part by someone trained to perceive flaws.

V-notch ball: In rotary ball valves, a ball-valve design that features a half-sphere with a V-shaped notch flow passage. In general, V-notch balls provide better rangeability than full-ball designs.

Volume booster: An actuator accessory used to increase the stroking speed when large input signal changes are received, decreasing stroking times by up to 90 percent (depending on actuator size and packing friction). Volume boosters also permit normal air flow from the positioner, with minor changes in the positioner input signal. Also referred to as *flow booster*.

Volume tank: A large tank attached to an actuator that is used to supply additional air to stroke the valve upon air failure.

V-ring packing: A packing style characterized by V-shaped (in cross-section) seamless rings. When axial force is applied to the ring, the radial forces cause the thin edges of the ring to press tightly against the wall of the packing box, as well as the stem or shaft. V-ring packing typically provides a strong seal with minimal friction in moderate temperatures. Also referred to as *chevron packing*.

Wafer body: In rotary valves, a body that has a short face-to-face that can be sandwiched between two flanged piping end connections, using special length studs and nuts. Also referred to as a *flangeless body*.

Wall thickness: The thickness of the pressure-retaining outside shell of a pipe or a valve body.

Water-hammer effect: In liquid services, the reaction caused when a valve is suddenly closed (or a pump is turned off), which causes a shock wave to be transmitted by the liquid throughout the piping system. Although water hammer causes noise, the primary disadvantage of allowing it to happen is the damage to piping and equipment installed in the piping system.

Weir: In diaphragm and enclosed-body pinch valves, an integral bar cast into the bottom of the metal body that is used to help compress the elastomeric diaphragm or liner.

Welded end connection: Another term for a *buttweld end connection*.

Wrench: Similar to a handle, a wrench is not attached to the valve stem and can be moved from valve to valve or removed altogether for security reasons.

X-ray testing: Another term for *radiography*.

Y body: A globe or check valve body where the linear action is inclined 45 to 60° from the axis of the inlet and outlet ports.

Yield strength: The force needed before a material begins to stretch, deform, or fracture.

Yoke: The part of an actuator that supports and secures the actuator to the top-works of a linear valve.

Zero: In control valves, the signal point when the valve begins to stroke.

Index

Index

About the Author

Philip L. Skousen ABC is director of corporate communications and training for Valtek International in Springville, Utah, a worldwide manufacturer of control valves, actuators, positioners, and intelligent systems for valves. He is also an adjunct instructor at Brigham Young University, where he teaches marketing and business communication courses. Mr. Skousen has written or contributed to dozens of technical articles for *The Valve Magazine, Chemical Engineering, Chemical Processing,* and other trade magazines.